Polynomial Identities
and Asymptotic Methods

Mathematical
Surveys
and
Monographs

Volume 122

Polynomial Identities and Asymptotic Methods

Antonio Giambruno
Mikhail Zaicev

American Mathematical Society

EDITORIAL COMMITTEE

Jerry L. Bona   Peter S. Landweber
Michael G. Eastwood Michael P. Loss

J. T. Stafford, Chair

2000 *Mathematics Subject Classification*. Primary 16R10, 16R20, 16R30, 16R40, 16R50, 16P90, 16W22, 16W55, 17B01.

For additional information and updates on this book, visit
www.ams.org/bookpages/surv-122

Library of Congress Cataloging-in-Publication Data
Giambruno, A.
  Polynomial identities and asymptotic methods / Antonio Giambruno, Mikhail Zaicev.
    p. cm. — (Mathematical surveys and monographs ; v. 122)
  Includes bibliographical references and index.
  ISBN 0-8218-3829-6 (alk. paper)
  1. PI-algebras. 2. Rings (Algebra). I. Zaicev, Mikhail. II. Title. III. Mathematical surveys and monographs ; no. 122.

QA251.G43   2005
512′.4–dc22   2005053010

Copying and reprinting. Individual readers of this publication, and nonprofit libraries acting for them, are permitted to make fair use of the material, such as to copy a chapter for use in teaching or research. Permission is granted to quote brief passages from this publication in reviews, provided the customary acknowledgment of the source is given.

Republication, systematic copying, or multiple reproduction of any material in this publication is permitted only under license from the American Mathematical Society. Requests for such permission should be addressed to the Acquisitions Department, American Mathematical Society, 201 Charles Street, Providence, Rhode Island 02904-2294, USA. Requests can also be made by e-mail to reprint-permission@ams.org.

© 2005 by the American Mathematical Society. All rights reserved.
The American Mathematical Society retains all rights
except those granted to the United States Government.
Printed in the United States of America.

∞ The paper used in this book is acid-free and falls within the guidelines
established to ensure permanence and durability.
Visit the AMS home page at http://www.ams.org/

10 9 8 7 6 5 4 3 2 1   10 09 08 07 06 05

Contents

Preface	ix
Chapter 1. Polynomial Identities and PI-Algebras	1
1.1. Basic definitions and examples	1
1.2. T-ideals and varieties of algebras	3
1.3. Homogeneous and multilinear polynomials	5
1.4. Stable identities and generic elements	10
1.5. Special types of identities	12
1.6. Symmetric functions	15
1.7. Identities of matrix algebras	16
1.8. A theorem of Lewin	20
1.9. Identities of block-triangular matrices	24
1.10. Central polynomials in matrix algebras	26
1.11. Structure theorems	29
1.12. Some applications of the structure theorems	35
1.13. The Gelfand-Kirillov dimension of a PI-algebra	36
Chapter 2. S_n-Representations	43
2.1. Finite dimensional representations	43
2.2. S_n-representations	46
2.3. Inducing S_n-representations	50
2.4. S_n-actions on multilinear polynomials	52
2.5. Hooks and symmetric and alternating sets of variables	57
Chapter 3. Group Gradings and Group Actions	61
3.1. Group-graded algebras	61
3.2. Abelian gradings and group actions	63
3.3. G-actions, G-gradings and free algebras	65
3.4. Wedderburn decompositions	69
3.5. Finite dimensional simple superalgebras	74
3.6. Involutions on matrix algebras	77
3.7. Superalgebras and Grassmann envelopes	80
3.8. Supercommutative envelopes	83
Chapter 4. Codimension and Colength Growth	87
4.1. Codimensions and colengths	87
4.2. An exponential upper bound for the codimensions	94
4.3. Identities of graded algebras	97
4.4. Robinson-Schensted correspondence	101
4.5. Cocharacters of PI-algebras	104

4.6.	Capelli polynomials and the strip theorem	107
4.7.	Amitsur polynomials and hooks	108
4.8.	Finitely generated superalgebras	110
4.9.	Colength growth: a polynomial upper bound	115

Chapter 5. Matrix Invariants and Central Polynomials — 119
- 5.1. S_n-action on tensor space — 119
- 5.2. Trace identities — 122
- 5.3. A primer of matrix invariants — 124
- 5.4. The discriminant — 125
- 5.5. Invariants and central polynomials — 128
- 5.6. Constructing S_k-maps — 131
- 5.7. Computing central polynomials — 132
- 5.8. Cocharacters and trace cocharacters — 135
- 5.9. Multialternating polynomials — 137
- 5.10. Asymptotics for the codimensions of $k \times k$ matrices — 139

Chapter 6. The PI-Exponent of an Algebra — 143
- 6.1. The exponential growth of the codimensions — 143
- 6.2. A candidate for the PI-exponent — 145
- 6.3. Graded identities and Grassmann envelopes — 151
- 6.4. Gluing Young tableaux — 155
- 6.5. Existence of the exponent — 160
- 6.6. Computing the exponent of some algebras — 161

Chapter 7. Polynomial Growth and Low PI-exponent — 165
- 7.1. The Grassmann algebra and standard polynomials — 165
- 7.2. Varieties of polynomial growth — 169
- 7.3. Locally noetherian varieties — 175
- 7.4. Polynomial growth and bounded multiplicities — 179
- 7.5. Types of polynomial growth — 185
- 7.6. Varieties of exponent two — 189

Chapter 8. Classifying Minimal Varieties — 193
- 8.1. Minimal superalgebras — 193
- 8.2. Some examples — 196
- 8.3. The superenvelope of a minimal superalgebra — 199
- 8.4. Products of verbally prime T-ideals — 205
- 8.5. Classifying minimal varieties of exponential growth — 207
- 8.6. Some consequences — 211

Chapter 9. Computing the Exponent of a Polynomial — 215
- 9.1. The exponent of standard and Capelli polynomials — 215
- 9.2. An upper bound for the exponent of a polynomial — 219
- 9.3. Powers of standard polynomials — 225
- 9.4. Essential hooks and reduced algebras — 238
- 9.5. The exponent of Amitsur polynomials — 242
- 9.6. The exponent of a Lie monomial — 245
- 9.7. Evaluating polynomials — 247
- 9.8. Asymptotics for the standard and the Capelli identities — 251

Chapter 10. G-Identities and $G \wr S_n$-Action   255
   10.1. G-identities, G-codimensions and $G \wr S_n$-action   255
   10.2. Decomposable monomials   259
   10.3. Essential G-identities. Amitsur's theorem on $*$-identities   261
   10.4. Representations of wreath products   264
   10.5. Graded identities and polynomial growth   267
   10.6. The $\mathbb{Z}_2 \wr S_n$-action   272
   10.7. Finite dimensional algebras with φ-action   274
   10.8. The $\mathbb{Z}_2$-exponent of a finite dimensional algebra   276
   10.9. Simple and semisimple φ-algebras   280

Chapter 11. Superalgebras, $*$-Algebras and Codimension Growth   283
   11.1. Notation and more   283
   11.2. $*$-varieties of almost polynomial growth   285
   11.3. Supervarieties of almost polynomial growth   289
   11.4. Capelli identities on superalgebras   292
   11.5. Superalgebras and polynomial growth   294
   11.6. $*$-algebras and the Nagata-Higman theorem   296
   11.7. Polynomial growth of the $*$-codimensions   298
   11.8. Supervarieties of exponent 2   301
   11.9. Further properties   304

Chapter 12. Lie Algebras and Non-associative Algebras   307
   12.1. Introduction to Lie algebras   307
   12.2. Identities of Lie algebras   309
   12.3. Codimension growth of Lie algebras   314
   12.4. Exponents of Lie algebras   323
   12.5. Overexponential codimension growth   327
   12.6. Lie superalgebras, alternative and Jordan algebras   328
   12.7. The general non-associative case   330

Appendix A. The Generalized-Six-Square Theorem   333
   A.1. The Theorem   333
   A.2. Basics   334
   A.3. Representations of integers   336
   A.4. A crucial lemma   338
   A.5. The proof of Theorem A.1.2   339

Bibliography   341

Index   349

Preface

One of the main objectives of this book is to show how one can combine methods of ring theory, combinatorics, and representation theory of groups with an analytical approach in order to study the polynomial identities satisfied by a given algebra. The idea of applying analytical methods to the theory of polynomial identities appeared in the early 1970s and nowadays this approach is one of the most powerful tools of the theory.

A polynomial identity of an algebra A is a polynomial in non-commuting indeterminates vanishing under all evaluations in A and the algebras having at least one such nontrivial relation are called PI-algebras. For instance, $xy - yx \equiv 0$ is a polynomial identity for any commutative algebra. Hence, in particular, the polynomial ring in one or several variables is an example of a PI-algebra. Another natural example is given by the exterior algebra of a vector space, or Grassmann algebra G which appears in algebra, analysis, geometry and other branches of modern mathematics. It is easy to see that G satisfies the polynomial identity $[[x, y], z] \equiv 0$ where $[x, y] = xy - yx$ is the Lie commutator of x and y. Any nilpotent algebra A such that $A^n = 0$ is a PI-algebra since it satisfies the polynomial identity $x_1 x_2 \cdots x_n \equiv 0$.

All subalgebras, homomorphic images and direct products of algebras satisfying a given identity $f \equiv 0$ still satisfy $f \equiv 0$. Hence the PI-algebras form a quite wide class including commutative algebras, finite dimensional algebras, algebraic algebras of bounded exponent and many more.

Non commutative polynomials vanishing on an algebra can be found in the early papers of Dehn ([**De**]) and Wagner ([**Wa**]). The general interest in PI-theory started after a paper of Kaplansky ([**K1**]) in 1948. In that paper it was proved that any primitive PI-algebra is a finite dimensional simple algebra suggesting that satisfying a polynomial identity is some finiteness condition on a given algebra. Most of the structure theory of PI-algebras was developed in the 1960s and the 1970s and an account of it can be found in the early books of Jacobson ([**Ja**]) and Procesi ([**Pr3**]). A comprehensive collection of the results on the structure theory of PI-algebras can be found in the book of Rowen ([**Ro2**]).

Two years after Kaplansky's theorem, Amitsur and Levitsky proved by purely combinatorial methods that a certain polynomial, called the standard polynomial of degree $2k$, is an identity of minimal degree for the algebra of $k \times k$ matrices ([**AL**]). This theorem was the beginning of a new approach to PI-theory, the main objective being the description of the polynomial identities satisfied by a given algebra. A few years later, Kostant ([**Ks**]) related the Amitsur-Levitsky theorem to cohomology theory and to the invariant theory of $k \times k$ matrices. In the 1970s the theory of PI-algebras was related to the more general theory of trace identities as developed by Procesi ([**Pr2**]) via invariant theory and independently by Razmyslov ([**R2**]).

Let $F\langle X\rangle$ denote the algebra of noncommutative polynomials in a given set of variables X over a field F, i.e., the free algebra on X over F. The polynomial identities satisfied by an algebra A form an ideal of $F\langle X\rangle$ invariant under all endomorphisms of the free algebra, called a T-ideal. Moreover, every T-ideal of $F\langle X\rangle$ is of this type. Hence describing the identities of an algebra means describing the T-ideals of the free algebra.

Since distinct algebras can have the same ideal of identities, the theory of T-ideals is linked to the theory of varieties of algebras. Recall that a variety of algebras is a class of algebras satisfying a given set of identities. The varieties were introduced by Birkhoff ([**Br**]) and Malcev ([**Ma1**]) in order to study the identities of algebraic structures and this seems to be the most natural language in the theory of identities.

The description of a T-ideal is in general a hard problem. Specht in 1950 ([**Sp**]) conjectured that over a field of characteristic zero every proper T-ideal of $F\langle X\rangle$ is finitely generated as a T-ideal. Many instances of this conjecture were proved in the following years but a complete proof was only given after a series of papers by Kemer in 1987 ([**Ke5**]). His proof is based on some basic structure theory of the T-ideals which has given a new impetus to the subject. It involves the study of superidentities of superalgebras and certain graded tensor products with the Grassmann algebra, called Grassmann envelopes. The theorems and techniques developed by Kemer are contained mostly in his monograph ([**Ke7**]) and have become in recent years some of the basic tools for studying the identities of a given algebra.

Even if every proper T-ideal is finitely generated, the polynomial identities of a given algebra, like the algebra of $n \times n$ matrices, are far from being understood. An important observation is that even if we know the generators of a given T-ideal, it is quite impossible in general to deduce from them, say, information on the polynomials of the T-ideal of a given degree. To overcome some of these difficulties it is natural to introduce a function measuring the growth of the identities of a T-ideal in some sense. This approach was introduced by Regev in 1972 ([**Re1**]) and nowadays it has become not only a main tool but also a major object of research in the theory of PI-algebras in characteristic zero.

Since the base field is of characteristic zero, by the well-known polarization process, every identity is equivalent to a finite set of multilinear ones. Then one can slice any T-ideal into subspaces P_n of polynomials in a given fixed set of n variables, $n = 1, 2, \ldots$, and the function defined by the codimensions of these spaces is the growth function associated to the given T-ideal. Since T-ideals are invariant under endomorphisms, the permutation action of the symmetric group S_n turns P_n into an S_n-module and the representation theory of the symmetric group, which is well-understood in characteristic zero, can be successfully applied.

One associates to each T-ideal a sequence of characters of the symmetric groups S_n, $n = 1, 2, \ldots$, called the sequence of cocharacters of the given PI-algebra and a numerical sequence, called the sequence of codimensions, given by the corresponding degrees measuring the growth of the T-ideal. The most significant results of the representation theory of the symmetric group and the corresponding combinatorial theory of Young diagrams such as the branching theorem, the Littlewood-Richardson rule, the hook formula etc., became an essential tool in the development of the theory.

The starting point in the investigation of the growth of T-ideals is a theorem of Regev ([**Re1**]) stating that the codimension sequence of a PI-algebra is exponentially bounded. By results of Kemer it turns out that this growth is either polynomial or exponential. Also in recent years ([**GZ1**], [**GZ2**]) it has been proved that the exponent of the growth rate for a proper T-ideal is an integer called the exponent of the T-ideal or of a corresponding PI-algebra. Having at hand an integer scale provided by the exponent, the theory has developed in the last years towards the classification of T-ideals according to the asymptotic behaviour of their sequence of codimensions and in this book we shall give an account of these results. Among them, the classification of maximal T-ideals of a given exponent (or minimal varieties of a given exponent) ([**GZ9**]), the prominent role played by the standard and the Capelli polynomials in combinatorial PI-theory and the precise relation between the growth of the corresponding T-ideals and the growth of the algebra of $n \times n$ matrices ([**GZ8**]), etc. This approach to the combinatorial theory of PI-algebras is also related to other theories of independent interest.

Since every T-ideal is multigraded by the degree, using standard methods the sequence of cocharacters is strictly related to the corresponding Hilbert or Poincaré series. In particular, the problem of decomposing the cocharacter sequence into irreducibles, translates into the problem of writing the Hilbert series as a sum of Schur functions. This relation establishes a precise link between the combinatorial theory of PI-algebras and the theory of symmetric functions ([**M**]) and many problems translate in a natural way into that setting.

Another strictly related theory is that of trace identities and the corresponding invariant theory of $n \times n$ matrices as developed by Procesi in [**Pr2**]. The methods of invariant theory and the development of the theory of trace identities obtained independently by Razmyslov in [**R2**] are one of the basic tools needed in order to develop the theory of PI-algebras.

It is well known that any field of characteristic zero is a splitting field for the symmetric group, hence the base field is usually not relevant when studying T-ideals. This seems to infer that no significant result of number theory should play a role. Nevertheless, as we shall see, a delicate extension of the well-known four squares theorem asserting that any integer is the sum of at most four squares, will be crucial in the computation of the growth of some polynomials.

For the reader interested in the general theory of polynomial identities, the first monographs devoted to the subject were published in the 1970s ([**Pr3**], [**Ja**], ([**Ro2**]). In several books concerning ring theory or other areas of algebra one can find some parts dedicated to the theory of polynomial identities. Among them we cite the books of Herstein [**H**], Cohn [**C**], Rowen [**Ro3**], Passman [**P**], Zhevlakov, Slinko, Shestakov and Shirshov [**ZSSS**], Formanek [**F4**] and Beidar, Martindale and Mikhalev [**BMM**]. Polynomial identities of Lie algebras are treated extensively in the books by Bahturin [**B1**] and Razmyslov [**R4**]. The solution of the Specht problem is contained in the important monograph by Kemer [**Ke7**]. The book by Drensky [**D10**] is a very good source for a first year graduate course in PI-theory. The recent achievements in PI-theory have also stimulated the appearance of new monographs and surveys devoted to polynomial identities ([**GRZ2**], [**DF**], [**BR**]). The book by Belov and Rowen [**BR**] appeared but there seems to be no significant overlap between the two books.

The general scheme of the book is as follows. The core of the book is Chapter 6 where we prove the integrality of the exponential growth of any proper variety or T-ideal. All the previous chapters contain the material needed for this purpose.

In the first chapter, we introduce the basic definitions and we give an account of the main results of the structure theory of PI-algebras. One of the main tools for computing the asymptotic behavior of the codimensions is the representation theory of the symmetric group and we give an account of this theory in Chapter 2. We present most of the classical results including the branching rules, the hook formula and the Littlewood-Richardson rule. We then study the permutation action of the symmetric group on the space of multilinear polynomials in a fixed number of variables and we derive most of the properties of this action that we shall use throughout.

In Chapter 3 we deal with group gradings and group actions. Group graded algebras and, in particular, superalgebras play an important role in different areas of mathematics and theoretical physics. The reason for studying superalgebras and their identities is twofold. It is an interesting fast growing subject. More important, there is a well understood connection between superidentities and ordinary identities that allows one to reduce some problems to the finite dimensional case, and this is one of the basic reductions in this book. In this chapter we generalize Wedderburn theorems to the case of superalgebras and algebras with involution. We also introduce the Grassmann envelope and the superenvelope of an algebra and prove their basic properties.

In Chapter 4 we define the basic notions of the theory, namely the sequences of codimensions and colengths and we prove the most important properties of their asymptotic behaviour. We also prove a basic structure theorem concerning the Grassmann envelope of a superalgebra and the well-known hook theorem and strip theorem.

Chapter 5 is devoted to the introduction of the invariant theory of $n \times n$ matrices and the consequent theory of trace polynomial identities. This subject is interesting on its own and is an important area of modern mathematics. In this chapter we apply results of invariant theory in order to prove the existence of suitable central polynomials for $n \times n$ matrices. Such polynomials are used in the subsequent chapter for finding the precise lower bound of the codimension growth. We also give the asymptotics of the codimensions of the algebra of $n \times n$ matrices.

Chapter 6 is the central chapter of the book and we prove that the sequence of codimensions of any PI-algebra (or proper variety) has an integral exponential growth, called the PI-exponent of the algebra. We also give a constructive way for determining it.

In the following chapters we apply the results obtained in order to further develop the theory. Chapter 7 is mainly devoted to the characterization of varieties having polynomial growth (or PI-exponent ≤ 1). The Grassmann algebra and its properties play a basic role in this description.

In Chapter 8 we classify all varieties minimal of given exponent. This leads to the notion of minimal superalgebra. We prove that such varieties have an ideal of identities which is the product of verbally prime T-ideals and are strictly related to the algebras of block triangular matrices. The classification of minimal varieties gives an effective way for computing the exponent of a variety. In fact in Chapter 9 we define the exponent of a polynomial, or set of polynomials, as the exponent of the corresponding variety and we compute it for some significant classes of polynomials such as standard polynomials, Capelli polynomials, and Amitsur polynomials. This leads in some significant cases to the determination of a generating algebra for the corresponding variety.

In Chapter 10 and Chapter 11 we extend our approach to graded algebras and to algebras with involution. We consider G-identities for an algebra A where G is a finite group of automorphisms and antiautomorphisms of A. We study such identities via the representation theory of the wreath product $G \wr S_n$ and we focus our attention to the case when G is a group of automorphisms or antiautomorphisms of A of order two. In this last case A has a structure of superalgebra or of algebra with involution and we prove that the corresponding G-codimensions have integral exponential growth in case A is finite dimensional. Chapter 11 is entirely devoted to superalgebras and algebras with involution and their identities. We characterize the corresponding varieties of polynomial growth and we prove that no intermediate growth is allowed for such varieties. We also relate the newly found invariants to the ordinary ones.

In the last chapter of the book we study our numerical invariants and their asymptotics in other classes of non-associative algebras. Even in algebras which are close to being associative, the sequences of codimensions and colengths show a wild behavior. We deal mostly with Lie algebras and the growth of their identities. In this setting the sequence of codimensions is no longer exponentially bounded and we give an account of the various phenomena that can occur. As an outcome, the combinatorial theory of PI-algebra seems to be much more developed in the associative case.

We are very grateful to A. Berele, V. Drensky, P. Koshlukov, S. Mishchenko, A. Regev and A. Valenti for reading and commenting on parts of the manuscript. Special thanks are due to A. Regev for preparing Appendix A. We would like to mention O. M. Di Vincenzo and I. Shestakov for useful discussions and remarks. We also thank F. Benanti and D. La Mattina for helping during the preparation of the manuscript.

This project was supported in part by the research grant PRIN 2003 "Algebras with polynomial identities and combinatorial methods", by the Istituto Nazionale di Alta Matematica of Italy and by the research grants RFBR No. 02-01-00219 and SSC-1910.2003.1 of Russia.

CHAPTER 1

Polynomial Identities and PI-Algebras

In this chapter we give the basic definitions and results of the theory of polynomial identities of associative algebras. After discussing the duality between T-ideals of the free algebra and varieties of algebras, we prove the basic properties of such ideals. Special attention is devoted to the standard polynomials and the Capelli polynomials since they play an important role in the book.

We prove some of the key results pertaining to the structure theory of PI-algebras such as Kaplansky's theorem, Amitsur-Levitzki's theorem, the existence of central polynomials of Formanek and Razmyslov and Posner's theorem. As an application we can study the ideal of identities of $k \times k$ matrices and prove some embedding theorems into an algebra of matrices. Even if some of these results are not essential for the development of the main subject of the book, we still feel that the reader should be acquainted with the main results and techniques of structure theory.

In this chapter we also present a proof of Lewin's theorem allowing to determine the structure of the T-ideal of identities of an algebra of block triangular matrices. We finally give a brief introduction to the theory of growth of finitely generated algebras and the Gelfand-Kirillov dimension.

1.1. Basic definitions and examples

Throughout this book we shall be dealing with algebras over a field and, except for the last chapter, all algebras will be associative. We start with the basic definition of free algebra. Let F be a field and X a set. The free associative algebra on X over F is the algebra $F\langle X \rangle$ of polynomials in the non-commuting indeterminates $x \in X$.

A linear basis of $F\langle X \rangle$ consists of all words in the alphabet X (including the empty word 1). Such words are called monomials and the product of two monomials is given by juxtaposition. We shall often call monomial also a non-zero scalar multiple of a word in X. The elements of $F\langle X \rangle$ are called polynomials and if $f \in F\langle X \rangle$, we write $f = f(x_1, \ldots, x_n)$ to indicate that $x_1, \ldots, x_n \in X$ are the only indeterminates occurring in f. For the elements of X we shall usually use the symbols x, x_i. We shall also use y, y_i, z, z_i, for new elements of X, when needed. Throughout we shall assume that X is an infinite set.

We define $\deg u$, the degree of a monomial u, as the length of the word u. Also $\deg_{x_i} u$, the degree of u in the indeterminate x_i, is the number of occurrences of x_i in u. Accordingly, the degree $\deg f$ of a polynomial $f = f(x_1, \ldots, x_n)$ is the maximum degree of a monomial in f; $\deg_{x_i} f$, the degree of f in x_i, is the maximum of $\deg_{x_i} u$, for u a monomial in f.

The algebra $F\langle X \rangle$ is defined, up to isomorphism, by the following universal property: given an associative F-algebra A, any map $X \to A$ can be uniquely extended to a homomorphism of algebras $F\langle X\rangle \to A$. The cardinality of X is called the rank of $F\langle X\rangle$. For most of the book we shall consider the free algebra $F\langle X\rangle$ of countable rank on the set $X = \{x_1, x_2, \ldots\}$.

DEFINITION 1.1.1. Let A be an F-algebra and $f = f(x_1, \ldots, x_n) \in F\langle X\rangle$. We say that $f \equiv 0$ is a polynomial identity of A if $f(a_1, \ldots, a_n) = 0$ for all $a_1, \ldots, a_n \in A$.

Let Φ denote the set of all homomorphisms $\varphi : F\langle X\rangle \to A$. Then it is clear that $f \equiv 0$ is a polynomial identity for A if and only if $f \in \bigcap_{\varphi \in \Phi} \ker \varphi$. We shall usually say that $f \equiv 0$ is an identity on A or that A satisfies $f \equiv 0$; sometimes we shall say that f itself is an identity of A.

Since the trivial polynomial $f = 0$ is an identity for any algebra A, we make the following:

DEFINITION 1.1.2. If A satisfies a non-trivial polynomial identity $f \equiv 0$, then we say that A is a PI-algebra.

For $a, b \in A$, let $[a, b] = ab - ba$ denote the Lie commutator of a and b. We next give some examples of PI-algebras.

EXAMPLE 1.1.3. If A is a commutative algebra, then A is a PI-algebra since it satisfies the identity $[x, y] \equiv 0$.

EXAMPLE 1.1.4. Any nilpotent algebra is a PI-algebra. In fact if, say, $A^n = 0$, for some $n \geq 1$, then $x_1 \cdots x_n \equiv 0$ is a polynomial identity for A.

EXAMPLE 1.1.5. Let A be an algebra nil of bounded index. This means that there exists an integer $n \geq 1$ such that $a^n = 0$, for all $a \in A$. Then clearly $x^n \equiv 0$ is a polynomial identity of A.

EXAMPLE 1.1.6. Let $UT_n(F)$ be the algebra of $n \times n$ upper triangular matrices over F. Then $UT_n(F)$ is a PI-algebra since it satisfies the identity
$$[x_1, x_2] \cdots [x_{2n-1}, x_{2n}] \equiv 0.$$
This is easily seen by noticing that the commutator of any two upper triangular matrices is a strictly upper triangular matrix. But the set of strictly upper triangular matrices forms a nilpotent two-sided ideal I of $UT_n(F)$ such that $I^n = 0$.

EXAMPLE 1.1.7. The algebra $M_2(F)$ of 2×2 matrices over F satisfies the identity $[[x, y]^2, z] \equiv 0$.

To see this, recall that if $a \in M_2(F)$, its characteristic polynomial is $x^2 - \operatorname{tr}(a)x + \det(a)E$, where $\operatorname{tr}(a)$ and $\det(a)$ are the trace and the determinant of a, respectively and E is the identity 2×2 matrix. In case a is a commutator, $\operatorname{tr}(a) = 0$ and, so, $a^2 + \det(a)E = 0$. This says that $a^2 = -\det(a)E$, i.e., the square of any commutator is a scalar matrix (hence central).

EXAMPLE 1.1.8. Let G be the Grassmann (exterior) algebra on a countable dimension vector space over a field F of characteristic different from 2. The algebra G can be constructed as follows. Let $F\langle X\rangle$ be the free algebra of countable rank on $X = \{x_1, x_2, \ldots\}$. If I is the two-sided ideal of $F\langle X\rangle$ generated by the set of polynomials $\{x_i x_j + x_j x_i \mid i, j \geq 1\}$, then $G = F\langle X\rangle/I$. If we write $e_i = x_i + I$ for $i = 1, 2, \ldots$, then G has the following presentation:
$$G = \langle 1, e_1, e_2, \ldots \mid e_i e_j = -e_j e_i, \text{ for all } i, j \geq 1\rangle.$$

Let S_n denote the symmetric group on $\{1, \ldots, n\}$. Notice that from $e_i e_j = -e_j e_i$ it easily follows that for any $1 \leq k < l \leq n$,

$$e_{i_1} \cdots e_{i_{k-1}} e_{i_k} e_{i_{k+1}} \cdots e_{i_{l-1}} e_{i_l} e_{i_{l+1}} \cdots e_{i_n}$$
$$= -e_{i_1} \cdots e_{i_{k-1}} e_{i_l} e_{i_{k+1}} \cdots e_{i_{l-1}} e_{i_k} e_{i_{l+1}} \cdots e_{i_n}.$$

Hence, by writing any permutation $\sigma \in S_n$ as a product of transpositions, we obtain

$$e_{\sigma(i_1)} \cdots e_{\sigma(i_n)} = (\operatorname{sgn} \sigma) e_{i_1} \cdots e_{i_n},$$

where $\operatorname{sgn} \sigma$ is the signature of the permutation σ (i.e., $\operatorname{sgn} \sigma = +1$ or -1 according to whether σ is an even or an odd permutation).

It follows that

$$B = \{1,\ e_{i_1} \cdots e_{i_k} \mid 1 \leq i_1 < \ldots < i_k\}$$

spans G over F. We claim that B is a basis of G. In fact, suppose that $h = \sum_{i=1}^{n} \alpha_i w_i = 0$ is a relation with the minimal number of non-zero coefficients α_i where $w_i \in B$. If the element e_m appears in w_1 but not in w_2, then $e_m w_1 = 0$ and $e_m h = \sum_{i=2}^{n} \alpha_i e_m w_i = 0$ is a relation with a smaller number of non-zero coefficients, a contradiction. Therefore B is a basis of G.

It is convenient to write G in the form $G = G^{(0)} \oplus G^{(1)}$ where

$$G^{(0)} = \operatorname{span}\{e_{i_1} \cdots e_{i_{2k}} \mid 1 \leq i_1 < \ldots < i_{2k},\ k \geq 0\},$$

$$G^{(1)} = \operatorname{span}\{e_{i_1} \cdots e_{i_{2k+1}} \mid 1 \leq i_1 < \ldots < i_{2k+1},\ k \geq 0\}.$$

It is easily checked that $G^{(0)} G^{(0)} + G^{(1)} G^{(1)} \subseteq G^{(0)}$ and $G^{(0)} G^{(1)} + G^{(1)} G^{(0)} \subseteq G^{(1)}$. As we shall see later this says that the decomposition $G = G^{(0)} \oplus G^{(1)}$ is a $\mathbb{Z}_2$-grading of G. Notice that $G^{(0)}$ coincides with the center of G.

We next claim that G satisfies the identity $[[x, y], z] \equiv 0$. In fact, since $G^{(0)}$ is central, any commutator of two elements of G turns out to be a linear combination of monomials in the e_i's of even length. Thus $[G, G] \subseteq G^{(0)}$ and the conclusion follows.

The Grassmann algebra is a very important example and we shall study G in more detail later.

1.2. T-ideals and varieties of algebras

Given an algebra A, we define

$$\operatorname{Id}(A) = \{f \in F\langle X \rangle \mid f \equiv 0 \text{ on } A\},$$

the set of polynomial identities of A. Clearly, $\operatorname{Id}(A)$ is a two-sided ideal of $F\langle X \rangle$. Moreover, if $f = f(x_1, \ldots, x_n)$ is any polynomial in $\operatorname{Id}(A)$, and $g_1, \ldots, g_n$ are arbitrary polynomials in $F\langle X \rangle$, it is clear that $f(g_1, \ldots, g_n) \in \operatorname{Id}(A)$. Since any endomorphism of $F\langle X \rangle$ is determined by mapping $x \mapsto g$, $x \in X$, $g \in F\langle X \rangle$, it follows that $\operatorname{Id}(A)$ is an ideal invariant under all endomorphisms of $F\langle X \rangle$. The ideals with this property are called T-ideals.

DEFINITION 1.2.1. An ideal I of $F\langle X \rangle$ is a T-ideal if $\varphi(I) \subseteq I$ for all endomorphisms φ of $F\langle X \rangle$.

Hence $\operatorname{Id}(A)$ is a T-ideal of $F\langle X \rangle$. On the other hand, it is easy to check that all T-ideals of $F\langle X \rangle$ are actually of this type. In fact, if I is a T-ideal, it is easily proved that $\operatorname{Id}(F\langle X \rangle / I) = I$.

We have seen that any algebra A determines a T-ideal of $F\langle X \rangle$. On the other hand, many algebras may correspond to the same T-ideal (as the ideal of its identities). In order to make this connection clearer, we need the notion of a variety of algebras.

DEFINITION 1.2.2. Given a non-empty set $S \subseteq F\langle X \rangle$, the class of all algebras A such that $f \equiv 0$ on A for all $f \in S$ is called the variety $\mathcal{V} = \mathcal{V}(S)$ determined by S.

A variety $\mathcal{V}$ is called non-trivial if $S \neq 0$ and $\mathcal{V}$ is proper if it is non-trivial and contains a non-zero algebra. For instance, the class of all commutative algebras forms a proper variety with $S = \{[x,y]\}$. Also, if $S = \{x^n\}$, then $\mathcal{V}(S)$ is the class of all algebras which are nil of exponent bounded by n.

Notice that if $\mathcal{V}$ is the variety determined by the set S and $\langle S \rangle_T$ is the T-ideal of $F\langle X \rangle$ generated by S, then $\mathcal{V}(S) = \mathcal{V}(\langle S \rangle_T)$ and $\langle S \rangle_T = \bigcap_{A \in \mathcal{V}} \mathrm{Id}(A)$. Let us write $\langle S \rangle_T = \mathrm{Id}(\mathcal{V})$. Thus to each variety corresponds a T-ideal of $F\langle X \rangle$; the converse is also true as we shall see later. Let us return to the basic properties of a variety.

In general, how can we decide whether a given class of algebras is a variety? A celebrated theorem of Birkhoff gives an answer to this question. Observe that a variety $\mathcal{V}$ is closed under taking homomorphic images, subalgebras and direct products. The theorem of Birkhoff shows that actually these properties characterize the varieties.

THEOREM 1.2.3 (Birkhoff). *A non-empty class $\mathcal{V}$ of algebras is a variety if and only if it satisfies the following properties:*

1) *if $A \in \mathcal{V}$, and $B \to A$ is a monomorphism, then $B \in \mathcal{V}$;*
2) *if $A \in \mathcal{V}$ and $A \to B$ is an epimorphism, then $B \in \mathcal{V}$;*
3) *if $\{A_\gamma\}_{\gamma \in \Gamma}$ is a family of algebras and $A_\gamma \in \mathcal{V}$, for all $\gamma \in \Gamma$, then $\prod_{\gamma \in \Gamma} A_\gamma \in \mathcal{V}$.*

The theorem of Birkhoff holds in a much more general setting and a proof can be found, for instance, in [**C**, Theorem 3.7, Vol. 3].

Now let $\mathcal{V}$ be a variety, $A \in \mathcal{V}$ an algebra and $Y \subseteq A$ a subset of A. We say that A is relatively free on Y (with respect to $\mathcal{V}$), if for any algebra $B \in \mathcal{V}$ and for every function $\alpha : Y \to B$, there exists a unique homomorphism $\beta : A \to B$ extending α. When $\mathcal{V}$ is the variety of all algebras, this is just the definition of a free algebra on Y. The cardinality of Y is called the rank of A.

Relatively free algebras are easily described in terms of free algebras.

THEOREM 1.2.4. *Let X be a non-empty set, $F\langle X \rangle$ a free algebra on X and $\mathcal{V}$ a variety with corresponding ideal $\mathrm{Id}(\mathcal{V}) \subseteq F\langle X \rangle$. Then $F\langle X \rangle / \mathrm{Id}(\mathcal{V})$ is a relatively free algebra on the set $\bar{X} = \{x + \mathrm{Id}(\mathcal{V}) \mid x \in X\}$. Moreover, any two relatively free algebras with respect to $\mathcal{V}$ of the same rank are isomorphic.*

PROOF. Let $B \in \mathcal{V}$ and let $\alpha : \bar{X} \to B$ be a set-theoretic map. Define a map $\beta : X \to B$ by setting $\beta(x) = \alpha(x + \mathrm{Id}(\mathcal{V}))$. Since $F\langle X \rangle$ is the free algebra on X, β can be extended to a homomorphism $\bar{\beta} : F\langle X \rangle \to B$. Now, if $f \in \mathrm{Id}(\mathcal{V})$, then f is an identity of B, hence $\bar{\beta}(f) = 0$. This shows that $\mathrm{Id}(\mathcal{V}) \subseteq \ker(\bar{\beta})$ and α extends to a homomorphism $F\langle X \rangle / \mathrm{Id}(\mathcal{V}) \to B$. Thus $F\langle X \rangle / \mathrm{Id}(\mathcal{V})$ is a relatively free algebra on $\bar{X}$.

Let $F_1, F_2 \in \mathcal{V}$ be relatively free algebras of the same rank on $X = \{x_i \mid i \in I\}$ and $Y = \{y_i \mid i \in I\}$, respectively. Since F_1 and F_2 are relatively free algebras with respect to $\mathcal{V}$, there exist homomorphisms $\alpha_1 : F_1 \to F_2$ and $\alpha_2 : F_2 \to F_1$ such that $\alpha_1(x_i) = y_i$ and $\alpha_2(y_i) = x_i$, for all $i \in I$. It is clear that $\alpha_1 \alpha_2$ and $\alpha_2 \alpha_1$ are the identity maps on Y and X, respectively. Hence they are isomorphisms. $\square$

The correspondence between T-ideals and varieties is well understood.

THEOREM 1.2.5. *There is a one-to-one correspondence between T-ideals of $F\langle X\rangle$ and varieties of algebras. In this correspondence a variety $\mathcal{V}$ corresponds to the T-ideal of identities $\mathrm{Id}(\mathcal{V})$ and a T-ideal I corresponds to the variety of algebras satisfying all the identities in I.*

PROOF. If I_1 and I_2 are two T-ideals, $I_1 \neq I_2$, then there exists, say, $f \in I_1 \setminus I_2$. But then $\mathcal{V}(I_1) \neq \mathcal{V}(I_2)$ since $F\langle X\rangle/I_2$ does not satisfy f and, so, $F\langle X\rangle/I_2 \in \mathcal{V}(I_2)$ but $F\langle X\rangle/I_2 \notin \mathcal{V}(I_1)$.

Now if $\mathcal{V}_1$ and $\mathcal{V}_2$ are two varieties, $\mathcal{V}_1 \neq \mathcal{V}_2$, there exists, say, $A \in \mathcal{V}_1 \setminus \mathcal{V}_2$. Hence there exists $f \in \mathrm{Id}(\mathcal{V}_2)$ such that $f \notin \mathrm{Id}(A)$. Since $\mathrm{Id}(A) \supseteq \mathrm{Id}(\mathcal{V}_1)$, it follows that $\mathrm{Id}(\mathcal{V}_2) \neq \mathrm{Id}(\mathcal{V}_1)$. $\square$

If $\mathcal{V}$ is a variety and A is an F-algebra such that $\mathrm{Id}(A) = \mathrm{Id}(\mathcal{V})$ (for instance, $A = F\langle X\rangle/\mathrm{Id}(\mathcal{V})$), then we say that $\mathcal{V}$ is the variety generated by A and we write $\mathcal{V} = \mathrm{var}(A)$. Also, we shall refer to $F\langle X\rangle/\mathrm{Id}(\mathcal{V})$ as the relatively free algebra of the variety $\mathcal{V}$ of rank $|X|$.

1.3. Homogeneous and multilinear polynomials

When the base field F is infinite, the study of the identities of a given algebra can be reduced to the study of homogeneous or multilinear polynomials. In this section we give the corresponding definitions and prove these reductions.

Let $F_n = F\langle x_1, \ldots, x_n\rangle$ be the free algebra of rank $n \geq 1$ over F. This algebra can be naturally decomposed as

$$F_n = F_n^{(1)} \oplus F_n^{(2)} \oplus \cdots$$

where, for every $k \geq 1$, $F_n^{(k)}$ is the subspace spanned by all monomials of total degree k. Since $F_n^{(i)} F_n^{(j)} \subseteq F_n^{(i+j)}$, for all $i, j \geq 1$, we say that F_n is graded by the degree or that it has a structure of graded algebra. The $F_n^{(i)}$'s are called the homogeneous components of F_n.

This decomposition can be further refined as follows: for every $k \geq 1$ write

$$F_n^{(k)} = \bigoplus_{i_1 + \cdots + i_n = k} F_n^{(i_1, \ldots, i_n)}$$

where $F_n^{(i_1, \ldots, i_n)}$ is the subspace spanned by all monomials of degree i_1 in $x_1, \ldots, i_n$ in x_n. Clearly $F_n^{(i_1, \ldots, i_n)} F_n^{(j_1, \ldots, j_n)} \subseteq F_n^{(i_1+j_1, \ldots, i_n+j_n)}$ and in this case we say that F_n is multigraded. Such decompositions extend in an obvious way to $F\langle X\rangle$ for X countable.

DEFINITION 1.3.1. *A polynomial f belonging to $F_n^{(k)}$ for some $k \geq 1$, will be called homogeneous of degree k. If f belongs to some $F_n^{(i_1, \ldots, i_n)}$, it will be called multihomogeneous of multidegree $(i_1, \ldots, i_n)$. We also say that a polynomial f is homogeneous in the variable x_i, if x_i appears with the same degree in every monomial of f.*

A useful property of T-ideals is that if F is an infinite field, they are homogeneous with respect to the above multigradings, i.e., they have a corresponding decomposition in multihomogeneous polynomials. If $f(x_1, \ldots, x_n) \in F\langle X \rangle$, we can always write
$$f = \sum_{i_1 \geq 0, \ldots, i_n \geq 0} f^{(i_1, \ldots, i_n)}$$
where $f^{(i_1, \ldots, i_n)} \in F_n^{(i_1, \ldots, i_n)}$ is the sum of all monomials in f where $x_1, \ldots, x_n$ appear at degree $i_1, \ldots, i_n$, respectively. The polynomials $f^{(i_1, \ldots, i_n)}$ which are non-zero are called the multihomogeneous components of f.

THEOREM 1.3.2. *Let F be an infinite field. If $f \equiv 0$ is a polynomial identity for the algebra A, then every multihomogeneous component of f is still a polynomial identity for A.*

PROOF. For every variable x_t, $1 \leq t \leq n$, we can decompose $f = \sum_{i=0}^m f_i$, where f_i is the sum of all monomials of f in which x_t appears at degree i and $m = \deg_{x_t} f$ is the degree of f in x_t. By an obvious induction argument, in order to prove the theorem, it is enough to prove that, for every variable x_t, $f_i \equiv 0$ for all $i \geq 0$.

Let $\alpha_0, \ldots, \alpha_m$ be distinct elements of F. Clearly, for every $j = 0, \ldots, m$,
$$f(x_1, \ldots, \alpha_j x_t, \ldots, x_n) \equiv 0$$
is still an identity for A. Since each f_i is homogeneous in x_t of degree i,
$$f_i(x_1, \ldots, \alpha_j x_t, \ldots, x_n) = \alpha_j^i f_i(x_1, \ldots, x_t, \ldots, x_n).$$
Hence
$$(1.1) \qquad f(x_1, \ldots, \alpha_j x_t, \ldots, x_n) = \sum_{i=0}^m \alpha_j^i f_i(x_1, \ldots, x_n) \equiv 0$$
on A, for all $j = 0, \ldots, m$.

Write the Vandermonde matrix
$$\Delta = \begin{pmatrix} 1 & 1 & \cdots & 1 \\ \alpha_0 & \alpha_1 & \cdots & \alpha_m \\ \vdots & \vdots & & \vdots \\ \alpha_0^m & \alpha_1^m & \cdots & \alpha_m^m \end{pmatrix}.$$

Then (1.1) says that for every $a_1, \ldots, a_n \in A$, if we write $f_i(a_1, \ldots, a_n) = \bar{f}_i$, then
$$(\bar{f}_0 \cdots \bar{f}_m)\Delta = 0.$$
Since the Vandermonde determinant $\det(\Delta) = \prod_{0 \leq i < j \leq m}(\alpha_j - \alpha_i)$ is non-zero, it follows that $f_0 \equiv 0, \ldots, f_m \equiv 0$ are identities of A. The proof is therefore complete. $\square$

Notice that the previous result is still valid if F is a finite field such that $|F| > \deg f$. On the other hand, if K is a field with q elements, K satisfies the identity $x^q - x \equiv 0$ but the homogeneous components of this identity do not vanish in K.

One of the most important consequences of Theorem 1.3.2 is that over an infinite field every T-ideal is generated by its multihomogeneous polynomials.

1.3. HOMOGENEOUS AND MULTILINEAR POLYNOMIALS

Among multihomogeneous polynomials a special role is played by the multilinear ones.

DEFINITION 1.3.3. A polynomial f is linear in the variable x_i if x_i occurs with degree 1 in every monomial of f. A polynomial which is linear in each of its variables is called multilinear.

In the language above we can say that a polynomial $f(x_1, \ldots, x_n) \in F\langle X \rangle$ is multilinear if it is multihomogeneous of multidegree $(1, \ldots, 1)$.

Since in a multilinear polynomial $f(x_1, \ldots, x_n)$ each variable appears in each monomial at degree 1, it is clear that this polynomial is always of the form

$$f(x_1, \ldots, x_n) = \sum_{\sigma \in S_n} \alpha_\sigma x_{\sigma(1)} \cdots x_{\sigma(n)}$$

where $\alpha_\sigma \in F$ and S_n is the symmetric group on $\{1, \ldots, n\}$. Notice that if $f(x_1, \ldots, x_n)$ is a polynomial linear in one variable, say x_1, then

$$f(\sum \alpha_i y_i, x_2, \ldots, x_n) = \sum \alpha_i f(y_i, x_2, \ldots, x_n)$$

for every $\alpha_i \in F, y_i \in F\langle X \rangle$. We shall use this property throughout. For instance, the most important feature of multilinear polynomials is given in the following remark.

REMARK 1.3.4. Let A be an F-algebra spanned by a set B over F. If a multilinear polynomial f vanishes on B, then f is a polynomial identity of A.

PROOF. Let $a_1 = \sum \alpha_{1i} u_i, \ldots, a_n = \sum \alpha_{ni} u_i$ be elements of A where the u_i's are elements of B. Then, since $f = f(x_1, \ldots, x_n)$ is linear in each of its variables, $f(a_1, \ldots, a_n) = \sum \alpha_{1i_1} \cdots \alpha_{ni_n} f(u_{i_1}, \ldots, u_{i_n}) = 0$. □

We are going to show how to reduce an arbitrary polynomial identity to a multilinear one. First some notation.

DEFINITION 1.3.5. Let S be a set of polynomials in $F\langle X \rangle$ and $f \in F\langle X \rangle$. We say that f is a consequence of the polynomials in S (or f follows from the polynomials in S) if $f \in \langle S \rangle_T$, the T-ideal generated by the set S.

DEFINITION 1.3.6. Two sets of polynomials are equivalent if they generate the same T-ideal.

In the proof of the next result we shall use the so-called process of multilinearization.

THEOREM 1.3.7. *If the algebra A satisfies an identity of degree k, then it satisfies a multilinear identity of degree $\leq k$.*

PROOF. Let $f(x_1, \ldots, x_n) \in F\langle X \rangle$ be a polynomial identity for the algebra A. If each variable x_i appears at degree ≤ 1, in each monomial of f, then by eventually specializing some of the variables to zero, we obtain a multilinear polynomial as desired. Hence we may assume that there exists a variable, say x_1, such that $\deg_{x_1} f = d > 1$.

Compute the polynomial

$$h(y_1, y_2, x_2, \ldots, x_n)$$
$$= f(y_1 + y_2, x_2, \ldots, x_n) - f(y_1, x_2, \ldots, x_n) - f(y_2, x_2, \ldots, x_n).$$

Notice that h is still a polynomial identity for A. Let us check that h is a non-zero polynomial. Suppose $h = 0$. Since any map $X \to X$ can be extended to an endomorphism of $F\langle X \rangle$, replacing y_1 and y_2 with x_1 in h we also get a zero polynomial, i.e.

$$h(x_1, x_1, x_2, \ldots, x_n) = f(2x_1, x_2, \ldots, x_n) - 2f(x_1, x_2, \ldots, x_n) = 0.$$

If we decompose f into the sum $f = f_0 + f_1 + \cdots + f_d$ where f_k is the sum of monomials of degree k in x_1, then the previous relations imply

$$-f_0 + (2^2 - 2)f_2 + \cdots + (2^d - 2)f_d = 0$$

which contradicts the inequality $d > 1$.

Since $\deg_{y_1} h = d - 1 < \deg_{x_1} f$, by an induction argument we obtain a multilinear polynomial which is an identity on A. $\square$

The method of multilinearization actually has a very important consequence.

THEOREM 1.3.8. *If $\operatorname{char} F = 0$, every non-zero polynomial $f \in F\langle X \rangle$ is equivalent to a finite set of multilinear polynomials.*

PROOF. We examine more closely the process of multilinearization. By Theorem 1.3.2, f is equivalent to the set of its multihomogeneous components. Hence we may assume that $f = f(x_1, \ldots, x_n)$ is multihomogeneous. We apply the process of multilinearization to f: if $\deg_{x_1} f = d > 1$, then we write

$$f(y_1 + y_2, x_2, \ldots, x_n) = \sum_{i=0}^{d} g_i(y_1, y_2, x_2, \ldots, x_n)$$

where $\deg_{y_1} g_i = i$, $\deg_{y_2} g_i = d - i$ and $\deg_{x_t} g_i = \deg_{x_t} f$, for all $t = 2, \ldots, n$.

Then all polynomials $g_i = g_i(y_1, y_2, x_2, \ldots, x_n)$, $i = 1, \ldots, d - 1$, are consequences of f.

Notice that for every i,

$$g_i(y_1, y_1, x_2, \ldots, x_n) = \binom{d}{i} f(y_1, x_2, \ldots, x_n).$$

Since $\operatorname{char} F = 0$, $\binom{d}{i} \neq 0$, hence f is a consequence of any g_i, $i = 1, \ldots, d-1$. We now apply induction in order to complete the proof. $\square$

We remark that the hypothesis of characteristic zero can be changed to $\operatorname{char} F > \deg f$ and the above result still holds.

The effect of the process of multilinearization applied to a multihomogeneous polynomial is also easily described. Let $f(x_1, \ldots, x_n)$ be such polynomial and write

$$f(x_1, \ldots, x_n) = \sum_{\sigma \in S_{d+1}} \alpha_\sigma a_{\sigma(1)} x_1 a_{\sigma(2)} x_1 \cdots a_{\sigma(d)} x_1 a_{\sigma(d+1)}$$

where $a_1, \ldots, a_{d+1}$ are (eventually trivial) monomials in the x_i's not involving x_1. Then the process of multilinearization applied to the variable x_1 leads to the polynomial

$$g(y_1, \ldots, y_d, x_2, \ldots, x_n) = \sum_{\sigma \in S_{d+1}} \sum_{\tau \in S_d} \alpha_\sigma a_{\sigma(1)} y_{\tau(1)} a_{\sigma(2)} y_{\tau(2)} \cdots a_{\sigma(d)} y_{\tau(d)} a_{\sigma(d+1)}.$$

We can record the previous result in the language of T-ideals.

COROLLARY 1.3.9. *If $\mathrm{char}\, F = 0$, every T-ideal is generated, as a T-ideal, by the multilinear polynomials it contains.*

In the proof of Theorem 1.3.7 we used in an essential way the fact that T-ideals are invariant under all endomorphisms of the free algebra. In the next lemma we show that this property can be exploited also in the case of subspaces of $F\langle X \rangle$.

Recall that a subspace L of the algebra A is a Lie ideal of A if $[L, A] \subseteq L$, where $[L, A]$ denotes the additive subgroup generated by all Lie commutators $[l, a]$, $l \in L, a \in A$. Also, if $f = f(x_1, \ldots, x_n) \in F\langle X \rangle$ and A is an F-algebra, let $f(A) = \mathrm{span}_F \{ f(a_1, \ldots, a_n) \mid a_1, \ldots, a_n \in A \}$.

LEMMA 1.3.10. *Let F be an infinite field.*
1) *If W is a subspace of $F\langle X \rangle$ such that $\varphi(W) \subseteq W$ for all endomorphisms φ of the algebra $F\langle X \rangle$, then W is a Lie ideal of $F\langle X \rangle$.*
2) *For any F-algebra A and for any $f \in F\langle X \rangle$, $f(A)$ is a Lie ideal of A.*

PROOF. Since W is invariant by endomorphisms of $F\langle X \rangle$, in order to prove 1) it is clearly enough to show that $[W, x] \subseteq W$ for $x \in X$.

Let $f = f(x_1, \ldots, x_n) \in W$, and write $f = \sum_i g_i$ as the sum of its multihomogeneous components. Since W is a subspace of $F\langle X \rangle$, by applying a Vandermonde determinant argument to $F\langle X \rangle / W$ as in the proof of Theorem 1.3.2, we obtain that every g_i lies in W. Also, if we show that $[g_i, x_{n+1}] \in W$, for all i, it will follow that $[f, x_{n+1}] \in W$. Therefore, without loss of generality, we may assume that f is multihomogeneous.

Denote $m_j = \deg_{x_j} f$, for all $j = 1, \ldots, n$. Let $g = f(x_1 + y_1, \ldots, x_n + y_n) \in W$ where $y_1, \ldots, y_n \in X$. If we consider the multihomogeneous component $g_i = g_i(x_1, \ldots, x_n, y_i)$ of g of multidegree $(m_1, \ldots, m_{i-1}, m_i - 1, m_{i+1}, \ldots, m_n)$ in $x_1, \ldots, x_i, \ldots, x_n$ and of multidegree $(0, \ldots, 0, 1, 0, \ldots, 0)$ in $y_1, \ldots, y_i, \ldots, y_n$, then, as above we obtain that $g_i \in W$. Since $\mathrm{ad}(x_{n+1}) : a \to [a, x_{n+1}]$ is a derivation of $F\langle X \rangle$,

$$[x_{i_1} \cdots x_{i_k}, x_{n+1}] = \sum_{j=1}^{k} x_{i_1} \cdots x_{i_{j-1}} [x_{i_j}, x_{n+1}] x_{i_{j+1}} \cdots x_{i_k}.$$

Hence
$$[f(x_1, \ldots, x_n), x_{n+1}] = \sum_i g_i(x_1, \ldots, x_n, [x_i, x_{n+1}]) \in W$$

and we are done.

The proof of the second part of the lemma will follow once we check the inclusion $[f(a_1, \ldots, a_n), b] \in f(A)$, for any $a_1, \ldots, a_n, b \in A$. Now, by the first part of the proof,
$$W = f(F\langle X \rangle) = \mathrm{span}\{ f(y_1, \ldots, y_n) \mid y_1, \ldots, y_n \in F\langle X \rangle \}$$
is a Lie ideal of $F\langle X \rangle$. Hence
$$[f(x_1, \ldots, x_n), x_{n+1}] = \sum_i \alpha_i f(y_1^{(i)}, \ldots, y_n^{(i)})$$

where the α_i's are scalars and $y_1^{(i)}, \ldots, y_n^{(i)} \in F\langle X \rangle$, for all i. By applying to this last equality the homomorphism $\varphi : F\langle X \rangle \to A$ such that $\varphi(x_1) = a_1, \ldots, \varphi(x_n) = a_n, \varphi(x_{n+1}) = b$, we obtain

$$[f(a_1, \ldots, a_n), b] = \sum_i \alpha_i f(a_1^{(i)}, \ldots, a_n^{(i)}) \in f(A),$$

for suitable $a_1^{(i)}, \ldots, a_n^{(i)} \in A$, as required. $\square$

1.4. Stable identities and generic elements

If A is an algebra over a field F, by extending the scalars, we obtain a new algebra over F whose identities are satisfied by A. In general the larger algebra has a different T-ideal of identities, nevertheless as we shall see, this essentially happens only in case F is a finite field.

Let C be a commutative algebra over F and consider the F-algebra $A \otimes_F C$. Some of the polynomial identities of A might still vanish on $A \otimes_F C$. We give them a special name.

DEFINITION 1.4.1. Let f be an identity for the F-algebra A. We say that f is a stable identity for A if for every commutative F-algebra C, f is still an identity for $A \otimes_F C$.

It is clear that if F is a finite field, not every identity of F is stable. For instance, if $|F| = q$, $x^q - x \equiv 0$ is an identity of F but it does not vanish on any proper field extension of F. We next show that for algebras over an infinite field, every identity is stable. As a consequence, for such algebras every identity is still an identity if we extend the base field.

LEMMA 1.4.2. *If F is an infinite field and A is an F-algebra, then every polynomial identity of A is stable.*

PROOF. Let $f(x_1, \ldots, x_n)$ be a polynomial identity for A, C a commutative algebra over F and $\bar{A} = A \otimes_F C$. By Theorem 1.3.2 we may assume that f is multihomogeneous of multidegree $(m_1, \ldots, m_n)$.

For $\bar{a}_1, \ldots, \bar{a}_n \in \bar{A}$ we need to show that $f(\bar{a}_1, \ldots, \bar{a}_n) = 0$. Suppose first that $\bar{a}_1 = a_1 \otimes c_1, \ldots, \bar{a}_n = a_n \otimes c_n$. Then

$$f(\bar{a}_1, \ldots, \bar{a}_n) = f(a_1, \ldots, a_n) \otimes c_1^{m_1} \cdots c_n^{m_n} = 0$$

and we are done in this case.

Now let $\bar{a}_1 = b_1 \otimes d_1 + b_2 \otimes d_2, \bar{a}_2 = a_2 \otimes c_2, \ldots, \bar{a}_n = a_n \otimes c_n$. Then

$$f(\bar{a}_1, \ldots, \bar{a}_n)$$
$$= f(b_1 \otimes d_1, a_2 \otimes c_2, \ldots, a_n \otimes c_n) + f(b_2 \otimes d_2, a_2 \otimes c_2, \ldots, a_n \otimes c_n)$$
$$+ \sum_{i=1}^{m_1-1} f_i(b_1 \otimes d_1, b_2 \otimes d_2, a_2 \otimes c_2, \ldots, a_n \otimes c_n)$$

where

(1.2) $\quad f(x_1 + y_1, x_2, \ldots, x_n) - f(x_1, x_2, \ldots, x_n) - f(y_1, x_2, \ldots, x_n)$
$$= \sum_{i=1}^{m_1-1} f_i(x_1, y_1, x_2, \ldots, x_n)$$

and $\deg_{x_1} f_i = i$. Since all polynomials f_i in (1.2) are multihomogeneous consequences of f, by the first part of the proof, it follows that $f(\bar{a}_1, \ldots, \bar{a}_n) = 0$ also in this case.

By generalizing this argument for arbitrary $\bar{a}_1 = \sum a_{1i} \otimes c_{1i}, \ldots, \bar{a}_n = \sum a_{ni} \otimes c_{ni} \in \bar{A}$, $a_{ij} \in A, c_{ij} \in C$, we write $f(\bar{a}_1, \ldots, \bar{a}_n)$ as a sum of expressions of the form

$$\bar{g} = g(a_{i_1 j_1} \otimes c_{i_1 j_1}, \ldots, a_{i_k j_k} \otimes c_{i_k j_k})$$

where $g = g(x_1, \ldots, x_k)$ is a multihomogeneous consequence of f. Again by the first part of the proof we obtain $\bar{g} = 0$. □

As an application of the previous lemma, we next find a useful explicit form for a relatively free algebra of a variety generated by a finite dimensional algebra. This algebra will be generated by "generic elements".

Let A be a finite dimensional algebra over the infinite field F. Set $\dim_F A = m$ and consider $\{u_1, \ldots, u_m\}$ a basis of A over F. Let $\xi_j^{(i)}$, $i \geq 1$, $1 \leq j \leq m$, be commutative indeterminates and let $F[\xi_j^{(i)} \mid i \geq 1, 1 \leq j \leq m]$ be the polynomial ring over F in these indeterminates. We construct $B = A \otimes_F F[\xi_j^{(i)}]$, the algebra tensor product of A and $F[\xi_j^{(i)}]$.

DEFINITION 1.4.3. The elements
$$\xi^{(i)} = \sum_{j=1}^m u_j \otimes \xi_j^{(i)}, \quad i = 1, 2, \ldots$$
are called generic elements. The subalgebra $\tilde{A}$ of B generated by $\xi^{(1)}, \xi^{(2)}, \ldots$ over F is called the algebra of generic elements of A.

THEOREM 1.4.4. *If F is infinite, the algebra $\tilde{A}$ is a relatively free algebra of countable rank of the variety $\mathrm{var}(A)$, i.e., $\tilde{A} \cong F\langle X \rangle / \mathrm{Id}(A)$, where X is countable.*

PROOF. Let $X = \{x_1, x_2, \ldots\}$ be countable and let $\psi : F\langle X \rangle \to \tilde{A}$ be the homomorphism induced by mapping $x_i \to \xi^{(i)}$, $i = 1, 2, \ldots$. We shall prove that $\ker \psi = \mathrm{Id}(A)$. By Lemma 1.4.2, $\mathrm{Id}(A \otimes F[\xi_j^{(i)}]) = \mathrm{Id}(A)$, hence $\ker \psi \supseteq \mathrm{Id}(A)$. Suppose now that $g = g(x_1, \ldots, x_n) \in \ker \psi$, i.e., $g(\xi^{(1)}, \ldots, \xi^{(n)}) = 0$ in $\tilde{A}$ and let $a_1, \ldots, a_n$ be arbitrary elements of A. Write each a_i as a linear combination of the basis $\{u_1, \ldots, u_m\}$ of A and let
$$a_i = \sum_{j=1}^m \lambda_j^{(i)} u_j$$
with $\lambda_1^{(i)}, \ldots, \lambda_m^{(i)} \in F$. Since $F[\xi_j^{(i)}]$ is the free commutative algebra of countable rank, any set-theoretical map $\xi_j^{(i)} \to \lambda_j^{(i)}$ extends to a homomorphism $F[\xi_j^{(i)}] \to F$. Hence, due to the universal property of the tensor product, the map
$$a \mapsto a, \ a \in A, \ \xi_j^{(i)} \mapsto \lambda_j^{(i)}, \ 1 \leq i \leq n, \ 1 \leq j \leq m,$$
extends to a homomorphism $\varphi : A \otimes F[\xi_j^{(i)}] \to A$ such that $\varphi(\xi^{(i)}) = a_i$, $1 \leq i \leq n$. Hence
$$0 = \varphi(g(\xi^{(1)}, \ldots, \xi^{(n)})) = g(\varphi(\xi^{(1)}), \ldots, \varphi(\xi^{(n)})) = g(a_1, \ldots, a_n).$$
Since $a_1, \ldots, a_n$ are arbitrary elements of A, $g(x_1, \ldots, x_n) \equiv 0$ is an identity of A and $\ker \psi = \mathrm{Id}(A)$ follows. □

A special case of interest is when $A = M_k(F)$ is the algebra of $k \times k$ matrices over F. In this case, choosing the matrix units e_{ij}'s as a basis for $M_k(F)$, we have the polynomial algebra $F[\xi_{ij}^{(t)}]$ in the variables $\xi_{ij}^{(t)}$, $t \geq 1$, $1 \leq i, j \leq k$. Then

$M_k(F) \otimes_F F[\xi_{ij}^{(t)}] \cong M_k(F[\xi_{ij}^{(t)}])$ and

$$\xi^{(t)} = \sum_{i,j=1}^{k} \xi_{ij}^{(t)} e_{ij}$$

is the matrix with entries $\xi_{ij}^{(t)}$. The elements $\xi^{(t)}$ are called $k \times k$ generic matrices and the algebra $F\{\xi\} = \widetilde{M_k(F)} = F\{\xi^{(1)}, \xi^{(2)}, \ldots\}$ is called the algebra of $k \times k$ generic matrices over F. Thus from the previous theorem we have the following.

COROLLARY 1.4.5. *The algebra $F\{\xi\}$ of $k \times k$ generic matrices over the infinite field F is the relatively free algebra of countable rank of the variety generated by $M_k(F)$.*

1.5. Special types of identities

In this section we introduce the notion of alternating polynomial and we describe some of the basic properties. Important examples will be the standard and the Capelli polynomials.

DEFINITION 1.5.1. Let $f = f(x_1, \ldots, x_n, y_1, \ldots, y_t)$ be a polynomial linear in each of the variables $x_1, \ldots, x_n$. We say that f is alternating in the variables $x_1, \ldots, x_n$ if, for any $1 \leq i < j \leq n$, the polynomial f becomes zero when we substitute x_i instead of x_j.

By the linearity of f in $x_1, \ldots, x_n$, it is easily seen that if f is alternating in $x_1, \ldots, x_n$, then for $1 \leq i < j \leq n$,

$$f(x_1, \ldots, x_i, \ldots, x_j, \ldots, x_n, y_1, \ldots, y_t) = -f(x_1, \ldots, x_j, \ldots, x_i, \ldots, x_n, y_1, \ldots, y_t).$$

This property is also equivalent to the definition of alternating polynomial in case char $F \neq 2$. Moreover, by writing any permutation of S_n as a product of transpositions, it follows that if $f(x_1, \ldots, x_n, y_1, \ldots, y_t)$ is alternating in $x_1 \ldots, x_n$, then

$$f(x_{\sigma(1)}, \ldots, x_{\sigma(n)}, y_1, \ldots, y_t) = (\operatorname{sgn} \sigma) f(x_1, \ldots, x_n, y_1, \ldots, y_t).$$

In case f is alternating in all of its variables, we simply say that f is alternating. The basic property of polynomials alternating in some set of variables is given in the following.

PROPOSITION 1.5.2. *Let $f(x_1, \ldots, x_n, y_1, \ldots, y_t)$ be a polynomial alternating in $x_1, \ldots, x_n$ and let A be an F-algebra. If $a_1, \ldots, a_n \in A$ are linearly dependent over F, then $f(a_1, \ldots, a_n, b_1, \ldots, b_t) = 0$ for all $b_1, \ldots, b_t \in A$.*

PROOF. By the hypothesis, one of the a_i's, say a_1, can be written as a linear combination of the others, $a_1 = \sum_{i=2}^{n} \alpha_i a_i, \alpha_i \in F$. But then

$$f(a_1, \ldots, a_n, b_1, \ldots, b_t) = \sum_{i=2}^{n} \alpha_i f(a_i, a_2, \ldots, a_n, b_1, \ldots, b_t) = 0$$

since f is alternating on $x_1, \ldots, x_n$ and in each term $f(a_i, a_2, \ldots, a_n, b_1, \ldots, b_t)$ two arguments coincide. $\square$

As an example of alternating polynomials we next define the Capelli polynomials.

1.5. SPECIAL TYPES OF IDENTITIES

DEFINITION 1.5.3. The polynomial

$$Cap_m(x_1,\ldots,x_m;y_1,\ldots,y_{m+1}) = \sum_{\sigma\in S_m}(\operatorname{sgn}\sigma)y_1 x_{\sigma(1)} y_2 x_{\sigma(2)}\cdots y_m x_{\sigma(m)} y_{m+1}$$

is called the Capelli polynomial of rank m or the mth Capelli polynomial.

The Capelli polynomial Cap_m is multilinear and alternating in $x_1,\ldots,x_m$. It plays a central role among alternating polynomials since, as we shall see below, every polynomial which is alternating on $x_1,\ldots,x_m$ can be written as a linear combination of Capelli polynomials obtained by specializing the y_i's.

It turns out that specialization of the y_i's is an important feature of Capelli polynomials. For algebras without 1, by abuse of notation, we make the following definition.

DEFINITION 1.5.4. An algebra A satisfies the Capelli identity of rank m (or the mth Capelli identity) if A satisfies all polynomials obtained from

$$Cap_m(x_1,\ldots,x_m;y_1,\ldots,y_{m+1})$$

by eventually setting the variables y_i equal to 1 in all possible ways.

Notice that there are 2^{m+1} such polynomials.

PROPOSITION 1.5.5. *If $f \in F\langle X\rangle$ is a polynomial alternating in $x_1,\ldots,x_m$, then*

$$f = \sum_{w_1,\ldots,w_{m+1}} \alpha_{w_1,\ldots,w_{m+1}} Cap_m(x_1,\ldots,x_m;w_1,\ldots,w_{m+1})$$

is a linear combination of Capelli polynomials where $w_1,\ldots,w_{m+1}$ are suitable (eventually trivial) monomials in $F\langle X\rangle$.

PROOF. Take any non-zero monomial $\beta w_1 x_{i_1} w_2 x_{i_2}\cdots x_{i_m} w_{m+1}$ of f, where the w_i's are monomials in the remaining variables. Since every permutation can be written as a product of transpositions, an easy induction shows that for all permutations $\sigma \in S_m$, $w_1 x_{\sigma(i_1)} w_2 x_{\sigma(i_2)}\cdots x_{\sigma(i_m)} w_{m+1}$ appears in f with coefficient $(\operatorname{sgn}\sigma)\beta$. It follows that $Cap_m(x_1,\ldots,x_m;w_1,\ldots,w_{m+1})$ is a summand of f with coefficient $\pm\beta$ and f is a linear combination of such polynomials. □

In light of Proposition 1.5.5, an algebra A satisfies the Capelli identity of rank m if every polynomial alternating in m variables vanishes in A.

The polynomial $Cap_m(x_1,\ldots,x_m;1,\ldots,1)$, obtained by specializing all the y_i's to 1 is called the standard polynomial of degree m and it has some special features that we shall explore next.

DEFINITION 1.5.6. The polynomial

$$St_m(x_1,\ldots,x_m) = \sum_{\sigma\in S_n}(\operatorname{sgn}\sigma)x_{\sigma(1)}\cdots x_{\sigma(m)}$$

is called the standard polynomial of degree m.

Next we state the basic properties of the standard polynomial. We adopt the convention that the symbol $\hat{}$ means omission. Hence $f(x_1,\ldots,\hat{x_i},\ldots,x_m)$ is a polynomial in which the variable x_i does not appear.

PROPOSITION 1.5.7.
1) If $f(x_1, \ldots, x_m)$ is a multilinear alternating polynomial of degree m, then $f = \alpha St_m(x_1, \ldots, x_m)$, for some $\alpha \in F$.
2) $St_{m+1}(x_1, \ldots, x_{m+1}) = \sum_{i=1}^{m+1} (-1)^{i+1} x_i St_m(x_1, \ldots, \hat{x}_i, \ldots, x_{m+1})$. Hence if $St_m \equiv 0$ is an identity for an algebra A, $St_{m+1} \equiv 0$ is also an identity for A.

PROOF. Part 1) is a special case of the previous proposition since in this case all the monomials $w_1, \ldots, w_{m+1}$ are trivial.

Part 2) is an easy corollary of part 1) since the polynomial St_{m+1} can be written as
$$St_{m+1} = x_1 f_1 + \ldots + x_{m+1} f_{m+1}$$
where $f_i = f_i(x_1, \ldots, \hat{x}_i, \ldots, x_{m+1})$ is a multilinear alternating polynomial in $x_1, \ldots, \hat{x}_i, \ldots, x_{m+1}$. Hence we just need to observe that
$$f_i = \beta_i St_m(x_1, \ldots, \hat{x}_i, \ldots, x_{m+1})$$
with $\beta_i = (-1)^{i+1}$. □

Recall that an algebra A is algebraic of bounded degree n over F if each element of A is algebraic over F of degree $\leq n$, i.e., for every $a \in A$, there exists a polynomial $f(x) \in F[x]$, $\deg f \leq n$, such that $f(a) = 0$.

THEOREM 1.5.8. Let A be an F-algebra.
1) If $\dim_F A = n < \infty$, then A satisfies the Capelli identity of rank $(n+1)$. In particular, $St_{n+1} \equiv 0$ on A.
2) If A is algebraic of bounded degree n, over F, then A satisfies
$$Cap_{n+1}(y, xy, x^2 y, \ldots, x^n y; z_1, \ldots, z_{n+2}) \equiv 0.$$
Hence $St_{n+1}(y, xy, x^2 y, \ldots, x^n y) \equiv 0$ on A.

PROOF. 1) Since Cap_{n+1} is multilinear, in order to prove that it vanishes on A it is enough to prove that it vanishes on a basis of A. Since $\dim_F A = n$ and Cap_{n+1} is alternating on $n+1$ elements, by Proposition 1.5.2, the claim follows.

2) Let $a \in A$ and $f(x) \in F[x]$, $\deg f \leq n$, be such that $f(a) = 0$. For all $b \in A$, $f(a)b = 0$ implies that the elements $b, ab, a^2 b, \ldots, a^n b$ are linearly dependent over F. Again by Proposition 1.5.2, it follows that
$$Cap_{n+1}(y, xy, x^2 y, \ldots, x^n y; z_1, \ldots, z_{n+2}) \equiv 0$$
on A. □

The previous theorem, when applied to matrices, says that the algebra $M_n(F)$ satisfies $Cap_{n^2+1} \equiv 0$ (and $St_{n^2+1} \equiv 0$). We shall see in the next sections that, while $n^2 + 1$ is actually the minimal degree of a Capelli polynomial satisfied by $M_n(F)$, the same is not true for the standard polynomial. In fact we shall prove that the minimal degree is $2n$. The second part of the previous theorem is of special interest for $M_n(F)$ in the case of standard polynomials. Notice that we can write
$$St_{n+1}(y, xy, x^2 y, \ldots, x^n y) = \sum_{\sigma \in S^*_{n+1}} (\text{sgn}\, \sigma) x^{\sigma(0)} y x^{\sigma(1)} y \cdots y x^{\sigma(n)} y$$
where S^*_{n+1} is the symmetric group acting on the set $\{0, 1, \ldots, n\}$. This polynomial is called the polynomial of algebraicity.

1.6. Symmetric functions

In this section we shall deduce some basic properties of the symmetric functions. Let $\mathbb{Q}$ be the field of rational numbers and let $\mathbb{Q}[\xi] = \mathbb{Q}[\xi_1, \ldots, \xi_n]$ be the polynomial algebra in the commutative indeterminates $\xi_1, \ldots, \xi_n$. The symmetric group S_n acts on $Q[\xi]$ by permuting the indeterminates: if $\sigma \in S_n$, the position $\sigma(\xi_i) = \xi_{\sigma(i)}$, $1 \leq i \leq n$, induces an automorphism of $\mathbb{Q}[\xi]$ that we shall still denote σ. The subring of invariants

$$\mathbb{Q}[\xi]^{S_n} = \{f \in \mathbb{Q}[\xi] \mid \sigma(f) = f, \text{ for all } \sigma \in S_n\}$$

is called the ring of symmetric polynomials in $\xi_1, \ldots, \xi_n$. This ring has been extensively studied and several sets of generators have been determined (see [**M**, p. 19ff]). Among them we are interested in the elementary symmetric polynomials and the power sums.

DEFINITION 1.6.1. The polynomials $e_k = \sum_{1 \leq i_1 < i_2 < \cdots < i_k \leq n} \xi_{i_1} \cdots \xi_{i_k}$, $1 \leq k \leq n$, are called the elementary symmetric functions on $\xi_1, \ldots, \xi_n$.

DEFINITION 1.6.2. The polynomials $p_r = \sum_{i=1}^n \xi_i^r$, $r \geq 1$, are called the power sums symmetric functions on $\xi_1, \ldots, \xi_n$.

The relation between these two types of symmetric functions is given in the following.

PROPOSITION 1.6.3 (Newton's formulas). *For $k = 1, \ldots, n$, we have*

$$ke_k = \sum_{r=1}^k (-1)^{r-1} p_r e_{k-r}$$

where $e_0 = 1$.

PROOF. Let x be an indeterminate and compute the formal power series

$$\sum_{r \geq 1} (-1)^{r-1} p_r x^{r-1} = \sum_{i=1}^n \sum_{r \geq 1} (-1)^{r-1} \xi_i^r x^{r-1} = \sum_{i=1}^n \frac{\xi_i}{1 + \xi_i x}$$

$$= \sum_{i=1}^n \frac{d}{dx} \log(1 + \xi_i x) = \frac{d}{dx} \log \prod_{i=1}^n (1 + \xi_i x)$$

$$= \frac{d}{dx} \log \sum_{k=0}^n e_k x^k = \sum_{k=1}^n k e_k x^{k-1} / \sum_{k=0}^n e_k x^k,$$

where we applied the obvious equality $\sum_{k=0}^n e_k x^k = \prod_{i=1}^n (1 + \xi_i x)$ and we developed $\frac{1}{1+\xi_i x} = 1 - \xi_i x + \xi_i^2 x^2 - \cdots$ as a formal power series in x.

Thus

$$\left(\sum_{r \geq 1} (-1)^{r-1} p_r x^{r-1}\right)\left(\sum_{k=0}^n e_k x^k\right) = \sum_{k=1}^n k e_k x^{k-1}.$$

By matching the coefficients of x^{k-1}, $k = 1, \ldots, n$, we get the desired formulas. □

Now let $n \geq 1$ be an integer. Recall that λ is a partition of n if $\lambda = (\lambda_1, \ldots, \lambda_r)$ where $\lambda_1 \geq \cdots \geq \lambda_r > 0$ are positive integers such that $\sum_{i=1}^r \lambda_i = n$. In this case we write $\lambda \vdash n$. For C a commutative algebra, let $M_n(C)$ denote the algebra of $n \times n$ matrices over C and let $\text{tr} : M_n(C) \to C$ be the usual trace function.

COROLLARY 1.6.4. *Let C be a commutative algebra over $\mathbb{Q}$ and $a \in M_n(C)$. If $f_a(x) = \sum_{i=0}^{n} \alpha_i x^{n-i}$ is the characteristic polynomial of a, then, for all $k = 1, \ldots, n$,*

$$\alpha_k = \sum_{\lambda = (\lambda_1, \ldots, \lambda_r) \vdash k} q_\lambda \operatorname{tr}(a^{\lambda_1}) \cdots \operatorname{tr}(a^{\lambda_r})$$

for some $q_\lambda \in \mathbb{Q}$ not depending on a.

PROOF. Suppose first that C is an integral domain. Then $a \in M_n(F)$, where F is the field of fractions of C. If $\varepsilon_1, \ldots, \varepsilon_n$ are the eigenvalues of a in some splitting field of $f_a(x)$, then

$$f_a(x) = (x - \varepsilon_1) \cdots (x - \varepsilon_n) = \sum_{i=0}^{n} (-1)^i e_i(\varepsilon_1, \ldots, \varepsilon_n) x^{n-i},$$

where $e_i(\varepsilon_1, \ldots, \varepsilon_n)$ are the elementary symmetric functions evaluated in $\varepsilon_1, \ldots, \varepsilon_n$. Also, for every $j = 1, \ldots, n$, since $\varepsilon_1^j, \ldots, \varepsilon_n^j$ are the eigenvalues of a^j, it follows that $\operatorname{tr}(a^j) = p_j(\varepsilon_1, \ldots, \varepsilon_n)$, is the power sum p_j evaluated in $\varepsilon_1, \ldots, \varepsilon_n$. The conclusion of the corollary now follows in this case from Newton's formulas.

In general, it is well known that there exists an epimorphism $\varphi : B \to C$, where B is an integral domain. Actually B is the polynomial algebra in some set of indeterminates. This φ induces an epimorphism $\bar{\varphi} : M_n(B) \to M_n(C)$ given by $\bar{\varphi}(a_{ij}) = (\varphi(a_{ij}))$ for $(a_{ij}) \in M_n(B)$. Since, for any $a' \in M_n(B)$, $\operatorname{tr}(\bar{\varphi}(a')) = \varphi(\operatorname{tr}(a'))$, the conclusion follows also in this case. □

1.7. Identities of matrix algebras

By Theorem 1.5.8 the algebra $M_k(F)$ of $k \times k$ matrices over F, being finite dimensional, is a PI-algebra and satisfies the Capelli identity of rank $(k^2 + 1)$. Hence it satisfies St_{k^2+1}. We shall next determine the minimal degree of a Capelli polynomial and a standard polynomial vanishing in $M_k(F)$.

PROPOSITION 1.7.1.
1) *The Capelli polynomial Cap_{k^2+1} is an identity for $M_k(F)$.*
2) *The Capelli polynomial Cap_{k^2} is not an identity for $M_k(F)$ but it vanishes on every proper subalgebra of $M_k(F)$.*

PROOF. 1) has been proved in the previous section. If A is a proper subalgebra of $M_k(F)$, $\dim_F A < k^2$, hence by Theorem 1.5.8, $Cap_{k^2} \equiv 0$ on A. In order to show that Cap_{k^2} does not vanish on $M_k(F)$, we need to find a non-zero valuation. Let the e_{ij}'s be the usual matrix units. Then we evaluate

$$Cap_{k^2}(e_{11}, e_{12}, \ldots, e_{1k}, e_{21}, \ldots, e_{k,k-1}, e_{kk}; e_{11}, e_{11}, e_{21}, \ldots, e_{k-1,k}, e_{k1}) = e_{11} \neq 0,$$

where for $x_1, \ldots, x_{k^2}$, we substituted all the e_{ij}'s ordered according to the left lexicographic order of the indices, the indeterminates y_1, y_{k^2+1} were replaced by e_{11}, e_{k1}, respectively and for all other y_i's we made the unique substitution making $y_1 x_1 y_2 x_2 y_3 \cdots x_{k^2} y_{k^2+1}$ the only monomial with non-zero evaluation. □

We now turn to the problem of determining polynomial identities of $M_k(F)$ of minimal degree. The first result in this direction is the following ([**AL**]).

THEOREM 1.7.2. *$M_k(F)$ does not satisfy any polynomial identity of degree $< 2k$.*

1.7. IDENTITIES OF MATRIX ALGEBRAS

PROOF. Let $f \equiv 0$ be a polynomial identity for $M_k(F)$ of degree $d < 2k$. By the multilinearization process we may assume that f is multilinear. Also, by eventually multiplying f by new variables on the right and by renaming the variables, we may assume that $\deg f = 2k - 1$ and

$$f(x_1, \ldots, x_{2k-1}) = \sum_{\sigma \in S_{2k-1}} \alpha_\sigma x_{\sigma(1)} \cdots x_{\sigma(2k-1)}$$

with $\alpha_1 \neq 0$. We compute $f(e_{11}, e_{12}, e_{22}, \ldots, e_{k-1,k}, e_{kk}) = \alpha_1 e_{1k} \neq 0$, since $x_1 \cdots x_{2k-1}$ is the only monomial with non-zero evaluation. Thus f is not an identity for $M_k(F)$. $\square$

A celebrated theorem of Amitsur and Levitzki states that St_{2k} is a polynomial identity for $M_k(F)$. The original proof ([**AL**]) is by induction and uses the combinatorial properties of the matrix units. There are several other subsequent different proofs; among the most significant ones here we present the one by Razmyslov ([**R2**]) and another by Rosset ([**R**]). Both of these proofs use arguments that we shall use in other parts of this book. We start with an easy reduction.

REMARK 1.7.3. In order to prove that St_{2k} vanishes in $M_k(F)$, it is enough to prove that it vanishes in $M_k(\mathbb{Q})$.

PROOF. If St_{2k} is an identity for $M_k(\mathbb{Q})$, then it is also an identity for $M_k(\mathbb{Z})$. Let $\mathbb{Z}_p$ be the field of p elements and $\varphi_p : M_k(\mathbb{Z}) \to M_k(\mathbb{Z}_p)$ the canonical map reducing the entries of a matrix modulo p. Clearly St_{2k} is still an identity for $M_k(\mathbb{Z}_p)$. Since St_{2k} is multilinear, by Remark 1.3.4, in order to prove that St_{2k} vanishes in $M_k(F)$, it is enough to prove that St_{2k} vanishes when evaluated on the matrix units e_{ij}. But this is the same as proving that St_{2k} is an identity for $M_k(P)$, where P is the prime field of F. $\square$

Next we need some lemmas on the properties of the trace. Recall that for C a commutative algebra, $\text{tr} : M_k(C) \to C$ is the usual trace.

LEMMA 1.7.4. *Let C be a commutative algebra over $\mathbb{Q}$. If $a \in M_k(C)$ is such that* $\text{tr}(a) = \text{tr}(a^2) = \cdots = \text{tr}(a^k) = 0$, *then* $a^k = 0$.

PROOF. By Corollary 1.6.4, a has a characteristic polynomial x^k. $\square$

Let G be the Grassmann algebra generated by $e_1, e_2, \ldots$ over $\mathbb{Q}$. Recall from Example 1.1.8 that $G = G^{(0)} \oplus G^{(1)}$ where $G^{(0)}$ ($G^{(1)}$) is the subspace spanned by the monomials in the e_i's of even length (odd length resp.).

LEMMA 1.7.5. *If $a, b \in M_k(G^{(1)})$, then* $\text{tr}(ab) = -\text{tr}(ba)$.

PROOF. We may clearly write $a = \sum_i a_i w_i, b = \sum_i b_i w_i$ where $a_i, b_i \in M_k(F)$ and for all i, $w_i \in G^{(1)}$ are monomials in the basis elements $e_1, e_2, \ldots$. Recalling that tr is a symmetric bilinear form on $M_k(F)$ and for any two monomials $w_i, w_j \in G^{(1)}$, $w_i w_j = -w_j w_i$, we get

$$\text{tr}(ab) = \sum_{i,j} \text{tr}(a_i b_j) w_i w_j = -\sum_{i,j} \text{tr}(b_j a_i) w_j w_i = -\text{tr}(ba).$$

$\square$

As an immediate consequence we obtain the following.

COROLLARY 1.7.6. *Let $r \geq 1$ and $a_1, \ldots, a_{2r} \in M_k(F)$, char $F \neq 2$. Then $\operatorname{tr}(St_{2r}(a_1, \ldots, a_{2r})) = 0$.*

PROOF. Let $a = \sum_{i=1}^{2r} a_i e_i \in M_k(G^{(1)})$. Recalling that for all $\sigma \in S_{2r}$,
$$e_{\sigma(1)} \cdots e_{\sigma(2r)} = (\operatorname{sgn} \sigma) e_1 \cdots e_{2r}$$
we get
$$a^{2r} = \sum_{\sigma \in S_{2r}} (\operatorname{sgn} \sigma) a_{\sigma(1)} \cdots a_{\sigma(2r)} e_1 \cdots e_{2r} = St_{2r}(a_1, \ldots, a_{2r}) e_1 \cdots e_{2r}.$$

Apply the previous lemma to a and $a^{2r-1} \in M_k(G^{(1)})$. We obtain $2\operatorname{tr}(a^{2r}) = 0$ and, since char $F \neq 2$, $0 = \operatorname{tr}(a^{2r}) = \operatorname{tr}(St_{2r}(a_1, \ldots, a_{2r})) e_1 \cdots e_{2r}$. Hence $\operatorname{tr}(St_{2r}(a_1, \ldots, a_{2r})) = 0$. □

THEOREM 1.7.7 (Amitsur and Levitzki [**AL**]). *The algebra $M_k(F)$ satisfies the standard identity $St_{2k} \equiv 0$.*

PROOF (Rosset). By the previous remark we need only to show that St_{2k} vanishes on $M_k(\mathbb{Q})$. Let $a_1, \ldots, a_{2k} \in M_k(\mathbb{Q})$ and let $G = G^{(0)} \oplus G^{(1)}$ be the Grassmann algebra over the field of rationals. Set $a = \sum_{i=1}^{2k} a_i e_i \in M_k(G^{(1)})$. As we noticed in the proof of the previous corollary,
$$a^{2k} = St_{2k}(a_1, \ldots, a_{2k}) e_1 \cdots e_{2k}.$$

Hence, in order to prove that $St_{2k} \equiv 0$ is an identity for $M_k(\mathbb{Q})$, it is enough to show that $a^{2k} = 0$. For all $i = 1, \ldots, k$, since $a, a^{2i-1} \in M_k(G^{(1)})$, by Lemma 1.7.5, $\operatorname{tr}(a^{2i}) = \operatorname{tr}(aa^{2i-1}) = -\operatorname{tr}(a^{2i-1}a) = -\operatorname{tr}(a^{2i})$. Hence $\operatorname{tr}(a^{2i}) = 0$. It follows that $a^2 \in M_k(G^{(0)})$ is such that $\operatorname{tr}(a^{2i}) = 0$ for all $i \geq 1$. By Lemma 1.7.4, $a^{2k} = 0$, the desired conclusion. □

Given any polynomial
$$f(x_1, \ldots, x_r, y_1, \ldots, y_n) \in F\langle X \rangle,$$
multilinear in $x_1, \ldots, x_r$, we define the operator of alternation $\mathcal{A}_{x_1, \ldots, x_r}$ on the variables $x_1, \ldots, x_r$ as
$$\mathcal{A}_{x_1, \ldots, x_r} f = \sum_{\sigma \in S_r} (\operatorname{sgn} \sigma) f(x_{\sigma(1)}, \ldots, x_{\sigma(r)}, y_1, \ldots, y_n).$$

For instance, it is clear that $\mathcal{A}_{x_1, \ldots, x_r} x_1 \cdots x_r = St_r(x_1, \ldots, x_r)$ and
$$\mathcal{A}_{x_1, \ldots, x_r} y_1 x_1 y_2 \cdots y_r x_r y_{r+1} = Cap_r(x_1, \ldots, x_r; y_1, \ldots, y_{r+1}).$$

Moreover, we have.

REMARK 1.7.8. $\mathcal{A}_{x_1, \ldots, x_{2k}} [x_1, x_2] \cdots [x_{2k-1}, x_{2k}] = 2^k St_{2k}(x_1, \ldots, x_{2k})$.

PROOF. By Proposition 1.5.7, $f = \mathcal{A}_{x_1, \ldots, x_{2k}} [x_1, x_2] \cdots [x_{2k-1}, x_{2k}] = \beta St_{2k}$ and we can determine β by computing the coefficient of the monomial $x_1 \cdots x_{2k}$ in f. Clearly $[x_{\sigma(1)}, x_{\sigma(2)}] \cdots [x_{\sigma(2k-1)}, x_{\sigma(2k)}]$ contains the monomial $x_1 \cdots x_{2k}$ if and only if σ is a product of some of the transpositions $(12), (23), \ldots, (2k-1, 2k)$. It follows that there are 2^k occurrences of the monomial $x_1 \cdots x_{2k}$ in f with coefficient 1. □

Next we give the proof of the Amitsur-Levizki theorem due to Razmyslov.

1.7. IDENTITIES OF MATRIX ALGEBRAS

PROOF (Razmyslov). As before we may assume that $F = \mathbb{Q}$. By Remark 1.7.3 and Corollary 1.4.5, it is enough to prove that St_{2k} vanishes on $\mathbb{Q}\{\xi\}$, the algebra of $k \times k$ generic matrices over $\mathbb{Q}$. Let $\xi^{(1)}$ be a generic matrix and let $f(x) = \sum_{r=0}^{k} \alpha_r x^{k-r}$ be its characteristic polynomial. By the Hamilton-Cayley theorem, $f(\xi^{(1)}) = 0$. Hence, by Corollary 1.6.4, any $k \times k$ matrix satisfies the relation

$$f(x) = \sum_{r=0}^{k} \sum_{\lambda=(\lambda_1,\ldots,\lambda_n) \vdash r} q_\lambda \operatorname{tr}(x^{\lambda_1}) \cdots \operatorname{tr}(x^{\lambda_n}) x^{k-r} \equiv 0,$$

for some $q_\lambda \in \mathbb{Q}$.

Since the trace is additive, we can apply to the previous expression the process of multilinearization. This is the obvious generalization of the process defined in Section 1.3, where we linearize all variables regardless of the trace function.

We obtain the polynomial with trace

$$g(x_1,\ldots,x_k) = \sum_{\sigma \in S_k} x_{\sigma(1)} \cdots x_{\sigma(k)} + \sum_{r=1}^{k} \sum_{\sigma \in S_k} \sum_{\lambda=(\lambda_1,\ldots,\lambda_n) \vdash r} q_\lambda \operatorname{tr}(x_{\sigma(1)} \cdots x_{\sigma(\lambda_1)})$$
$$\cdots \operatorname{tr}(x_{\sigma(\lambda_1 + \cdots + \lambda_{n-1}+1)} \cdots x_{\sigma(r)}) x_{\sigma(r+1)} \cdots x_{\sigma(k)}.$$

Clearly $g(\xi^{(1)},\ldots,\xi^{(k)}) = 0$ for all generic matrices $\xi^{(1)},\ldots,\xi^{(k)} \in \mathbb{Q}\{\xi\}$. We now substitute, for all $i = 1,\ldots,k$, x_i with $[x_{2i-1}, x_{2i}]$. We then apply the operator $\mathcal{A}_{x_1,\ldots,x_{2k}}$ to the resulting polynomial. By the previous remark we obtain the polynomial with trace

$$g'(x_1,\ldots,x_{2k}) = 2^k k! St_{2k}(x_1,\ldots,x_{2k}) + \sum_{r=1}^{k} \sum_{\lambda \vdash r} q'_\lambda h_\lambda$$

where $q'_\lambda \in \mathbb{Q}$ and the h_λ's are polynomials with trace of the form

$$\operatorname{tr}(St_{2\lambda_1}(x_{i_1},\ldots,x_{i_{2\lambda_1}})) \cdots \operatorname{tr}(St_{2\lambda_n}(x_{j_1},\ldots,x_{j_{2\lambda_n}})) St_{2(k-r)}(x_{l_1},\ldots,x_{l_{2(k-r)}}).$$

We must have $g'(\xi^{(1)},\ldots,\xi^{(2k)}) = 0$ for $\xi^{(1)},\ldots,\xi^{(2k)}$ generic matrices. Moreover, since by Corollary 1.7.6, $\operatorname{tr}(St_{2j}(\xi^{(1)},\ldots,\xi^{(2j)})) = 0$, for all $j \geq 1$ we have $h_\lambda(\xi^{(1)},\ldots,\xi^{(2k)}) = 0$ for all $\lambda \vdash r$. It follows that $St_{2k}(\xi^{(1)},\ldots,\xi^{(2k)}) = 0$ and the proof is completed. □

We conclude this section by showing that actually the only identities of degree $2k$ for $M_k(F)$ are the scalar multiples of St_{2k}.

PROPOSITION 1.7.9. *Let char $F \neq 2$. If f is a polynomial identity of $M_k(F)$ and $\deg(f) = 2k$, then $f = \alpha St_{2k}$, for some $\alpha \in F$.*

PROOF. Suppose first that f is multilinear and write

$$f = f(x_1,\ldots,x_{2k}) = \sum_{\sigma \in S_{2k}} \alpha_\sigma x_{\sigma(1)} \cdots x_{\sigma(2k)}.$$

Without loss of generality, we may assume that α_1, the coefficient of $x_1 \cdots x_{2k}$ in f, is equal to 1. Let e_{ij}, $1 \leq i \leq 2k$, be the usual matrix units. Then

$$0 = f(e_{11}, e_{12}, e_{22}, e_{23}, \ldots, e_{i-1,i}, e_{ii}, e_{ii}, e_{i,i+1}, \ldots, e_{k-1,k}, e_{kk}) = (1 - \alpha_{\sigma_i}) e_{1k}$$

for all $i = 1,\ldots,k$, where $\sigma_i \in S_k$ is the transposition $(2i-1, 2i)$. Hence

$$\alpha_{(2i-1,2i)} = -1.$$

We now replace in f, $x_1, \ldots, x_{2k}$ with the elements

$$e_{12}, e_{22}, e_{23}, \ldots, e_{i-1,i}, e_{ii}, e_{ii}, e_{i,i+1}, \ldots, e_{k-1,k}, e_{kk}, e_{k1}$$

respectively, and we compute the coefficient of e_{1k} in the resulting evaluation. We obtain

$$\alpha_{(2i-2, 2i-1)} = -\alpha_1 = -1,$$

for all $i = 1, \ldots, k$. The outcome of this is that $\alpha_\tau = -1$ in f, for all transpositions of the type $(i, i+1) \in S_{2k}$.

Since f is an identity of $M_k(F)$, for any $\rho \in S_{2k}$ the polynomial

$$f(x_{\sigma(1)}, \ldots, x_{\sigma(2k)}) = \sum_{\rho \in S_{2k}} \alpha_\rho x_{\sigma\rho(1)} \cdots x_{\sigma\rho(2k)} = \sum_{\pi \in S_{2k}} \alpha_{\sigma^{-1}\pi} x_{\pi(1)} \cdots x_{\pi(2k)}$$

is still a polynomial identity of $M_k(F)$ of degree $2k$. Replacing σ^{-1} with σ and comparing the coefficients of the monomials $x_1 \cdots x_{2k}$ and $x_{\pi(1)} \cdots x_{\pi(2k)}$ with $\pi = (i-1, i)$ we obtain

$$\alpha_{\sigma(i-1,i)} = -\alpha_\sigma$$

for all $i = 1, \ldots, 2k-1$. Since the transpositions $(12), (23), \ldots, (2k-1, 2k)$ generate S_{2k}, we conclude that $\alpha_\sigma = \operatorname{sgn} \sigma$ and $f = St_{2k}$ as expected.

In general, let g be the multilinear polynomial obtained from f by the process of multilinearization. Clearly g is symmetric in at least two variables. On the other hand, by the first part of the proof, g is a scalar multiple of St_{2k}. Hence, since $\operatorname{char} F \neq 2$, g being symmetric and alternating in two variables must be the zero polynomial. This proves that there are no identities of degree $2k$ of $M_k(F)$ which are not multilinear. $\square$

We close this section with the following general problem: can one find a set of polynomials generating $\operatorname{Id}(M_k(F))$ as a T-ideal? From a celebrated theorem of Kemer (see [**Ke7**]) it is know that in characteristic zero every T-ideal is finitely generated. In the case of $M_k(F)$, a complete answer is known only for $k = 2$. In fact, Razmyslov in [**R1**] exhibited 9 polynomials generating $\operatorname{Id}(M_2(F))$ as a T-ideal in case $\operatorname{char} F = 0$. Drensky in [**D2**] proved that St_4 and $[[x_1, x_2]^2, x_3]$ is a minimal set of generators of $\operatorname{Id}(M_2(F))$. Recently Koshlukov has extended this result to any characteristic different from 2 ([**Ko**]). It turns out that if $\operatorname{char} F \neq 2, 3$ the above two identities generate $\operatorname{Id}(M_2(F))$ but one extra identity is needed when $\operatorname{char} F = 3$.

1.8. A theorem of Lewin

Let A and B be two F-algebras with T-ideals of identities $\operatorname{Id}(A)$ and $\operatorname{Id}(B)$, respectively. The product of the T-ideals $\operatorname{Id}(A)\operatorname{Id}(B)$ is still a T-ideal and a general problem is that of constructing an F-algebra C containing A and B such that $\operatorname{Id}(C) = \operatorname{Id}(A)\operatorname{Id}(B)$. One such construction was performed by Lewin in [**Le**] and we prove this result in this section. We shall apply Lewin's theorem in the next section in order to relate the identities of upper block triangular matrices to those of matrices.

For $F\langle X \rangle$, the free associative algebra of countable rank, let us denote by R the free $F\langle X \rangle$-bimodule with a countable set of free generators $r_1, r_2, \ldots$. There exists a uniquely defined linear map $\delta : F\langle X\rangle \to R$ such that $\delta(x_i) = r_i$, for all $i \geq 1$ and $\delta(fg) = \delta(f)g + f\delta(g)$ for any $f, g \in F\langle X\rangle$. Clearly δ is defined on monomials by

$$\delta(x_{i_1}\cdots x_{i_n}) = \sum_{j=1}^n x_{i_1}\cdots x_{i_{j-1}} r_{i_j} x_{i_{j+1}}\cdots x_{i_n}$$

and then extended by linearity to all $F\langle X\rangle$. For short we shall call any linear map satisfying $\delta(ab) = \delta(a)b + a\delta(b)$ a derivation. Consider the linear map

$$\psi : F\langle X\rangle \to \begin{pmatrix} F\langle X\rangle & R \\ 0 & F\langle X\rangle \end{pmatrix} \quad \text{such that} \quad \psi(f) = \begin{pmatrix} f & \delta(f) \\ 0 & f \end{pmatrix}.$$

Notice that ψ is an algebra homomorphism. In fact, if we assume δ to be only an additive map, then ψ is multiplicative if and only if δ is a derivation.

If we consider two ideals U, V of $F\langle X\rangle$, the quotient space $R/(UR + RV)$ has a natural structure of free $(F\langle X\rangle/U, F\langle X\rangle/V)$-bimodule with basis $\{r_i + (UR+RV), i = 1, 2, \ldots\}$.

Let π_U, π_V be the canonical homomorphisms of $F\langle X\rangle$ onto $F\langle X\rangle/U$ and $F\langle X\rangle/V$ respectively, and let $\overline{\delta} : F\langle X\rangle \to R/(UR + RV)$ be defined by $\overline{\delta}(f) = \delta(f) + (UR + RV)$. Then $\overline{\delta}$ is also a derivation since it satisfies

$$\overline{\delta}(fg) = \delta(fg) + UR + RV = \delta(f)g + f\delta(g) + UR + RV$$

$$= (\delta(f) + UR + RV)g + f(\delta(g) + UR + RV) = \overline{\delta}(f)g + f\overline{\delta}(g)$$

as a linear map from $F\langle X\rangle$ to its bimodule $R/(UR + RV)$.

This implies that the map

$$\varphi : F\langle X\rangle \to \begin{pmatrix} F\langle X\rangle/U & R/(UR+RV) \\ 0 & F\langle X\rangle/V \end{pmatrix}$$

such that

$$\varphi(f) = \begin{pmatrix} \pi_U(f) & \overline{\delta}(f) \\ 0 & \pi_V(f) \end{pmatrix}$$

is a homomorphism of algebras. The idea of constructing such homomorphisms arose to Magnus and allows to construct some matrix realization of the algebra $F\langle X\rangle/\ker\varphi$ starting from some realization of $F\langle X\rangle/U$ and $F\langle X\rangle/V$. Next we shall prove that $\ker\varphi = U \cdot V$.

THEOREM 1.8.1 (Lewin [**Le**]). *Let R be the free $(F\langle X\rangle, F\langle X\rangle)$-bimodule with free generators $r_1, r_2, \ldots$. For non-trivial two-sided ideals U and V of $F\langle X\rangle$, let*

$$\pi_U : F\langle X\rangle \to F\langle X\rangle/U \quad \text{and} \quad \pi_V : F\langle X\rangle \to F\langle X\rangle/V$$

be the corresponding canonical epimorphisms. Also let $\delta : F\langle X\rangle \to R$ be a derivation with $\delta(x_i) = r_i$, $i = 1, 2, \ldots$ and let $\overline{\delta}(f) = \delta(f) + (UR + RV)$. Then the linear map

$$\varphi : F\langle X\rangle \to \begin{pmatrix} F\langle X\rangle/U & R/(UR+RV) \\ 0 & F\langle X\rangle/V \end{pmatrix}$$

such that $\varphi(f) = \begin{pmatrix} \pi_U(f) & \overline{\delta}(f) \\ 0 & \pi_V(f) \end{pmatrix}$ is a homomorphism of algebras and $\ker\varphi = U \cdot V$.

PROOF. First note that for any $f \notin U$, $\pi_U(f) \neq 0$ and therefore $f \notin \ker \varphi$. This says that $\ker \varphi \subseteq U$. Similarly, $\ker \varphi \subseteq V$. On the other hand, for all $f \in U$, $g \in V$ we have that $\varphi(fg) = \varphi(f)\varphi(g) = 0$ and we obtain $U \cap V \supseteq \ker \varphi \supseteq U \cdot V$. Therefore in order to prove the theorem it is enough to verify the inclusion $V \cap \ker \varphi \subseteq U \cdot V$.

Consider the basis B of $F\langle X \rangle$ consisting of all monomials in elements of X. We consider X linearly ordered by requiring that $x_1 < x_2 < \ldots$. We then define a linear order on B by setting, for any two monomials $a, b \in B$, $a > b$ if $\deg a > \deg b$ or $\deg a = \deg b$ and a is strictly greater than b in the right lexicographic order.

Now consider the linear order on the set $R_0 = \{r_1, r_2, \ldots\}$ of generators of R given by $r_1 < r_2 < \ldots$. Then an F-basis of R consists of all products ar_ib with $a, b \in B$. This basis can also be ordered by setting $ar_ib < cr_jd$ if and only if $b < d$ or $b = d$ and $r_i < r_j$ or $b = d$, $r_i = r_j$ and $a < c$.

Given any polynomial $f \in F\langle X \rangle$, f has a leading term (greatest monomial in the given order) which we shall denote by $\overline{f}$. We now define a subset $C \subseteq B$ in the following way: a monomial c belongs to C if $c = \overline{v}$ for some $v \in V$ and c cannot be written as $c = b\overline{w}$ for some $w \in V$, $b \neq 1$.

For any $c \in C$ let us fix some $a_c \in V$ with $\overline{a}_c = c$. Then any $v \in V$ has the leading term of the type $bc = \overline{ba_c}$, $c \in C$, and hence $\overline{v - ba_c} < \overline{v}$ and $v - ba_c \in V$. Therefore all $a_c, c \in C$ generate V as a left $F\langle X \rangle$-module.

Since V is a non-trivial ideal of $F\langle X \rangle$, any $c \in C$ is a monomial of positive degree and it can be written as $c = x_id_c$ where $d_c \in B$ and $x_i \in X$. Note that $d_c \notin C$ by the definition of C. For convenience we shall denote such x_i by x_c and we shall also write $r_i = \delta(x_c) = r_c$. Using this convention we decompose the basis $R_0 B$ of the free left $F\langle X \rangle$-module $T = RF\langle X \rangle$ into three parts

$$B_1 = \{r_c d_c | c \in C\}, \ B_2 = \{rbc | r \in R_0, b \in B, c \in C\},$$
$$B_3 = \{rb | r \in R_0, b \in B, rb \notin B_1 \cup B_2\}.$$

Notice that $B_1 \cap B_2 = \emptyset$. In fact, if $rbc_1 = r_c d_c$, then $bc_1 = d_c$ and this implies $x_c bc_1 = x_c d_c \in C$. But since $c_1 \in C$, $x_c bc_1 \notin C$, a contradiction.

Using the above decomposition we define a new basis of T. First note that $\overline{a}_c = x_c d_c$ and $\delta(\overline{a}_c) = \delta(x_c)d_c + x_c\delta(d_c)$, hence $\overline{\delta(a_c)}$, the leading term of $\delta(a_c)$, is equal to $\delta(x_c)d_c = r_c d_c$. Now we replace B_1 with $K_1 = \{\delta(a_c) | c \in C\}$.

For any monomial bc with $b \in B$, $c \in C$ we fix $v_{bc} = ba_c \in V$ with $\overline{v}_{bc} = bc$ and take $K_2 = \{rv_{bc} | r \in R_0, b \in B, c \in C\}$.

Then there exists a one-to-one correspondence $b \longleftrightarrow b'$ between elements of $B_1 \cup B_2$ and $K_1 \cup K_2$ such that $\overline{b'} = b$. Let $K_3 = B_3$, standard arguments with reduction of leading terms show that $K = K_1 \cup K_2 \cup K_3$ generates T as a left $F\langle X \rangle$-module. On the other hand, any non-trivial relation among $b'_1, \ldots, b'_n \in K$ leads to a non-trivial relation among the basic elements $\overline{b'_1}, \ldots, \overline{b'_n} \in R_0 F\langle X \rangle$. Therefore K is a basis of T.

Note that the elements $v_{bc}, b \in B, c \in C$, form a linear basis of V. Hence K_2 is an F-basis of $R_0 V = \mathrm{span}\{K_2\}$ and

$$RV = F\langle X \rangle R_0 V = F\langle X \rangle K_2.$$

Since K is a basis of $T = RF\langle X \rangle$, we have

$$UR = UF\langle X \rangle R_0 F\langle X \rangle = UR_0 F\langle X \rangle = UK = UK_1 \oplus UK_2 \oplus UK_3$$

and

(1.3) $$UR + RV = UK_1 \oplus F\langle X \rangle K_2 \oplus UK_3.$$

Now let $v \in V \cap \ker \varphi$. Then $\delta(v) \in U \cdot R + R \cdot V$. Using the derivation property of δ and the decomposition of v into the sum $v = \sum_{c \in C} b_c a_c$ with $b_c \in F\langle X \rangle$ we can write $\delta(v)$ as the sum $\delta(v) = A_v + B_v$ where

$$A_v = \sum_c \delta(b_c) a_c, \quad B_v = \sum_c b_c \delta(a_c).$$

Clearly, $A_v \in RV$, $B_v \in F\langle X \rangle \cdot K_1$ and the inclusion $\delta(v) \in UR + RV$ implies

$$B_v \in (UR + RV) \cap F\langle X \rangle \cdot K_1.$$

By (1.3) we conclude that $B_v \in UK_1$. But since the elements $\delta(a_c)$ of K_1 are left-independent, it follows that all b_c from the expression for B_v lie in U. Hence $v = \sum b_c a_c \in UV$, and the proof of the theorem is complete. $\square$

Theorem 1.8.1 has an important application in case U and V are T-ideals.

COROLLARY 1.8.2. *Let A, B be two F-algebras and let M be an (A, B)-bimodule. Suppose that*

1) *A contains a relatively free subalgebra $\widetilde{A}$ with free generators $a_1, a_2, \ldots$ and $U = \mathrm{Id}(\widetilde{A})$ is the T-ideal of identities of $\widetilde{A}$;*
2) *B contains a relatively free subalgebra $\widetilde{B}$ with free generators $b_1, b_2, \ldots$ and $V = \mathrm{Id}(\widetilde{B})$;*
3) *M contains a free $(\widetilde{A}, \widetilde{B})$-bimodule with free generators $w_1, w_2, \ldots$.*

Then the elements $\begin{pmatrix} a_i & w_i \\ 0 & b_i \end{pmatrix}$, $i = 1, 2, \ldots$, are free generators of a relatively free subalgebra $\widetilde{C}$ of the algebra

$$\begin{pmatrix} A & M \\ 0 & B \end{pmatrix}$$

and $\mathrm{Id}(\widetilde{C}) = UV$.

PROOF. Let $\widetilde{A} \cong F\langle X \rangle / U$, $\widetilde{B} \cong F\langle X \rangle / V$, $W = \{w_1, w_2, \ldots\}$. The isomorphisms $F\langle X \rangle / U \to \widetilde{A}$, $F\langle X \rangle / V \to \widetilde{B}$ are defined by the set theoretical maps $x_i + U \to a_i$, $x_i + V \to b_i$ respectively, $i = 1, 2, \ldots$. Also if R is a free $(F\langle X \rangle, F\langle X \rangle)$-bimodule with generators $r_1, r_2, \ldots$, then the map $r_i + (UR + RV) \to w_i$ defines a corresponding bimodule isomorphism $R/(UR + RV) \to AWB$.

Therefore we obtain an algebra isomorphism

$$\psi : \begin{pmatrix} F\langle X \rangle / U & R/(UR + RV) \\ 0 & F\langle X \rangle / V \end{pmatrix} \to \begin{pmatrix} \widetilde{A} & \widetilde{A} W \widetilde{B} \\ 0 & \widetilde{B} \end{pmatrix} \subseteq \begin{pmatrix} A & M \\ 0 & B \end{pmatrix}$$

such that

$$\psi\left(\begin{pmatrix} \overline{x}_i & r_i \\ 0 & \widetilde{x}_i \end{pmatrix}\right) = \begin{pmatrix} a_i & w_i \\ 0 & b_i \end{pmatrix} = a_i + w_i + b_i$$

where $\overline{x}_i = x_i + U$, $\widetilde{x}_i = x_i + V$, $i = 1, 2, \ldots$. Clearly $\ker \psi = 0$. Denote by φ the homomorphism

$$F\langle X \rangle \to \begin{pmatrix} F\langle X \rangle / U & R/(UR + RV) \\ 0 & F\langle X \rangle / V \end{pmatrix}$$

defined in Theorem 1.8.1 and let $\overline{\varphi} = \psi \varphi$. Then by the previous theorem $\mathrm{Im}(\overline{\varphi})$ is the F-algebra generated by the elements

$$\begin{pmatrix} a_i & w_i \\ 0 & b_i \end{pmatrix}, \quad i = 1, 2, \ldots$$

and
$$\mathrm{Im}(\bar\varphi) \cong F\langle X\rangle/\ker\bar\varphi = F\langle X\rangle/UV.$$
□

1.9. Identities of block-triangular matrices

A natural generalization of the algebra of $k \times k$ matrices is given by the algebra of block-triangular matrices. As we shall see in a later chapter, this algebra plays an important role in PI-theory in particular in the study of the asymptotic behaviour of the sequence of codimensions.

For an ordered set of positive integers $(d_1, \ldots, d_m)$ let us denote by
$$UT(d_1, \ldots, d_m)$$
the subalgebra of the matrix algebra $M_{d_1+\cdots+d_m}(F)$ consisting of all matrices of the type
$$\begin{pmatrix} A_1 & \cdots & * \\ \vdots & \ddots & \vdots \\ 0 & \cdots & A_m \end{pmatrix}$$
where $A_1 \in M_{d_1}(F), \ldots, A_m \in M_{d_m}(F)$. Any such matrix has a block-triangular structure and we call $UT(d_1, \ldots, d_m)$ the algebra of upper block-triangular matrices of size $d_1, \ldots, d_m$. In particular, for $m = 1$ we have $UT(d_1) = M_{d_1}(F)$.

In order to describe the T-ideal of identities of $UT(d_1, \ldots, d_m)$ in terms of the T-ideals of the building blocks $M_{d_1}(F), \ldots, M_{d_m}(F)$ we shall apply Lewin's theorem from the previous section. We prove

THEOREM 1.9.1 ([**GZ7**]). *If $UT(d_1, \ldots, d_m)$ is an upper block-triangular matrix algebra over an infinite field F, then*
$$\mathrm{Id}(UT(d_1, \ldots, d_m)) = \mathrm{Id}(M_{d_1}(F)) \cdots \mathrm{Id}(M_{d_m}(F)).$$

PROOF. Let $I_1 = \mathrm{Id}(M_{d_1}(F)), \ldots, I_m = \mathrm{Id}(M_{d_m}(F))$. If $g_1 \in I_1, \ldots, g_m \in I_m$, it is easy to check that $g_1 \cdots g_m$ is an identity of $UT(d_1, \ldots, d_m)$. Hence $I_1 \cdots I_m \subseteq \mathrm{Id}(UT(d_1, \ldots, d_m))$.

In order to prove the other inclusion we consider
$$R = UT(d_1, \ldots, d_m) \otimes F[\xi_{ij}^{(t)} | i, j, t \geq 1],$$
the algebra tensor product of $UT(d_1, \ldots, d_m)$ with the polynomial algebra in the commutative indeterminates $\xi_{ij}^{(t)}$. Since F is infinite, by Lemma 1.4.2,
$$\mathrm{Id}(UT(d_1, \ldots, d_m)) = \mathrm{Id}(R).$$
In the next lemma we shall prove that R contains a subalgebra C with $\mathrm{Id}(C) = I_1 \cdots I_m$. As a consequence $\mathrm{Id}(UT(d_1, \ldots, d_m)) = \mathrm{Id}(R) \subseteq \mathrm{Id}(C) = I_1 \cdots I_m$ and the proof of the theorem will be complete. □

LEMMA 1.9.2. *The tensor product $UT(d_1, \ldots, d_m) \otimes F[\xi_{ij}^{(t)} | i, j, t \geq 1]$ contains as a subalgebra a relatively free algebra of countable rank with T-ideal $I = I_1 \cdots I_m$.*

PROOF. We shall prove the lemma by induction on m. Since the case $m = 1$ follows from Corollary 1.4.5, we may assume that $m > 1$. For convenience denote

$A = UT(d_1, \ldots, d_{m-1})$, $B = M_{d_m}(F)$, $t = d_1 + \cdots + d_{m-1}$, $s = d_m$. Decompose the set $\{\xi_{ij}^{(t)} | i, j, t \geq 1\}$ of commuting indeterminates into three disjoint subsets,

$$\{\xi_{ij}^{(t)} | i, j, t \geq 1\} = \{\rho_{ij}^{(t)} | i, j, t \geq 1\} \cup \{\mu_{ij}^{(t)} | i, j, t \geq 1\} \cup \{\nu_{ij}^{(t)} | i, j, t \geq 1\}.$$

Consider the embedding $A \to M_{r+s}(F)$, $B \to M_{r+s}(F)$,

$$A = \begin{pmatrix} M_{d_1}(F) & \cdots & * & 0 \\ \vdots & \ddots & \vdots & \vdots \\ 0 & \cdots & M_{d_{m-1}}(F) & 0 \\ 0 & \cdots & 0 & 0 \end{pmatrix}, \quad B = \begin{pmatrix} 0 & \cdots & 0 & 0 \\ \vdots & \ddots & \vdots & \vdots \\ 0 & \cdots & 0 & 0 \\ 0 & \cdots & 0 & M_{d_m}(F) \end{pmatrix}.$$

Then $B = \mathrm{span}\{e_{ij} | r+1 \leq i, j \leq r+s\}$ and A is the linear span of some of the matrix units e_{ij} with $1 \leq i, j \leq r$. For $t = 1, 2, \ldots$ denote

$$a_t = \sum_{i,j} \rho_{ij}^{(t)} e_{ij} \quad \text{where} \quad e_{ij} \in A,$$

$$b_t = \sum_{i,j=1}^{s} \nu_{ij}^{(t)} e_{r+i, r+j}.$$

As we mentioned above, the algebra $A = UT(d_1, \ldots, d_{m-1})$ satisfies all identities of the form $g = g_1 \cdots g_{m-1}$, $g_1 \in I_1, \ldots, g_{m-1} \in I_{m-1}$, i.e., $\mathrm{Id}(A) \supseteq I_1 \cdots I_{m-1}$. On the other hand, by induction on m, $A \otimes F[\rho_{ij}^{(t)}]$ contains a relatively free algebra with T-ideal $I_1 \ldots I_{m-1}$. Hence,

$$\mathrm{Id}(A) = I_1 \cdots I_{m-1}$$

and by Theorem 1.4.4 the algebra $\widetilde{A} = F\{a_1, a_2, \ldots\}$ is relatively free with $\mathrm{Id}(\widetilde{A}) = I_1 \cdots I_{m-1}$. Also $\widetilde{B} = F\{b_1, b_2, \ldots\}$ is relatively free and $\mathrm{Id}(\widetilde{B}) = I_m$.

Denote also

$$u_t = \sum_{i=1}^{r} \sum_{j=1}^{s} \mu_{ij}^{(t)} e_{i, r+j}.$$

Next we shall prove that $u_1, u_2, \ldots$ generate in $UT(d_1, \ldots, d_m) \otimes F[\xi_{ij}^{(t)} | i, j, t \geq 1]$ a free $(\widetilde{A}, \widetilde{B})$-bimodule.

Let $f_1, f_2, \ldots$ be some linearly independent elements of $\widetilde{A}$. We need to verify that

$$\sum_{i,j} f_i u_j g_{ij} \neq 0$$

whenever all g_{ij} are non-zero elements of $\widetilde{B}$. Since $u_1, u_2, \ldots$ depend on disjoint sets of commutative indeterminates $\mu_{ij}^{(t)}$, we need to check this inequality only for one u_i, say, u_1. Suppose

(1.4) $$f_1 u_1 g_1 + \cdots + f_n u_1 g_n = 0,$$

for some $n \geq 1$ and $g_1, \ldots, g_n \neq 0$ in $\widetilde{B}$.

Since g_1 is a non-zero element of the relatively free algebra $\widetilde{B}$ it has a non-zero value in B under some specialization of the variables $\nu_{ij}^{(t)}$. This value is a linear combination of matrix units e_{ij}, $r+1 \leq i, j \leq r+s$, with some non-zero coefficient. Denote by $\sum \lambda_{ij}^{(k)} e_{ij}$ the value of g_k under this specialization. Let the coefficient of $e_{i_0 j_0}$ be non-zero, i.e., $\lambda_{i_0 j_0}^{(1)} \neq 0$.

The left-hand side of (1.4) is zero for any specialization of the variables $\nu_{ij}^{(t)}$. Hence, we obtain

$$\sum_{i,j=r+1}^{r+s} (\sum_k \lambda_{ij}^{(k)} f_k) u_1 e_{ij} = 0 \tag{1.5}$$

in $UT(d_1, \ldots, d_m) \otimes F[\rho_{ij}^{(t)}, \mu_{ij}^{(t)}, \nu_{ij}^{(t)}]$ and $\lambda_{i_0 j_0}^{(1)} \neq 0$. Now consider a specialization of the variables $\mu_{ij}^{(1)} \to F$ such that $u_1 \to e_{r i_0}$. Then from (1.5) it follows that

$$\sum_j \sum_k \lambda_{i_0 j}^{(k)} f_k e_{rj} = 0. \tag{1.6}$$

But for any fixed j the product $f_k e_{rj}$ is a matrix whose first, $\ldots, (j-1)$th, $(j+1)$th, $\ldots, (r+s)$th columns are all zero. Hence the equality (1.5) holds for any j, in particular, for $j = j_0$ we have

$$(\sum_k \lambda_{i_0 j_0}^{(k)} f_k) e_{rj_0} = 0. \tag{1.7}$$

Note that the coefficient of e_{rj_0} is a linear combination of the linearly independent elements $f_1, \ldots, f_n \in \widetilde{A}$ with $\lambda_{i_0 j_0}^{(1)} \neq 0$. Therefore

$$f = \sum_k \lambda_{i_0 j_0}^{(k)} f_k$$

is a non-zero element of $\widetilde{A}$, that is, f is not an identity of A. But from (1.7) it follows that for any specialization of the variables $\rho_{ij}^{(t)} \to F$, i.e., for any homomorphism $\varphi : \widetilde{A} \to A$ one has $\varphi(f) e_{rj_0} = 0$. This means that the rth column of the matrix $\varphi(f)$ is zero. From Lemma 1.3.10 it follows that the subspace of A generated by all values $\varphi(f)$ of f in A is a Lie ideal in A. On the other hand, it is easy to observe that A has no non-zero Lie ideals whose rth column is zero. For instance, if $C = (c_{ij}) \in A$ is a matrix whose $(k+1)$th, $\ldots, r$th columns are zero and $c_{ik} \neq 0$ for some i, k, then $[C, e_{kr}]$ has a non-zero element in the (i, r) entry. This contradiction proves that $u_1, u_2, \ldots$ generates a free $(\widetilde{A}, \widetilde{B})$-bimodule.

Now we apply Corollary 1.8.2 and we conclude that $F\{a_i + u_i + b_i | i = 1, 2, \ldots\}$ is the relatively free algebra corresponding to the T-ideal

$$I_1 \cdots I_{m-1} \cdot I_m = \mathrm{Id}(M_{d_1}(F)) \cdots \mathrm{Id}(M_{d_m}(F)).$$

$\square$

1.10. Central polynomials in matrix algebras

In this section we prove the existence of central polynomials for matrix algebras, i.e., polynomials which do not vanish but take central values in the algebra. The existence of these polynomials is of great interest not only for the development of the structure theory of PI-algebras, but also, as we shall see in later chapters also for the combinatorial PI-theory.

DEFINITION 1.10.1. Let A be an F-algebra with non-zero center. A polynomial $f(x_1, \ldots, x_n) \in F\langle X \rangle$ is a central polynomial for A if
  (1) for all $a_1, \ldots, a_n \in A$, $f(a_1, \ldots, a_n)$ lies in the center of A;
  (2) f is not a polynomial identity of A;
  (3) f has zero constant term.

The first example of a central polynomial for $M_2(F)$ was given by Wagner and Hall as follows.

EXAMPLE 1.10.2. The polynomial $[x_1, x_2]^2$ is a central polynomial for $M_2(F)$. This is readily seen since $\operatorname{tr}[a, b] = 0$ and when a and b are 2×2 matrices, this implies that $[a, b]^2$ is a scalar matrix (see also Example 1.1.7).

Another interesting example is the following.

EXAMPLE 1.10.3. The polynomial $[x_1, x_2]$ is a central polynomial for the Grassmann algebra G.

For many years the Wagner-Hall polynomial was the only known central polynomial for matrix algebras. Nevertheless, Kaplansky in [**K2**] conjectured the existence of central polynomials for matrix algebras of any size. This conjecture was proved independently by Formanek in [**F1**] and Razmyslov in [**R1**]. Here we shall give Formanek construction of a central polynomial for $k \times k$ matrices. Throughout, F will be an arbitrary field and $F\{\xi\} = F\{\xi^{(1)}, \xi^{(2)}, \ldots\}$ the algebra of $k \times k$ generic matrices.

We start by proving the following fact due to Procesi.

LEMMA 1.10.4. *The eigenvalues of the generic matrix* $\xi^{(1)} = \sum_{i,j=1}^{k} \xi_{ij}^{(1)} e_{ij}$ *are distinct and algebraically independent over* F.

PROOF. If we specialize $\xi^{(1)}$ to the matrix $a = \sum_{i=1}^{k} \xi_{ii}^{(1)} e_{ii} \in M_k(F(\xi_{ij}^{(1)}))$, the characteristic polynomial $\det(x - \xi^{(1)})$ of $\xi^{(1)}$ specializes to the characteristic polynomial $\det(x - a)$ of a. Since the latter has distinct and algebraically independent roots, so does the characteristic polynomial of $\xi^{(1)}$. □

Next we make a reduction that we shall use in the proof of the main theorem.

REMARK 1.10.5. Let $f(x_1, \ldots, x_n) \in F\langle X \rangle$ be such that $f(\xi_0^{(1)}, \xi^{(2)}, \ldots, \xi^{(n)})$ is a non-zero scalar matrix where $\xi_0^{(1)} = \sum_{i=1}^{k} \xi_{ii}^{(1)} e_{ii}$ is a diagonal matrix. Then $f(x_1, \ldots, x_n)$ is a central polynomial of $M_k(F)$.

PROOF. Let K be the algebraic closure of the rational function field $F(\xi_{ij}^{(t)})$, $1 \leq t \leq n$. By the previous lemma there exists $a \in M_k(K)$ such that $\eta^{(1)} = a^{-1} \xi^{(1)} a$ is a diagonal matrix. Define $\eta^{(t)} = a^{-1} \xi^{(t)} a$, for all $t = 2, \ldots, n$. Now,

$$a^{-1} f(\xi^{(1)}, \xi^{(2)}, \ldots, \xi^{(n)}) a = f(\xi_0^{(1)}, a^{-1} \xi^{(2)} a, \ldots, a^{-1} \xi^{(n)} a) = f(\eta^{(1)}, \ldots, \eta^{(n)}),$$

is a non-zero scalar matrix, hence $f(\xi^{(1)}, \ldots, \xi^{(n)})$ is a scalar matrix. □

THEOREM 1.10.6 (Formanek [**F1**], Razmyslov [**R1**]). *For every* $k \geq 1$, $M_k(F)$ *has a central polynomial.*

PROOF. Let $F[\xi_1, \ldots, \xi_{k+1}]$ be the polynomial ring in the commuting indeterminates $\xi_1, \ldots, \xi_{k+1}$, and consider $F\langle x, y_1, \ldots, y_k \rangle$, the free algebra on $x, y_1, \ldots, y_k$. Define the F-linear map $\varphi : F[\xi_1, \ldots, \xi_{k+1}] \to F\langle x, y_1, \ldots, y_k \rangle$ induced by setting

$$\varphi(\xi_1^{a_1} \cdots \xi_{k+1}^{a_{k+1}}) = x^{a_1} y_1 x^{a_2} y_2 \cdots y_{k-1} x^{a_k} y_k x^{a_{k+1}}.$$

Then set

(1.8) $$g(\xi_1, \ldots, \xi_{k+1}) = \prod_{2 \leq i \leq k} (\xi_1 - \xi_i)(\xi_{k+1} - \xi_i) \prod_{2 \leq j < l \leq k} (\xi_j - \xi_l)^2$$

and
$$G(x, y_1, \ldots, y_k) = \varphi(g(\xi_1, \ldots, \xi_{k+1})).$$
We shall prove next that
$$f(x, y_1, \ldots, y_k)$$
$$= G(x, y_1, \ldots, y_k) + G(x, y_2, \ldots, y_k, y_1) + \ldots + G(x, y_k, y_1, \ldots, y_{k-1})$$
is a central polynomial for $M_k(F)$. By the previous remark, it is enough to show that $f(X, \xi^{(2)}, \ldots, \xi^{(k+1)})$ is a non-zero scalar matrix where $X = \sum_{i=1}^{k} \xi_i e_{ii}$ is a diagonal matrix. Moreover, since f is linear in each variable $y_1, \ldots, y_k$, it is enough to evaluate $\xi^{(2)}, \ldots, \xi^{(k+1)}$ on the matrix units e_{ij}.

Now, the monomial $x^{a_1} y_1 x^{a_2} y_2 \cdots y_{k-1} x^{a_k} y_k x^{a_{k+1}}$ evaluates to
$$X^{a_1} e_{i_1 j_1} X^{a_2} e_{i_2 j_2} \cdots X^{a_k} e_{i_k j_k} X^{a_{k+1}} = \xi_{i_1}^{a_1} \xi_{i_2}^{a_2} \cdots \xi_{i_k}^{a_k} \xi_{j_k}^{a_{k+1}} e_{i_1 j_1} e_{i_2 j_2} \cdots e_{i_k j_k}.$$
This implies that
$$G(X, e_{i_1 j_1}, \ldots, e_{i_k j_k}) = g(\xi_{i_1}, \ldots, \xi_{i_k}, \xi_{j_k}) e_{i_1 j_1} e_{i_2 j_2} \cdots e_{i_k j_k}.$$
Recalling the form of g given in (1.8), we obtain that $g(\xi_{i_1}, \ldots, \xi_{i_k}, \xi_{j_k}) = 0$ unless $(i_1, \ldots, i_k)$ is a permutation of $(1, \ldots, k)$ and $i_1 = j_k$. Also, in this case $g(\xi_{i_1}, \ldots, \xi_{i_k}, \xi_{j_k}) = \prod_{1 \leq s < t \leq k} (\xi_s - \xi_t)^2$.

Recalling the multiplication rules of the e_{ij}'s, we obtain that
$$G(X, e_{i_1 j_1}, \ldots, e_{i_k j_k}) = \prod_{1 \leq s < t \leq k} (\xi_s - \xi_t)^2 e_{i_1 i_1}$$
when $j_1 = i_2, j_2 = i_3, \ldots, j_{k-1} = i_k, j_k = i_1$ and $G(X, e_{i_1 j_1}, \ldots, e_{i_k j_k}) = 0$ in all other cases.

It follows that
$$f(X, e_{i_1 j_1}, \ldots, e_{i_k j_k}) = \prod_{1 \leq s < t \leq k} (\xi_s - \xi_t)^2 I_{k \times k}$$
where $I_{k \times k}$ is the $k \times k$ identity matrix, when $j_1 = i_2, j_2 = i_3, \ldots, j_{k-1} = i_k, j_k = i_1$ and $G(X, e_{i_1 j_1}, \ldots, e_{i_k j_k}) = 0$ in all other cases. Thus f takes central values and, since $\xi_1, \ldots, \xi_k$ are distinct, f is a central polynomial of $M_k(F)$. □

An important feature of central polynomials is the following observation due to Procesi [**Ro2**].

LEMMA 1.10.7. *If f is a central polynomial of $M_k(F)$, then f is a polynomial identity of $M_{k-1}(F)$.*

PROOF. Embed $M_{k-1}(F)$ into the upper left corner of $M_k(F)$. Since f is a central polynomial of $M_k(F)$, for any $a_1, \ldots, a_n \in M_{k-1}(F)$, $f(a_1, \ldots, a_n) \in M_{k-1}(F) \cap F I_{k \times k} = 0$. □

As in the case of polynomial identities, one can ask what is the minimal degree of a central polynomial for $M_k(F)$, say, when char $F = 0$. The central polynomial of Formanek is of degree k^2. Even if the Razmyslov polynomial is of higher degree, Halpin ([**Hl**]) deduced from it a new central polynomial of degree k^2. It was believed for some time that k^2 was the answer to the above problem for $k \geq 3$. But in [**DK**] Drensky and Kasparian constructed a new central polynomial for 3×3 matrices of degree 8 and proved that 8 is the minimal degree. In [**F4**] Formanek conjectured that when char $F = 0$, the minimal degree of a central polynomial of $M_k(F)$, $k \geq 3$ is $\frac{1}{2}(k^2 + 3k - 2)$. Notice that this is the only quadratic function that takes values 1, 4 and 8 for $k = 1, 2$ and 3, respectively. In [**D9**] Drensky constructed central polynomials of degree $(k-1)^2 + 1$, for any $k \geq 3$ and these are the central polynomials of smallest degree so far constructed.

1.11. Structure theorems

In this section we state some basic results of the structure theory of PI-algebras. We start with Kaplansky's theorem giving the structure of a primitive PI-algebra. This theorem was proved in 1948 ([**K1**]) and was actually the starting point of the theory of PI-algebras.

Throughout we assume that the reader is familiar with the basic notions and results of ring theory that can be found for instance in the first chapters of Herstein's book [**H**]. These include the Jacobson radical of a ring, the Wedderburn theorem on semisimple artinian rings ([**H**, Theorem 1.4.4]), and the Wedderburn-Artin theorem on simple artinian rings ([**H**, Theorem 2.1.6]). We recall these last two theorems.

THEOREM 1.11.1 (Wedderburn, Wedderburn-Artin). *Let A be a ring. Then*
1) *A is simple artinian if and only if $A \cong M_k(D)$ for some division ring D, $k \geq 1$;*
2) *A is semisimple artinian if and only if $A = I_1 \oplus \cdots \oplus I_n$ where $I_1, \ldots, I_n$ are simple artinian rings and are all minimal two-sided ideals of A.*

Let A be a ring and let M be a left A-module. Let us denote by $\text{End}_A(M)$ the commuting ring of A in M, i.e., the ring of endomorphisms of the A-module M. If the module M is irreducible, then any non-zero endomorphism must be an automorphism. Hence we have the following.

LEMMA 1.11.2 (Schur). *If M is an irreducible A-module, then $\text{End}_A(M)$ is a division ring.*

Recall that a left A-module M is faithful if, for $a \in A$, $aM = 0$ implies $a = 0$. Then a ring A is said to be left primitive if it has a faithful irreducible left module. Even though there are examples of left primitive rings that are not right primitive (see [**B**]), one simply speaks of primitive rings meaning left primitive rings. Let A be a primitive ring and let M be a faithful irreducible A-module. By Schur's lemma, $D = \text{End}_A(M)$ is a division ring. Hence A can be regarded as a ring of D-linear transformations of M. We write $A \subseteq \text{End}_D(M)$, where $\text{End}_D(M)$ is the ring of D-linear transformations of M. The important feature of this embedding is given by Jacobson's theorem. First we need the following notion.

DEFINITION 1.11.3. Let V be a left vector space over the division ring D and S a non-empty subset of $\text{End}_D(V)$. We say that S is a dense set of D-linear transformations on V if, given any finite number of D-linearly independent elements $v_1, \ldots, v_n$ of V and any arbitrary elements $w_1, \ldots, w_n$ of V, there exists $a \in S$ with $a(v_i) = w_i$ for all $i = 1, \ldots, n$.

Notice that in case V is finite dimensional over D, then $\text{End}_D(V)$ is the only dense set of linear transformations of V over D.

The connection between the above concept and primitive rings is given in the following theorem whose proof can be found for instance in [**H**, Theorem 2.1.2].

THEOREM 1.11.4 (Jacobson). *Let A be a primitive ring with faithful irreducible module M. If $D = \text{End}_A(M)$, then A is a dense ring of D-linear transformations on M.*

As an immediate consequence one gets the following characterization of primitive rings.

THEOREM 1.11.5. *If A is a primitive ring, there exists a division ring D such that one of the following holds:*
 (1) $A \cong M_k(D)$, *for some* $k \geq 1$.
 (2) *for every $k \geq 1$ there exists a subring $A^{(k)}$ of A and an epimorphism $\varphi_k : A^{(k)} \to M_k(D)$.*

PROOF. Let A be a dense ring of D-linear transformations of the vector space V. If V is finite dimensional over D then, as we remarked above, $A \cong \operatorname{End}_D(V) \cong M_k(D)$, where $k = \dim_D V$.

If V is infinite dimensional over D, pick $v_1, v_2, \ldots$ countable many linearly independent elements of V. For every $k \geq 1$, set $V_k = \operatorname{span}_D\{v_1, \ldots, v_k\}$ and $A^{(k)} = \{a \in A \mid aV_k \subseteq V_k\}$. Then, by the density theorem, the restriction map $\varphi_k : A^{(k)} \to \operatorname{End}_D(V_k)$ is an epimorphism and the proof is complete. □

Recall that given a division ring D a maximal subfield of D is a subfield of D not contained in any other subfield of D. The existence of a maximal subfield is guaranteed by Zorn's lemma. If K is a maximal subfield of the division ring D, then $C_D(K) = \{a \in D \mid ak = ka, \text{ for all } k \in K\}$, the centralizer of K in D, must coincide with K. This property actually characterizes maximal subfields. One of the important features of maximal subfields is that they are splitting fields as shown below. The proof of the next lemma is direct (see [**P**]) and makes no use of the property of tensor products of simple algebras.

LEMMA 1.11.6. *Let D be a division ring with center Z and let K be a maximal subfield of D. Then $D \otimes_Z K$ is a dense ring of linear transformations on the vector space D over the field K.*

PROOF. For every $d \in D$, let $L_d : D \to D$ denote left multiplication by d and denote $D_l = \{L_d \mid d \in D\}$. Then $L : D \to D_l \subseteq \operatorname{End}_Z(D)$ such that $L(d) = L_d$, is a Z-algebra homomorphism. Also, for every $k \in K$, let $T_k : D \to D$ denote right multiplication by k and denote $K_r = \{T_k \mid k \in K\}$. Then, since K is commutative, the map $T : K \to K_r \subseteq \operatorname{End}_Z(D)$ such that $T(k) = T_k$, is also a Z-algebra homomorphism. Since left and right multiplications on D commute, it follows that the map $D \otimes_Z K \to D_l K_r$ defined by $\sum d_i \otimes k_i \to \sum L_{d_i} T_{k_i}$ is well defined and an epimorphism of Z-algebras. This in turn makes D a $D \otimes_Z K$-module with respect to the action $(\sum d_i \otimes k_i)x = \sum d_i x k_i$, for $x, d_i \in D, k_i \in K$.

Since for every $x \in D$, $x \neq 0$ we have that $Dx = D$, it follows that $(D \otimes_Z K)x = D$. This says that D is irreducible as a $D \otimes_Z K$-module. Next we prove that D is a faithful $D \otimes_Z K$-module. Since every element of $D \otimes_Z K$ can always be written in the form $\sum d_i \otimes k_i$ with all the k_i's linearly independent over Z, we only need to show that if $\sum_{i=1}^n d_i D k_i = 0$ for some Z-linearly independent elements $k_1, \ldots, k_n \in K$, then $d_1 = \ldots = d_n = 0$. We prove this by induction on n. The case $n = 1$ is clear. Suppose that the result holds for $n-1$ and suppose by contradiction that not all the d_i's are zero. Let $d_n \neq 0$. By multiplying $\sum d_i D k_i$ on the left by d_n^{-1}, we may assume that $d_n = 1$. For all $x, y \in D$, we have

$$x\left(\sum_{i=1}^n d_i y k_i\right) = 0 \quad \text{and} \quad \sum_{i=1}^n d_i x y k_i = 0.$$

By subtracting these two identities we obtain

$$\sum_{i=1}^{n-1}(xd_i - d_i x)yk_i = 0,$$

for all $x \in D$. Thus by induction, $d_i x = x d_i$ for all i and for all $x \in D$. This says that $d_1,\ldots,d_n \in Z$, the center of D. From the equality $\sum_{i=1}^{n} d_i k_i = \sum_{i=1}^{n} d_i 1 k_i = 0$ we obtain that $k_1,\ldots,k_n$ are linearly dependent over Z, a contradiction. This proves that D is a faithful $D \otimes_Z K$-module.

Thus $D \otimes_Z K$ is a primitive ring with faithful irreducible module D. By Jacobson's theorem in order to complete the proof, it is enough to show that the commuting ring $\mathrm{End}_{D \otimes_Z K}(D) \cong K$. To see this, let $\alpha \in \mathrm{End}_{D \otimes_Z K}(D)$. For every $x \in D$, we have $\alpha(x) = \alpha(x \cdot 1) = x\alpha(1)$, since α commutes with left multiplication by D. Also, if $x \in K$, then $x\alpha(1) = \alpha(x \cdot 1) = \alpha(1 \cdot x) = \alpha(1)x$, since α commutes with right multiplication by K. Hence $\alpha(1) \in D$ centralizes K and, since K is a maximal subfield of D, we get $\alpha(1) \in K$. Thus α is the right multiplication by $\alpha(1)$ namely, $\alpha = T_{\alpha(1)} \in K_r$ and by the defined action of $D \otimes_Z K$ on D, this says that $\mathrm{End}_{D \otimes_Z K}(D) \cong K$. $\square$

For an integer n let us denote by $[n/2]$ the integer part of $n/2$. For an algebra A let $Z = Z(A)$ denote the center of A. In case A is a finite dimensional central simple algebra, it is well known that $\dim_{Z(A)} A = n^2$ is a square and we call n the degree of the algebra A ([**H**, Theorem 4.2.2]).

THEOREM 1.11.7 (Kaplansky [**K1**]). *Let A be a primitive algebra satisfying a polynomial identity of degree d. Then A is a finite dimensional central simple algebra and $\dim_{Z(A)} A \leq [d/2]^2$.*

PROOF. Let f be the polynomial identity of degree d satisfied by A and assume, as we may, that f is multilinear. Being A primitive, it has a faithful irreducible A-module V with commuting ring D. If $\dim_D V = \infty$, then by Theorem 1.11.5, for each positive integer n there exists a subalgebra $A^{(n)}$ of A with $M_n(D)$ as a homomorphic image. Since polynomial identities are inherited by subalgebras and homomorphic images, it follows that $M_n(Z(D)) \subseteq M_n(D)$ satisfies f. Since by Theorem 1.7.2, $M_n(Z(D))$ does not satisfy identities of degree less than $2n$, it follows that $n \leq d/2$ and this is a contradiction. Thus $\dim_D V = n < \infty$ and again, by Theorem 1.11.5, $A \cong M_n(D)$. Write $Z = Z(A)$.

Since $M_n(D)$ satisfies f and D is a subring, D also satisfies f. Moreover, if K is a maximal subfield of D, since f is multilinear and K is commutative, we obtain that $D \otimes_Z K$ still satisfies f. Since, by the previous lemma, $D \otimes_Z K$ has a faithful irreducible module with commuting ring K, by the first part of the proof we obtain that $D \otimes_Z K \cong M_m(K)$, for some integer $m \geq 1$. The outcome of this is that

$$A \subseteq A \otimes_Z K \cong M_n(D) \otimes_Z K \cong M_n(D \otimes_Z K) \cong M_n(M_m(K)) \cong M_{nm}(K).$$

Since f is multilinear, $A \otimes_Z K \cong M_{nm}(K)$ still satisfies f. Recalling that $M_{nm}(K)$ does not satisfy identities of degree less that $2nm$, we obtain that $nm \leq d/2$. Now, $\dim_Z A = \dim_K A \otimes_Z K = (nm)^2 \leq [d/2]^2$ and the proof is complete. $\square$

The next step is to relate the identities of finite dimensional simple algebras to the identities of matrices. We do this in the following lemma.

LEMMA 1.11.8. *Let A be a central simple algebra of dimension n^2 over its center F. Then $\mathrm{Id}(A) = \mathrm{Id}(M_n(F))$. Moreover, f is a central polynomial of $M_n(F)$ if and only if f is a central polynomial of A.*

PROOF. By Theorem 1.11.1, $A \cong M_t(D)$, for some finite dimensional central division algebra D over F. If F is finite, D is a finite division algebra and by a theorem of Wedderburn, D is a field. Hence $D = F$, since D is central over F. It follows that $A \cong M_t(F)$ in this case and we are done.

Suppose F is infinite and let K be a maximal subfield of D. By Lemma 1.11.6, $D \otimes_F K \cong M_m(K)$ for some m. Hence $A \otimes_F K \cong M_{tm}(K)$. As above, by comparing dimensions we see that $A \otimes_F K \cong M_n(K)$. Notice that if f is a central polynomial of A, $[f,y]$ is an identity of A. Since F is infinite, by Lemma 1.4.2 every polynomial identity of A is stable. Hence A and $M_n(F)$ have the same identities and the same central polynomials. $\square$

A ring is semiprime if it does not contain non-zero nilpotent ideals. Notice that if I is a non-zero nilpotent right ideal of the semiprime ring A, then AI is a non-zero two-sided nilpotent ideal of A and this is not allowed. Thus a semiprime ring has no non-zero nilpotent one-sided ideals.

For a polynomial $f \in F\langle X \rangle$ and a subset S of an algebra A, let $f(S)$ denote the set of all values $f(s_1, \ldots, s_n)$ with $s_1, \ldots, s_n \in S$.

THEOREM 1.11.9. *If A is a semiprime PI-algebra, then A has no non-zero nil ideals.*

PROOF. Let f be a multilinear polynomial satisfied by A and suppose that A has a non-zero nil ideal N. Since $f(N) = 0$, then also $f(N)N = 0$. Let $g \in F\langle X \rangle$ be a multilinear polynomial of minimal degree with the property that $g(I)I = 0$, for some non-zero right ideal $I \subseteq N$. Such g and I clearly exist by the above. We shall reach a contradiction by proving that $I^2 = 0$ and this is not allowed in a semiprime algebra.

Suppose that there exists $a \in I$ such that $a^2 I = 0$. We may clearly assume that the monomial $x_1 \cdots x_n$ appears in g with non-zero coefficient and write g in the form
$$g(x_1, \ldots, x_n) = h(x_1, \ldots, x_{n-1})x_n + h'(x_1, \ldots, x_n)$$
where x_n never appears as the rightmost variable in every monomial of h'. Since $a^2 I = 0$ we see that, for any $r_1, \ldots, r_{n-1} \in I$,
$$g(ar_1, \ldots, ar_{n-1}, a) = h(ar_1, \ldots, ar_{n-1})a.$$
Recalling that $aI \subseteq I$ and $g(I)I = 0$, we obtain that
$$h(ar_1, \ldots, ar_{n-1})aI = g(ar_1, \ldots, ar_{n-1}, a)I \subseteq g(I)I = 0,$$
for all $r_1, \ldots, r_{n-1} \in I$. Thus $h(aI)aI = 0$. But then, h is a non-zero multilinear polynomial of smaller degree that g such that $h(aI)aI = 0$ where aI is a right ideal of A contained in N. By the minimality of the degree of g, aI must be zero. We have proved that if $a \in I$ is such that $a^2 I = 0$, then $aI = 0$. But, if b is any element of I, b is nilpotent, say $b^k = 0$; hence $b^k I = 0$. Since $(b^{k-1})^2 = 0$, by the above, $b^{k-1}I = 0$ and by induction we finally get $bI = 0$. This says that $I^2 = 0$ and, being A semiprime, we conclude that $I = 0$. $\square$

THEOREM 1.11.10. *If A has no non-zero nil ideals, then the polynomial algebra $A[x]$ is semisimple.*

PROOF. Let $J = J(A[x])$ be the Jacobson radical of $A[x]$ and let I be the ideal of A consisting of the leading coefficients of the polynomials of J. We shall prove that I is a nil ideal and since by hypothesis A has no non-zero nil ideals, this will imply that $I = 0$ and so, $J = 0$.

Let $f = f(x) = \sum_{i=0}^{n} a_i x^i \in J$ be non-zero. Since $xf \in J$, the element $1 - xf$ is invertible in $A[x]$ and let $g \in A[x]$ be such that $(1 - xf)g = 1$. We claim that, for every $m \geq 1$,

$$(1.9) \qquad g = x^m f^m g + \sum_{i=0}^{m-1} x^i f^i.$$

The proof is by induction on m. Since $(1 - xf)g = 1$, then $g = xfg + 1$ and the case $m = 1$ is proved. Suppose the above formula holds for $m - 1$. Then we have

$$g = x^{m-1} f^{m-1} g + \sum_{i=0}^{m-2} x^i f^i = x^{m-1} f^{m-1}(xfg + 1) + \sum_{i=0}^{m-2} x^i f^i = x^m f^m g + \sum_{i=0}^{m-1} x^i f^i$$

and the claim is proved.

Write $g = \sum_{i=0}^{r} b_i x^i$ and take $m > r$. Recalling that $f = \sum_{i=0}^{n} a_i x^i \in J$ and equating in (1.9) the coefficient of x^{m+nm+i}, $1 \leq i \leq r$, we obtain $0 = a_n^m b_i$. It follows that $a_n^m g = 0$ and, since g is invertible in $A[x]$, we get $a_n^m = 0$ and the leading coefficient of f is nilpotent. Thus I is a nil ideal and this completes the proof of the theorem. □

Recall that, given a collection of algebras $\{A_\gamma \mid \gamma \in \Gamma\}$, we say that an algebra A is a subdirect product of the algebras A_γ, $\gamma \in \Gamma$, if A can be embedded into the algebra direct product $\Pi_{\gamma \in \Gamma} A_\gamma$ so that $\pi_\gamma(A) = A_\gamma$ for all $\gamma \in \Gamma$ where $\pi_\gamma : A \to A_\gamma$ is the projection onto A_γ. It is readily seen that if $\{I_\gamma \mid \gamma \in \Gamma\}$ is a collection of ideals of A, then A is a subdirect product of the algebras A/I_γ, $\gamma \in \Gamma$, if and only if $\bigcap_{\gamma \in \Gamma} I_\gamma = 0$. For instance, since the Jacobson radical of an algebra A is the intersection of the primitive ideals of A, it follows that a semisimple algebra is a subdirect product of primitive algebras.

Next result is due to Rowen ([**Ro1**]) and is an important application of the existence of central polynomials for matrices.

THEOREM 1.11.11. *Let A be a semiprime PI-algebra. If I is a non-zero ideal of A, then $I \cap Z(A) \neq 0$.*

PROOF. Let f be an identity of A of degree d. Suppose first that A is semisimple. Then A is a subdirect product of primitive algebras A_γ, $\gamma \in \Gamma$. Each A_γ still satisfies f hence, by Kaplansky's theorem, it is a simple algebra of dimension $n_\gamma^2 \leq [d/2]^2$ over its center. Let I_γ be the image of I under the canonical projection $A \to A_\gamma$. Since A_γ is simple, then either $I_\gamma = A_\gamma$ or $I_\gamma = 0$. Consider the set $\Gamma' = \{\gamma \in \Gamma \mid I_\gamma = A_\gamma\}$. Since I is a non-zero ideal of A, the set Γ' is nonempty and let n_0^2 be the largest dimension of a central simple algebra A_γ with $\gamma \in \Gamma'$. Let h_{n_0} be a central polynomial of $n_0 \times n_0$ matrices. By Lemma 1.11.8 for each algebra A_γ with $\dim_{Z(A_\gamma)} A_\gamma = n_0^2$,

$$0 \neq h_{n_0}(I_\gamma) = h_{n_0}(A_\gamma) \subseteq Z(A_\gamma).$$

For any other algebra A_γ of smaller dimension over its center, by Lemmas 1.10.7 and 1.11.8, we must have $h_{n_0}(I_\gamma) = 0$. The outcome of this is that $0 \neq h_{n_0}(I) \subseteq I \cap Z(A)$ and the theorem is proved in case A is a semisimple algebra.

Now let A be any semiprime algebra. By Theorem 1.11.9, A has no non-zero nil ideals and, so, by Theorem 1.11.10, the polynomial ring $A[x]$ is semisimple. Since $I[x]$ is a non-zero ideal of $A[x]$, by the first part of the proof we have that $0 \neq I[x] \cap Z(A[x]) = I[x] \cap Z(A)[x] = (I \cap Z(A))[x]$. Thus $I \cap Z(A) \neq 0$. □

An immediate consequence of the previous theorem is the following.

COROLLARY 1.11.12. *If A is a semiprime PI-algebra and $Z(A)$ is a field, then A is a finite dimensional central simple algebra.*

PROOF. If I is a non-zero ideal of A, by the previous theorem I intersects $Z(A)$ non-trivially. Since $Z(A)$ is a field, I contains an invertible element and, so, coincides with A. This shows that A is simple. But then we can apply Kaplansky's theorem. □

A ring A is prime if the product of any two non-zero ideals of A is still non-zero. A useful equivalent formulation is the following. A is prime if $aAb = 0$ for $a, b \in A$ implies either $a = 0$ or $b = 0$. From the definition it also follows that the center of a prime ring, if non-zero, is an integral domain whose elements are not zero divisors in A.

Suppose that A is a prime ring with non-zero center Z and let K be the field of fractions of Z. Consider the tensor product $A \otimes_Z K$ and let $c = \sum a_i \otimes z'_i z_i^{-1}$ be an element of $A \otimes_Z K$. By taking common denominators, c can be written in the form $a \otimes z^{-1}$ for a suitable $a \in A, z \in Z \setminus \{0\}$. In other words, $A \otimes_Z K$ is the central localization of A. One usually writes $a \otimes z^{-1}$ as az^{-1} and $A \otimes_Z K$ as A_Z. Thus $A_Z = \{az^{-1} \mid a \in A, z \in Z, z \neq 0\}$.

It is readily seen that since A is prime, the central localization A_Z is still a prime ring. In fact, if $az_1^{-1} A_Z bz_2^{-1} = 0$ with $az_1^{-1}, bz_2^{-1} \in A_Z$, then in particular, $aAbz_1^{-1} z_2^{-1} = 0$ and, by multiplying by $z_1 z_2$ on the right, we obtain $aAb = 0$. But then the primeness of A forces $a = 0$ or $b = 0$ and, so, either $az_1^{-1} = 0$ or $bz_2^{-1} = 0$. Note also that if A is a prime PI-algebra, then by Theorem 1.11.11 $Z(A) \neq 0$.

THEOREM 1.11.13 (Posner [**Po**]). *Let A be a prime algebra satisfying a polynomial identity of degree d. Then the central localization A_Z of A is a central simple algebra of dimension $n^2 \leq [d/2]^2$ over its center K. Moreover, $\mathrm{Id}(A) = \mathrm{Id}(M_n(K))$.*

PROOF. Suppose first that Z is infinite. By what we remarked above, the central localization A_Z is still a prime algebra. Moreover, $A_Z = A \otimes_Z K$, where K is the field of fractions of Z, has the same identities of A, being Z infinite. It can be easily checked that actually K is the center of A_Z and, being a field, by Corollary 1.11.12, A_Z is a finite dimensional central simple algebra. In case Z is finite, then $Z = K$ and one applies Corollary 1.11.12. Since by Lemma 1.11.8, $\mathrm{Id}(A_Z) = \mathrm{Id}(M_n(K))$ we have $\mathrm{Id}(A) = \mathrm{Id}(M_n(K))$. By Kaplansky's theorem we also have $n^2 \leq [d/2]^2$. □

The formulation of Posner's theorem given above is a combination of the original one ([**Po**]) and the above consequences of the existence of central polynomials for matrices.

Next we state a theorem which is of great interest in general PI-theory but will not be used in any essential way in this book. We do not present the proof of this result here since it requires among other things, a deep combinatorial result known as the Shirshov height theorem, which is of independent interest but will not be used in this book. For the interested reader, a proof of these results can be found in [**Ro3**, Vol. 2, p. 141ff] and in the very recent [**BR**, p. 44ff].

THEOREM 1.11.14 (Razmyslow-Kemer-Braun). *Let A be a finitely generated PI-algebra over an arbitrary field. Then the Jacobson radical of A is a nilpotent ideal.*

1.12. Some applications of the structure theorems

Our first application concerns the ideal of identities of $k \times k$ matrices. We start with the following theorem due to Amitsur that follows from the proof of Theorem 1.11.11.

THEOREM 1.12.1. *If A is a semiprime PI-algebra over an infinite field F, then $\mathrm{Id}(A) = \mathrm{Id}(M_k(F))$, for some $k \geq 1$.*

PROOF. Since by Lemma 1.4.2, over an infinite field all identities are stable, by tensoring A with the algebraic closure of F, we may assume that F is algebraically closed.

Since A is semiprime, by Theorem 1.11.10 the polynomial algebra $A[x]$ is semisimple. Now, $A[x] \cong A \otimes_F F[x]$ and by Lemma 1.4.2 again, A and $A[x]$ satisfy the same identities. Thus we may assume that A is semisimple. By the proof of Theorem 1.11.11, A can be embedded into the direct product of simple algebras A_γ of dimension $\leq [d/2]^2$ over their center. Since F is algebraically closed, for each γ, $A_\gamma \cong M_k(F)$, for some $k \leq d/2$. Clearly A satisfies the same polynomial identities as $M_n(F)$, where n is the largest integer such that $A_\gamma \cong M_n(F)$, for some $\gamma \in \Gamma$. □

Combining Theorem 1.11.14 with the results on upper block triangular matrix algebras proved in Section 1.9 one can prove the following surprising result.

THEOREM 1.12.2. *Let A be a finitely generated PI-algebra over an infinite field F. Then A satisfies all the identities of $M_k(F)$, for some $k \geq 1$.*

PROOF. Let $J = J(A)$ be the Jacobson radical of A. Since the quotient algebra A/J is semisimple, by Theorem 1.12.1 it satisfies all identities of $M_k(F)$ for some k. Write $I = \mathrm{Id}(A/J) = \mathrm{Id}(M_k(F))$. For any $f = f(x_1, \ldots, x_n) \in I$, we have that $f(a_1, \ldots, a_n) \in J$ for all $a_1, \ldots, a_n \in A$. Since by Theorem 1.11.14 $J^t = 0$ for some t, we obtain that $I^t \subseteq \mathrm{Id}(A)$. Now by Theorem 1.9.1, I^t is the T-ideal of identities of the algebra $UT(d_1, \ldots, d_m)$ where $d_1 = \ldots = d_m = k$. Thus
$$\mathrm{Id}(UT(d_1, \ldots, d_m)) \subseteq \mathrm{Id}(A)$$
and A satisfies all identities of the algebra $M_{tk}(F)$. □

In the proof of Theorem 1.12.2 we used the nilpotency of the Jacobson radical of a finitely generated PI-algebra. Later in the book in Section 4.8 we shall prove a much stronger result over a field of characteristic zero, without the help of the Razmyslow-Kemer-Braun theorem but based on a result of Kemer.

It is useful in general to know when a PI-algebra can be embedded into $k \times k$ matrices over a field or more generally over a commutative ring for some k. Not every PI-algebra can be embedded into matrices over a commutative ring. For instance, since in characteristic zero the Grassmann algebra does not satisfy any standard identity, it cannot be embedded into matrices of any size (see also [**Ro3**, vol. 1, p. 371]). Nevertheless, as an application of Kaplansky's theorem one can give a positive result for algebras with no non-zero nil ideals.

THEOREM 1.12.3. *Let A be a PI-algebra with no non-zero nil ideals over an infinite field F. Then A can be embedded into $M_k(C)$ for some commutative F-algebra C and $k \geq 1$.*

PROOF. First extend the scalars by embedding A into $\bar{A} = A \otimes \bar{F}$, where $\bar{F}$ is the algebraic closure of F. Clearly, in order to prove the theorem it is enough to embed $\bar{A}$ into $M_k(C)$, for some commutative $\bar{F}$-algebra C. Hence, without loss of generality, we may assume F to be algebraically closed. Since A has no non-zero nil ideals, by Theorem 1.11.10 the polynomial ring $A[x]$ is semisimple. Hence by passing to $A[x]$, we may assume that A is semisimple.

As we remarked before, A is then a subdirect product of primitive algebras A_γ, $\gamma \in \Gamma$. Each primitive algebra A_γ still satisfies all the identities of A, hence by Kaplansky's theorem, is a simple algebra of dimension $\leq k^2$ over its center, for some fixed $k \geq 1$. Since F is algebraically closed each primitive algebra A_γ is isomorphic to $M_n(F)$ for some $n \leq k$.

If we denote by C the direct product of all the centers Z_γ of A_γ, then the direct product $\Pi_\gamma A_\gamma$ embeds into $M_k(C)$. Hence also A, being embedded into $\Pi_\alpha A_\alpha$, embeds into $M_k(C)$. □

An ideal I of a ring A is prime if A/I is a prime ring. As an application of Posner's theorem we next detect all the prime ideals of the free algebra $F\langle X \rangle$. We start by proving that the algebra of generic matrices $F\{\xi\} = F\{\xi^{(1)}, \xi^{(2)}, \dots\}$ is an integral domain. Both results are due to Amitsur ([**A2**]).

THEOREM 1.12.4. *If F is an infinite field, the algebra $F\{\xi\}$ of $k \times k$ generic matrices is a domain.*

PROOF. It is well known (see for instance [**L**]), that there exists a division algebra D over F of dimension k^2 over its center Z. By Lemma 1.11.8 the identities of D over Z are the same as the identities of $M_k(Z)$.

Consider the Z-algebra $\tilde{D}$ of generic elements of D (see Theorem 1.4.4) generated by $\eta^{(1)}, \eta^{(2)}, \dots$, over Z. By Lemma 1.11.8 we have that $\tilde{D}$ has the same identities as $M_k(Z)$. Hence by Corollary 1.4.5 and Theorem 1.2.4, $\tilde{D}$ is isomorphic to the algebra of $k \times k$ generic matrices $Z\{\xi\} = Z\{\xi^{(1)}, \xi^{(2)}, \dots\}$. On the other hand, $\tilde{D}$ is a domain being a subalgebra of $D \otimes Z[\eta_j^{(i)} \mid i \geq 1,\ 1 \leq j \leq k^2]$. Hence $F\{\xi\} \subseteq Z\{\xi\}$ being embedded into a domain is a domain. □

THEOREM 1.12.5. *Let F be an infinite field and let I be a non-zero T-ideal of $F\langle X \rangle$. Then I is a prime ideal if and only if $I = \mathrm{Id}(M_k(F))$, for some $k \geq 1$.*

PROOF. First suppose that $F\langle X \rangle / I$ is a prime PI-algebra. Then, by Theorem 1.12.1, $\mathrm{Id}(F\langle X \rangle / I) = \mathrm{Id}(M_k(F))$, for some k. Since $\mathrm{Id}(F\langle X \rangle / I) = I$, the conclusion follows.

Conversely, by the previous theorem $F\langle X \rangle / \mathrm{Id}(M_k(F)) \cong F\{\xi\}$ is a domain. Hence $I = \mathrm{Id}(M_k(F))$ is a prime ideal. □

1.13. The Gelfand-Kirillov dimension of a PI-algebra

We conclude this chapter by introducing the notion of the Gelfand-Kirillov dimension of a finitely generated algebra. We shall prove some of the basic properties of this dimension and the applications to PI-algebras. In chapter 6 we shall introduce the exponent of a PI-algebra and we shall compare it to the Gelfand-Kirillov dimension in some significant cases. For an account on the Gelfand-Kirillov dimension we refer the reader to the book [**KL**].

Let A be a finitely generated algebra over a field F and let $\{a_1, \ldots, a_k\}$ be a set of generators. For every $n \geq 1$ define

$$A_n = \text{span}_F\{a_{i_1} \cdots a_{i_t} \mid 1 \leq t \leq n, 1 \leq i_i, \ldots, i_t \leq k\},$$

the subspace of A spanned by all possible products of at most n generators.

DEFINITION 1.13.1. The Gelfand-Kirillov dimension of A is

$$\text{GKdim}(A) = \limsup_{n \to \infty} \log_n(\dim A_n).$$

The first basic properties are the following.

1) The GK-dimension does not depend on the particular choice of the generating set ([**KL**, Lemma 1.1]).
2) If B is a subalgebra or a homomorphic image of A, then $\text{GKdim}(B) \leq \text{GKdim}(A)$ ([**KL**, Lemma 3.1]).
3) For any finitely generated algebra A either $\text{GKdim}(A) = 0$ or $\text{GKdim}(A) = 1$ or $\text{GKdim}(A) \geq 2$ ([**KL**, Theorem 2.5]).

The above definition means that if $\text{GKdim}(A) = s < \infty$, then the sequence of dimensions $\{\dim A_n\}_{n \geq 1}$ asymptotically behaves like a polynomial of degree s. It is clear that $\text{GKdim}(A) = 0$ if and only if A is a finite dimensional algebra over F. Moreover, given any real number $\alpha \geq 2$, it can be shown that there exist algebras whose GK-dimension is α ([**KL**, Theorem 2.9]).

We start by giving two significant examples.

EXAMPLE 1.13.2. Let $A = F[X_1, \ldots, X_k]$ be the algebra of polynomials in the commuting variables $X_1, \ldots, X_k$ over F. Notice that the number of monomials in $X_1, \ldots, X_k$ of degree i is equal to $\binom{k+i-1}{i}$. Hence

$$\dim A_n = \sum_{i=0}^{n} \binom{k+i-1}{i} = \binom{k+n}{n}.$$

It follows that $\text{GKdim}(A) = k$.

EXAMPLE 1.13.3. Let $k > 1$ and let $A = F\{x_1, \ldots, x_k\}$ be the free algebra on the set $\{x_1, \ldots, x_k\}$ over F. In this case $\dim A_n = \sum_{i=0}^{n} k^i$. It follows that A has exponential growth and $\text{GKdim}(A) = \infty$.

Given any finite dimensional algebra A over an infinite field, Example 1.13.2 above can be applied to get information on the GK-dimension of any relatively free algebra of finite rank of $\text{var}(A)$ as follows. Recall that if $F[\xi_j^{(i)} \mid i \geq 1, j = 1, \ldots, m]$ is the polynomial ring over F in the commutative indeterminates $\xi_j^{(i)}$, one constructs $B = A \otimes_F F[\xi_j^{(i)}]$, the algebra tensor product of A and $F[\xi_j^{(i)}]$. Then, if $\{u_1, \ldots, u_m\}$ is a basis of A, the elements

$$\xi^{(i)} = \sum_{j=1}^{m} u_j \otimes \xi_j^{(i)}, \quad i = 1, 2, \ldots$$

are the generic elements of B and the algebra $F\{\xi^{(1)}, \xi^{(2)}, \ldots\} = F\{\xi\}$ generated by $\xi^{(1)}, \xi^{(2)}, \ldots$ over F is the relatively free algebra of countable rank of $\operatorname{var}(A)$ (see Theorem 1.4.4).

Denote by $F_k\{\xi\}$ the subalgebra of $F\{\xi\}$ generated by $\xi^{(1)}, \ldots, \xi^{(k)}$, i.e., the relatively free algebra of $\operatorname{var}(A)$ of rank k.

We have

PROPOSITION 1.13.4. *Let A be a finite dimensional algebra over the infinite field F and let $F_k\{\xi\}$ be the relatively free algebra of rank k of $\operatorname{var}(A)$. If $\dim_F A = m$, then the number of linearly independent monomials in $\xi^{(1)}, \ldots, \xi^{(k)}$ of degree at most n is less or equal to $m(n+k)^{km}$. In particular, $\operatorname{GKdim}(F_k\{\xi\}) \leq km$.*

PROOF. Any monomial in the generic elements $\xi^{(i)} = \sum_{j=1}^m u_j \otimes \xi_j^{(i)}$, $1 \leq i \leq k$, of degree at most n belongs to $A \otimes F[\xi_j^{(i)}]_n$ where $F[\xi_j^{(i)}]_n$ is the subspace of $F[\xi_j^{(i)}]$ consisting of all polynomials in the elements $\xi_j^{(i)}$, $1 \leq i \leq k$, $1 \leq j \leq m$, of total degree at most n. By Example 1.13.2, $\dim F[\xi_j^{(i)}]_n = \binom{km+n}{n} \leq (n+k)^{km}$; hence $\dim A \otimes F[\xi_j^{(i)}]_n \leq m(n+k)^{km}$ and the claim follows. $\square$

Since any algebra in $\operatorname{var}(A)$ is a homomorphic image of a relatively free algebra, from the above proposition it follows that any finitely generated algebra in $\operatorname{var}(A)$ has finite GK-dimension. Next we show that this is a special case of a much more general result due to Berele ([**Be1**]).

THEOREM 1.13.5. *If A is a finitely generated PI-algebra over an infinite field, then $\operatorname{GKdim}(A) < \infty$.*

PROOF. By Theorem 1.12.2, A satisfies all identities of $M_k(F)$, for some $k \geq 1$. If A is generated by m elements, then A is a homomorphic image of the relatively free algebra of rank m of the variety $\operatorname{var}(M_k(F))$. Since the GK-dimension does not increase by passing to homomorphic images, it follows that it is enough to prove our theorem for a relatively free algebra A of finite rank of a variety $\operatorname{var}(B)$ where $\dim B < \infty$. But this has already been proved in Proposition 1.13.4. $\square$

Our next objective is to present a deep result of Procesi which allows to compute explicitly the GK-dimension of the algebra of $k \times k$ generic matrices. We start by proving the following property of the GK-dimension.

LEMMA 1.13.6. *Let A be a finitely generated algebra over a field F. If A is a finitely generated module over a central subalgebra B, then $\operatorname{GKdim}(A) = \operatorname{GKdim}(B)$.*

PROOF. First we claim that B is a finitely generated algebra. In fact, let $A = \sum_{i=1}^n Ba_i$ and, for every $1 \leq i, j \leq n$, write $a_i a_j = \sum_{k=1}^n b_{ijk} a_k$ with $b_{ijk} \in B$. If $c_1, \ldots, c_m$ generate A as an F-algebra, then, for $1 \leq i \leq m$, write $c_i = \sum_{j=1}^m b_{ij} a_j$ with $b_{ij} \in B$.

Let C be the (commutative) subalgebra of B generated by the elements b_{ijk}, b_{ij} over F. Then C, being finitely generated is noetherian. Also, A is a finitely generated C-module and so, B being a submodule is also finitely generated over C. Now recalling that C is a finitely generated subalgebra of B, it follows that B is finitely generated as an F-algebra and the claim is established.

Since the GK-dimension does not change by adding a unit element, we may assume that A is an algebra with 1. Hence, by considering the right regular representation of A, we can embed A into $M_n(B)$, the algebra of $n \times n$ matrices over B. Now, if S is a finite generating set for B and $E = \{e_{ij} \mid 1 \le i, j \le n\}$, then $(S \otimes 1) \cup (1 \otimes E)$ is a generating set for $B \otimes M_n(F) \cong M_n(B)$. Since $\dim M_n(F) < \infty$, it follows that $\operatorname{GKdim}(B) = \operatorname{GKdim}(M_n(B))$ and, so, $\operatorname{GKdim}(A) = \operatorname{GKdim}(B)$. □

Let A be a finitely generated commutative algebra over F. By the Noether Normalization Theorem (see [**Mt**, Corollary 14.G]) A is a finitely generated module over the polynomial algebra $F[X_1, \ldots, X_m]$ and $m = \operatorname{tr\,deg}(A)$ is the transcendence degree of A. Moreover, m equals the Krull dimension of A, i.e., the maximal length of a chain of prime ideals of A. By the previous lemma, recalling Example 1.13.2 we obtain

THEOREM 1.13.7. *If A is a finitely generated commutative algebra over a field F, then $\operatorname{GKdim}(A) = \operatorname{tr\,deg}(A/F)$.*

We now define the GK-dimension for not necessarily finitely generated algebras.

DEFINITION 1.13.8. *If A is an arbitrary F-algebra, we define*
$$\operatorname{GKdim}(A) = \sup_B \operatorname{GKdim}(B)$$
where the supremum is taken over all finitely generated subalgebras B of A.

Clearly, if K is an algebraic extension field of F, we have that $\operatorname{GKdim}(K) = 0$. On the other hand, for a purely transcendental extension $K = F(q_1, \ldots, q_m)$ of F and for any finitely generated F-subalgebra B of K we have that $\operatorname{GKdim}(B) \le m$ and $\operatorname{GKdim}(F[q_1, \ldots, q_m]) = m$. It follows that, for any field extension $K \supseteq F$, $\operatorname{GKdim}(K) = \operatorname{tr\,deg}(K/F)$.

More generally, for a prime PI-algebra A the computation of the GK-dimension can also be reduced to that of the transcendence degree of the center of A over F.

THEOREM 1.13.9. *Let A be a finitely generated prime PI-algebra over F and $Z = Z(A)$ its center. Then $\operatorname{GKdim}(A) = \operatorname{tr\,deg}(Z/F)$.*

PROOF. By Posner's theorem the central localization A_Z of A is a finite dimensional central simple algebra over K, the field of fractions of Z. Let $\{q_1, \ldots, q_r\}$ be a maximal set of algebraically independent elements of K over F, i.e., $F \subseteq F(q_1, \ldots, q_r)$ is a purely transcendental extension and $F(q_1, \ldots, q_r) \subseteq K$ is algebraic. Then clearly $\operatorname{tr\,deg}(K/F) = r$.

Since A is a finitely generated F-algebra, there exists a finite set of elements $b_1, \ldots, b_m \in A$ such that $b_1, \ldots, b_m$ generate A as an F-algebra and any $a \in A$ can be written as a K-linear combination of $b_1, \ldots, b_m$ in A_Z. Let
$$b_i b_j = \sum_k \lambda_{ij}^k b_k, \quad 1 \le i, j \le m,$$
for some $\lambda_{ij}^k \in K$. Denote by A' the algebra generated over F by the set
$$\{b_1, \ldots, b_m, q_1, \ldots, q_r, \lambda_{ij}^k, \ 1 \le i, j, k \le m\}$$
and by C the algebra generated over F by the set $\{q_1, \ldots, q_r, \lambda_{ij}^k, \ 1 \le i, j, k \le m\}$. By Lemma 1.13.6, $\operatorname{GKdim}(A') = \operatorname{GKdim}(C)$ and $\operatorname{GKdim}(C) = q = \operatorname{tr\,deg}(K/F) = \operatorname{GKdim}(K)$.

Next we prove that $\operatorname{GKdim}(A) = \operatorname{GKdim}(A_Z)$.

Let B be a finitely generated subalgebra of A_Z and let $a_1, \ldots, a_t$ be a generating set of B. Then one can find $z \in Z$ such that $a_1 = c_1 z^{-1}, \ldots, a_t = c_t z^{-1}$. Since we are interested in computing $\sup_B \mathrm{GKdim}(B)$, without loss of generality, we may assume that $c_1, \ldots, c_t$ generate A as an F-algebra.

By the Lemma 1.13.6, $\mathrm{GKdim}(A_Z) = \mathrm{GKdim}(K)$. Since by the previous theorem $\mathrm{GKdim}(K) = \mathrm{tr\,deg}(K/F)$, we get the desired conclusion. $\square$

In light of the previous theorem, in order to compute the GK-dimension of the algebra of $n \times n$ generic matrices, one needs to study the center of this algebra.

Recall that if the e_{ij}'s are the usual matrix units and $\xi_{ij}^{(t)}$, $t \geq 1, i,j = 1, \ldots, n$, are commutative variables, $F\{\xi\} = F\{\xi^{(1)}, \xi^{(2)}, \ldots\}$ is the algebra of $n \times n$ generic matrices $\xi^{(t)} = \sum_{i,j=1}^{n} \xi_{ij}^{(t)} e_{ij}$ over F.

Since by Theorem 1.12.4 the ring of generic matrices $F\{\xi\}$ is a PI-domain, by Posner's theorem the central localization of $F\{\xi\}$ is a division ring of dimension n^2 over its center. Clearly this property still holds if we restrict ourselves to a finite number m of generic matrices.

DEFINITION 1.13.10. The central localization of the ring of $n \times n$ generic matrices $F\{\xi^{(1)}, \ldots, \xi^{(m)}\}$ is called the $n \times n$ generic division ring and is denoted by $F(\xi^{(1)}, \ldots, \xi^{(m)})$.

Let us denote by Z the center of $F\{\xi^{(1)}, \ldots, \xi^{(m)}\}$ and by C its field of fractions that is the center of the corresponding generic division ring.

We close this section by stating a theorem of Procesi allowing to compute the transcendence degree of Z over F. We shall not use this result in other parts of the book but its proof can be found in [**Pr3**, Theorem 6.3]. An alternative approach is given in [**F4**, Theorem 28].

THEOREM 1.13.11. *Let $u_1, \ldots, u_n$ be the eigenvalues of $\xi^{(1)}$ and consider the field $K = C(u_1, \ldots, u_n)$. Then:*
  (1) *K is a splitting field for the generic division ring $F(\xi^{(1)}, \ldots, \xi^{(m)})$.*
  (2) *K is a Galois extension of C with Galois group S_n.*
  (3) *K is a pure transcendental extension of F of transcendence degree $(m-1)n^2 + 1$.*

Recall that a field L is unirational over a subfield M if it has finite index in a rational function field over M. An immediate consequence of the previous theorem is the following.

COROLLARY 1.13.12. *The $n \times n$ generic division ring $F(\xi^{(1)}, \ldots, \xi^{(m)})$ has center C unirational over F of transcendence degree $(m-1)n^2 + 1$. Hence*
$$\mathrm{GKdim}(F\{\xi^{(1)}, \ldots, \xi^{(m)}\}) = \mathrm{tr\,deg}(C/F) = (m-1)n^2 + 1.$$

It is not known if C is actually a rational function field over F. For 2×2 generic matrices this property holds ([**Pr3**, Theorem 6.5]). We conclude this section with an explicit computation of a transcendence basis of C over F in the case of two 2×2 generic matrices ([**FHL**]). In order to simplify the notation, write $\xi^{(1)} = \xi$ and $\xi^{(2)} = \eta$.

THEOREM 1.13.13. *The center of the 2×2 generic division ring $F(\xi, \eta)$ coincides with the rational function field over F in the five variables*
$$\mathrm{tr}(\xi), \mathrm{tr}(\eta), \det(\xi), \det(\eta), \mathrm{tr}(\xi\eta).$$

1.13. THE GELFAND-KIRILLOV DIMENSION

PROOF. Set $K = F(\mathrm{tr}(\xi), \mathrm{tr}(\eta), \det(\xi), \det(\eta), \mathrm{tr}(\xi\eta))$. Let E be the 2×2 identity matrix and let A be the algebra generated by the elements $E, \xi, \eta, \xi\eta$ over K. It is not hard to prove that $E, \xi, \eta, \xi\eta$ are linearly independent over K. Moreover, the relations
$$\xi^2 = \mathrm{tr}(\xi)\xi - \det(\xi), \ \eta^2 = \mathrm{tr}(\eta)\eta - \det(\eta),$$
$$\eta\xi = (\mathrm{tr}(\xi\eta) - \mathrm{tr}(\xi)\mathrm{tr}(\eta))E + \mathrm{tr}(\eta)\xi + \mathrm{tr}(\xi)\eta - \xi\eta$$
imply that A is a four-dimensional algebra over K with basis $\{E, \xi, \eta, \xi\eta\}$. Since it easily follows that A is central simple over K, we obtain that $A = F(\xi, \eta)$. Thus K coincides with the center C of $F(\xi, \eta)$. Since by the previous corollary, $\mathrm{tr}\deg(C/F) = 5$, the elements $\mathrm{tr}(\xi), \mathrm{tr}(\eta), \det(\xi), \det(\eta), \mathrm{tr}(\xi\eta)$ are algebraically independent over F. $\square$

CHAPTER 2

S_n-Representations

In the first section of this chapter we recall the basic definitions and results of the representation theory of finite groups over an algebraically closed field of characteristic zero. Our aim is to get the reader familiar with the basic terminology and facts used throughout the book. The second section is devoted to the classical ordinary representation theory of the symmetric group through the theory of Young tableaux. We present some of the most interesting results of the theory such as the hook formula, the branching theorem and the more general Littlewood-Richardson rule. These will be some of the basic tools in our investigation throughout the book. We give no proofs of the results of the first two sections. We feel that the representation theory of finite groups is a quite standard subject nowadays and can be found in several textbooks of algebra (see for instance [**CuR**], [**H**], [**Ja2**]). For the ordinary representation theory of the symmetric group we refer the reader to the books of Curtis and Reiner ([**CuR**]), Boerner ([**Bo**]), Robinson ([**Rob**]). The article of Henke and Regev in [**GRZ2**] is an excellent source. For a complete treatment of the theory in arbitrary characteristic we refer the reader to the books of James ([**Jm**]), James and Kerber ([**JK**]) and Sagan ([**Sa**]).

In the last two sections of this chapter we exploit the permutation action of the symmetric group S_n on the space of multilinear polynomials in n variables. This action has the basic property of leaving T-ideals invariant and will be a main tool throughout the book. We realize the generators of the irreducible S_n-representations as polynomials and we prove some basic properties on their symmetric and alternating sets of variables.

2.1. Finite dimensional representations

Let V be a vector space over a field F and let $GL(V)$ be the group of invertible endomorphisms of V. Even if most of the results in this chapter hold over a field of arbitrary characteristic, we shall assume, throughout the chapter starting with Section 2, that char $F = 0$. Recall the following.

DEFINITION 2.1.1. A representation of a group G on V is a homomorphism of groups $\rho : G \to GL(V)$.

Let us denote by End(V) the algebra of F-endomorphisms of V. If FG is the group algebra of G over F and ρ is a representation of G on V, it is clear that ρ induces a homomorphism of F-algebras $\rho' : FG \to \text{End}(V)$ such that $\rho'(1_G) = 1$.

Throughout we shall be dealing only with the case when $\dim_F V = n < \infty$, i.e., with finite dimensional representations. In this case n is called the dimension or the degree of the representation ρ. Now, a representation of a group G uniquely determines a finite dimensional FG-module (or G-module) in the following way. If $\rho : G \to GL(V)$ is a representation of G, V becomes a (left) G-module by defining

$gv = \rho(g)(v)$ for all $g \in G$, $v \in V$. It is also clear that if M is a G-module which is finite dimensional as a vector space over F, then $\rho : G \to GL(M)$, such that $\rho(g)(m) = gm$, for $g \in G$, $m \in M$, defines a representation of G on M.

DEFINITION 2.1.2. If $\rho : G \to GL(V)$ and $\rho' : G \to GL(W)$ are two representations of a group G, we say that ρ and ρ' are equivalent, and we write $\rho \sim \rho'$, if V and W are isomorphic as G-modules.

DEFINITION 2.1.3. A representation $\rho : G \to GL(V)$ is irreducible if V is an irreducible G-module. ρ is completely reducible if V is the direct sum of its irreducible submodules.

The basic tool for studying the representations of a finite group in the ordinary case (i.e., char $F = 0$) is Maschke's theorem. Recall that an algebra A is semisimple if $J(A) = 0$, where $J(A)$ is the Jacobson radical of A.

THEOREM 2.1.4 (Maschke). *Let G be a finite group and let char $F = 0$ or char $F = p > 0$ and $p \nmid |G|$. Then the group algebra FG is semisimple.*

As a consequence of Wedderburn's theorem (Theorem 1.11.1), it follows that, under the hypotheses of Maschke's theorem,

$$(2.1) \qquad FG \cong M_{n_1}(D^{(1)}) \oplus \cdots \oplus M_{n_k}(D^{(k)})$$

where $D^{(1)}, \ldots, D^{(k)}$ are finite dimensional division algebras over F. In light of these results one can classify all the irreducible representations of G: M is an irreducible G-module if and only if M is an irreducible $M_{n_i}(D^{(i)})$-module, for some i. On the other hand, $M_{n_i}(D^{(i)})$ has (up to isomorphisms) only one irreducible module, isomorphic to $\sum_{j=1}^{n_i} D^{(i)} e_{ji}$.

From the above it can also be deduced that every G-module V is completely reducible. Hence, if $\dim_F V < \infty$, V is the direct sum of a finite number of irreducible G-modules. We record this fact in the following.

COROLLARY 2.1.5. *Let G be a finite group and F a field of characteristic zero or $p > 0$ and $p \nmid |G|$. Then every representation of G is completely reducible and the number of inequivalent irreducible representations of G equals the number of simple components in the Wedderburn decomposition of the group algebra FG.*

The decomposition of the group algebra FG can be stated in terms of representations by means of the regular representation. Recall that the (left) regular representation of G is given by $\tau : FG \to \text{End}(FG)$ where $\tau(a)(b) = ab$, for all $a, b \in FG$. From the decomposition given in (2.1) it follows that the regular representation has the following decomposition

$$FG \cong n_1 J_1 \oplus \cdots \oplus n_k J_k$$

where $n_i J_i = J_i \oplus \cdots \oplus J_i$ (n_i-times), n_i is called the multiplicity of J_i in FG, and $J_i \cong \sum_{l=1}^{n_i} D^{(i)} e_{li}$ is a minimal left ideal of $M_{n_i}(D^{(i)})$. Note that $n_i = \dim J_i$ is the degree of the representation J_i. We can now state this in the following.

PROPOSITION 2.1.6. *Under the hypotheses of Maschke's theorem, every irreducible representation of G (up to equivalence) appears in the regular representation of G with multiplicity equal to its degree.*

Recall that an element $e \in FG$ is an idempotent if $e^2 = e$. It is well known that since FG is finite dimensional semisimple, every one-sided ideal of FG is generated by an idempotent. Moreover every two-sided ideal is generated by a central idempotent. We say that an idempotent is minimal (central resp.) if it generates a minimal one-sided (two-sided resp.) ideal. We record this in the following.

PROPOSITION 2.1.7. *If M is an irreducible representation of G, then $M \cong J_i$, a minimal left ideal of $M_{n_i}(D^{(i)})$, for some $i \in 1, \ldots, k$. Hence there exists a minimal idempotent $e \in FG$ such that $M \cong FGe$.*

When the field F is a splitting field for the group G, e.g., F is algebraically closed, then the following properties hold.

PROPOSITION 2.1.8. *Let F be a splitting field for G. Then the number of non-equivalent irreducible representations of G equals the number of conjugacy classes of G.*

Since by Corollary 2.1.5 this number equals the number of simple components of FG, it follows that when F is a splitting field for G, it equals the dimension of the center of FG over F.

A basic tool in representation theory is provided by the theory of characters. From now on assume that F is a splitting field for G of characteristic zero and let $\text{tr} : \text{End}(V) \to F$ be the trace function on $\text{End}(V)$.

DEFINITION 2.1.9. *Let $\rho : G \to GL(V)$ be a representation of G. Then the map $\chi_\rho : G \to F$ such that $\chi_\rho(g) = \text{tr}(\rho(g))$ is called the character of the representation ρ and $\dim V = \deg \chi_\rho$ is called the degree of the character χ_ρ.*

We say that the character χ_ρ is irreducible if ρ is irreducible. Since $\chi_\rho(g) = \chi_\rho(h)$ provided g is conjugate to h in G, χ_ρ is constant on the conjugacy classes of G, i.e., χ_ρ is a class function of G. Notice that $\chi_\rho(1) = \deg \chi_\rho$.

Let $C(G)$ be the vector space of class functions on G. One can define an inner product $(\ ,\ )$ on $C(G)$ by setting $(\chi, \psi) = \frac{1}{|G|} \sum_{g \in G} \chi(g) \psi(g^{-1})$ for $\chi, \psi \in C(G)$. The basic property of this inner product is that the irreducible characters of the group G form an orthonormal basis of $C(G)$.

We next state the basic properties of the characters of a group G.

PROPOSITION 2.1.10. *Suppose that F is an algebraically closed field of characteristic zero and let $J_1, \ldots, J_k$ be a complete list of non-equivalent irreducible representations of G with characters $\chi_1, \ldots, \chi_k$, respectively. Let $\rho : G \to GL(V)$ be a representation of G and write $V \cong m_1 J_1 \oplus \cdots \oplus m_k J_k$ with $m_i \geq 0$. Then:*

(1) $\chi_\rho = \sum_{i=1}^{k} m_i \chi_i$.
(2) $(\chi_\rho, \chi_i) = m_i$, *for all i.*
(3) $(\chi_\rho, \chi_\rho) = \sum_{i=1}^{k} m_i^2$.
(4) χ_ρ *is irreducible if and only if $(\chi_\rho, \chi_\rho) = 1$.*
(5) *If ρ' is another representation of G, then $\rho \sim \rho'$ if and only if $\chi_\rho = \chi_{\rho'}$.*

When the field F is algebraically closed and of characteristic zero, one can write all the minimal central idempotents of the group algebra FG in terms of the irreducible characters of G.

REMARK 2.1.11. Under the hypotheses of the previous proposition, the group algebra FG decomposes as $FG = \oplus_{i=1}^{k} e_i FG$ where, for every $i = 1, \ldots, k$, $e_i FG \cong M_{n_i}(F)$ and

$$e_i = \frac{\chi_i(1)}{|G|} \sum_{g \in G} \overline{\chi_i(g)} g$$

is a minimal central idempotent of FG.

Now let H be a subgroup of the group G and let V be a G-module. Then V can be regarded, by restriction, as an H-module; we denote this module by $V \downarrow H$ and we call it the induced module on H.

Conversely, if V is an H-module, the space $FG \otimes_{FH} V$ has a natural structure of G-module. We denote this module by $V \uparrow G$ and we call it the G-module induced by V.

Accordingly, if χ is the character of the G-module (H-module) V, then $\chi \downarrow H$ ($\chi \uparrow G$, resp.) represents the character of the induced module. When an irreducible module is induced up or down, it remains no longer irreducible in general. In characteristic zero, a basic tool relating induced modules is given by Frobenius reciprocity theorem.

THEOREM 2.1.12 (Frobenius). *Let G be a finite group with subgroup H. If χ is an irreducible character of G and ψ is an irreducible character of H, then*

$$(\chi \downarrow H, \psi) = (\psi \uparrow G, \chi).$$

2.2. S_n-representations

In this section we describe the ordinary representation theory of the symmetric group S_n, $n \geq 1$. Since $\mathbb{Q}$, the field of rational numbers, is a splitting field for S_n, for any field F of characteristic zero, the group algebra FS_n has a decomposition into simple components which are algebras of matrices over the field F itself. Moreover, by Proposition 2.1.8, the number of irreducible non-equivalent representations equals the number of conjugacy classes of S_n. Recall the following.

DEFINITION 2.2.1. Let $n \geq 1$ be an integer. A partition λ of n is a finite sequence of integers $\lambda = (\lambda_1, \ldots, \lambda_r)$ such that $\lambda_1 \geq \cdots \geq \lambda_r > 0$ and $\sum_{i=1}^{n} \lambda_i = r$. In this case we write $\lambda \vdash n$ or $|\lambda| = n$.

If $r = 1$, then $\lambda_1 = n$ and we write $\lambda = (n)$. For the partition λ with $\lambda_1 = \ldots = \lambda_n = 1$ the notation $\lambda = (1^n)$ is usually used. More generally, we write $\lambda = (k^d)$ as soon as $\lambda = (k, \ldots, k)$ and $n = kd$.

It is well known that the conjugacy classes of S_n are indexed by the partitions of n: if $\sigma \in S_n$, we decompose σ into the product of disjoint cycles, including 1-cycles. This decomposition is unique if we require that

$$\sigma = \pi_1 \pi_2 \cdots \pi_r$$

with $\pi_1, \ldots, \pi_r$ cycles of length $\lambda_1 \geq, \ldots, \geq \lambda_r \geq 1$, respectively. Then the partition $\lambda = (\lambda_1, \ldots, \lambda_r)$ uniquely determines the conjugacy class of σ.

Since, as we mentioned above, all the irreducible characters of S_n are indexed by the partitions of n, let us denote by χ_λ the irreducible S_n-character corresponding to $\lambda \vdash n$.

It is standard to use the notation $d_\lambda = \chi_\lambda(1)$ for the degree of χ_λ. It follows that FS_n has the following decomposition

$$FS_n = \bigoplus_{\lambda \vdash n} I_\lambda \cong \bigoplus_{\lambda \vdash n} M_{d_\lambda}(F),$$

where $I_\lambda = e_\lambda FS_n \cong M_{d_\lambda}(F)$ is the minimal two-sided ideal of FS_n corresponding to $\lambda \vdash n$, and $e_\lambda = \sum_{\sigma \in S_n} \chi_\lambda(\sigma)\sigma$ is the essential central idempotent deduced from Remark 2.1.11. In other words, the character of the regular representation τ of S_n has the decomposition

$$\chi_\tau = \sum_{\lambda \vdash n} d_\lambda \chi_\lambda.$$

We record this in the following.

PROPOSITION 2.2.2. *Let F be any field of characteristic zero and $n \geq 1$. Then there is a one-to-one correspondence between irreducible S_n-characters and partitions of n. Let $\{\chi_\lambda \mid \lambda \vdash n\}$ be a complete set of irreducible characters of S_n and let $d_\lambda = \chi_\lambda(1)$ be the degree of χ_λ, $\lambda \vdash n$. Then*

$$FS_n = \bigoplus_{\lambda \vdash n} I_\lambda \cong \bigoplus_{\lambda \vdash n} M_{d_\lambda}(F),$$

where $I_\lambda = e_\lambda FS_n$ and $e_\lambda = \sum_{\sigma \in S_n} \chi_\lambda(\sigma)\sigma$ is up to a scalar, the unit element of I_λ.

Next we describe a formula for computing the degrees d_λ.

DEFINITION 2.2.3. If $\lambda = (\lambda_1, \ldots, \lambda_r) \vdash n$, the Young diagram associated to λ is the finite subset of $\mathbb{Z} \times \mathbb{Z}$ defined as $D_\lambda = \{(i,j) \in \mathbb{Z} \times \mathbb{Z} \mid i = 1, \ldots, r, \ j = 1, \ldots, \lambda_i\}$.

There are two standard notations. In one notation, a Young diagram D_λ is denoted as an array of boxes corresponding to the points (i,j). In the other notation, and this is the one we shall adopt, the array of boxes denoting D_λ is such that the first coordinate i (the row index) increases from top to bottom and the second coordinate j (the column index) increases from left to right. For instance, the diagram $D_{(4,2,2,1)}$ is represented by

$$D_{(4,\,2,\,2,\,1)} = \quad \begin{array}{|c|c|c|c|} \hline & & & \\ \hline & \\ \cline{1-2} & \\ \cline{1-2} \\ \cline{1-1} \end{array}$$

For a partition $\lambda \vdash n$ we shall denote by λ' the conjugate partition of λ; $\lambda' = (\lambda'_1, \ldots, \lambda'_s)$ is the partition such that $\lambda'_1, \ldots, \lambda'_s$ are the lengths of the columns of D_λ. Hence D'_λ is obtained from D_λ by flipping D_λ along its main diagonal.

DEFINITION 2.2.4. Let $\lambda \vdash n$. A Young tableau T_λ of the diagram D_λ is a filling of the boxes of D_λ with the integers $1, 2, \ldots, n$. We shall also say that T_λ is a tableau of shape λ.

Of course there are $n!$ distinct tableaux. Among these a prominent role is played by the so-called standard tableaux.

DEFINITION 2.2.5. A tableau T_λ of shape λ is standard if the integers in each row and in each column of T_λ increase from left to right and from top to bottom, respectively.

For instance, the tableau

$$T_{(5,\,2,\,1)} = \begin{array}{|c|c|c|c|c|} \hline 1 & 4 & 3 & 7 & 8 \\ \hline 5 & 2 \\ \cline{1-2} 6 \\ \cline{1-1} \end{array}$$

is not standard. The tableau

$$T_{(4,\,4,\,1)} = \begin{array}{|c|c|c|c|} \hline 1 & 2 & 4 & 6 \\ \hline 3 & 5 & 8 & 9 \\ \hline 7 \\ \cline{1-1} \end{array}$$

is standard.

There is a strict connection between standard tableaux and degrees of the irreducible S_n-characters.

THEOREM 2.2.6. *Given a partition $\lambda \vdash n$, the number of standard tableaux of shape λ equals d_λ, the degree of χ_λ, the irreducible character corresponding to λ.*

Next we give two different formulas for the degree d_λ of the irreducible character χ_λ: the hook formula ([**JK**, Theorem 2.3.21]) and the Young-Frobenius formula ([**Bo**, Theorem 4.2]). First we need some further terminology.

Given a diagram D_λ, $\lambda \vdash n$, we identify a box of D_λ with the corresponding point (i, j). For instance, the third box of the first row has coordinate $(1, 3)$.

DEFINITION 2.2.7. For any box $(i, j) \in D_\lambda$, we define the hook number of (i, j) as $h_{ij} = \lambda_i + \lambda'_j - i - j + 1$, where $\lambda' = (\lambda'_1, \ldots, \lambda'_s)$ is the conjugate partition of λ.

Note that h_{ij} counts the number of boxes in the "hook" with edge in (i,j), i.e., the boxes to the right and below (i, j). In the following example we have written inside each box its hook number:

$$\begin{array}{|c|c|c|c|} \hline 6 & 4 & 3 & 1 \\ \hline 4 & 2 & 1 \\ \cline{1-3} 1 \\ \cline{1-1} \end{array}$$

PROPOSITION 2.2.8 (The Hook Formula).

$$d_\lambda = \frac{n!}{\prod_{i,j} h_{ij}}$$

where the product runs over all boxes of D_λ.

The next formula is obtained as follows. Let $\lambda = (\lambda_1, \ldots, \lambda_r) \vdash n$ and define $l_1 = \lambda_1 + k - 1$, $l_2 = \lambda_2 + k - 2, \ldots, l_k = \lambda_k$. Then

PROPOSITION 2.2.9 (The Young-Frobenius Formula).

$$d_\lambda = \frac{n!}{l_1! \cdots l_k!} \prod_{1 \leq i < j \leq k} (l_i - l_j),$$

where the product runs over all boxes of D_λ.

Next we describe a complete set of minimal left ideals of FS_n. Given any tableau T_λ of shape $\lambda \vdash n$, let us denote by $T_\lambda = D_\lambda(a_{ij})$, where a_{ij} is the integer in the (i,j) box. Then

2.2. S_n-REPRESENTATIONS

DEFINITION 2.2.10. The row-stabilizer of T_λ is

$$R_{T_\lambda} = S_{\lambda_1}(a_{11}, a_{12}, \ldots, a_{1\lambda_1}) \times \cdots \times S_{\lambda_r}(a_{r1}, a_{r2}, \ldots, a_{r\lambda_r})$$

where $S_{\lambda_i}(a_{i1}, a_{i2}, \ldots, a_{i\lambda_i})$ denotes the symmetric group acting on the integers $a_{i1}, a_{i2}, \ldots, a_{i\lambda_i}$.

Hence R_{T_λ} is the subgroup of S_n consisting of all permutations stabilizing the rows of T_λ.

DEFINITION 2.2.11. The column-stabilizer of $T_\lambda = D_\lambda(a_{ij})$ is

$$C_{T_\lambda} = S_{\lambda'_1}(a_{11}, a_{21}, \ldots, a_{\lambda'_1 1}) \times \cdots \times S_{\lambda'_s}(a_{1\lambda_1}, a_{2\lambda_1}, \ldots, a_{\lambda'_s \lambda_1})$$

where $\lambda' = (\lambda'_1, \ldots, \lambda'_s)$ is the conjugate partition of λ.

Hence C_{T_λ} is the subgroup of S_n consisting of all permutations stabilizing the columns of T_λ.

DEFINITION 2.2.12. For a given tableau T_λ, we define

$$e_{T_\lambda} = \sum_{\substack{\sigma \in R_{T_\lambda} \\ \tau \in C_{T_\lambda}}} (\mathrm{sgn}\,\tau)\sigma\tau.$$

It can be shown that $e_{T_\lambda}^2 = a e_{T_\lambda}$, where $a = \frac{n!}{d_\lambda} = \prod_{i,j} h_{ij}$ is a non-zero integer, i.e., e_{T_λ} is an essential idempotent of FS_n.

Given a partition $\lambda \vdash n$, the symmetric group S_n acts on the set of Young tableaux of shape λ as follows: if $\sigma \in S_n$ and $T_\lambda = D_\lambda(a_{ij})$, then $\sigma T_\lambda = D_\lambda(\sigma(a_{ij}))$. This action has the property that

$$R_{\sigma T_\lambda} = \sigma R_{T_\lambda} \sigma^{-1} \quad \text{and} \quad C_{\sigma T_\lambda} = \sigma C_{T_\lambda} \sigma^{-1}.$$

It follows that $\sigma e_{T_\lambda} \sigma^{-1} = e_{\sigma T_\lambda}$.

We record the most important facts about e_{T_λ} in the following.

PROPOSITION 2.2.13. *For every Young tableau T_λ of shape $\lambda \vdash n$, the element e_{T_λ} is a minimal essential idempotent of FS_n and $FS_n e_{T_\lambda}$ is a minimal left ideal of FS_n with character χ_λ. If T_λ and T_λ^* are Young tableaux of the same shape, then e_{T_λ} and $e_{T_\lambda^*}$ are conjugated in FS_n through some $\sigma \in S_n$; moreover, $\sigma e_{T_\lambda} \sigma^{-1} = e_{\sigma T_\lambda}$.*

The above proposition says that for any two tableaux T_λ, T_λ^* of the same shape λ, $FS_n e_{T_\lambda} \cong FS_n e_{T_\lambda^*}$, as S_n-modules. Whereas if T_λ and T_μ are of a different shape, $FS_n e_{T_\lambda} \not\cong FS_n e_{T_\mu}$. Notice that if $e_{T_\lambda}^2 = a e_{T_\lambda}$, then $\frac{e_{T_\lambda}}{a}$ is an idempotent which generates the same left ideal $FS_n e_{T_\lambda} = FS_n \frac{e_{T_\lambda}}{a}$.

The standard tableaux come into play if one wants to find, among the $n!$ essential idempotents arising from tableaux of shape λ, some orthogonal ones. In fact, we have

PROPOSITION 2.2.14. *If $T_1, \ldots, T_{d_\lambda}$ are all the standard tableaux of shape λ, then I_λ, the minimal two-sided ideal of FS_n corresponding to λ, has the decomposition*

$$I_\lambda = \bigoplus_{i=1}^{d_\lambda} FS_n e_{T_i}.$$

One should beware of the above decomposition; it is only a decomposition as left modules. In general the e_{T_i} are not orthogonal. It can be shown that there exists a natural order on the set of standard tableaux $T_1 > \cdots > T_{d_\lambda}$ such that $e_{T_i} e_{T_j} = 0$ if $i > j$ ([**JK**, Theorem 3.1.24]).

We will use the following property of essential idempotents ([**CuR**, Lemma 28.13]).

LEMMA 2.2.15 (Von Neumann). *Let $\lambda \vdash n$, $x \in FS_n$ and $\sigma x \tau = (sgn\,\tau)x$ for all $\sigma \in R_{T_\lambda}, \tau \in C_{T_\lambda}$. Then $x = \gamma e_{T_\lambda}$ for some rational number γ.*

2.3. Inducing S_n-representations

In this section we regard the group S_n embedded in S_{n+1} as the subgroup of all permutations fixing the integer $n+1$. The next theorem gives a decomposition into irreducibles of any S_n-module induced up to S_{n+1}.

Let us denote by M_λ an irreducible S_n-module corresponding to the partition $\lambda \vdash n$. We have ([**JK**, Theorem 2.4.3])

THEOREM 2.3.1 (Branching Theorem). *Let the group S_n be embedded into S_{n+1} as the subgroup fixing the integer $n+1$. Then*
 (1) *if $\lambda \vdash n$, then $M_\lambda \uparrow S_{n+1} \cong \sum_{\mu \in \lambda^+} M_\mu$ where λ^+ is the set of all partitions of $n+1$ whose diagram is obtained from D_λ by adding one box;*
 (2) *if $\mu \vdash n+1$, then $M_\mu \downarrow S_n \cong \sum_{\lambda \in \mu^-} M_\lambda$ where μ^- is the set of partitions of n whose diagram is obtained from D_μ by deleting one box.*

Note that by Frobenious reciprocity (1) and (2) are equivalent formulations.

We go one step further and we state a more general result. First we need some definitions.

We embed the group $S_n \times S_m$ into S_{n+m} by letting S_m act on $\{n+1, \ldots, n+m\}$. Recall that if M is an S_n-module and N is an S_m-module, then $M \otimes_F N$ has a natural structure of $S_n \times S_m$-module.

DEFINITION 2.3.2. If M is an S_n-module and N is an S_m-module, then the outer tensor product of M and N is defined as
$$M \widehat{\otimes} N = (M \otimes N) \uparrow S_{n+m}.$$

Recall that (n) denotes a one-row partition $\mu \vdash n$, i.e., $\mu_1 = n$. We have ([**JK**, Theorem 2.8.2])

THEOREM 2.3.3 (Young's Rule). *Let $\lambda \vdash n$ and $m \geq 1$. Then*
$$M_\lambda \widehat{\otimes} M_{(m)} \cong \sum M_\mu$$
where the sum runs over all partitions μ of $n+m$ such that we have $\mu_1 \geq \lambda_1 \geq \mu_2 \geq \cdots \geq \mu_{n+m} \geq \lambda_{n+m}$.

The above rule says that μ is obtained from λ by adding m boxes to the diagram D_λ in such a way that no two new boxes appear in the same column of the new diagram. In order to introduce the most general Littlewood-Richardson rule, which gives a decomposition of $M_\lambda \widehat{\otimes} M_\mu$ for all λ, μ, we need some further definitions.

DEFINITION 2.3.4. An unordered partition of n is a finite sequence of positive integers $\alpha = (\alpha_1, \ldots, \alpha_t)$ such that $\sum_{i=1}^t \alpha_i = n$. In this case we write $\alpha \models n$.

2.3. INDUCING S_n-REPRESENTATIONS

DEFINITION 2.3.5. Let $\lambda \vdash n$ and $\alpha \models n$. A (generalized) Young tableau of shape λ and content α is a filling of the diagram D_λ by positive integers in such a way that the integer i occurs exactly α_i times.

For instance,

1	2	2	2	3	4
2	1	1			
3					

is a tableau of shape $\lambda = (6, 3, 1)$ and content $\alpha = (3, 4, 2, 1)$. Note that a Young tableau T_λ, $\lambda \vdash n$ is of content (1^n).

DEFINITION 2.3.6. A Young tableau is semistandard if the numbers are non-decreasing along the rows and strictly increasing down the columns.

We now consider the obvious partial order on the set of partitions. Let $\lambda = (\lambda_1, \ldots, \lambda_p) \vdash n$ and $\mu = (\mu_1, \ldots, \mu_q) \vdash m$, then $\lambda \geq \mu$ if and only if $p \geq q$ and $\lambda_i \geq \mu_i$ for all $i = 1, \ldots, p$. In the language of Young diagrams $\lambda \geq \mu$ means that D_μ is a subdiagram of D_λ.

Let $\lambda \vdash n$, $\mu \vdash m$. We say that $\lambda \geq \mu$ if $\lambda_i \geq \mu_i$ for all $i \geq 1$, i.e., $D_\lambda \supseteq D_\mu$. If $\lambda \geq \mu$, we define the skew-partition $\lambda \setminus \mu = (\lambda_1 - \mu_1, \lambda_2 - \mu_2, \ldots)$; the corresponding diagram $D_{\lambda \setminus \mu}$ is the set of boxes of D_λ which do not belong to D_μ. For instance, if $\lambda = (4, 2, 2, 1)$, $\mu = (3, 2)$, then $\lambda \setminus \mu = (1, 0, 2, 1)$ and

$$D_{\lambda \setminus \mu} = \begin{array}{|c|c|c|c|} \hline & & & * \\ \hline * & * & & \\ \hline * & * & & \\ \hline & & & \\ \hline \end{array}$$

DEFINITION 2.3.7. A skew-tableau $T_{\lambda \setminus \mu}$ is a filling of the boxes of the skew-diagram $D_{\lambda \setminus \mu}$ with distinct natural numbers. If repetitions occur, then we have the notion of (generalized) skew-tableau. We also have the natural notions of standard and semistandard skew-tableaux.

DEFINITION 2.3.8. Let $\alpha = (\alpha_1, \ldots, \alpha_t) \models n$. We say that α is a lattice permutation if for each j the number of i's which occur among $\alpha_1, \ldots, \alpha_j$ is greater than or equal to the number of $(i+1)$'s for each i.

We can now formulate ([**JK**, Theorem 2.8.13])

THEOREM 2.3.9 (Littlewood-Richardson Rule). Let $\lambda \vdash n$ and $\mu \vdash m$. Then

$$M_\lambda \widehat{\otimes} M_\mu \cong \sum_{\nu \vdash n+m} k^\mu_{\nu \setminus \lambda} M_\nu$$

where $k^\mu_{\nu \setminus \lambda}$ is the number of semistandard tableau of shape $\nu \setminus \lambda$ and content μ which yield lattice permutations when we read their entries from right to left and downwards.

An algorithm for determining the decomposition in the Littlewood-Richardson-rule is the following: consider $T_\mu = D_\mu(a_{ij})$ where the a_{ij}'s are symbols. Then:

(1) Add to D_λ all the boxes with the symbols a_{1j}; after the additions, no row of the new tableau may contain more boxes than a preceding row.
(2) Next add all boxes with the symbols a_{2j}, according to the same rule, and so on, until all the boxes with the symbols have been added.

(3) These additions must be such that for all i, if $y < j$, a_{iy} goes in a later column than a_{ij}, and for all j, if $x < i$, a_{xj} goes in an earlier row than a_{ij}.

2.4. S_n-actions on multilinear polynomials

In this section we introduce an action of the symmetric group S_n on the space of multilinear polynomials in n fixed variables. This action is the main tool for proving most of the results in this book. We assume throughout this and the next section that $\operatorname{char} F = 0$. We start with a remark about arbitrary irreducible S_n-modules.

LEMMA 2.4.1. *Let M be an irreducible left S_n-module with character $\chi(M) = \chi_\lambda$, $\lambda \vdash n$. Then M can be generated as an S_n-module by an element of the form $e_{T_\lambda} f$ for some $f \in M$ and some Young tableau T_λ of shape λ. Moreover, for any Young tableau T_λ^* of shape λ there exists $f' \in M$ such that $M = FS_n e_{T_\lambda^*} f'$.*

PROOF. Recall that $FS_n = \bigoplus_{\mu \vdash n} I_\mu$, where I_μ is the two-sided ideal of FS_n corresponding to μ and by Proposition 2.2.14,

$$FS_n = \bigoplus_{\substack{\mu \vdash n \\ T_\mu \, standard}} FS_n e_{T_\lambda}.$$

Since $M = FS_n M$, there exists $\mu \vdash n$, T_μ standard tableau and $f \in M$ such that $0 \neq FS_n e_{T_\mu} f \subseteq M$. By the irreducibility of M we get $FS_n e_{T_\mu} f = M$. Also, since $\chi(M) = \chi_\lambda$, we obtain that $\lambda = \mu$. Finally, if T_λ^* is another Young tableau of the same shape then, by Proposition 2.2.13, $e_{T_\lambda} = \sigma e_{T_\lambda^*} \sigma^{-1}$ and $g = \sigma e_{T_\lambda^*} f'$ where $f' = \sigma^{-1} f$. □

The previous lemma says that, given a partition $\lambda \vdash n$ and a Young tableau T_λ of shape λ, any irreducible S_n-module M such that $\chi(M) = \chi_\lambda$ can be generated by an element of the form $e_{T_\lambda} f$ for some $f \in M$. By the definition of R_{T_λ}, for any $\sigma \in R_{T_\lambda}$ we have that $\sigma e_{T_\lambda} f = e_{T_\lambda} f$, i.e., $e_{T_\lambda} f$ is stable under the R_{T_λ}-action. The number of R_{T_λ}-stable elements in an arbitrary S_n-module M is closely related to the number of irreducible S_n-submodules of M having character χ_λ.

LEMMA 2.4.2. *Let T_λ be a Young tableau corresponding to $\lambda \vdash n$ and let M be an S_n-module such that $M = M_1 \oplus \cdots \oplus M_m$ where $M_1, \ldots, M_m$ are irreducible S_n-submodules with character χ_λ. Then m is equal to the maximal number of linearly independent elements $g \in M$ such that $\sigma g = g$ for all $\sigma \in R_{T_\lambda}$.*

PROOF. Clearly it is enough to prove the lemma when $m = 1$, i.e., $M = M_1$ and $\chi(M) = \chi_\lambda$. By the previous lemma, M can be generated by an element of the type $g = e_{T_\lambda} f$, for some $f \in M$. Obviously, $\sigma g = g$ for all $\sigma \in R_{T_\lambda}$. On the other hand, suppose $0 \neq h \in M$ is also stable under the R_{T_λ}-action. Then $h = ae_{T_\lambda} f$ for some $a \in FS_n$, $0 \neq f \in M$, and we have the equality $\sigma ae_{T_\lambda} f = ae_{T_\lambda} f$ for any $\sigma \in R_{T_\lambda}$. It follows that $b = \sigma ae_{T_\lambda} - ae_{T_\lambda} \in J \cap FS_n e_{T_\lambda}$, where $J = \{x \in FS_n \mid xf = 0\}$ is the left annihilator of f. By irreducibility, $J \cap FS_n e_{T_\lambda}$ is either 0 or $FS_n e_{T_\lambda}$. In this last case $J \supseteq FS_n e_{T_\lambda}$, so $FS_n e_{T_\lambda} f = 0$, a contradiction. It follows that $J \cap FS_n e_{T_\lambda} = 0$. Hence $b = 0$ and $\sigma ae_{T_\lambda} = ae_{T_\lambda}$ is stable under the left R_{T_λ}-action. Clearly, $ae_{T_\lambda} \tau = (\operatorname{sgn} \tau) ae_{T_\lambda}$ for any $\tau \in C_{T_\lambda}$. Hence $\sigma ae_{T_\lambda} \tau = (\operatorname{sgn} \tau) ae_{T_\lambda}$ for all $\sigma \in R_{T_\lambda}, \tau \in C_{T_\lambda}$ and by Lemma 2.2.15 we obtain $ae_{T_\lambda} = \gamma e_{T_\lambda}$ with $\gamma \in F$. Hence $h = \gamma g$ and the proof is complete. □

2.4. S_n-ACTIONS ON MULTILINEAR POLYNOMIALS

Now let A be a PI-algebra and $\mathrm{Id}(A)$ its T-ideal of identities. As we showed in Corollary 1.3.9, in characteristic zero, $\mathrm{Id}(A)$ is determined by its multilinear polynomials.

We introduce
$$P_n = \mathrm{span}\{x_{\sigma(1)} \cdots x_{\sigma(n)} \mid \sigma \in S_n\},$$
the vector space of multilinear polynomials in $x_1, \ldots, x_n$ in the free algebra $F\langle X \rangle$. We define a map
$$\varphi : FS_n \longrightarrow P_n$$
by setting
$$\varphi : \sum_{\sigma \in S_n} \alpha_\sigma \sigma \mapsto \sum_{\sigma \in S_n} \alpha_\sigma x_{\sigma(1)} \cdots x_{\sigma(n)}.$$

It is clear that φ is a linear isomorphism. Hence, when no confusion arises, we shall use the same notation for an element $f \in FS_n$ and its image in P_n. This isomorphism turns P_n into an S_n bimodule; if $\sigma, \tau \in S_n$, then
$$\sigma(x_{\tau(1)} \cdots x_{\tau(n)}) = x_{\sigma\tau(1)} \cdots x_{\sigma\tau(n)} = (x_{\sigma(1)} \cdots x_{\sigma(n)})\tau.$$

The interpretation of the left S_n-action on a polynomial $f(x_1, \ldots, x_n) \in P_n$, for $\sigma \in S_n$, is
$$\sigma f(x_1, \ldots, x_n) = f(x_{\sigma(1)}, \ldots, x_{\sigma(n)}),$$
that is, of permuting the variables according to σ.

In fact, by linearity it is enough to check it on a monomial. Hence if we write
$$x_{\pi(1)} \cdots x_{\pi(n)} = M_\pi(x_1, \ldots, x_n),$$
we need to show that $\sigma M_\pi(x_1, \ldots, x_n) = M_\pi(x_{\sigma(1)}, \ldots, x_{\sigma(n)})$. Now, if we rename the variables $x_{\sigma(1)} = y_1, \ldots, x_{\sigma(n)} = y_n$, then
$$M_\pi(x_{\sigma(1)}, \ldots, x_{\sigma(n)}) = M_\pi(y_1, \ldots, y_n) = y_{\pi(1)} \cdots y_{\pi(n)}$$
$$= x_{\sigma\pi(1)} \cdots x_{\sigma\pi(n)} = \sigma M_\pi(x_1, \ldots, x_n)$$
since $y_{\sigma(i)} = x_{\sigma\pi(i)}$, $i = 1, \ldots, n$.

The right action of τ on $f(x_1, \ldots, x_n)$ is that of changing the places in each monomial $x_{\sigma(1)} \cdots x_{\sigma(n)}$ according to the permutation τ and is independent of σ. This means that the ith factor $x_{\sigma(i)}$ will be placed in the $\tau^{-1}(i)$ place of the new monomial, $i = 1, \ldots, n$. In other words, if
$$\sigma = \begin{pmatrix} 1 & 2 & \ldots & n \\ i_1 & i_2 & \ldots & i_n \end{pmatrix},$$
then $x_{\sigma(1)} x_{\sigma(2)} \cdots x_{\sigma(n)} = x_{i_1} x_{i_2} \cdots x_{i_n}$ and
$$(x_{i_1} x_{i_2} \cdots x_{i_n})\tau = x_{i_{\tau(1)}} x_{i_{\tau(2)}} \cdots x_{i_{\tau(n)}}.$$

For example, let $n = 4$ and $x_{i_1} \cdots x_{i_4} = x_2 x_3 x_1 x_4$, i.e., $i_1 = 2, i_2 = 3, i_3 = 1, i_4 = 4$. Then $\tau = (1234)$ acts from the right as follows:
$$(x_2 x_3 x_1 x_4)\tau = x_{i_2} x_{i_3} x_{i_4} x_{i_1} = x_3 x_1 x_4 x_2.$$

Note that T-ideals are not invariant in general under the right action.

Denote $P_n = P_n(x_1, \ldots, x_n) = P_n(x)$. If $y_1, \ldots, y_n$ are other variables, one can also consider $P_n(y_1, \ldots, y_n) = P_n(y)$. If A is a PI-algebra and $\mathrm{char}\, F = 0$, it suffices to study the multilinear identities of A. Namely, one should study $P_n(x) \cap \mathrm{Id}(A)$, $P_n(y) \cap \mathrm{Id}(A)$, etc. However, the correspondence $x_i \to y_i$ yields the isomorphism $P_n(x) \cong P_n(y) \cong FS_n$ and it suffices to study just $P_n(x) \cap \mathrm{Id}(A)$.

Since T-ideals are invariant under permutations of the variables, we obtain that $P_n \cap \mathrm{Id}(A)$ is a left S_n-submodule of P_n. Hence
$$P_n(A) = \frac{P_n}{P_n \cap \mathrm{Id}(A)}$$
has an induced structure of left S_n-module. If $F\langle X \rangle$ is the free algebra of countable rank on $X = \{x_1, x_2, \ldots, \}$, then $P_n(A)$ is the space of multilinear elements in the first n variables of the relatively free algebra $F\langle X \rangle/\mathrm{Id}(A)$. If $\mathcal{V} = \mathrm{var}(A)$, we also write $P_n(\mathcal{V}) = P_n(A)$.

DEFINITION 2.4.3. For $n \geq 1$, the S_n-character of $P_n(A) = P_n/(P_n \cap \mathrm{Id}(A))$ is called the nth cocharacter of A (or of the T-ideal $\mathrm{Id}(A)$) and is denoted $\chi_n(A)$.

In the following chapters we shall study the asymptotic behaviour of the sequence of codimensions $\{c_n(A)\}_{n \geq 1}$ where $c_n(A) = \chi_n(A)(1)$ and the sequence of cocharacters $\{\chi_n(A)\}_{n \geq 1}$ for any PI-algebra A.

If we decompose the nth cocharacter into irreducibles, we obtain
$$(2.2) \qquad \chi_n(A) = \sum_{\lambda \vdash n} m_\lambda \chi_\lambda,$$
where χ_λ is the irreducible S_n-character associated to the partition $\lambda \vdash n$ and $m_\lambda \geq 0$ is the corresponding multiplicity.

Using Lemma 2.4.2 we can characterize the multiplicities m_λ as follows. Recall that by Theorem 1.2.4, if $X = \{x_1, x_2, \ldots\}$ and $Y = \{y_1, y_2, \ldots\}$ are countable sets, then $F\langle X \rangle/\mathrm{Id}(A) \cong F\langle Y \rangle/\mathrm{Id}(A)$ is the relatively free algebra of countable rank of $\mathrm{var}(A)$.

THEOREM 2.4.4. *Let A be a PI-algebra with nth cocharacter $\chi_n(A)$ given in (2.2). Let $\lambda = (\lambda_1, \ldots, \lambda_k) \vdash n$ and let W_λ be a submodule of $P_n(A)$ with character $\chi(W_\lambda) = m_\lambda \chi_\lambda$. If*
$$\varphi : \frac{F\langle X \rangle}{\mathrm{Id}(A)} \to \frac{F\langle Y \rangle}{\mathrm{Id}(A)}$$
is a homomorphism such that
$$\varphi(\bar{x}_1) = \cdots = \varphi(\bar{x}_{\lambda_1}) = \bar{y}_1,$$
$$\varphi(\bar{x}_{\lambda_1+1}) = \cdots = \varphi(\bar{x}_{\lambda_1+\lambda_2}) = \bar{y}_2,$$
$$\vdots$$
$$\varphi(\bar{x}_{\lambda_1+\cdots+\lambda_{k-1}+1}) = \cdots = \varphi(\bar{x}_n) = \bar{y}_k,$$
where $\bar{x}_i = x_i + \mathrm{Id}(A)$ and $\bar{y}_i = y_i + \mathrm{Id}(A)$, $i = 1, \ldots, k$, then $m_\lambda = \dim \varphi(W_\lambda)$.

PROOF. Consider the decomposition
$$W_\lambda = M_1 \oplus \cdots \oplus M_q$$
where $M_1 \cong \cdots \cong M_q$ are irreducible S_n-modules and $q = m_\lambda$.

Let T_λ be the Young tableau filled up with the integers $1, \ldots, n$ canonically as follows. The integers $1, \ldots, \lambda_1$ are inserted from left to right into the boxes of the first row of the diagram D_λ, $\lambda_1 + 1, \ldots, \lambda_1 + \lambda_2$ are inserted into the boxes of the second row and so on. By Lemma 2.4.1, the modules $M_1, \ldots, M_q$ are generated by elements of the form $e_{T_\lambda} f_1, \ldots, e_{T_\lambda} f_q$, respectively, where $f_i \in M_i, i = 1, \ldots, q$.

By the construction of T_λ any $e_{T_\lambda} f_i$ is symmetric on each of the sets
$$X_1 = \{\bar{x}_1, \ldots, \bar{x}_{\lambda_1}\}, \ldots, X_k = \{\bar{x}_{\lambda_1 + \ldots + \lambda_{k-1}+1}, \ldots, \bar{x}_n\}.$$
We set $g_i = \varphi(e_{T_\lambda} f_i)$ and denote by $f_i^* = f_i^*(x_1, \ldots, x_n)$ the complete linearization of g_i where $\bar{y}_j$ is linearized into the elements of the set X_j, $j = 1, \ldots, k$. In other words, if $\psi : F\langle Y \rangle/\mathrm{Id}(A) \to F\langle X \rangle/\mathrm{Id}(A)$ is the homomorphism defined by
$$\psi(\bar{y}_1) = \bar{x}_1 + \cdots + \bar{x}_{\lambda_1},$$
$$\psi(\bar{y}_2) = \bar{x}_{\lambda_1+1} + \cdots \bar{x}_{\lambda_1+\lambda_2},$$
$$\vdots$$
$$\psi(\bar{y}_k) = \bar{x}_{\lambda_1 + \cdots + \lambda_{k-1}+1} + \cdots + \bar{x}_n,$$
then
$$f_i^* = (\lambda_1)! \cdots (\lambda_k)! e_{T_\lambda} f_i, \quad i = 1, \ldots, q,$$
and $e_{T_\lambda} f_i$ is, up to a scalar, the multilinearization of the polynomial $\psi(g_i)$.

In particular, the polynomials $g_1, \ldots, g_q$ are non-zero. Moreover, $g_1, \ldots, g_q$ are linearly independent (mod $\mathrm{Id}(A)$) since an equality of the type $\alpha_1 g_1 + \cdots + \alpha_k g_q = 0$ with $\alpha_1, \ldots, \alpha_k \in F$, implies that $\alpha_1 e_{T_\lambda} f_1 + \cdots + \alpha_k e_{T_\lambda} f_k \equiv 0$ is an identity of A. Hence $\dim W_\lambda \geq q$.

On the other hand, let $h_1, \ldots, h_l \in W_\lambda$ and suppose that $\varphi(h_1), \ldots, \varphi(h_l)$ are linearly independent. By applying to $h_1, \ldots, h_l$ the complete linearization as above, we get $h_1^*, \ldots, h_l^* \in P_n(A)$. Since each h_i^* is a consequence of h_i as an identity, it follows that $h_i^* \in W_\lambda$. Moreover, h_i^* is symmetric on any set $X_j, j = 1, \ldots, k$. Hence $\sigma h_i^* = h_i^*$ for all $\sigma \in R_{T_\lambda}$ where R_{T_λ} is the subgroup of S_n stabilizing the rows of T_λ. Note that
$$\varphi(h_i^*) = (\lambda_1)! \cdots (\lambda_k)! \varphi(h_i), \quad i = 1, \ldots, l,$$
and therefore $h_1^*, \ldots, h_l^*$ are also linearly independent. By Lemma 2.4.2, $l \leq q$ and the proof is complete. $\square$

Another useful tool in computing the nth cocharacter is the following.

THEOREM 2.4.5. *Let A be a PI-algebra with nth cocharacter $\chi_n(A)$ given in (2.2). For a partition $\mu \vdash n$, the multiplicity m_μ is equal to zero if and only if for any Young tableau T_μ of shape μ and for any polynomial $f = f(x_1, \ldots, x_n) \in P_n$, the algebra A satisfies the identity $e_{T_\mu} f \equiv 0$.*

PROOF. Consider the decompositions
$$FS_n = \bigoplus_{\lambda \vdash n} I_\lambda, \qquad P_n = Q \oplus J$$
where $Q = P_n \cap \mathrm{Id}(A)$ and $J \simeq P_n(A) = P_n/(P_n \cap \mathrm{Id}(A))$. Fix some $\mu \vdash n$. Then $m_\mu = 0$ in (2.2) if and only if $I_\mu J = 0$. On the other hand, the equality $I_\mu J = 0$ is equivalent to the inclusion $I_\mu P_n \subseteq Q$. Since I_μ is the sum of all left ideals $FS_n e_{T_\mu}$, the inclusion $I_\mu P_n \subseteq Q$ takes place if and only if $e_{T_\mu} f \in Q$ for any $f \in P_n$, i.e., $e_{T_\mu} f \equiv 0$ is an identity of A and the proof is complete. $\square$

Recall that in the case of characteristic zero any identity is equivalent to a finite set of multilinear identities. Next we describe the structure of the space of multilinear identities in the language of S_n-action.

Let $F\langle X\rangle^*$ denote the free algebra on X without unit element. $F\langle X\rangle^*$ is the subalgebra of $F\langle X\rangle$ consisting of all polynomials in the elements of X without constant term. If I is a proper T-ideal of $F\langle X\rangle$, then $I \subseteq F\langle X\rangle^*$; in fact, if $f(x_1,\ldots,x_n) \in I$, $f(0,\ldots,0) \in I$ implies that f has zero constant term. Then one says that two sets of polynomials are equivalent in $F\langle X\rangle^*$ if they generate the same T-ideal of $F\langle X\rangle^*$.

The following interesting result holds.

THEOREM 2.4.6. *Let $f_i = f_i(x_1,\ldots,x_n)$, $i = 1,\ldots,r$, and $g_j = g_j(x_1,\ldots,x_n)$, $j = 1,\ldots,s$, be multilinear polynomials of $F\langle X\rangle^*$. Then the two sets $\{f_1,\ldots,f_r\}$ and $\{g_1,\ldots,g_s\}$ are equivalent in $F\langle X\rangle^*$ if and only if $FS_n f_1 + \cdots + FS_n f_r = FS_n g_1 + \cdots + FS_n g_s$.*

PROOF. Suppose first that $\sum_i FS_n f_i = \sum_j FS_n g_j$. Clearly, for any $\sigma \in S_n$ there exists an endomorphism $\varphi : F\langle X\rangle^* \to F\langle X\rangle^*$ such that $\varphi(x_k) = x_{\sigma(k)}$, $k = 1,\ldots,n$. Hence, for any multilinear $f = f(x_1,\ldots,x_n)$, $FS_n f$ lies in the T-ideal $\langle f\rangle_T$ generated by $f \in F\langle X\rangle^*$. In particular, $g_1,\ldots,g_s \in \langle f_1,\ldots,f_r\rangle_T$. Similarly, $f_1,\ldots,f_r \in \langle g_1,\ldots,g_s\rangle_T$. Hence $\{f_1\ldots,f_r\}$ and $\{g_1,\ldots,g_s\}$ are equivalent.

Now let $\{f_1\ldots,f_r\}$ and $\{g_1\ldots,g_s\}$ be equivalent sets of identities. The inclusion $f_i \in \langle g_1\ldots,g_s\rangle_T$ means that f_i can be written as a sum of elements of the type
$$\varphi(g_j),\ a\varphi(g_j),\ \varphi(g_j)b,\ a\varphi(g_j)b$$
where $a,b \in F\langle X\rangle$ and φ is an endomorphism of $F\langle X\rangle^*$. Since f_i and g_j are multilinear of the same degree it follows that f_i is a sum of some of the $\varphi(g_j)$'s. Let
$$f_i = \sum_{i,j} \varphi_{ij}(g_j).$$

If
$$\varphi_{ij}(x_l) = \sum_k \alpha_{ijl}^k x_k + h_{ijl}$$
where h_{ijl} is a sum of monomials of degree at least 2, then also
$$(2.3) \qquad f_i = \sum_{i,j} \psi_{ij}(g_j)$$
where
$$\psi_{ij}(x_l) = \sum_k \alpha_{ijl}^k x_k.$$

Since any g_j is multilinear, (2.3) implies that f_i is a linear combination of values $g_j(x_{i_1},\ldots,x_{i_n})$. Moreover, all $i_1,\ldots,i_n$ should be distinct and $\{i_1,\ldots,i_n\} = \{1,\ldots,n\}$, that is, $g_j(x_{i_1},\ldots,x_{i_n}) = g_j(x_{\sigma(1)},\ldots,x_{\sigma(n)})$ for some $\sigma \in S_n$ and $f_i \in FS_n g_1 + \cdots + FS_n g_s$. We get the inclusion
$$FS_n f_i \subseteq FS_n g_1 + \cdots + FS_n g_s.$$
Similarly, $FS_n g_j \subseteq FS_n f_1 + \cdots + FS_n f_r$ and $FS_n f_1 + \cdots + FS_n f_r = FS_n g_1 + \cdots + FS_n g_s$. $\square$

As a special case of the previous theorem, we see that two multilinear identities $f = f(x_1,\ldots,x_n) \equiv 0$ and $g = g(x_1,\ldots,x_n) \equiv 0$ of the same degree are equivalent in $F\langle X\rangle^*$ if and only if $FS_n f = FS_n g$ in P_n.

Another useful consequence of Lemma 2.4.1 is the following.

THEOREM 2.4.7. *For any multilinear polynomial $f \in P_n$ there exist a finite set of polynomials $g_1, \ldots, g_r \in P_n$ and partitions $\lambda(1), \ldots, \lambda(r)$ of n such that $FS_n f = FS_n e_{T_{\lambda(1)}} g_1 + \cdots + FS_n e_{T_{\lambda(r)}} g_r$.*

PROOF. Write $M = FS_n f$ and consider the decomposition $M = M_1 \oplus \cdots \oplus M_r$ into the sum of irreducible S_n-modules. By Lemma 2.4.1 there exist $g_1 \in M_1, \ldots, g_r \in M_r$ and Young tableaux $T_{\lambda(1)}, \ldots, T_{\lambda(r)}$ such that $M_1 = FS_n e_{T_{\lambda(1)}} g_1$, $\ldots, M_r = FS_n e_{T_{\lambda(r)}} g_r$. □

2.5. Hooks and symmetric and alternating sets of variables

In this section we prove some technical results about the structure of multilinear polynomials of the type $e_{T_\lambda} f$ where λ is a partition of n and T_λ is a Young tableau of shape λ. Recall that R_{T_λ} and C_{T_λ} are the row-stabilizer and the column-stabilizer subgroups of S_n, respectively.

LEMMA 2.5.1. *Let H be a subgroup of C_{T_λ}, M an S_n-module and $e_{T_\lambda} u \neq 0$ for some $u \in M$. Then*

$$\left(\sum_{\sigma \in H} (\operatorname{sgn} \sigma) \sigma \right) e_{T_\lambda} u \neq 0.$$

PROOF. Write $C_{T_\lambda} = a_1 H \cup a_2 H \cup \ldots \cup a_m H$ where $a_1 = 1, a_2, \ldots, a_m$ is a left transversal of H in C_{T_λ} and let $r = \sum_{\sigma \in H} (\operatorname{sgn} \sigma) \sigma$. If $r e_{T_\lambda} u = 0$, then $a_i r e_{T_\lambda} u = 0$ in M for all $i = 1, \ldots, m$. Hence

$$e_{T_\lambda}^2 u = \left(\sum_{\tau \in R_{T_\lambda}} \tau \right) \left(\sum_{\sigma \in C_{T_\lambda}} (\operatorname{sgn} \sigma) \sigma \right) e_{T_\lambda} u$$

$$= \left(\sum_{\tau \in R_{T_\lambda}} \tau \right) (a_1 r e_{T_\lambda} u \pm \cdots \pm a_m r e_{T_\lambda} u) = 0,$$

a contradiction since $e_{T_\lambda}^2 = \gamma e_{T_\lambda}$ for some integer $\gamma \neq 0$. □

A slight modification of the previous argument gives the proof of the following.

LEMMA 2.5.2. *Let H be a subgroup of R_{T_λ}, M an S_n-module and $e_{T_\lambda} u \neq 0$ for some $u \in M$. Then*

$$\left(\sum_{\sigma \in H} \sigma \right) \left(\sum_{\tau \in C_{T_\lambda}} (\operatorname{sgn} \tau) \tau \right) e_{T_\lambda} u \neq 0.$$

Let $\lambda \vdash n$ and T_λ a tableau associated to λ. We make the following.

DEFINITION 2.5.3. We say that a multilinear polynomial $f = f(x_1, \ldots, x_n)$ *corresponds to the tableau* T_λ if $f = e_{T_\lambda} f_0$ for some multilinear polynomial $f_0 \in P_n$.

Given integers $d, l, t \geq 0$ we define the partition

$$h(d, l, t) = (\underbrace{l+t, \ldots, l+t}_{d}, \underbrace{l, \ldots, l}_{t}).$$

It is clear that the diagram associated with $h(d, l, t)$ is hook shaped (see picture).

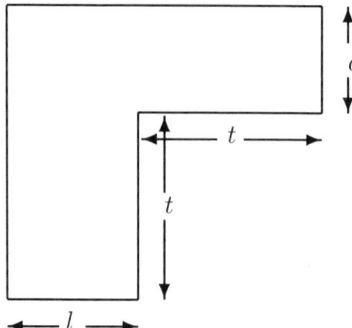

We also define an infinite hook $H(d,l)$ as follows:

$$H(d,l) = \bigcup_{n \geq 1} \{\lambda = (\lambda_1, \lambda_2, \ldots) \vdash n \mid \lambda_{d+1} \leq l\}.$$

Thus $H(d,l)$ can be realized as the set of all diagrams whose shape lies in the hook shaped region given in the picture below

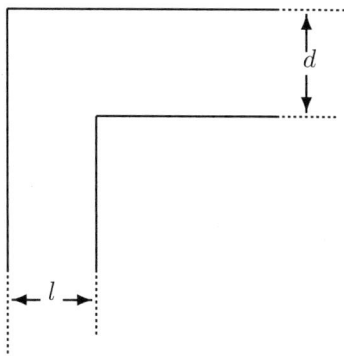

The integer d is called the hand and l the foot of the hook. In what follows we shall say that a partition λ lies in the hook $H(d,l)$ and we shall write $\lambda \in H(d,l)$ if the corresponding Young diagram D_λ is contained in $H(d,l)$. Analogously if V is an S_n-module with character $\chi(V) = \sum_{\lambda \vdash n} m_\lambda \chi_\lambda$, we shall write $\chi(V) \subseteq H(d,l)$ if $\lambda \in H(d,l)$ for all partitions λ such that $m_\lambda \neq 0$.

Recall the partial order on partitions given in Section 2.3. If $\lambda = (\lambda_1, \ldots, \lambda_p) \vdash n$ and $\mu = (\mu_1, \ldots, \mu_q) \vdash m$, then $\lambda \geq \mu$ if and only if $p \geq q$ and $\lambda_i \geq \mu_i$ for all $i = 1, \ldots, p$. Hence $\lambda \geq \mu$ means that D_μ is a subdiagram of D_λ.

The following two lemmas record important properties of the hooks.

LEMMA 2.5.4. *Let $\lambda \vdash n$ be such that $\lambda \geq h(d,l,t)$ for some d,l,t, and let $f(x_1, \ldots, x_n)$ be a multilinear polynomial corresponding to a tableau T_λ. Then there exist $r \in FS_n$ such that $rf \neq 0$, and a subset Y of $\{x_1, \ldots, x_n\}$ such that*

(1) *$Y = Y_1 \cup \ldots \cup Y_d$, rf is symmetric in each set of variables $Y_i, i = 1, \ldots, d$, and $|Y_i| = t + l$;*
(2) *rf can be decomposed into the sum of multilinear polynomials $rf = f_1 + f_2 + \cdots + f_k$ such that for every f_i there is a partition $Y = Y_1' \cup \cdots \cup Y_{t+l}'$ with $|Y_j'| = d$ and f_i is alternating in each set of variables $Y_j', j = 1, \ldots, t+l$.*

2.5. HOOKS AND SYMMETRIC AND ALTERNATING SETS

PROOF. By hypothesis the tableau T_λ contains a rectangular tableau T_0 with d rows and $t+l$ columns. Let $N_j, j = 1,\ldots,d$, be the set of integers in the jth row of T_0, $N'_i, i = 1,\ldots,t+l$, the set of integers in the ith column of T_0 and $N = N_1 \cup \ldots \cup N_d$. Set $H = \{\sigma \in R_{T_\lambda} |\ \sigma(i) = i \text{ for any } i \notin N\}$ and

$$r_0 = \sum_{\tau \in C_{T_\lambda}} (\operatorname{sgn} \tau)\tau,$$

$$r = \left(\sum_{\sigma \in H} \sigma\right) r_0.$$

Let $Y_j = \{x_i|\ i \in N_j\}$, $Z_i = \{x_s|\ s \in N'_i\}$.

Clearly the element $g = rf$ is symmetric in the variables from Y_j for each j. On the other hand, the polynomial $r_0 f$ is alternating in the variables of each Z_i, therefore for any $\sigma \in H$ the element $\sigma r_0 f$ is alternating in the variables of each $Y'_i = \sigma(Z_i)$ for every $i = 1,\ldots,t+l$. Hence, rf is the required multilinear polynomial. By Lemma 2.5.2, rf is non-zero and the proof is complete. □

LEMMA 2.5.5. *Let $f(x_1,\ldots,x_n)$ correspond to the Young tableau T_λ and suppose that $\lambda \geq h(d,l,t)$. Then for some $r \in FS_n$, $rf \neq 0$ and there is a subset Y of $\{x_1,\ldots,x_n\}$ such that*

(1) *$Y = Y_1 \cup \ldots \cup Y_l$, rf is alternating in each set of variables $Y_i, i = 1,\ldots,l$, and $|Y_i| = t + d$;*
(2) *rf can be decomposed into the sum of multilinear polynomials $rf = f_1 + f_2 + \cdots + f_k$ such that for every f_i there is a partition $Y = Y'_1 \cup \cdots \cup Y'_{t+d}$ with $|Y'_j| = l$ and f_i is symmetric in each set of variables $Y'_j, j = 1,\ldots,t+d$.*

PROOF. As in the previous lemma T_λ contains a rectangular tableau T_0 with $d+t$ rows and l columns. Let $N_j, j = 1,\ldots,l$, be the set of integers in the jth column of T_0, $N'_i, i = 1,\ldots,t+d$, the set of integers in the ith row of T_0 and $N = N_1 \cup \ldots \cup N_l$. Denote $H = \{\sigma \in C_{T_\lambda}|\ \sigma(i) = i \text{ for any } i \notin N\}$ and

$$r = \left(\sum_{\sigma \in H} (\operatorname{sgn} \sigma)\sigma\right).$$

Let $Y_j = \{x_i|\ i \in N_j\}$, $Z_i = \{x_s|\ s \in N'_i\}$.

By definition the element f is symmetric in the variables from Z_i. Hence, for any $\sigma \in H$ the polynomial σf is symmetric in the variables from $Y'_i = \sigma(Z_i)$. This implies the second part of the lemma. Also, the polynomial rf is non-zero by Lemma 2.5.1 and is alternating on the variables from Y_j for any $j = 1,\ldots,l$. □

We remark that Lemma 2.5.5 can also be deduced from Lemma 2.5.4 by considering the linear map $\varphi : FS_n \to FS_n$ defined by setting $\varphi(\sigma) = (\operatorname{sgn} \sigma)\sigma$, for $\sigma \in S_n$. It is easily checked that φ is an isomorphism of S_n-modules and for $\lambda \vdash n$, $\varphi(I_\lambda) = I_{\lambda'}$, where λ' is the conjugate partition of λ. Notice that for any tableau T_λ,

$$\varphi(e_{T_\lambda}) = \varphi\left(\left(\sum_{\sigma \in R_{T_\lambda}} \sigma\right)\left(\sum_{\tau \in C_{T_\lambda}} (\operatorname{sgn} \tau)\tau\right)\right) = \left(\sum_{\sigma \in C_{T_{\lambda'}}} (\operatorname{sgn} \sigma)\sigma\right)\left(\sum_{\tau \in R_{T_{\lambda'}}} \tau\right).$$

We now consider the structure of the multilinear polynomials corresponding to Young tableau T_λ provided that λ lies in an infinite hook $H(d,l)$.

LEMMA 2.5.6. *Let $\lambda \vdash n$, T_λ a Young tableau of shape λ and $f = e_{T_\lambda} g$ where $g = g(x_1,\ldots,x_n)$ is some multilinear polynomial in $x_1,\ldots,x_n$. If $\lambda \in H(d,l)$, then there exists a decomposition of $X_n = \{x_1,\ldots,x_n\}$ into a disjoint union*

$$(2.4) \qquad X_n = X_1 \cup \ldots \cup X_{d'} \cup Y_1 \ldots \cup Y_{l'}$$

with $d' \leq d, l' \leq l$ and a multilinear polynomial $f' = f'(x_1,\ldots,x_n)$ such that
- *f' is symmetric on the variables in each set $X_i, 1 \leq i \leq d'$;*
- *f' is alternating on the variables in each set $Y_j, 1 \leq j \leq l'$;*
- *$FS_n f = FS_n f'$;*
- *the integers $d', l', |X_1|, \ldots, |X_{d'}|, |Y_1|, \ldots, |Y_{l'}|$ are uniquely determined by λ and do not depend on the choice of the tableau T_λ;*
- *the decomposition (2.4) is uniquely defined by T_λ and does not depend on g.*

PROOF. Let $\lambda = (\lambda_1,\ldots,\lambda_t)$. Denote by Y_1 the set consisting of all x_i with i appearing in the 1st column of T_λ, by Y_2 the set of all x_j with j appearing in the 2nd column of T_λ, and so on. Denote by H the subgroup in S_n of all permutations stabilizing the first l columns of T_λ (in case $\lambda_1 < l$, we consider all $\lambda_1 = l'$ columns of T_λ). Then by Lemma 2.5.1

$$f' = \left(\sum_{\tau \in H} (\operatorname{sgn}\tau)\tau\right) e_{T_\lambda} g \neq 0.$$

Hence $FS_n f' = FS_n f$ since f generates an irreducible S_n-module. In addition, f' is alternating on the variables in any set $Y_j, 1 \leq j \leq \min(l,l')$. If $\lambda_1 = l' \leq l$, i.e., T_λ has no more than l columns, then $\{x_1,\ldots,x_n\} = Y_1 \cup \ldots \cup Y_{l'}$ and the proof is complete. Suppose that T_λ has at least $l+1$ columns. In this case denote by X_1 all variables indexed by the integers appearing in the first row of T_λ which do not lie in $Y_1 \cup \ldots \cup Y_l$. If $X_1 \cup Y_1 \cup \ldots \cup Y_l \neq X_n$, then denote by X_2 all variables indexed by the integers in the second row of T_λ which do not lie in $Y_1 \cup \ldots \cup Y_l$, and so on.

Since $\lambda \in H(d,l)$, we finally get the decomposition $X_n = X_1 \cup \ldots \cup X_{d'} \cup Y_1 \ldots \cup Y_{l'}$ with $d' \leq d$. Obviously, f' is symmetric on the variables from any set X_j, since f was symmetric on the variables from the jth row of T_λ. Clearly, $d', l', |X_1|, \ldots, |X_{d'}|, |Y_1|, \ldots, |Y_{l'}|$ are defined only by the shape of the diagram D_λ and do not depend on the choice of the tableau T_λ. By the construction all $X_1, \ldots, Y_l$ do not depend on g. □

CHAPTER 3

Group Gradings and Group Actions

If an algebra A has an additional structure of graded algebra, or if a finite group G acts on A as a group of automorphisms and antiautomorphisms, one can try to relate the polynomial identities of A to more general identities (graded identities, G-identities) by taking into account the additional structure. We have in mind two reasons for pursuing this study: the important role played by the algebra of $k \times k$ matrices over the base field whose gradings, automorphisms and involutions are well understood, and a basic result of Kemer asserting that the identities of any algebra are the same as those of some graded tensor product of the Grassmann algebra with a suitable finite dimensional $\mathbb{Z}_2$-graded algebra or superalgebra.

In this chapter we study algebras which are graded by a finite group or on which a finite group acts, and we prove that in the case of abelian groups if the base field has enough roots of unity, there is a well understood duality between group gradings and group actions. We introduce the notions of free graded algebra, free algebra with G-action, and the corresponding notions of graded identity and G-identity. We then specialize our study to the case when an automorphism or an antiautomorphism of order two acts on the algebra A leading to a structure of superalgebra or of algebra with involution on A.

In this setting we prove a generalization of the Wedderburn-Malcev theorem on the structure of the finite dimensional algebras over a field of characteristic zero. In the case of superalgebras, we classify all simple superalgebras over an algebraically closed field of characteristic zero. We also introduce the Grassmann envelope and the supercommutative envelope of a superalgebra and we establish a useful connection between varieties of algebras and varieties of superalgebras. Finally, in the case of the algebra of $k \times k$ matrices over an infinite field, we prove that the ideals of identities with involution are actually determined by the identities with transpose or symplectic involution.

3.1. Group-graded algebras

In this section we introduce the notion of an algebra graded by a group and we give several examples. Let A be an algebra over a field F and let G be any group.

DEFINITION 3.1.1. The algebra A is G-graded if A can be written as the direct sum of subspaces $A = \oplus_{g \in G} A^{(g)}$ such that for all $g, h \in G$, $A^{(g)} A^{(h)} \subseteq A^{(gh)}$.

From the definition it is clear that any $a \in A$ can be uniquely written as a finite sum $a = \sum_{g \in G} a_g$ with $a_g \in A^{(g)}$. The subspaces $A^{(g)}$ are called the homogeneous components of A. Accordingly, an element $a \in A$ is homogeneous (or homogeneous of degree g) if $a \in A^{(g)}$. A subspace $B \subseteq A$ is graded or homogeneous if $B = \oplus_{g \in G}(B \cap A^{(g)})$. In other words, B is graded if, for any $b \in B$, $b = \sum_{g \in G} b_g$ implies that $b_g \in B$ for all $g \in G$. Similarly, one can define graded subalgebras, graded ideals, etc. Notice that if H is a subgroup of G, then clearly $B = \oplus_{h \in H} A^{(h)}$ is a graded subalgebra of A. In particular, if e is the unit of G, $A^{(e)}$ is a subalgebra of A. Next we give some examples of graded algebras.

EXAMPLE 3.1.2. Any algebra A can be graded by any group G by setting $A = A^{(e)}$ and $A^{(g)} = 0$ for any $g \neq e$. This grading is called trivial.

EXAMPLE 3.1.3. The free associative algebra of countable rank $A = F\langle X \rangle$ has a natural $\mathbb{Z}$-grading by setting $A = \oplus_{n \in \mathbb{Z}} A^{(n)}$ where $A^{(n)} = 0$ if $n \leq 0$, and $A^{(n)}$ is the linear span of all monomials of total degree n, in case $n > 0$.

If $X' = \{x_1, \ldots, x_k\}$ is a finite set, then $A = F\langle X' \rangle$, the free associative algebra of rank k, can be graded by the group $\mathbb{Z}^k = \mathbb{Z} \oplus \cdots \oplus \mathbb{Z}$ by setting

$$A^{(n_1, \ldots, n_k)} = \{f \in F\langle X' \rangle \mid f \text{ is multihomogeneous, } \deg_{x_i} f = n_i, \ i = 1, \ldots, k\}.$$

EXAMPLE 3.1.4. The group algebra $A = FG$ of a group G is naturally graded by G by setting $A^{(g)} = \operatorname{span}_F \{g\}$.

EXAMPLE 3.1.5. Let $A = M_k(F)$ be the algebra on $k \times k$ matrices over F and let G be an arbitrary group. Given any k-tuple $(g_1, \ldots, g_k) \in G^k$, one can define a G-grading of A by setting

$$A^{(g)} = \operatorname{span}_F \{e_{ij} \mid g_i^{-1} g_j = g\},$$

where e_{ij} are the usual matrix units. It is not difficult to show that this G-grading can be obtained by identifying $M_k(F)$ with the algebra $\operatorname{End}(V)$ of all linear transformations of a G-graded k-dimensional vector space V.

EXAMPLE 3.1.6. Let $A = M_2(F)$ and let $G = \langle a \rangle \times \langle b \rangle$ be the direct product of two cyclic groups of order two. Then define

$$A^{(1)} = \operatorname{span}_F \left\{ \begin{pmatrix} 1 & 0 \\ 0 & 1 \end{pmatrix} \right\}, \ A^{(a)} = \operatorname{span}_F \left\{ \begin{pmatrix} 1 & 0 \\ 0 & -1 \end{pmatrix} \right\},$$

$$A^{(b)} = \operatorname{span}_F \left\{ \begin{pmatrix} 0 & 1 \\ 1 & 0 \end{pmatrix} \right\}, \ A^{(ab)} = \operatorname{span}_F \left\{ \begin{pmatrix} 0 & 1 \\ -1 & 0 \end{pmatrix} \right\}.$$

A direct inspection shows that $A = \oplus_{g \in G} A^{(g)}$ and $A^{(g)} A^{(h)} = A^{(gh)}$, for any $g, h \in G$.

Notice that the above grading on $M_2(F)$ cannot be obtained as in Example 3.1.5. In the next example we generalize the construction of Example 3.1.6 to arbitrary $n \geq 2$.

EXAMPLE 3.1.7. Let F be a field containing ε, a primitive kth root of 1. Consider the following two $k \times k$ matrices

$$A = \begin{pmatrix} \varepsilon^{k-1} & 0 & \cdots & 0 \\ 0 & \varepsilon^{k-2} & \cdots & 0 \\ \vdots & \vdots & \ddots & \vdots \\ 0 & 0 & \cdots & 1 \end{pmatrix}, \ B = \begin{pmatrix} 0 & 1 & 0 & \cdots & 0 \\ 0 & 0 & 1 & \cdots & 0 \\ \vdots & \vdots & \vdots & \ddots & \vdots \\ 0 & 0 & 0 & \cdots & 1 \\ 1 & 0 & 0 & \cdots & 0 \end{pmatrix}.$$

Then A and B satisfy the relations

(3.1) $$AB = \varepsilon BA, \ A^k = B^k = I,$$

where I is the identity $k \times k$ matrix.

We claim that the matrices $A^i B^j$, $1 \le i, j \le k$, are linearly independent over F. In fact, for $1 \le j \le k$, denote by L_j the subspace of $M_k(F)$ generated by the elements $B^j, AB^j, \ldots, A^{k-1}B^j$. Since $L_j \subseteq \text{span}\{e_{1,j+1}, e_{2,j+2}, \ldots, e_{kj}\}$ it is clear that
$$L_1 + \cdots + L_k = L_1 \oplus \cdots \oplus L_k.$$
On the other hand, if $z_1, \ldots, z_k$ are indeterminates, then $z_1 A + z_2 A^2 + \cdots + z_k A^k = 0$ gives a system of k linear equations in $z_1, \ldots, z_k$. Since ε is a primitive kth root of 1, the determinant of this system is a non-zero Vandermonde. It follows that for any (not all zero) scalars $\alpha_1, \ldots, \alpha_k$, the matrix $\alpha_1 A + \alpha_2 A^2 + \cdots + \alpha_k A^k$ is non-zero; hence since B is invertible, $\alpha_1 A B^j + \cdots + \alpha_k A^k B^j \ne 0$.

We have shown that the matrices $A^i B^j$, $1 \le i, j \le k$ are linearly independent over F. Since they are k^2 in number, they form a basis of $M_k(F)$. Now consider $H = \langle a \rangle_k \times \langle b \rangle_k$, the group direct product of two cyclic groups of order k and, for $h = a^i b^j \in H$, denote by $M_k(F)^{(h)}$ the one-dimensional subspace $\text{span}_F\{A^i B^j\}$. From relations (3.1) it follows that, for all $h, s \in H$, $M_k(F)^{(h)} M_k(F)^{(s)} \subseteq M_k(F)^{(hs)}$, so,

$$(3.2) \qquad M_k(F) = \bigoplus_{h \in H} M_k(F)^{(h)}$$

is an H-grading on $M_k(F)$.

In the next chapters we shall be mostly dealing with $\mathbb{Z}_2$-gradings. Nevertheless we shall also be using some of the properties of G-graded algebras. Next we examine the gradings by finite abelian groups since in this case there is a close connection with group actions.

3.2. Abelian gradings and group actions

In this section we describe the duality between G-actions and G-gradings on an associative algebra A over a field of characteristic zero in case G is a finite abelian group of automorphisms.

Let G be an arbitrary finite abelian group, $|G| = k$, let A be an algebra over the field F of characteristic zero and suppose that F contains a primitive kth root of 1.

Let $\text{Aut}(A)$ denote the group of automorphisms of the algebra A and assume that $G \subseteq \text{Aut}(A)$. We shall mostly use the exponential notation and write $g(a) = a^g$ for $a \in A$, $g \in G$. For convenience we extend the action of G to an action of the group algebra FG by setting $a^{\alpha_1 g_1 + \cdots + \alpha_k g_k} = \alpha_1 a^{g_1} + \cdots + \alpha_k a^{g_k}$, where $g_1, \ldots, g_k \in G$ and $\alpha_1, \ldots, \alpha_k \in F$.

Let
$$\hat{G} = \{\chi_1, \ldots, \chi_k\}$$
denote the set of all irreducible characters for the group G. Since G is abelian, each χ_i is linear, hence a homomorphism. This says that $\hat{G}$ is a group under multiplication defined by $(\chi_i \chi_j)(g) = \chi_i(g) \chi_j(g)$ and $\hat{G} \cong G$.

The group algebra FG decomposes into the direct sum of fields and let $f_1, \ldots, f_k$ be the minimal idempotents of FG. By Remark 2.1.11, for every $i = 1, \ldots, k$,

$$f_i = \frac{1}{k} \sum_{j=1}^{k} \chi_i(g_j^{-1}) g_j.$$

An easy computation shows that for any $g \in G$, $gf_i = \chi_i(g)f_i$ and

(3.3) $$\chi_i(f_j) = \delta_{ij},$$

where δ_{ij} is the Kronecker delta.

Then one defines

(3.4) $$A^{(\chi_i)} = \{a \in A \mid a^g = \chi_i(g)a, \text{ for all } g \in G\}.$$

We claim that $A^{(\chi_i)}$ is the subspace of A spanned by the elements of the form a^{f_i}, $a \in A$. In fact, since $1 = f_1 + \cdots + f_k$, for every $a \in A$, we can write $a = a^{f_1} + \cdots + a^{f_k}$. Then $a^g = a^{f_1 g} + \cdots a^{f_k g} = \chi_1(g)a^{f_1} + \cdots + \chi_k(g)a^{f_k}$ and if $a \in A^{(\chi_i)}$, $a^g = \chi_i(g)a$ implies that $(\chi_j(g) - \chi_i(g))a^{f_j} = 0$, for all $j = 1, \ldots, k$, $g \in G$. Clearly this says that $a^{f_j} = 0$ for all $j \neq i$ and $a = a^{f_i}$ follows. This proves the claim.

Since for $a \in A$, $a = a^{f_1} + \cdots + a^{f_k}$ it also follows that

(3.5) $$A = \oplus_{\chi \in \hat{G}} A^{(\chi)}.$$

Now let $a \in A^{(\chi)}$, $b \in A^{(\psi)}$. Then $a^g = \chi(g)a$, $b^g = \psi(g)b$, so, $(ab)^g = a^g b^g = (\chi\psi)(g)ab$. This implies that

(3.6) $$A^{(\chi)}A^{(\psi)} \subseteq A^{(\chi\psi)}$$

and (3.5) is a $\hat{G}$-grading on A.

We have proved that any algebra A with G-action can be considered as a G-graded algebra (since $\hat{G} \cong G$). For instance, if $G = \{1, \varphi\} \cong \mathbb{Z}_2$ is the cyclic group of order 2, then $A = A^{(0)} \oplus A^{(1)}$ where $A^{(0)} = \{a \in A | \varphi(a) = a\} = A^G$, the subalgebra of fixed elements under the G-action, and $A^{(1)} = \{a \in A | \varphi(a) = -a\}$.

Now let $A = \oplus_{g \in G} A_g$ be an associative algebra graded by a finite abelian group G of order k. We define an action of $\hat{G}$ on A by setting

$$\chi(a) = \sum_{g \in G} \chi(g) a_g$$

where χ is an irreducible G-character and $a \in A$ is of the form $a = \sum_{g \in G} a_g$, $a_g \in A_g$. Since $\hat{G}$ is isomorphic to G, this proves that any G-grading on an algebra A gives rise to a G-action on A.

As an example, consider $G = \mathbb{Z}_k$, the cyclic group of order k. Then the map

$$a_0 + a_1 + \cdots + a_{k-1} \mapsto a_0 + \epsilon a_1 + \cdots + \epsilon^{k-1} a_{k-1}$$

where $a_0 \in A_0, \ldots, a_{k-1} \in A_{k-1}$, and ϵ is a primitive kth root of 1, defines an automorphism of the algebra A of order k.

We record the previous discussion in the following.

THEOREM 3.2.1. *Let G be a finite abelian group and suppose that F contains a primitive $|G|$th root of one. Then any G-grading on an algebra A defines a $\hat{G}$-action on A by automorphisms and vice versa. In this action, a subspace $V \subseteq A$ is a graded subspace of A if and only if V is invariant under the $\hat{G}$-action. An element $a \in A$ is homogeneous in the G-grading if and only if a is an eigenvector for any $\chi \in \hat{G}$.*

PROOF. For the $\hat{G}$-action on A defined above, if $V = \oplus_{g \in G} V^{(g)}$ is a graded subspace of A, then $\chi(V^{(g)}) = V^{(g)}$ for all $\chi \in \hat{G}$, $g \in G$. Hence $\chi(V) = V$. Suppose now that V is a subspace of A stable under the $\hat{G}$-action. If V is not a graded subspace then there exists $v = v_{g_1} + \cdots + v_{g_t} \in V$ with $g_1, \ldots, g_t \in G$

distinct and $v_{g_1}, \ldots, v_{g_t} \notin V$. Let $\chi \in \hat{G}$ be such that $\chi(g_1) = \lambda, \chi(g_2) = \mu$ and $\lambda \neq \mu$. Then
$$u = \lambda v - \chi(v) = (\lambda - \mu)v_{g_2} + \cdots \in V$$
and applying the same procedure to u we obtain that at least one of the components $v_{g_2}, \ldots, v_{g_t}$ lies in V. This contradiction shows that any $\hat{G}$-stable subspace of A is graded. In particular, if $\dim V = 1$, we get the required property for the homogeneous elements and the $\hat{G}$-eigenvectors. $\square$

Recall that $\mathbb{Z}_2$-graded algebras are usually called superalgebras. If $A = A^{(0)} \oplus A^{(1)}$ is a superalgebra, then the subspaces $A^{(0)}$ and $A^{(1)}$ are called the even and the odd component of A, respectively. Accordingly, the elements of $A^{(0)}$ are called even and the elements of $A^{(1)}$ are called odd. Notice that, if $G = \mathbb{Z}_2$, any field of characteristic zero satisfies the hypothesis of the previous theorem and $\hat{G} = \langle 1, \phi \rangle$ where
$$\phi(a) = \begin{cases} a, & \text{if } a \in A^{(0)}, \\ -a, & \text{if } a \in A^{(1)}. \end{cases}$$
In this case Theorem 3.2.1 takes the following form.

COROLLARY 3.2.2. *Let $A = A^{(0)} \oplus A^{(1)}$ be a superalgebra over a field F, char $F = 0$. Then the linear map $\phi : A \to A$ defined by $\phi(a_0 + a_1) = a_0 - a_1$, where $a_0 \in A^{(0)}, a_1 \in A^{(1)}$, is an automorphism of A of order 1 (if $A = A^{(0)}$) or 2 (if $A^{(1)} \neq 0$). A subspace $V \subseteq A$ is graded if and only if $\phi(V) = V$. An element $a \in A$ is even (odd) if and only if $\phi(a) = a$ ($\phi(a) = -a$, respectively).*

The duality between G-gradings and $\hat{G}$-actions will be used in the construction of a structure theory of graded algebras.

3.3. G-actions, G-gradings and free algebras

Recall that for an F-algebra A, a linear map $\varphi : A \to A$ is an antiautomorphism of A if $\varphi(ab) = \varphi(b)\varphi(a)$ for all $a, b \in A$. The set of automorphisms and antiautomorphisms of A form a group under the composition of functions, denoted $\text{Aut}^*(A)$. Since the product of any two antiautomorphisms is an automorphism, $\text{Aut}(A)$, the group of automorphisms of A, is a subgroup of $\text{Aut}^*(A)$ of index at most 2.

Let G be a finite subgroup of $\text{Aut}^*(A)$. We shall be studying in the next chapters the identities with automorphisms and antiautomorphisms of A (called G-identities), and to this end here we introduce an universal object called the free algebra with G-action.

Let X be a set and G a finite group with a subgroup H of index ≤ 2. We construct $F\langle X|G\rangle$, the free algebra on X with G-action as follows. The algebra $F\langle X|G\rangle$ is the algebra freely generated by the set $\{x^g = g(x) \mid x \in X, g \in G\}$. We let G act on $F\langle X|G\rangle$ in a natural way by requiring that if $g_1, g_2 \in G$, then $(x^{g_1})^{g_2} = x^{g_2 g_1}$, and then by extending this action to all of $F\langle X|G\rangle$ as follows: if v, w are monomials and $g \in G$, then define $(vw)^g = v^g w^g$, if $g \in H$ and $(vw)^g = w^g v^g$, if $g \in G \setminus H$. Then extend this action by linearlity to all of $F\langle X|G\rangle$. Thus G acts on $F\langle X|G\rangle$ with H as automorphisms and $G \setminus H$ as antiautomorphisms. The elements of $F\langle X|G\rangle$ are called G-polynomials.

DEFINITION 3.3.1. *$F\langle X|G\rangle$ is called the free algebra on X with G-action.*

Now let A be an F-algebra and G a finite group acting on it ($G \subseteq \mathrm{Aut}^*(A)$). Notice that $H = G \cap \mathrm{Aut}(A)$, is a subgroup of G of index at most 2. $F\langle X|G\rangle$ has the following universal property: any set theoretical map $\varphi : X \to A$ extends uniquely to a homomorphism $\bar\varphi : F\langle X|G\rangle \to A$ such that $\bar\varphi(f^g) = \bar\varphi(f)^g$, for any $f \in F\langle X|G\rangle, g \in G$.

If we let $\bar\Phi$ be the set of all such homomorphisms, then $\mathrm{Id}^G(A) = \bigcap_{\bar\phi \in \bar\Phi} \ker \bar\phi$ is the ideal of G-polynomial identities of A. Hence, a G-polynomial $f(x_1^{g_1}, \ldots, x_n^{g_n})$ is a G-identity for A if $f(a_1^{g_1}, \ldots, a_n^{g_n}) = 0$ for all $a_1, \ldots, a_n \in A$. We write $f \equiv 0$ on A, in this case.

DEFINITION 3.3.2. $\mathrm{Id}^G(A) = \{f \in F\langle X|G\rangle \mid f \equiv 0 \text{ on } A\}$ is the ideal of G-polynomial identities of A.

By naturally extending the ordinary case (G trivial), the degree of a monomial M in a variable $x \in X$, is defined as the number of times the variables x^g appear in M (regardless of the exponent $g \in G$).

We next introduce another universal object called the free G-graded algebra. Let $F\langle X\rangle$ be the free algebra over F on a countable set X and let G be a finite group. We write X in the form $X = \bigcup_{g \in G} X_g$, where $X_g = \{x_1^{(g)}, x_2^{(g)}, \ldots\}$ are disjoint sets. The indeterminates from X_g are said to be of homogeneous degree g. The homogeneous degree of a monomial $x_{i_1}^{(g_1)} \cdots x_{i_t}^{(g_t)} \in F\langle X\rangle$ is defined to be $g_1 g_2 \cdots g_t$, as opposed to its total degree, which is defined to be t. Denote by $F\langle X\rangle^{(g)}$ the subspace of the algebra $F\langle X\rangle$ generated by all the monomials having homogeneous degree g. Notice that $F\langle X\rangle^{(g)} F\langle X\rangle^{(h)} \subseteq F\langle X\rangle^{(gh)}$, for every g, h in G. It follows that
$$F\langle X\rangle = \bigoplus_{g \in G} F\langle X\rangle^{(g)}$$
is a G-grading. We denote by $F\langle X\rangle^{gr}$ the algebra $F\langle X\rangle$ with this grading.

DEFINITION 3.3.3. $F\langle X\rangle^{gr}$ is called *the free G-graded algebra of countable rank over F.*

The algebra $F\langle X\rangle^{gr}$ has the following universal property: given any G-graded algebra A, any set theoretical map $\psi : X \to A$ such that $\psi(X_g) \subseteq A_g$ extends uniquely to a homomorphism $\bar\psi : F\langle X\rangle^{gr} \to A$ of G-graded algebras. If we let $\bar\Psi$ be the set of all such homomorphisms, $\mathrm{Id}^{gr}(A) = \bigcap_{\bar\psi \in \bar\Psi} \ker \bar\psi$ is called the ideal of G-graded polynomial identities of A. This means that a graded polynomial $f(x_1^{(g_1)}, \ldots, x_n^{(g_n)}) \in F\langle X\rangle^{gr}$ is a graded identity for the algebra A, and we write $f \equiv 0$ on A, if $f(a_1^{(g_1)}, \ldots, a_n^{(g_n)}) = 0$ for all $a_1^{(g_1)} \in A^{(g_1)}, \ldots, a_n^{(g_n)} \in A^{(g_n)}$.

DEFINITION 3.3.4. $\mathrm{Id}^{gr}(A) = \{f \in F\langle X\rangle^{gr} \mid f \equiv 0 \text{ on } A\}$ is the ideal of graded identities of A.

The above two notions are related if G is abelian and $\mathrm{char}\, F = 0$. We next describe this relation between $F\langle X|G\rangle$ and $F\langle X\rangle^{gr}$.

Let $G \subseteq \mathrm{Aut}^*(A)$ be a finite abelian group acting on A and let $f_1, \ldots, f_k$ be the minimal idempotents of the group algebra FG. Let $\hat G = \{\chi_1, \ldots, \chi_k\}$ be the dual group of G and recall that $\chi_i(f_j) = \delta_{ij} f_j$.

For any $x \in X$ we can write $x = x^1 = x^{f_1 + \cdots + f_k} = x^{f_1} + \cdots + x^{f_k}$; hence, for $g \in G$, $x^g = x^{g(f_1 + \cdots + f_k)} = x^{gf_1} + \cdots + x^{gf_k} = \chi_1(g) x^{f_1} + \cdots + \chi_k(g) x^{f_k}$. This says

that
$$\operatorname{span}_F\{x^g | g \in G\} = \operatorname{span}_F\{x^{f_1}, \ldots, x^{f_k}\}.$$

It follows that $F\langle X|G\rangle$ is freely generated, as an algebra, by the set $\{x^{f_1}, \ldots, x^{f_k} | x \in X\}$.

Now, the space $W = \operatorname{span}_F\{x^g | g \in G, x \in X\}$ can be decomposed as
$$W = W^{(\chi_1)} \oplus \cdots \oplus W^{(\chi_k)}$$
where $W^{(\chi_i)} = \operatorname{span}\{x^{f_i} | x \in X\}$. This in turn induces in the free algebra $F\langle X|G\rangle$ the following grading as a vector space

$$(3.7) \qquad F\langle X|G\rangle = \bigoplus_{\chi \in \hat{G}} F\langle X|G\rangle^{(\chi)}.$$

Notice that if $G \subseteq \operatorname{Aut}(A)$ acts as a group of automorphisms on A, then the one in (3.7) is a $\hat{G}$-grading as an algebra and it is easy to see that it is the one described in Theorem 3.2.1. It follows that in this case $F\langle X|G\rangle$ is the free $\hat{G}$-graded algebra on $\tilde{X} = \{x^{f_i} \mid x \in X, i = 1, \ldots, k\}$.

An important property of the above G-grading of the free algebra $F\langle X|G\rangle$ is given in the following.

LEMMA 3.3.5. *Let A be an algebra, $G \subseteq \operatorname{Aut}(A)$ a finite abelian group and let $\varphi : F\langle X|G\rangle \to A$ be a homomorphism. Then the following conditions are equivalent:*

1) *φ commutes with the G-action, i.e., $\varphi(f^g) = \varphi(f)^g$ for all $f \in F\langle X|G\rangle$, $g \in G$;*
2) *φ is a homomorphism of $\hat{G}$-graded algebras.*

PROOF. Let $A = \oplus_{i=1}^n A^{(\chi_i)}$ and suppose first that φ is a homomorphism commuting with the G-action. Let $\varphi(x_i) = a_i, i \geq 1$. Then $\varphi(x_i^g) = a_i^g$ for any $g \in G$, so,

$$(3.8) \qquad \varphi(x_i^{f_j}) = a_i^{f_j}$$

for all minimal idempotents $f_j \in FG$. Since $x_i^{f_j} \in F\langle X|G\rangle^{(\chi_j)}$ and $a_i^{f_j} \in A^{(\chi_j)}$, from (3.8) it follows that φ is a homomorphism of graded algebras.

If, on the other hand, $\varphi : F\langle X|G\rangle \to A$ preserves the $\hat{G}$-grading, for $x \in X$ write

$$(3.9) \qquad \varphi(x^{f_1}) = b_1 \in A^{(\chi_1)}, \ldots, \varphi(x^{f_k}) = b_k \in A^{(\chi_k)}.$$

Then, recalling that $\chi_i(f_j) = \delta_{ij}f_j$, it follows that $b_i^{f_j} = \chi_i(f_j)b_i = \delta_{ij}b_i$ and, if we set $b = b_1 + \cdots + b_k$, we get $b^{f_i} = b_i, i = 1, \ldots, k$.

Now, if $g \in G$, write g as a linear combination of $f_1, \ldots, f_k$ in FG. We get $g = \lambda_1 f_1 + \cdots + \lambda_k f_k$, and $x^g = x^{\lambda_1 f_1 + \cdots + \lambda_k f_k} = \lambda_1 x^{f_1} + \cdots + \lambda_k x^{f_k}$.

From (3.9), we then obtain
$$\varphi(x^g) = \lambda_1 \varphi(x^{f_1}) + \cdots + \lambda_k \varphi(x^{f_k}) = \lambda_1 b_1 + \cdots + \lambda_k b_k = b^{\lambda_1 f_1 + \cdots + \lambda_k f_k} = b^g.$$

Thus, for any $x \in X$ and $g \in G$, $\varphi(x) = b$ implies $\varphi(x^g) = (\varphi(x))^g$ and φ commutes with the G-action. $\square$

From the above lemma we obtain.

PROPOSITION 3.3.6. *Let A be an algebra over a field F and $G \subseteq \mathrm{Aut}(A)$ a finite abelian group. Suppose that $\mathrm{char}\, F = 0$ and F contains a primitive $|G|$th root of one. Then $F\langle X|G\rangle = \oplus_{\chi \in \hat{G}} F\langle X|G\rangle^{(\chi)}$ is the free $\hat{G}$-graded algebra of countable rank and $\mathrm{Id}^G(A) = \mathrm{Id}^{gr}(A)$.*

PROOF. As above, let $|G| = k$ and let $f_1, \ldots, f_k$ be the minimal idempotents of FG. We shall prove that the set $\tilde{X} = \{x^{f_i} \mid x \in X, i = 1, \ldots, k\}$ is a set of free graded generators of the $\hat{G}$-graded algebra $F\langle X|G\rangle$.

Let $A = A^{(\chi_1)} \oplus \cdots \oplus A^{(\chi_k)}$ be a $\hat{G}$-graded algebra and let $\varphi : \tilde{X} \to A$ be any map with $\varphi(x^{f_i}) = b_i \in A^{(\chi_i)}$ for all $x \in X$, $1 \leq i \leq k$. Recall that by Theorem 3.2.1 the group $\hat{G}$ acts on the algebra A. Since $F\langle X|G\rangle$ is the free algebra on X with G-action, there exists a homomorphism $\bar{\varphi} : F\langle X|G\rangle \to A$ such that $\bar{\varphi}(x) = \bar{\varphi}(x^{f_1} + \cdots + x^{f_k}) = b_1 + \cdots + b_k$. By Lemma 3.3.5, $\bar{\varphi}$ is a homomorphism of graded algebras extending φ, i.e., $F\langle X|G\rangle$ is a free G-graded algebra.

In order to prove the equality $\mathrm{Id}^G(A) = \mathrm{Id}^{gr}(A)$, consider a G-polynomial $f = f(x_1^{g_1}, \ldots, x_n^{g_n}) \in \mathrm{Id}^G(A)$. As we remarked before Lemma 3.3.5 we can write

$$x_i^{g_i} = \sum_j \chi_j(g_i) x_i^{f_i},$$

so that f can be rewritten as a polynomial in graded indeterminates

$$f(x_1^{g_1}, \ldots, x_n^{g_n}) = h(x_1^{f_1}, \ldots, x_1^{f_k}, x_n^{f_1}, \ldots, x_n^{f_k})$$

with $x_i^{f_j} \in F\langle X|G\rangle^{(\chi_j)}$.

Next we check that h is a graded identity of A. Let $c_1^1, \ldots, c_1^k, \ldots, c_n^1, \ldots, c_n^k$ be arbitrary homogeneous elements of A, $c_i^j \in A^{(\chi_j)}$ and set

$$a_i = (\sum_j \chi_j(g_i) c_i^j)^{g_i^{-1}}.$$

Clearly,

$$a_i^{g_i} = \sum_j \chi_j(g_i) c_i^j$$

and $f(a_1^{g_1}, \ldots, a_n^{g_n}) = 0$ since f is a G-identity of A. On the other hand,

$$f(a_1^{g_1}, \ldots, a_n^{g_n}) = h(c_1^1, \ldots, c_1^k, \ldots, c_n^1, \ldots, c_n^k)$$

and h is a graded identity of A. Thus

$$f(x_1^{g_1}, \ldots, x_n^{g_n}) = h(x_1^{f_1}, \ldots, x_n^{f_k}) \in \mathrm{Id}^{gr}(A),$$

i.e., $\mathrm{Id}^G(A) \subseteq \mathrm{Id}^{gr}(A)$.

For the inverse inclusion, let $G = \{g_1, \ldots, g_k\}$ and for every $i = 1, \ldots, n$, $j = 1, \ldots, k$, write

$$\chi_1(g_j) x_i^{f_1} + \cdots + \chi_k(g_j) x_i^{f_k} = x_i^{g_j}.$$

Since the $k \times k$ matrix $(\chi_i(g_j))$ is non-degenerate, there exists a $k \times k$ matrix T such that

$$(x_i^{f_j}) = T(x_i^{g_j}).$$

The same argument as before shows that if

$$f = f(x_1^{f_1}, \ldots, x_1^{f_k}, \ldots, x_n^{f_1}, \ldots, x_n^{f_k}) \equiv 0$$

is a graded identity of A, then $f = h(x_1^{g_1}, \ldots, x_1^{g_k}, \ldots, x_n^{g_1}, \ldots, x_n^{g_k})$ for some polynomial h and $h \equiv 0$ is a G-identity of A. Hence $f \in \mathrm{Id}^G(A)$ and $\mathrm{Id}^{gr}(A) \subseteq \mathrm{Id}^G(A)$. This completes the proof of the proposition. □

In what follows, when G is a finite abelian group of automorphisms of A and F contains a primitive $|G|$th root of one, $\mathrm{char}\, F = 0$, we shall identify G and $\hat{G}$ and we shall say that the algebra A has G-action or is G-graded.

Next we consider the special case when G has order 2. We first give a definition.

DEFINITION 3.3.7. Let A be an F-algebra. A linear mapping $*: A \to A$ is an involution of A if, for all $a, b \in A$, we have

$$(ab)^* = b^* a^* \quad \text{and} \quad a^{**} = a.$$

Let A be an algebra over F with an automorphism or an antiautomorphism φ of order 2 and assume that the field F is of characteristic different from 2. Then A has a structure of superalgebra or algebra with involution, respectively.

Write $G = \langle \varphi \rangle$. Then $\frac{1+\varphi}{2}$ and $\frac{1-\varphi}{2}$ are the minimal idempotents of the group algebra FG. By applying the above discussion, we see that the free algebra with G-action is freely generated by the elements

$$x_i + x_i^\varphi \quad \text{and} \quad x_i - x_i^\varphi, \quad i = 1, 2, \ldots.$$

For convenience, for every i, let us write $x_i + x_i^\varphi = y_i$ and $x_i - x_i^\varphi = z_i$. Then $F\langle X|G\rangle = F\langle Y, Z\rangle$ is the free associative algebra on the two sets $Y = \{y_1, y_2, \ldots\}$ and $Z = \{z_1, z_2, \ldots\}$. The difference between the case when φ is an automorphism or an antiautomorphism of order 2 lies in the following.

In the first case, as we remarked above, $F\langle Y, Z\rangle$ has a structure of $\mathbb{Z}_2$-graded algebra (or superalgebra) where the variables from Y have homogeneous degree 0 and the variables from Z have homogeneous degree 1. Notice that in this case $F\langle Y, Z\rangle^{(0)}$ is spanned by all monomials on $Y \cup Z$ having an even number of variables from Z, and $F\langle Y, Z\rangle^{(1)}$ is spanned by all monomials with an odd number of variables from Z.

DEFINITION 3.3.8. $F\langle Y, Z\rangle$ is called the free superalgebra on Y and Z over F.

If φ is an involution, we write $\varphi = *$ and $F\langle Y, Z\rangle$ has an induced structure of algebra with involution where the variables from Y are symmetric, i.e., $y_i^* = y_i$, and the variables from Z are skew-symmetric, i.e., $z_i^* = -z_i$, $i = 1, 2, \ldots$. In this case one writes $F\langle Y, Z\rangle = F\langle X, *\rangle$.

DEFINITION 3.3.9. $F\langle X, *\rangle$ is called the free algebra with involution $*$ on X over F.

3.4. Wedderburn decompositions

In this section we deal with finite dimensional algebras. We wish to extend Wedderburn's theorems on simple, and semisimple algebras (Theorem 1.11.1) and the Wedderburn-Malcev theorem on finite dimensional algebras (that we shall prove below) to the setting of superalgebras and algebras with involution.

Let A be an unitary algebra with an ideal I and let $\pi : A \to A/I = \bar{A}$ be the canonical projection. We say that the elements $\bar{a}_1, \ldots, \bar{a}_n \in \bar{A}$ are lifted to the elements $a_1, \ldots, a_n \in A$ if $\pi(a_i) = \bar{a}_i$, $1 \leq i \leq n$. With this terminology in mind we can prove the following.

LEMMA 3.4.1. *Let I be a nil ideal of the algebra A. Then:*
1) *Every idempotent of A/I can be lifted to an idempotent of A.*
2) *Every finite set of orthogonal idempotents of A/I can be lifted to a set of orthogonal idempotents of A.*
3) *Every set of $n \times n$ matrix units of A/I can be lifted to a set of $n \times n$ matrix units of A.*

PROOF. 1) Let $e \in A$ with $\bar{e}$ an idempotent of $\bar{A}$. Then $e - e^2 \in I$ and, since I is nil, $(e - e^2)^k = 0$, for some $k \geq 1$. This says that $e^k - ke^{k+1} + \cdots = 0$, so, $e^k = e^{k+1}a$ where a is a polynomial expression in e over F. It easily follows that $e^k = e^{k+t}a^t$, for all $t \geq 1$, and
$$(ea)^{2k} = (e^{2k}a^k)a^k = e^k a^k = (ea)^k.$$
Thus $(ea)^k$ is the desired idempotent since from $\bar{e}^k = \bar{e}^{2k}$ we get $\overline{(ea)^k} = \bar{e}^k \bar{a}^k = \overline{e^{2k}a^k} = \bar{e}^k = \bar{e}$.

2) Let $f_1, \ldots, f_n \in \bar{A}$ be orthogonal idempotents. Suppose that $f_1, \ldots, f_{n-1}$ have been lifted to orthogonal idempotents $e_1, \ldots, e_{n-1} \in A$ and we want to lift f_n to an idempotent e_n orthogonal to $e_1, \ldots, e_{n-1}$.

Set $e' = \sum_{i=1}^{n-1} e_i$. By the first part of the lemma, f_n can be lifted to an idempotent $g \in A$ and $\overline{e'g} = (\sum_{i=1}^{n-1} f_i)f_n = 0$ implies that $e'g \in I$. If we set $g_1 = (1 - e'g)^{-1}g(1 - e'g)$, then $g_1^2 = g_1$ and $\bar{g}_1 = f_n$. Moreover, since $(1-e'g)g = g - e'g = g(1-e')$, we have that $g_1 e' = 0$. It follows that
$$e_n = (1 - e')g_1$$
is an idempotent, $\bar{e}_n = f_n$ and $e_n e' = e' e_n = 0$.

But for all $1 \leq i \leq n-1$, we have that $e' e_i = e_i e' = e_i$, hence $e_n e_i = e_n e' e_i = 0$ and $e_i e_n = e_i e' e_n = 0$ and e_n is the required idempotent.

3) Let f_{ij}, $1 \leq i, j \leq n$ be matrix units of $\bar{A}$. By 2) above $f_{11}, \ldots, f_{nn}$ can be lifted to orthogonal idempotents $e_{11}, \ldots, e_{nn}$ and we need to lift f_{ij} to e_{ij} for $i \neq j$. Let $a_{ij} \in A$ with $\overline{a_{ij}} = f_{ij}$ and define $e_{i1} = e_{ii}a_{i1}e_{11}$. If $b_i = e_{11} - a_{1i}e_{i1}$, then $\bar{b}_i = f_{11} - f_{1i}f_{i1} = 0$ and $b_i \in I$. Thus $1 - b_i$ is invertible and set $d_i = (1 - b_i)^{-1}$. Then $e_{1i} = d_i a_{1i} e_{ii}$ lifts f_{1i} and
$$e_{1i}e_{i1} = d_i a_{1i} e_{ii} e_{i1} = d_i a_{1i} e_{i1} e_{11} = d_i(e_{11} - b_i)e_{11} = d_i(1 - b_i)e_{11} = e_{11}.$$
It follows that $(e_{ii} - e_{i1}e_{1i})^2 = e_{ii}^2 - e_{ii}e_{i1}e_{1i} - e_{i1}e_{1i}e_{ii} + (e_{i1}e_{1i})^2 = e_{ii} - e_{i1}e_{1i}$ is an idempotent of I. Since I is nil, it must be zero and $e_{i1}e_{1i} = e_{ii}$ follows. Now the elements $e_{ij} = e_{i1}e_{1j}$ form a set of matrix units lifting the f_{ij}'s. □

Recall that if A is an algebra and N is an A-bimodule, then a linear map $f : A \to N$ is called a (generalized) derivation if $f(ab) = af(b) + f(a)b$, for all $a, b \in A$.

LEMMA 3.4.2. *Let A be a finite dimensional semisimple algebra over the field F and let N be an A-bimodule. If F is a splitting field for A, then any derivation $f : A \to N$ is inner, i.e., $f(a) = [a, u]$ for some $u \in N$.*

PROOF. We claim that there exist bases $\{a_1, \ldots, a_m\}$ and $\{a'_1, \ldots, a'_m\}$ of A over F such that $\sum_{i=1}^m a_i a'_i = 1$ and for every $a \in A$, the equality $a_i a = \sum_{i=1}^m \alpha_{ij} a_j$ implies $aa'_i = \sum_{i=1}^m a'_j \alpha_{ji}$, for some $\alpha_{ij} \in F$.

Since A is isomorphic to the direct sum of matrix algebras over F, it is clearly enough to prove the claim if A coincides with one of the summands, i.e., $A = M_n(F)$.

In this algebra we consider the first basis consisting of the usual matrix units e_{ij} and the second consisting of the elements $\frac{1}{n}e_{ij}$. It is straightforward to check that these bases satisfy the given properties and the claim is established.

Now if $f : A \to N$ is a derivation, it is enough to define $u = \sum_{i=1}^{m} a'_i f(a_i)$. In fact, take $a \in A$ and let $a_i a = \sum_{i=1}^{m} \alpha_{ij} a_j$. Then

$$[a, u] = au - ua = \sum_{i=1}^{m} aa'_i f(a_i) - \sum_{i=1}^{m} a'_i f(a_i) a$$

$$= \sum_i aa'_i f(a_i) - \sum_i a'_i f(a_i a) + \sum_i a'_i a_i f(a)$$

$$= \sum_{i,j} a'_j \alpha_{ji} f(a_i) - \sum_{i,j} a'_i \alpha_{ij} f(a_j) + f(a) = f(a)$$

and f is inner. $\square$

THEOREM 3.4.3 (Wedderburn-Malcev). *Let A be a finite dimensional algebra over a field F of characteristic zero and let $J(A)$ be its Jacobson radical. Then there exists a maximal semisimple subalgebra B such that*

$$A = B + J(A).$$

Moreover, if B and B' are semisimple subalgebras such that $A = B + J(A) = B' + J(A)$, then there exists $x \in J(A)$ such that $B' = (1+x)B(1+x)^{-1}$.

PROOF. We start by proving the existence of the decomposition. Write $\bar{A} = A/J$ and suppose first that F is a splitting field for $\bar{A}$, e.g., F is algebraically closed. Then $\bar{A} \cong \oplus_{i=1}^{k} M_{n_i}(F)$, for some $n_i \geq 1$ and $k \geq 1$. Let $f_1, \ldots, f_k \in \bar{A}$ be the orthogonal central idempotents corresponding to the unit elements of $M_{n_1}(F), \ldots, M_{n_k}(F)$ respectively. Since J is nilpotent, by Lemma 3.4.1 the f_i's can be lifted to orthogonal idempotent $e_1, \ldots, e_k \in A$.

For $1 \leq t \leq k$ let $A_t = e_t A e_t$. Then $\bar{A}_t = f_t \bar{A} f_t \cong M_{n_t}(F)$ and by Lemma 3.4.1 the matrix units $e_{ij} \in M_{n_t}(F)$ can be lifted to a set of matrix units $x_{ij}^{(t)}$ of A_t, $1 \leq i, j \leq n_t$. Set $B_t = \sum_{i,j} F x_{ij}^{(t)} \cong M_{n_t}(F) \cong \bar{A}_t$. Then $B = \oplus_{i=1}^{k} B_i \cong \oplus_{i=1}^{k} \bar{A}_i \cong \bar{A}$ is a subalgebra of A isomorphic to $\bar{A}$. Notice that since $B \cap J = 0$, it follows that $(B+J)/J \simeq B/B \cap J = B \simeq \bar{A} = A/J$. Thus $A = B + J$ and the desired decomposition is proved in this case.

Now let F be an arbitrary field and suppose that $J^2 = 0$. Let $\{a_1, \ldots, a_n\}$ be a basis of A over F such that $\{\bar{a}_1, \ldots, \bar{a}_m\}$ is a basis of $\bar{A}$ over F and $\{a_{m+1}, \ldots, a_n\}$ is a basis of J over F. For $1 \leq i, j \leq n$ write $a_i a_j = \sum_{k=1}^{n} \alpha_{ij}^{k} a_k$ where the $\alpha_{ij}^{k} \in F$ are the structure constants of A. Then we clearly have that $\bar{a}_i \bar{a}_j = \sum_{k=1}^{m} \alpha_{ij}^{k} \bar{a}_k$.

In order to find a subalgebra $S \subseteq A$ with $S \cap J = 0$ and $S \cong \bar{A}$ we need to find scalars x_{ij} such that the elements

$$b_i = a_i + \sum_{j=m+1}^{n} x_{ij} a_j, \ 1 \leq i \leq m,$$

form a basis of S over F. We will find b_i's of this form such that the algebras $\bar{A}$ and S have the same structure constants, i.e., $b_i b_j = \sum_{k=1}^{m} \alpha_{ij}^{k} b_k$ for $1 \leq i, j \leq m$.

This last equality translates into

$$(a_i + \sum_{t=m+1}^{n} x_{it}a_t)(a_j + \sum_{t=m+1}^{n} x_{jt}a_t) = \sum_k \alpha_{ij}^k (a_k + \sum_{t=m+1}^{n} x_{kt}a_t).$$

Since $J^2 = 0$, comparing the coefficients of the basis elements a_i, we get the following system of linear equations in the unknowns x_{ij},

(3.10) $$\alpha_{ij}^k + \sum_{t=m+1}^{n} \alpha_{it}^k x_{jt} + \sum_{t=m+1}^{n} \alpha_{tj}^k x_{it} = \sum_{s=1}^{m} \alpha_{ij}^s x_{ts},$$

$$1 \leq i,j \leq n, \ m+1 \leq k \leq n.$$

Let K be a splitting field for the algebra $\bar{A}$. Then $A_1 = A \otimes_F K$ is such that $\dim_K A_1 = \dim_F A = n$ and K is a splitting field for the semisimple subalgebra $\bar{A} \otimes K$. By the first part of the proof it follows that $A_1 = S_1 + J(A_1)$ for some semisimple subalgebra S_1 such that $S_1 \cap J(A_1) = 0$. Since by extending the scalars the structure constants do not change, the algebras A and A_1 have the same structure constants. Hence the decomposition of the algebra A_1 says that the system (3.10) has a solution in the field K. Since the coefficients of the system lie in F, the solution lies in F. This says that there exists a semisimple subalgebra S of A with $A = S + J$ and $S \cap J = 0$. Hence the wanted decomposition is proved when $J^2 = 0$.

In the general case we prove the theorem by induction on $\dim_F A$. By the previous part we may assume that $J^2 \neq 0$ and write $B = A/J^2$. Since $\dim_F B < \dim_F A$, by induction $B = S_1 + J_1$ for a suitable semisimple subalgebra $S_1 \cong B/J_1 \cong A/J$. Let $S = \pi^{-1}(S_1)$ where $\pi : A \to B$ is the canonical projection. Then $S_1 \cong S/(S \cap J^2)$ which implies that $J(S) = S \cap J^2$. By the inductive hypothesis we have that $S = T + (S \cap J^2)$ for some semisimple subalgebra $T \cong S/(S \cap J^2) \cong S_1$. Hence $T + J = T + (S \cap J^2) + J = S + J = A$ and $T \cap J = T \cap (S \cap J^2) = 0$. This completes the proof of the first part of the theorem.

Suppose now that B_1 and B_2 are maximal semisimple subalgebras of A and $A = B_1 + J = B_2 + J$. We wish to find an element $u \in J$ such that $B_1 = (1 + u)B_2(1 + u)^{-1}$. First assume that F is a splitting field for the algebra A.

Let $\pi : A \to \bar{A}$ be the canonical projection and let $\varphi_1 : \bar{A} \to A$ and $\varphi_2 : \bar{A} \to A$ be the homomorphisms such that $\pi\varphi_i = \mathrm{id}_{\bar{A}}$ and $\varphi_i(\bar{A}) = B_i$, $i = 1, 2$. We make J into an $\bar{A}$-bimodule by setting $xa = x\varphi_2(a)$ and $ax = \varphi_1(a)x$, for $x \in J, a \in \bar{A}$. Consider the function $f : \bar{A} \to A$ defined by setting $f(a) = \varphi_1(a) - \varphi_2(a)$. Since $\pi f(a) = \pi\varphi_1(a) - \pi\varphi_2(a) = a - a = 0$, we get that $f(a) \in J$, for all $a \in \bar{A}$. Moreover,

$$f(ab) = \varphi_1(ab) - \varphi_2(ab) = \varphi_1(a)(\varphi_1(b) - \varphi_2(b)) + (\varphi_1(a) - \varphi_2(a))\varphi_2(b)$$
$$= \varphi_1(a)f(b) + f(a)\varphi_2(b) = af(b) + f(a)b.$$

Hence f is a derivation and by Lemma 3.4.2 there exists $u \in J$ such that

(3.11) $$\varphi_1(a) - \varphi_2(a) = f(a) = au - ua = \varphi_1(a)u - u\varphi_2(a),$$

for all $a \in \bar{A}$.

We rewrite the last equality in the form $\varphi_1(a)(1 - u) = (1 - u)\varphi_2(a)$. Then, since $1 - u$ is invertible, we obtain $\varphi_1(a) = (1 - u)\varphi_2(a)(1 - u)^{-1}$, for all $a \in \bar{A}$, i.e., $B_1 = (1 - u)B_2(1 - a)^{-1}$.

Now let F be an arbitrary field. As before, let $\pi : A \to \bar{A}$ be the canonical projection and let $\varphi_1, \varphi_2 : \bar{A} \to A$ be the homomorphisms of $\bar{A}$ such that $\varphi_i(\bar{A}) =$

$B_i, \pi\varphi_i = \mathrm{id}_{\bar{A}}$, $i = 1, 2$. By choosing bases of $\bar{A}$ and J as above, we can write u with indeterminate coefficients $u = \sum x_i a_i$. Clearly π, φ_1 and φ_2 can be lifted from $\bar{A}$ and A to $\bar{A} \otimes K$ and $A \otimes K$, respectively, for any extension field $K \supseteq F$. Hence (3.11) can be viewed as a system of linear equations on x_{ij}'s both over F and K. If K is a splitting field for A, the system (3.11) has a solution in K as it was shown before. Since all the coefficients of (3.11) lie in F, it also has a solution in F, i.e., there exists $u \in J$ such that $\varphi_1(a) = (1-u)\varphi_2(a)(1-u)^{-1}$, for all $a \in \bar{A}$ and $B_1 = (1-u)B_2(1-a)^{-1}$. □

In order to have a unified approach to the two cases of involution and $\mathbb{Z}_2$-grading, according to Corollary 3.2.2, we view a superalgebra A as an algebra with φ-action where $\varphi \in \mathrm{Aut}(A)$, $\varphi^2 = 1$. Hence, given an algebra A with $\varphi \in \mathrm{Aut}^*(A)$, $\varphi^2 = 1$, we regard A as a superalgebra if φ is an automorphism and as an algebra with involution φ if φ is an antiautomorphism.

We say that the algebra A is φ-simple if $A^2 \neq 0$ and A has no non-zero φ-ideals (or φ-stable ideals) i.e., ideals I such that $I^\varphi = I$. If A is a superalgebra, we also say that A is a simple superalgebra. For example, any matrix algebra over a division ring endowed with any $\mathbb{Z}_2$-grading or involution is a simple superalgebra or a $*$-simple algebra with involution, respectively. On the other hand, the two-dimensional commutative algebra $F \oplus cF$ where $c^2 = 1$ (i.e., the group algebra of the cyclic group of order 2 generated by c) is also a simple superalgebra and a $*$-simple algebra. Next we prove the generalization of the Wedderburn and Wedderburn-Malcev theorems.

THEOREM 3.4.4. *Let A be a finite dimensional algebra over a field F with an automorphism or antiautomorphism φ of order two. Then:*
  (1) *$J(A)$ is a φ-ideal of A;*
  (2) *if A is a φ-simple algebra, then either A is simple or $A = B \oplus B^\varphi$ for some simple subalgebra B;*
  (3) *if A is semisimple, then A is a finite direct sum of φ-simple algebras;*
  (4) *if $\mathrm{char}\, F = 0$, there exists a maximal semisimple subalgebra B of A such that $B^\varphi = B$.*

PROOF. Since the Jacobson radical $J = J(A)$ is stable under the action of $\mathrm{Aut}^*(A)$, it follows that $J^\varphi = J$ and (1) is proved. Suppose that A is φ-simple but not simple. Since J is a φ-ideal of A, $J = 0$ and A is semisimple. If B is a minimal ideal of A, B^φ is still a minimal ideal and $B^\varphi \neq B$ by the φ-simplicity of A. Now, $B \oplus B^\varphi$ is a φ-ideal of A. Hence $A = B \oplus B^\varphi$ and we are done with (2).

Suppose now that A is a semisimple algebra. By Wedderburn's theorem, we can write $A = B_1 \oplus \cdots \oplus B_n$ where $B_1, \ldots, B_n$ are all the minimal two-sided ideals of A. Since for every i, B_i^φ is also a minimal ideal of A, then $B_i^\varphi = B_j$, for some $1 \leq j \leq m$. We now rename $B_1, \ldots, B_n$ and we write

$$A = C_1 \oplus \cdots \oplus C_m$$

where either $C_i = B_j$ with $B_j^\varphi = B_j$ or $C_i = B_j \oplus B_j^\varphi$, with $B_j^\varphi \neq B_j$. Thus $C_1, \ldots, C_m$ are minimal two-sided φ-ideals of A and the proof of (3) is complete.

Next we prove (4). Suppose that $\mathrm{char}\, F = 0$. If $J = 0$, then $B = A$ and there is nothing to prove. Let $J \neq 0$ and suppose first that $J^2 = 0$. Consider a maximal semisimple subalgebra $B \subseteq A$ and suppose that $B^\varphi \neq B$. Then B^φ is also a maximal semisimple subalgebra of A and, by the Wedderburn-Malcev theorem, it

is conjugated to B through an inner automorphism of A. Hence, there exists $x \in J$ such that
$$(1+x)^{-1}B^\varphi(1+x) = B.$$
Notice that since $J^2 = 0$, $(1+x)^{-1} = 1 - x$. Therefore the map $\tau : B \to B$ defined by $\tau(b) = (1-x)\varphi(b)(1+x)$, is an automorphism or an antiautomorphism of B and

(3.12) $$b^\varphi = (1+x)\tau(b)(1-x).$$

Suppose first that φ is an automorphism. Since $\varphi^2 = 1$, recalling that $J^2 = 0$, we obtain that
$$b = ((1+x)\tau(b)(1-x))^\varphi = (1+x^\varphi)(1+x)\tau^2(b)(1-x)(1-x^\varphi)$$
(3.13) $$= \tau^2(b) + (x+x^\varphi)\tau^2(b) - \tau^2(b)(x+x^\varphi).$$

Hence $\tau^2(b) = b$ and $(x+x^\varphi)b = b(x+x^\varphi)$. Write $x = x^+ + x^-$ where $x^+ = \frac{x+x^\varphi}{2}$ and $x^- = \frac{x-x^\varphi}{2}$. From the above it follows that $(1+x)b(1-x) = (1+x^-)b(1-x^-)$, for all $b \in B$.

Consider the subalgebra $B' = (1+\frac{x^-}{2})B(1-\frac{x^-}{2})$. For arbitrary $a = (1+\frac{x^-}{2})b(1-\frac{x^-}{2}) \in B'$ we have
$$a^\varphi = (1+\frac{(x^-)^\varphi}{2})\varphi(b)(1-\frac{(x^-)^\varphi}{2}) = (1-\frac{x^-}{2})(1+x^-)\tau(b)(1-x^-)(1+\frac{x^-}{2})$$
$$= (1+\frac{x^-}{2})\tau(b)(1-\frac{x^-}{2}) = a' \in B'.$$

Hence B' is a φ-invariant subalgebra and, being isomorphic to B, it is a maximal semisimple φ-invariant subalgebra of A.

If φ is an involution, from a computation similar to (3.13), we obtain that
$$b = \tau^2(b) + (x-x^\varphi)\tau^2(b) - \tau^2(b)(x-x^\varphi).$$

Hence $\tau^2(b) = b$ and $(x-x^\varphi)b = b(x-x^\varphi)$ follows. The outcome of this is that $(1+x)b(1-x) = (1+x^+)b(1-x^+)$, for all $b \in B$. As above it is easy to show that the subalgebra $B' = (1+\frac{x^+}{2})B(1-\frac{x^+}{2})$ is a maximal semisimple φ-invariant subalgebra of A. With this the theorem is proved when $J^2 = 0$.

Suppose now that $J^2 \neq 0$ and choose $m \geq 2$ such that $J^m \neq 0$, $J^{m+1} = 0$. Clearly, J^m is a φ-ideal of A. Therefore A/J^m is an algebra with induced φ-action and we can apply induction on m. Therefore there exists a maximal semisimple subalgebra of A/J^m which is φ-invariant. Denote by B its preimage in A. Then the Jacobson radical of B is $J(B) = J^m$ and $B = B' + J^m$ where B' is a maximal semisimple subalgebra of A. Since $(J^m)^2 \subseteq J^{2m} = 0$, by applying again induction on m we can assume $(B')^\varphi = B'$, that is, B' is a maximal semisimple φ-invariant subalgebra of A, and the proof of the theorem is complete. □

3.5. Finite dimensional simple superalgebras

In this section we describe the finite dimensional simple superalgebras over an algebraically closed field of characteristic zero. First we introduce three finite dimensional superalgebras which are simple over any field of characteristic zero.

(1) $A = M_n(F)$ with trivial grading $A = A^{(0)}$, $A^{(1)} = 0$.

(2) $A = M_{k,l}(F) = \left\{\begin{pmatrix} P & Q \\ R & S \end{pmatrix}\right\}$, $k \geq l > 0$, where P, Q, R, S are $k \times k$, $k \times l$, $l \times k$ and $l \times l$ matrices, respectively. A is endowed with the grading $A^{(0)} = \left\{\begin{pmatrix} P & 0 \\ 0 & S \end{pmatrix}\right\}$, $A^{(1)} = \left\{\begin{pmatrix} 0 & Q \\ R & 0 \end{pmatrix}\right\}$.

(3) $A = M_n(F \oplus cF)$, where $c^2 = 1$ with grading $A^{(0)} = M_n(F)$, $A^{(1)} = cM_n(F)$.

Since matrix algebras are simple in a non-graded sense, the algebras of type (1) and (2) above are simple as superalgebras. The algebras of type (3) are semisimple in the non-graded sense but they do not have any graded ideals. Note that $M_n(F \oplus cF)$ is isomorphic to $M_n(F) \otimes_F F\mathbb{Z}_2$. It is not difficult to see that all the above superalgebras are not isomorphic among themselves.

Our main objective in this section is to prove that over an algebraically closed field of characteristic zero the above are the only examples of simple superalgebras. We start with the following two lemmas.

LEMMA 3.5.1. *Let A be a finite dimensional simple superalgebra over an algebraically closed field F of characteristic zero. Then $A^{(0)}$ is semisimple.*

PROOF. If $A^{(0)}$ is not semisimple, being finite dimensional over F, its Jacobson radical is nilpotent. Hence $A^{(0)}$ contains a non-zero ideal I such that, $I^2 = 0$. Let AIA be the ideal of A generated by I. We claim that $(AIA)^3 = 0$. In fact, we shall prove that for any homogeneous b_1, $b_2 \in A$ and for any a_1, a_2, $a_3 \in I$, $b = a_1 b_1 a_2 b_2 a_3 = 0$. If b_1 or b_2 lies in $A^{(0)}$, then either $a_1 b_1 a_2 = 0$ or $a_2 b_2 a_3 = 0$ since I is an ideal of $A^{(0)}$. On the other hand, if both b_1 and b_2 are odd, then $b_1 a_2 b_2 \in A^{(0)}$ and $b = 0$ in this case also.

Since A is simple, $AIA = 0$ and since $1 \in A$ we conclude that $I = 0$, a contradiction. $\square$

LEMMA 3.5.2. *Let $A \cong M_n(F)$ with unit element E and let $B = M_r(F)$ be a subalgebra of A such that $E \in B$. Then $A = BC$ where C is the centralizer of B in A.*

PROOF. Write $E = e_{11} + \cdots + e_{rr}$ for the unit of B where the e_{ij}'s are the usual matrix units. Given $x \in A$, we set
$$x_{ij} = \sum_s e_{si} x e_{js}.$$
Then for any k, l,
$$x_{ij} \cdot e_{kl} = \sum_s e_{si} x e_{js} e_{kl} = e_{ki} x e_{jk} e_{kl} = e_{ki} x e_{jl},$$
$$e_{kl} \cdot x_{ij} = \sum_s e_{kl} e_{si} x e_{js} = e_{kl} e_{li} x e_{jl} = e_{ki} x e_{jl},$$
that is, x_{ij} commutes with B. On the other hand,
$$\sum_{i,j} x_{ij} e_{ij} = \sum_{i,j,s} e_{si} x e_{js} e_{ij} = \sum_{i,j} e_{ii} x e_{jj} =$$
$$= (e_{11} + \cdots + e_{rr}) x (e_{11} + \cdots + e_{rr}) = ExE = x.$$
Hence, $A = BC$, and the proof is complete. $\square$

THEOREM 3.5.3. *Let A be a finite dimensional simple superalgebra over an algebraically closed field F of characteristic zero. Then A is isomorphic to one of the superalgebras $M_n(F)$, $M_{k,l}(F)$, $k \geq l > 0$, and $M_n(F \oplus cF)$, $c^2 = 1$.*

PROOF. Let $\varphi \in \mathrm{Aut}(A)$ be the automorphism inducing the $\mathbb{Z}_2$-grading of A. Then A is φ-simple and, by Theorem 3.4.4, either A is simple or $A = B \oplus B^\varphi$, for some simple subalgebra B. Since F is algebraically closed, any finite dimensional simple algebra is isomorphic to $M_k(F)$, for some $k \geq 1$. Hence either A is isomorphic to a matrix algebra over F or $A = B \oplus B^\varphi$ and $B \cong M_k(F)$. In this latter case

$$A^{(0)} = \{a + a^\varphi \mid a \in B\}, \ A^{(1)} = \{a - a^\varphi \mid a \in B\}$$

and we obtain that $A \cong M_k(F \oplus cF)$ where $c^2 = 1$. Hence in this second case we are done.

Suppose now that A is a matrix algebra. We shall next describe, up to isomorphism, all $\mathbb{Z}_2$-gradings on A. Let $A = A^{(0)} \oplus A^{(1)}$. Since $A^{(0)}$ is semisimple, we can decompose $A^{(0)}$ into the direct sum of its simple components, $A = A_1 \oplus \cdots \oplus A_m$ where, since F is algebraically closed, $A_1, \ldots, A_m$ are matrix algebras over F. Denote by E the unit matrix of A and by $e_1, \ldots, e_m$ the units of $A_1, \ldots, A_m$, respectively. If $E = E^{(0)} + E^{(1)}$ with $E^{(0)} \in A^{(0)}$ and $E^{(1)} \in A^{(1)}$, then, since the grading is direct, for any homogeneous $a \in A$, we obtain $E^{(0)}a = aE^{(0)} = a$, i.e., $E = E^{(0)} \in A^{(0)}$. Recalling that $E = e_1 + \cdots + e_m$, then we can write

$$A = \oplus_{1 \leq i,j \leq m} e_i A e_j.$$

It is easy to check that for all i, j, $e_i A e_j$ is a graded subspace. Moreover, since

$$A^{(0)} = A_1 \oplus \cdots \oplus A_m \subseteq e_1 A e_1 \oplus \cdots \oplus e_m A e_m,$$

we must have that $e_i A e_j \subseteq A^{(1)}$, for all $i \neq j$. Now, for every i, $A e_i^2 A = A e_i A = A$, since A is simple. Hence, if $m \geq 3$, as a consequence of the above we get that $e_1 A e_3 = e_1 A e_2 \cdot e_2 A e_3 \subseteq A^{(0)}$ a contradiction. Thus $m \leq 2$.

Consider first the case when $m = 1$. Since $A^{(0)}$ is simple, $A^{(0)} \cong M_r(F)$, for some $r \geq 1$ and $A^{(0)}$ contains the unit E of $A \cong M_n(F)$. Set $B = A^{(0)}$. Notice also that the centralizer of any homogeneous element of A is also homogeneous in the $\mathbb{Z}_2$-grading. Hence, by applying Lemma 3.5.2, we have that $A = BC$, where $C = C^{(0)} \oplus C^{(1)}$ and $C^{(0)} \subseteq B = A^{(0)}$ is the center of B. Hence, $C^{(0)}$ consists of scalar matrices, $\dim C^{(0)} = 1$. Since any non-trivial graded ideal I of C generates a non-trivial graded ideal BI of A, C must be a simple superalgebra. Then, by induction on the dimension, C is isomorphic to one of the algebras $M_s(F)$, $M_{k,l}(F)$, $M_k(F \oplus cF)$ with one-dimensional even component, i.e., $C = F$ or $C = F \oplus cF$. Notice that $BC \cong B \otimes C$, since $B \cap C = C^{(0)} = \{\alpha E \mid \alpha \in F\}$. Hence $\dim BC = \dim B \cdot \dim C$ and C cannot be isomorphic to $F \oplus cF$ since otherwise $n^2 = \dim A = 2 \cdot \dim B = 2r^2$, a contradiction. Thus $C = F$, $A = B = A^{(0)}$ and we are done in the case $m = 1$.

Suppose now that $m = 2$. We have $A^{(0)} = A_1 \oplus A_2$, $E = e_1 + e_2$ where e_1 and e_2 are the unit elements of the matrix algebras A_1 and A_2, respectively. Also,

$$A = e_1 A e_1 \oplus e_1 A e_2 \oplus e_2 A e_1 \oplus e_2 A e_2 \oplus$$

as a vector space. The subalgebra $e_1 A e_1$ is a simple superalgebra with even component A_1. Hence, by the previous case, $e_1 A e_1 = A_1 \subseteq A^{(0)}$. Similarly, $e_2 A e_2 = A_2 \subseteq A^{(0)}$. It is easy to see that the algebra $A = M_n(F)$ with this structure is isomorphic, as a $\mathbb{Z}_2$-graded algebra, to $M_{k,l}(F)$ where $A_1 \cong M_k(F)$,

$A_2 \cong M_l(F)$ and $n = k + l$. Since $M_{k,l}(F)$ and $M_{l,k}(F)$ are isomorphic, we may assume $k \geq l$. This completes the proof. □

We conclude this section by stating a structure theorem for finite dimensional superalgebras over an algebraically closed field of charactersitc zero. Its proof is an obvious consequence of Theorem 3.4.4 in light of Theorem 3.5.3.

THEOREM 3.5.4. *Let A be a finite dimensional superalgebra over an algebraically closed field F of characteristic zero. Then there exists a maximal semisimple superalgebra $B \subseteq A$ such that*
$$A = B + J(A).$$
Moreover, B is a finite direct sum of simple superalgebras each isomorphic to either $M_k(F)$ or $M_{k,l}(F)$ with $k \geq l > 0$ or $M_k(F \oplus cF)$.

3.6. Involutions on matrix algebras

In this section we want to classify all involutions $*$ on the matrix algebra $M_k(F)$ when F is an infinite field, char $F \neq 2$. We first give a few examples of involutions.

EXAMPLE 3.6.1. Let $M_k(F)$ be the algebra of $k \times k$ matrices over F. For a matrix $A = (a_{ij}) \in M_k(F)$ let $A^t = (a_{ji})$ be the usual transpose. Then $* = t$ is an involution of $M_k(F)$, called the transpose involution.

If $k = 2n$ is even, let $s : M_{2n}(F) \to M_{2n}(F)$ be defined as follows: write $A = \begin{pmatrix} B & C \\ D & E \end{pmatrix} \in M_{2n}(F)$ where $B, C, D, E \in M_n(F)$. Then
$$A^s = \begin{pmatrix} E^t & -C^t \\ -D^t & B^t \end{pmatrix},$$
where t denotes the transpose. It is easy to check that $* = s$ is an involution called the symplectic involution of $M_k(F)$.

We shall be studying the polynomial expressions, involving an involution, which vanish on $M_k(F)$ and we shall see that, when F is infinite of characteristic different from 2, t and s are the only two involutions that we need to consider.

EXAMPLE 3.6.2. Given an algebra A denote by A^{op} the opposite algebra of A. Recall that A^{op} has the same additive structure of A and multiplication defined as follows: if $a, b \in A^{op}$, then $a \cdot b = ba$ where ba is the product of b and a in A. The algebra $A \oplus A^{op}$ has an involution $*$, denoted the exchange involution, defined as follows: if $a, b \in A$, then $(a, b)^* = (b, a)$.

Recall that, if A is an algebra with involution $*$, then $A^+ = \{a + a^* \mid a \in A\}$ and $A^- = \{a - a^* \mid a \in A\}$ are the sets of symmetric and skew-symmetric elements of A, respectively. Notice that for $a, b \in A^-$, $[a, b] \in A^-$, and A^- is a Lie subalgebra of A under the bracket operation. More generally, the following inclusions hold:
$$[A^+, A^+] + [A^-, A^-] \subseteq A^- \quad \text{and} \quad [A^+, A^-] \subseteq A^+.$$
This says that A has a structure of $\mathbb{Z}_2$-graded Lie algebra (see [**BMPZ**]).

For an algebra A denote by $U(A)$ the group of units of A. For any $u \in U(A)$, let us denote by $\gamma_u : A \to A$ the inner automorphism induced by u. Hence $\gamma_u(a) = uau^{-1}$ for all $a \in A$. Notice that if $u \in U(A) \cap (A^+ \cup A^-)$, i.e., if u is either a symmetric or a skew unit, and $*$ is an involution of A, then also $*\gamma_u$ is an involution. In fact, for $a \in A$,

$$*\gamma_u * \gamma_u(a) = *\gamma_u((u^{-1})^* a^* u^*) = (u^*)^{-1} u a u^{-1} u^* = a.$$

Conversely, if $*\gamma_u$ is an involution, for some $u \in U(A)$, what can we say about u? If the center Z of A is an integral domain (e.g., A is a prime algebra) and Z is fixed by the involution, it is easy to prove that u must be either a symmetric or a skew element of A.

LEMMA 3.6.3. *Let A be a prime algebra with involution $*$ and suppose that $*$ fixes the center Z of A elementwise. If $u \in U(A)$, then $*\gamma_u$ is still an involution of A if and only if $u = u^*$ or $u = -u^*$.*

PROOF. Suppose that $*\gamma_u$ is an involution. Then for $a \in A$, we must have

$$a = *\gamma_u * \gamma_u(a) = *\gamma_u((u^{-1})^* a^* u^*) = (u^{-1})^* u a u^{-1} u^*.$$

This says that $(u^{-1})^* u$ commutes with all elements $a \in A$. Hence it lies in the center Z of A. Write $u^* = u\alpha$, for some $\alpha \in Z$. Since Z is fixed by the involution, from $u^{**} = u$, we get $u\alpha^2 = u$. Hence $(1 - \alpha^2)u = 0$. Since u is a unit, this implies that $1 - \alpha^2 = 0$ and, since Z is a domain we get $\alpha = \pm 1$. This says that u is either symmetric or skew. □

Next we need to use the following well-known theorem whose proof can be found, for instance, in [**Ro2**].

THEOREM 3.6.4 (Skolem-Noether). *Let R be a finite dimensional central simple algebra over F and let A, B be isomorphic simple F-subalgebras of R. Then every isomorphism $\psi : A \to B$ of F-algebras can be extended to an inner automorphism of R.*

We can now classify all the involutions of a finite dimensional central simple algebra over F.

THEOREM 3.6.5. *Let A be a finite dimensional central simple algebra over F with an involution $*$. Then J is an involution of A if and only if $J = *\gamma_u$ for some unit u such that $u = u^*$ or $u = -u^*$.*

PROOF. If J is an involution, the composition $*J$ is an automorphism of A which is the identity on the center F. By Theorem 3.6.4, then $*J = \gamma_u$ for some unit $u \in A$. Hence by multiplying by $*$ on the left, we get $J = *\gamma_u$ and, by the previous lemma, $u \in A^+ \cup A^-$. The converse has been shown above. □

We remark that in the matrix algebra $M_{2n}(F)$ the transpose involution t and the symplectic involution s are related by

$$t = s\gamma_{I_0}$$

where $I_0 = \begin{pmatrix} 0 & I \\ -I & 0 \end{pmatrix} \in M_{2n}(F)$ and I is the identity $n \times n$ matrix. Notice that $I_0 = -I_0^t = -I_0^s$ is a skew element.

Next we show that the transpose and the symplectic involution have a prominent role in the study of the identities with involution of $M_k(F)$.

We start by proving a technical lemma.

LEMMA 3.6.6. *Let F be a field, char $F \neq 2$ and let $a \in GL_k(F)$. If F contains the square roots of all the eigenvalues of a, then there exists $p(x) \in F[x]$ such that $a = p(a)^2$.*

PROOF. Let $\alpha_1, \ldots, \alpha_n \in F$ be the distinct eigenvalues of a and consider the decomposition of the vector space $V = F^{(k)}$ as a direct sum of its eigenspaces with respect to a. Thus $V = V_1 \oplus \cdots \oplus V_n$ where, for every $i = 1, \ldots, n$ and $m_i \geq 1$, $V_i = \{v \in V \mid v(a - \alpha_i)^{m_i} = 0\}$.

We claim that in order to prove the lemma, it is enough to find polynomials $p_i(x) \in F[x]$ such that $V_i(a - p_i(a)^2) = 0$ for all $i = 1, \ldots, n$. In fact, since the polynomials $(x - \alpha_1)^{m_1}, \ldots, (x - \alpha_n)^{m_n}$ are relatively prime, by the Chinese remainder theorem, there exists $f(x) \in F[x]$ such that
$$f(x) \equiv p_i(x) \pmod{(x - \alpha_i)^{m_i}}, \quad i = 1, \ldots, m.$$
Hence $V_i(a - f(a)^2) = 0$ for all i, leading to $V(a - f(a)^2) = 0$ and $a - f(a)^2 = 0$.

Therefore, by replacing V with V_i, without loss of generality, we may assume that $V(a - \alpha)^m = 0$ for some $\alpha \in F$ and $m \geq 1$. Hence $(a - \alpha)^m = 0$. Write $b = a - \alpha$ and consider the formal binomial expansion
$$(3.14) \qquad (\alpha + b)^{\frac{1}{2}} = \alpha^{\frac{1}{2}} + \frac{1}{2}\alpha^{-\frac{1}{2}}b - \frac{1}{8}\alpha^{-\frac{3}{2}}b^2 + \cdots.$$
Since $b^m = 0$, the above expression has at most $m + 1$ non-zero terms. Now substitute in (3.14) b with $a - \alpha$ and then a with x. The resulting polynomial $p(x)$ has the desired property $p(a)^2 = a$. □

LEMMA 3.6.7. *Let $*$ be an involution of $M_k(F)$, char $F \neq 2$. If $a = -a^* \in M_k(F)$ has characteristic polynomial $f(x) = \sum_{i=0}^{k} \alpha_i x^{k-i}$, then $\alpha_1 = \alpha_3 = \alpha_5 = \ldots = 0$. In particular, if k is odd, $\det(a) = 0$.*

PROOF. Since a is a skew matrix,
$$\operatorname{tr}(a^{2r+1}) = \operatorname{tr}((a^{2r+1})^*) = \operatorname{tr}(-a^{2r+1}) = -\operatorname{tr}(a^{2r+1}),$$
for any $r \geq 1$ and $\alpha_1 = \operatorname{tr}(a) = 0$ follows. By applying Newton's formulas (Proposition 1.6.3), for $i = 0, 1, 2, \ldots$, we have
$$(2i + 1)\alpha_{2i+1} = \sum_{r=1}^{i} \operatorname{tr}(a^{2r})\alpha_{2i+1-2r}$$
and the proof in completed by induction on i. □

For an algebra A with involution $*$, let us denote by $\operatorname{Id}(A, *)$ the ideal of polynomial identities with involution of $F\langle X, *\rangle$.

THEOREM 3.6.8. *Let F be an infinite field, char $F \neq 2$ and let $*$ be an involution of $M_k(F)$. Then either*
$$\operatorname{Id}(M_k(F), *) = \operatorname{Id}(M_k(F), t) \quad or \quad \operatorname{Id}(M_k(F), *) = \operatorname{Id}(M_k(F), s).$$
The second possibility can occur only if k is even.

PROOF. Suppose first that F is algebraically closed. By Theorem 3.6.5, $* = t\gamma_u$ for some symmetric or skew unit $u \in M_k(F)$. Suppose first that $u = u^t$ is symmetric. Since F contains the square roots of all eigenvalues of u, by Lemma 3.6.6 there exists $v \in M_k(F)$ such that $u = v^2$. Moreover, since v is a polynomial expression in u and $u = u^t$, then $v = v^t$ holds also. We claim that $\gamma_v : (M_k(F), t) \to$

$(M_k(F), *)$ is an isomorphism of algebras with involution. In fact, for $a \in M_k(F)$, recalling that $* = t\gamma_u$, we have

$$\gamma_v(a^*) = va^*v^{-1} = v(uau^{-1})^t v^{-1}$$
$$= vu^{-1}a^t u v^{-1} = v^{-1}a^t v = (vav^{-1})^t = (\gamma_v(a))^t.$$

This proves the theorem when u is a symmetric element.

Suppose now that $u = -u^t$ is a skew element. By Lemma 3.6.7 if k is odd, $\det(u) = 0$. Hence, since u is a unit, k must be even. Now, recalling that $t = s\gamma_{I_0}$, it follows that $* = t\gamma_u = s\gamma_{I_0}\gamma_u = s\gamma_{I_0 u}$. Since $I_0^t = -I_0$ and $u^t = -u$, we obtain

$$(I_0 u)^s = t\gamma_{I_0}^{-1}(I_0 u) = (I_0^{-1} I_0 u I_0)^t = I_0^t u^t = I_0 u.$$

Thus $I_0 u$ is an s-symmetric unit. As in the first part of the proof one can construct an isomorphism $(M_k(F), s) \to (M_k(F), *)$ of algebras with involution. This completes the proof when F is algebraically closed.

Now let F be an arbitrary infinite field and let $\bar{F}$ be its algebraic closure. Since F is infinite, as in Lemma 1.4.2, it can be easily proved that every $*$-identity is stable. Hence $(M_k(F), *)$ and $(M_k(\bar{F}), *)$ have the same $*$-identities over F. Similarly, $(M_k(F), j)$ and $(M_k(\bar{F}), j)$ have the same $*$-identities over F where $j = t$ or s.

By the first part $(M_k(\bar{F}), *) \cong (M_k(\bar{F}), j)$ where $j = t$ or s. Hence $(M_k(\bar{F}), *)$ has the same $*$-identities as $(M_k(\bar{F}), j)$ over F. This completes the proof. □

3.7. Superalgebras and Grassmann envelopes

As in the case of ordinary algebras, given a non-empty set of polynomials in two sets of variables, one can define the notion of variety of superalgebras or supervariety. Throughout this section we assume that char $F = 0$.

Let $F\langle Y, Z \rangle$ be the free superalgebra over F of countable rank. Hence the variables from Y have homogeneous degree zero and those from Z have homogeneous degree one. Recall that, given a superalgebra $A = A^{(0)} \oplus A^{(1)}$, a polynomial $f(y_1, \ldots, y_n, z_1, \ldots, z_m) \in F\langle Y, Z \rangle$ is a graded identity or superidentity of A if

$$f(a_1, \ldots, a_n, b_1, \ldots, b_m) = 0$$

for all $a_1, \ldots, a_n \in A^{(0)}, b_1, \ldots, b_m \in A^{(1)}$. Also $\mathrm{Id}^{gr}(A)$ is the ideal of graded identities of A.

DEFINITION 3.7.1. A graded ideal $I = I^{(0)} \oplus I^{(1)}$ of $F\langle Y, Z \rangle$ is called a T_2-ideal if $\varphi(I) \subseteq I$ for all graded endomorphisms φ of $F\langle Y, Z \rangle$.

Notice that $\mathrm{Id}^{gr}(A)$ is a T_2-ideal of $F\langle Y, Z \rangle$. It is also clear that any T_2-ideal of $F\langle Y, Z \rangle$ is of the form $\mathrm{Id}^{gr}(A)$ for some superalgebra A.

As in the case of ordinary algebras any non-empty set $S \subseteq F\langle Y, Z \rangle$ of homogeneous elements defines a variety.

DEFINITION 3.7.2. Given a non-empty set $S = S^{(0)} \cup S^{(1)}$, $S^{(0)} \subseteq F\langle Y, Z \rangle^{(0)}$, $S^{(1)} \subseteq F\langle Y, Z \rangle^{(1)}$, the class of all superalgebras $A^{(0)} \oplus A^{(1)}$ such that $f \equiv 0$ on A for all $f \in S$ is called the supervariety determined by S.

For any supervariety $\mathcal{V}$, the set $\mathrm{Id}^{gr}(\mathcal{V})$ of all its graded identities is clearly a T_2-ideal. Also $F\langle Y, Z \rangle / \mathrm{Id}^{gr}(\mathcal{V})$ is the relatively free algebra of countable rank of $\mathcal{V}$ with even free generators Y and odd free generators Z.

It is well known that Birkhoff's theorem also holds for supervarieties if we replace ordinary homomorphisms with graded homomorphisms.

Among superalgebras a prominent role in this book is played by the Grassmann algebra G. Recall that G is the algebra with 1 generated by the countable set $\{e_1, e_2, \ldots\}$ satisfying the relations

$$e_i e_j + e_j e_i = 0, \quad i, j \geq 1.$$

A basis of G is given by 1 and by all monomials of the form

(3.15) $$e_{i_1} e_{i_2} \cdots e_{i_m}; \quad i_1 < i_2 < \ldots < i_m; \quad m \geq 1.$$

Hence G has a natural $\mathbb{Z}_2$-grading $G = G^{(0)} \oplus G^{(1)}$ where $G^{(0)}$ is the subspace of G spanned by all monomials (3.15) of even length m and $G^{(1)}$ is the subspace spanned by all monomials of odd length.

Given any superalgebra A one can form a new superalgebra with the help of G called the Grassmann envelope of A. This is of importance in PI-theory. Here is the definition.

DEFINITION 3.7.3. Let $A = A^{(0)} \oplus A^{(1)}$ be a superalgebra. The algebra

$$G(A) = (A^{(0)} \otimes G^{(0)}) \oplus (A^{(1)} \otimes G^{(1)})$$

is called the Grassmann envelope of A.

Clearly the Grassmann envelope $G(A)$ has a natural $\mathbb{Z}_2$-grading, $G(A) = G(A)^{(0)} \oplus G(A)^{(1)}$, where $G(A)^{(0)} = A^{(0)} \otimes G^{(0)}$, $G(A)^{(1)} = A^{(1)} \otimes G^{(1)}$. Next we exploit the relation between the multilinear graded identities of A and of $G(A)$.

Denote by $P_{k,m}$ the space of all multilinear polynomials of $F\langle Y, Z\rangle$ in the variables $y_1, \ldots, y_k, z_1, \ldots, z_m$. The intersection $P_{k,m} \cap \mathrm{Id}^{gr}(A)$ consists of all multilinear graded identities of the superalgebra A having degree k on even variables and degree m on odd variables.

We define a linear isomorphism $\sim : P_{k,m} \to P_{k,m}$ by the following rule: let $f \in P_{k,m}$ and write f as

$$f = \sum_{\substack{\sigma \in S_m \\ W=(w_0,w_1,\ldots,w_m)}} \alpha_{\sigma,W} w_0 z_{\sigma(1)} w_1 \cdots w_{m-1} z_{\sigma(m)} w_m$$

where $w_0, w_1, \ldots, w_m$ are monomials in $y_1, \ldots, y_k$ and $\alpha_{\sigma,W} \in F$. Then

$$\widetilde{f} = \sum_{\substack{\sigma \in S_m \\ W=(w_0,w_1,\ldots,w_m)}} (\mathrm{sgn}\,\sigma) \alpha_{\sigma,W} w_0 z_{\sigma(1)} w_1 \cdots w_{m-1} z_{\sigma(m)} w_m.$$

The map $\sim$ has the following basic properties.

LEMMA 3.7.4. Let $f \in P_{k,m}$. Then:
1) f is a graded identity of $G(A)$ if and only if $\widetilde{f}$ is a graded identity of A;
2) $\widetilde{\widetilde{f}} = f$.

PROOF. The second statement is obvious. In order to prove 1), let $\bar{y}_1, \ldots, \bar{y}_k \in A^{(0)}$, $\bar{z}_1, \ldots, \bar{z}_m \in A^{(1)}$ be arbitrary homogeneous elements of A and $g_1, \ldots, g_k \in G^{(0)}$, $h_1, \ldots, h_m \in G^{(1)}$ be arbitrary homogeneous elements of G. Fix a monomial

$$w = a_0(y_1, \ldots, y_k) z_{\sigma(1)} \cdots z_{\sigma(m)} a_m(y_1, \ldots, y_k)$$

and compute its value $\bar{w}$ on $\bar{y}_1 \otimes g_1, \ldots, \bar{y}_k \otimes g_k, \ldots, \bar{z}_1 \otimes h_1, \ldots, \bar{z}_m \otimes h_m$. Since $g_1, \ldots, g_k$ lie in the center of G and $h_1, \ldots, h_m$ skew-commute, we obtain

$$\bar{w} = a_0(\bar{y}_1, \ldots, \bar{y}_k)\bar{z}_{\sigma(1)} \cdots \bar{z}_{\sigma(m)} a_m(\bar{y}_1, \ldots, \bar{y}_k) \otimes g_1 \cdots g_k h_{\sigma(1)} \cdots h_{\sigma(m)}$$
$$= (\operatorname{sgn}\sigma) a_0(\bar{y}_1, \ldots, \bar{y}_k)\bar{z}_{\sigma(1)} \cdots \bar{z}_{\sigma(m)} a_m(\bar{y}_1, \ldots, \bar{y}_k) \otimes g_1 \cdots g_k h_1 \cdots h_m.$$

From this equality, by the definition of $\widetilde{\ }$ it follows that

$$f(\bar{y}_1 \otimes g_1, \ldots, \bar{y}_k \otimes g_k, \bar{z}_1 \otimes h_1, \ldots, \bar{z}_m \otimes h_m)$$
$$(3.16) \qquad = \widetilde{f}(\bar{y}_1, \ldots, \bar{y}_k, \bar{z}_1, \ldots, \bar{z}_m) \otimes g_1 \cdots g_k h_1 \cdots h_m.$$

Since $\bar{y}_1, \ldots, \bar{z}_m, g_1, \ldots, h_m$ are arbitrary homogeneous elements, we conclude that f is equal to zero identically on $G(A)$ if and only if $\widetilde{f} \equiv 0$ on A. This completes the proof of the lemma. $\square$

Now let $\mathcal{V}$ be a variety of algebras. Denote by $\mathcal{V}^*$ the class of all superalgebras $A = A^{(0)} \oplus A^{(1)}$ such that $G(A) \in \mathcal{V}$. Recalling that char $F = 0$, we have

THEOREM 3.7.5. *For any variety of algebras $\mathcal{V}$ the class $\mathcal{V}^*$ is a supervariety.*

PROOF. Denote by S the set of all multilinear identities of $\mathcal{V}$. Since char $F = 0$, some algebra B lies in $\mathcal{V}$ if and only if $g \equiv 0$ on B for all $g \in S$.

Now let $A = A^{(0)} \oplus A^{(1)}$ be a superalgebra and $B = G(A) = (A^{(0)} \otimes G^{(0)}) \oplus (A^{(1)} \otimes G^{(1)})$. As we remarked above, B is also $\mathbb{Z}_2$-graded, $B^{(0)} = A^{(0)} \otimes G^{(0)}$, $B^{(1)} = A^{(1)} \otimes G^{(1)}$. Hence B satisfies a multilinear identity $g \equiv 0$ of degree n if and only if B satisfies the family of 2^n graded identities $f \equiv 0$ where any f is obtained from g by replacing some k ($0 \leq k \leq n$) variables with even indeterminates and all other $n - k$ variables with odd indeterminates.

By Lemma 3.7.4, $B = G(A)$ satisfies a multilinear graded identity $f \equiv 0$ if and only if A satisfies a graded identity $\widetilde{f} \equiv 0$. This means that we can start with multilinear identities of the variety $\mathcal{V}$ and construct the set of graded identities W such that $G(A) \in \mathcal{V}$ if and only if $h \equiv 0$ on A for all $h \in W$. This completes the proof of the theorem. $\square$

For later use we record the following obvious consequence.

REMARK 3.7.6. If the superalgebras $A = A^{(0)} \oplus A^{(1)}$ and $B = B^{(0)} \oplus B^{(1)}$ have the same graded identities, then $G(A)$ and $G(B)$ have the same ordinary identities.

The Grassmann envelope plays an important role in PI-theory since it allows to relate ordinary identities and superidentities in a precise way. It will be clear from the following chapters that it is a basic tool for studying varieties that cannot be generated by any finitely generated algebra. A basic theorem in this setting is a theorem of Kemer stating that any variety can be generated by the Grassmann envelope of a finite dimensional superalgebra. We shall prove a weaker version of this theorem in the next chapter (Theorem 4.8.2).

A connection between the classification of the finite dimensional simple superalgebras and the Grassmann envelope is due to the following classification of Kemer of verbally prime T-ideals.

DEFINITION 3.7.7. A T-ideal I is verbally prime if for any T-ideals I_1, I_2 such that $I_1 I_2 \subseteq I$ we must have $I_1 \subseteq I$ or $I_2 \subseteq I$.

It is clear that a T-ideal I is verbally prime if whenever f and g are two polynomials in distinct variables and $fg \in I$, then either $f \in I$ or $g \in I$. A variety will be called prime if the corresponding T-ideal is verbally prime. Also an algebra A will be called verbally prime if it generates a prime variety. With this terminology at hand, we can state the following.

THEOREM 3.7.8 (Kemer [**Ke3**]). *Let $\mathcal{V}$ be a proper variety of algebras over a filed of characteristic zero. Then $\mathcal{V}$ is prime if and only if $\mathcal{V} = \mathrm{var}(G(A))$ where A is a finite dimensional simple superalgebra.*

In light of Theorem 3.5.4 it turns out that a proper prime variety must be generated by one of the algebras $M_k(F)$ or $M_k(G)$ or $M_{k,l}(G)$, $(0 < l \leq [k/2])$, where

$$M_{k,l}(G) = \begin{matrix} k \\ l \end{matrix} \begin{pmatrix} \overset{k}{G^{(0)}} & \overset{l}{G^{(1)}} \\ G^{(1)} & G^{(0)} \end{pmatrix}.$$

3.8. Supercommutative envelopes

In this section we shall define the superenvelope or S-envelope of a superalgebra and we shall compare it to the Grassmann envelope of the superalgebra. These are useful tools that will be used in the next chapters. Suppose throughout that char $F = 0$.

DEFINITION 3.8.1. A superalgebra $A = A^{(0)} \oplus A^{(1)}$ is said to be supercommutative if for all homogeneous elements $a, b \in A$, $ab - (-1)^{(dega)(degb)}ba = 0$.

The Grassmann algebra with its usual grading $G = G^{(0)} \oplus G^{(1)}$ is an example of supercommutative superalgebra.

One defines in an obvious way a free object S of countable rank in the variety of supercommutative superalgebras as follows. Take $U = \{u_1, u_2, \ldots\}$ and $V = \{v_1, v_2, \ldots\}$, be two countable sets and define $S = F[U, V]$ to be the algebra with 1, generated by $U \cup V$ over F, subject to the conditions that the elements of U are central and the elements of V anticommute. In symbols

$$S = \langle 1, u_1, v_1, u_2, v_2, \ldots \mid u_i u_j = u_j u_i, v_i v_j = -v_j v_i, u_i v_j = v_j u_i, \ i,j = 1, 2, \ldots \rangle.$$

The algebra $S = F[U, V]$ has a natural $\mathbb{Z}_2$-grading $S = S^{(0)} \oplus S^{(1)}$ if we require the variables of U to be even and those of V to be odd.

Also S has the following universal property: given any supercommutative algebra A, any map $\varphi : \{U, V\} \to A$ such that $\varphi(U) \subseteq A^{(0)}$ and $\varphi(V) \subseteq A^{(1)}$ can be uniquely extended to a homomorphism of superalgebras $\bar{\varphi} : S \to A$.

DEFINITION 3.8.2. $S = F[U, V]$ is called the free supercommutative algebra over F on the countable sets of commuting variables U and anticommuting variables V.

If we identify the generating elements $e_1, e_2, \ldots$, of the Grassman algebra G with the elements $v_1, v_2, \ldots$, of S respectively, then G embeds into S with induced $\mathbb{Z}_2$-grading. Thus $G^{(0)}$ is spanned by all monomials in the v_i's of even length and $G^{(1)}$ is spanned by all monomials in the v_i's of odd length. We remark that $S \cong F[U] \otimes_F G$.

As in the case of the Grassmann envelope, one can define the so-called superenvelope or S-envelope of any superalgebra.

DEFINITION 3.8.3. If $A = A^{(0)} \oplus A^{(1)}$ is a superalgebra, the superenvelope of A is the superalgebra
$$S(A) = (A^{(0)} \otimes S^{(0)}) \oplus (A^{(1)} \otimes S^{(1)}).$$

Next we want to prove that for any superalgebra A, $G(A)$ and $S(A)$ have the same identities.

PROPOSITION 3.8.4. *For any superalgebra A over a field of characteristic zero, $S(A)$ and $G(A)$ satisfy the same polynomial identities.*

PROOF. As we remarked above, $G \subseteq S$ is the subalgebra generated by V. Hence $S(A) \supseteq G(A)$ and $G(A)$ satisfies all the identities of $S(A)$.

Now let $f(x_1, \ldots, x_n)$ be a multilinear polynomial which is not an identity of $S(A)$. Then there exist $a_1, \ldots, a_n \in A^{(0)} \cup A^{(1)}$ and $p_1, \ldots, p_n \in S^{(0)} \cup S^{(1)}$ such that
$$f(a_1 \otimes p_1, \ldots, a_n \otimes p_n) \neq 0.$$
We may clearly assume that $a_1, \ldots, a_r \in A^{(0)}$, $p_1, \ldots, p_r \in S^{(0)}$, $a_{r+1}, \ldots, a_n \in A^{(1)}$, $p_{r+1}, \ldots, p_n \in S^{(1)}$. Hence, recalling that the p_i's commute or anticommute between them, we can write
$$f(a_1 \otimes p_1, \ldots, a_n \otimes p_n) = b \otimes p_1 \cdots p_n$$
with $0 \neq b \in A$ and $0 \neq p_1 \cdots p_n \in S$. It is clear that the same computations show that
$$f(a_1 \otimes u_1, \ldots, a_r \otimes u_r, a_{r+1} \otimes v_1, \ldots, a_n \otimes v_{n-r}) = b \otimes u_1 \cdots u_r v_1 \cdots v_{n-r} \neq 0.$$
Since S is the free supercommutative algebra, the map $\varphi : A \otimes \{U, V\} \to A \otimes G$ such that
$$\varphi(a \otimes v_i) = a \otimes e_i, \quad \varphi(a \otimes u_i) = a \otimes e_{n+2i-1} e_{n+2i}, \quad i = 1, 2, \ldots$$
for all $a \in A$, can be extended to a homomorphism $\bar\varphi : A \otimes S \to A \otimes G$ such that $\bar\varphi(S(A)) \subseteq G(A)$. We obtain
$$\bar\varphi(f(a_1 \otimes u_1, \ldots, a_r \otimes u_r, a_{r+1} \otimes v_1, \ldots, a_n \otimes v_{n-r}))$$
$$= f(a_1 \otimes v_{n+1} v_{n+2}, \ldots, a_r \otimes v_{n+2r-1} v_{n+2r}, a_{r+1} \otimes v_1, \ldots, a_n \otimes v_{n-r})$$
$$= b \otimes v_{n+1} v_{n+2} \cdots v_{n+2r-1} v_{n+2r} v_1 \cdots v_{n-r} \neq 0.$$
Hence f is not an identity of $G(A)$ and the proof is complete. $\square$

A useful feature of the supercommutative envelope is that one can often define in a natural way generic elements in the following sense.

PROPOSITION 3.8.5. *Let $A = A^{(0)} \oplus A^{(1)}$ be a finite dimensional superalgebra over the field F. If $\{a_1, \ldots, a_k\}$ and $\{b_1, \ldots, b_t\}$ are bases of $A^{(0)}$ and $A^{(1)}$ respectively, then the subalgebra of $S(A)$ generated by the elements*
$$\xi_i = a_1 \otimes u_{i1} + \cdots + a_k \otimes u_{ik} + b_1 \otimes v_{i1} + \cdots + b_t \otimes v_{it}, \quad i = 1, 2, \ldots$$
is a relatively free algebra of the variety $var(G(A))$ with free generators $\xi_1, \xi_2, \ldots$.

PROOF. Let $f(x_1, \ldots, x_n)$ be a non-zero polynomial such that $f(\xi_1, \ldots, \xi_n) = 0$. If $c_1, \ldots, c_n \in S(A)$, then, for all $1 \le i \le n$, we can write
$$c_i = a_1 \otimes p_{i1} + \cdots + a_k \otimes p_{ik} + b_1 \otimes q_{i1} + \cdots + b_t \otimes q_{it}$$
for suitable $p_{ij} \in S^{(0)}, q_{ij} \in S^{(1)}$. As in the previous proof, there exists a homomorphism $\bar\varphi : A \otimes S \to A \otimes S$ such that
$$\bar\varphi(a_i) = a_i, \; \bar\varphi(b_j) = b_j, \; \bar\varphi(u_{ij}) = p_{ij}, \; \bar\varphi(v_{ij}) = q_{ij}, \quad 1 \le i \le k, 1 \le j \le t.$$
It follows that $f(c_1, \ldots, c_n) = \bar\varphi(f(\xi_1, \ldots, \xi_n)) = 0$ and f is an identity for the algebra $S(A)$.

This says that the algebra generated by the elements $\xi_1, \xi_2, \ldots$ is a relatively free algebra of the variety $\mathrm{var}(S(A))$. Since $\mathrm{var}(S(A)) = \mathrm{var}(G(A))$, the proof is complete. □

CHAPTER 4

Codimension and Colength Growth

In Chapter 2 we defined a natural action of the symmetric group on the space of multilinear identities in a fixed number of variables and our main objective in this setting is to understand the decomposition of the corresponding module into irreducibles.

In this chapter we introduce two numerical sequences: the sequence of codimensions and the sequence of colengths measuring, for each representation, the dimension and the number of irreducible summands. As we shall see, the explicit computation of such sequences for a given algebra is in general a difficult problem and we present here some significant examples.

The key result about the sequence of codimensions is that it is exponentially bounded (Theorem 4.2.4). Besides being interesting on its own, this can be used in general, as a tool to prove existence theorems in PI-theory. We give two examples here. The first is a celebrated theorem of Regev stating that the tensor product of two PI-algebras is a PI-algebra (Theorem 4.2.6). The second asserts that a group graded algebra whose 1-component satisfies an ordinary identity, must be a PI-algebra (Theorem 4.3.3). The best known exponential bound for the codimensions is obtained by counting certain "good" permutations. We do this by means of a well-known theorem of Dilworth. On the other hand, this notion is also encountered in the Robinson-Schensted algorithm (Theorem 4.4.2) and we present it here together with some combinatorial consequences related to our investigation.

Concerning the cocharacter sequence of a PI-algebra we present the so-called hook theorem (Theorem 4.5.1) stating the existence of a hook shaped area of the plane containing all diagrams appearing with non-zero multiplicity. This theorem becomes the so-called strip theorem (Theorem 4.6.1) when the algebra A satisfies a Capelli polynomial, e.g., when A is a finitely generated algebra.

Concerning the sequence of colengths the most important result states that this sequence is polynomially bounded (Theorem 4.9.3). This result is obtained as a consequence of the hook-theorem as well as of the theorem stating that any variety is generated by the Grassmann envelope of a finitely generated superalgebra (Theorem 4.8.2).

4.1. Codimensions and colengths

Let F be a field of characteristic zero, $F\langle X \rangle$ the free associative algebra of countable rank on $X = \{x_1, x_2, \ldots\}$ and A a PI-algebra over F. By Corollary 1.3.9, we know that $\mathrm{Id}(A)$, the T-ideal of identities of A, is generated by its multilinear polynomials. Hence, recalling that P_n denotes the space of multilinear polynomials in the first n variables $x_1, \ldots, x_n$, it follows that $\mathrm{Id}(A)$ is generated by the subspace

$$(P_1 \cap \mathrm{Id}(A)) \oplus (P_2 \cap \mathrm{Id}(A)) \oplus \cdots \oplus (P_n \cap \mathrm{Id}(A)) \oplus \cdots$$

in the free associative algebra $F\langle X \rangle$. It is clear that if A satisfies all the identities of some other PI-algebra B, then $P_n \cap \mathrm{Id}(A) \supseteq P_n \cap \mathrm{Id}(B)$ and $\dim(P_n \cap \mathrm{Id}(A)) \geq \dim(P_n \cap \mathrm{Id}(B))$, for all $n = 1, 2, \ldots$. Therefore, the dimensions of the spaces $P_n \cap \mathrm{Id}(A)$ give us, in some sense, the growth of the identities of the algebra A. For some obvious reason that will be clear in the following sections (see Theorem 4.2.4), it is more convenient to consider the codimension of $P_n \cap \mathrm{Id}(A)$ in P_n.

DEFINITION 4.1.1. The non-negative integer
$$c_n(A) = \dim \frac{P_n}{P_n \cap \mathrm{Id}(A)}$$
is called the nth codimension of the algebra A.

Clearly, $\dim(P_n \cap \mathrm{Id}(A)) = n! - c_n(A)$.

REMARK 4.1.2. Let A be an algebra. Then A is a PI-algebra if and only if $c_n(A) < n!$, for some $n \geq 1$.

If $\mathcal{V}$ is a variety of algebras and $\mathcal{V} = \mathrm{var}(A)$ then we define $c_n(\mathcal{V}) = c_n(A)$. Next we give some examples of algebras for which the sequence of codimensions can be easily computed.

EXAMPLE 4.1.3. Let A be a nilpotent algebra and let $A^m = 0$. Then $c_n(A) = 0$, for all $n \geq m$.

EXAMPLE 4.1.4. If A is commutative then $c_n(A) \leq 1$, for all $n \geq 1$.

We next present one more example in which we can compute explicitly the sequence of codimensions ([**Ma2**]).

THEOREM 4.1.5. *For the algebra $UT_2(F)$ of 2×2 upper triangular matrices over the field F, $\mathrm{char}\, F = 0$, the following conditions hold.*

1) *The T-ideal of identities of $UT_2(F)$ is generated by the polynomial*
$$[x_1, x_2][x_3, x_4].$$
2) $c_n(A) = 2^{n-1}(n-2) + 2.$

PROOF. First notice that $A = UT_2(F)$ satisfies the identity
$$[x_1, x_2][x_3, x_4] \equiv 0. \tag{4.1}$$
We shall prove that $[x_1, x_2][x_3, x_4]$ generates $\mathrm{Id}(A)$ as a T-ideal and we shall compute the sequence of codimensions of A by exhibiting a basis of P_n (mod $P_n \cap \mathrm{Id}(A)$).

For any $r \geq 2$ define $[x_1, x_2, \ldots, x_r] = [[[x_1, x_2], x_3], \ldots, x_r]$, the left normed Lie commutator of $x_1, \ldots, x_r$.

Consider the set of polynomials in P_n of the type
$$x_{i_1} \cdots x_{i_m}[x_k, x_{j_1}, \ldots, x_{j_{n-m-1}}] \tag{4.2}$$
where
$$\{i_1, \ldots, i_m, j_1, \ldots, j_{n-m-1}, k\} = \{1, 2, \ldots, n\},$$
and
$$i_1 < \ldots < i_m, \quad j_1 < \ldots < j_{n-m-1}, \quad k > j_1, \quad m \neq n-1. \tag{4.3}$$
We claim that such polynomials span P_n (mod $P_n \cap Q$) where $Q = \langle [x_1, x_2][x_3, x_4] \rangle_T$, the T-ideal generated by $[x_1, x_2][x_3, x_4]$.

First note that for any $a_1, \ldots, a_k, b_1, \ldots, b_m, x, y \in F\langle X\rangle$ we have

$$a_{\sigma(1)} \cdots a_{\sigma(k)}[x,y]b_{\tau(1)} \cdots b_{\tau(m)} \equiv a_1 \cdots a_k[x,y]b_1 \cdots b_m \pmod{Q}.$$

It follows that, modulo Q, any multilinear polynomial can be written as a linear combination of polynomials of the type

$$x_{i_1} \cdots x_{i_k}[x_{j_1}, \ldots, x_{j_m}]$$

with $i_1 < \ldots < i_k$, $m + k = n$. Note also that any left normed commutator satisfies the relation

$$[x_{j_1}, x_{j_2}, \ldots, x_{j_r}, x_{j_{r+1}}, \ldots] \equiv [x_{j_1}, x_{j_2}, \ldots, x_{j_{r+1}}, x_{j_r}, \ldots] \pmod{Q}$$

since $[[a,b],[c,d]] = [a,b][c,d] - [c,d][a,b] \in Q$ for any $a, b, c, d \in F\langle X\rangle$. This proves the claim.

We next show that the elements (4.2) are linearly independent modulo the T-ideal $\mathrm{Id}(A)$. In fact, let $I = \{i_1, \ldots, i_m\}$, $J = \{j_1, \ldots, j_{n-m-1}\}$, $X_{I,J,k} = x_{i_1} \cdots x_{i_m}[x_k, x_{j_1}, \ldots, x_{j_{n-m-1}}]$ and suppose that a linear combination of elements of type (4.2) is zero (mod $\mathrm{Id}(A)$). Then, for some $\alpha_{I,J,k} \in F$,

$$f = f(x_1, \ldots, x_n) = \sum_{I,J,k} \alpha_{I,J,k} X_{I,J,k} \in \mathrm{Id}(A)$$

and we wish to show that all coefficients $\alpha_{I,J,k}$ are zero.

If we replace $x_{i_1}, \ldots, x_{i_m}$ with the unit matrix $E \in UT_2(F)$, x_k with the matrix unit e_{12} and all remaining $x_{j_1}, \ldots, x_{j_{n-m-1}}$ with e_{22}, then $X_{I,J,k}$ evaluates to e_{12}. All other $X_{I,J,k}$ with $(I', J', k') \neq (\{i_1, \ldots, i_m, j_1, \ldots, j_{n-m-1}, k\})$ take zero value. Hence $f \equiv 0$ is an identity of $UT_2(F)$ only if all coefficients $\alpha_{I,J,k}$ are zero. This shows that the elements (4.2) are linearly independent modulo $P_n \cap \mathrm{Id}(A)$. Since

$$P_n \supseteq P_n \cap \mathrm{Id}(A) \supseteq P_n \cap Q,$$

it follows that $\mathrm{Id}(A) = Q$ and the elements (4.2) form a basis of P_n modulo $P_n \cap \mathrm{Id}(A)$.

We now count for any fixed n, the total number of elements in (4.2) subject to condition (4.3), i.e., the nth codimension $c_n(A)$. If $0 \leq m \leq n-2$, then this number is equal to

$$\binom{n}{m}(n-m-1) = \binom{n}{n-m}(n-m-1).$$

If $m = n$, we have exactly one monomial $x_1 \cdots x_n$. Hence

$$c_n(A) = \sum_{j=2}^{n} \binom{n}{j}(j-1) + 1 = \sum_{j=2}^{n} j\binom{n}{j} - \sum_{j=2}^{n} \binom{n}{j} + 1$$

$$= n2^{n-1} - \binom{n}{1} - 2^n + \binom{n}{1} + \binom{n}{0} + 1 = n2^{n-1} - 2^n + 2 = 2^{n-1}(n-2) + 2.$$

$\square$

As a final example, in the next theorem we shall compute the codimensions of the Grassmann algebra.

It is in general a very difficult technical task to compute the precise values of the codimensions for a given algebra and, so far, they have been computed only for very few PI-algebras.

Since the sequence of codimensions measures, in some way, the rate of growth of the identities of a given algebra, it makes sense to try to estimate the codimensions asymptotically. In the next chapter we shall do this for the codimensions of the matrix algebra $M_k(F)$.

There is another numerical sequence that can be attached to the identities of an algebra A and this is the sequence of colengths defined as follows. Recall that, given a PI-algebra A, the symmetric group S_n acts on the space of multilinear polynomials P_n by permuting the variables and this in turn induces a structure of S_n-module to the space $P_n(A) = P_n/(P_n \cap \mathrm{Id}(A))$ (see Section 2.4). In characteristic zero, the character of this module is the nth cocharacter $\chi_n(A)$ of A. Thus assume that F has characteristic zero and let

$$\chi_n(A) = \sum_{\lambda \vdash n} m_\lambda \chi_\lambda$$

be the decomposition of $\chi_n(A)$ into irreducible S_n-characters. Then we have

DEFINITION 4.1.6.

$$l_n(A) = \sum_{\lambda \vdash n} m_\lambda$$

is called the nth colength of A.

In other words, $l_n(A)$ counts the number of irreducible S_n-modules appearing in the decomposition of $P_n(A)$. As in the case of codimensions, if $\mathcal{V}$ is a variety of algebras, we write $l_n(\mathcal{V}) = l_n(A)$ where A is an algebra generating $\mathcal{V}$.

EXAMPLE 4.1.7. If A is a commutative non-nilpotent algebra, then

$$P_n = \mathrm{span}\{x_1 \cdots x_n\} \pmod{P_n \cap \mathrm{Id}(A)}.$$

Hence $c_n(A) = 1$ and $l_n(A) = 1$, for all n.

We now present one more instance in which we can compute precisely the sequences of codimensions and the colengths of an algebra. Recall that for integers $d, l \geq 0$, $H(d, l)$ denotes the infinite hook whose hand is d and foot is l (see Section 2.5). In particular, if λ is any partition of $n \geq 1$, then $\lambda \subset H(1,1)$ if

$$\lambda = (p, \underbrace{1, \ldots, 1}_{n-p}) = (p, 1^{n-p}), \quad p \geq 1.$$

THEOREM 4.1.8 ([KR], [OR]). *For the infinite dimensional Grassmann algebra G over a field of characteristic zero the following conditions hold:*

1) *The T-ideal of identities of G is generated by the polynomial $[[x_1, x_2], x_3]$;*
2) $c_n(G) = 2^{n-1}$;
3) $\chi_n(G) = \sum_{\substack{\lambda \vdash n \\ \lambda \subset H(1,1)}} \chi_\lambda$;
4) $l_n(G) = n$.

PROOF. Recall that G is generated by the countable set $\{e_1, e_2, \ldots\}$ and G has a structure of superalgebra $G = G^{(0)} \oplus G^{(1)}$ where $G^{(0)}$ (resp. $G^{(1)}$) is the space generated by all monomials in the e_i's of even (resp. odd) length. Also, $G^{(0)}$ is the center of G, $[G, G] \subseteq G^{(0)}$, so G satisfies the identity

(4.4) $$[[x_1, x_2], x_3] \equiv 0.$$

Denote by $\mathcal{V}$ the variety determined by the identity (4.4). Then $G \in \mathcal{V}$. We claim that $\mathcal{V}$ also satisfies the identity

(4.5) $$[x,y][z,y] \equiv 0.$$

Indeed, by replacing in (4.4) x_1 with xy, x_2 with y and x_3 with z and by using the obvious equality $[xy,y] = [x,y]y$ we obtain

$$[[x,y]y,z] = [x,y]yz - z[x,y]y \equiv [x,y]yz - zy[x,y] \equiv 0.$$

On the other hand,

$$[[x,y],yz] = [x,y]yz - yz[x,y] \equiv 0.$$

Therefore $yz[x,y] \equiv zy[x,y]$ and we get (4.5). The linearization of (4.5) leads to the identity

(4.6) $$[x,y_1][z,y_2] \equiv -[x,y_2][z,y_1]$$

in $\mathcal{V}$.

From the relations (4.4), (4.6) it follows that any multilinear polynomial in $x_1, \ldots, x_n$ is a linear combination of products of the type

(4.7) $$x_{i_1} \cdots x_{i_k} [x_{j_1}, x_{j_2}] \cdots [x_{j_{2m-1}}, x_{j_{2m}}]$$

with

(4.8) $$i_1 < \ldots < i_k; \quad j_1 < j_2 \ldots < j_{2m}; \quad 2m + k = n.$$

The total number of polynomials in (4.7) subject to condition (4.8) is equal to

$$s_0 = \binom{n}{0} + \binom{n}{2} + \cdots + \binom{n}{2q}$$

where $q = [\frac{n}{2}]$ is the integer part of $\frac{n}{2}$. Denote by s_1 the sum of all binomial coefficients $\binom{n}{i}$ with odd $i \leq n$. Then

$$2^n = (1+1)^n = s_0 + s_1, \quad 0 = (1-1)^n = s_0 - s_1.$$

Hence $s_0 = s_1 = 2^{n-1}$ and we obtain the following upper bound for $c_n(\mathcal{V})$:

$$c_n(\mathcal{V}) = \dim P_n(\mathcal{V}) \leq 2^{n-1}.$$

Note that since $G \in \mathcal{V}$, then also $c_n(G) \leq c_n(\mathcal{V}) \leq 2^{n-1}$.

In order to find a lower bound for $c_n(\mathcal{V})$ we shall prove that any irreducible S_n-character χ_λ associated to a partition $\lambda \subset H(1,1)$ appears with non-zero multiplicity in $\chi_n(G)$. Indeed, let $\lambda = (k, 1^{n-k})$ be such a partition and let D_λ be the corresponding Young diagram (whose first row has length k and first column has length $n - k + 1$). We construct the Young tableau T_λ by inserting the integers $1, 2, \ldots, k$ into the boxes of the first row of D_λ from left to right and all remaining integers $k+1, \ldots, n$ into the boxes of the first column starting from the second box from top to bottom (see picture).

1	2	$\cdots$	k
$k+1$			
$\vdots$			
$n-1$			
n			

Then
$$R_{T_\lambda} = S_k \text{ and } C_{T_\lambda} = S_{n-k+1}\{1, k+1, \ldots, n\}$$
where $S_m\{i_1, \ldots, i_m\}$ denotes the symmetric group action on the set $\{i_1, \ldots, i_m\}$. Now we apply the essential idempotent e_{T_λ} to the monomial $w = x_1 \cdots x_n$. We obtain
$$w_k = e_{T_\lambda} w$$
$$= \left(\sum_{\sigma \in S_k} \sigma\right) \left(\sum_{\tau \in S_{n-k+1}\{1,k+1,\ldots,n\}} (\text{sgn } \tau) x_{\tau(1)} x_2 \cdots x_k x_{\tau(k+1)} \cdots x_{\tau(n)}\right)$$
and we claim that w_k is not an identity of G.

In fact, we replace the indeterminates $x_1, \ldots, x_k$ with
$$\bar{x}_1 = e_1, \ \bar{x}_{k+1} = e_2, \ \ldots, \bar{x}_n = e_{n-k+1}$$
and
$$\bar{x}_2 = e_{n-k+2}e_{n-k+3}, \ \bar{x}_3 = e_{n-k+4}e_{n-k+5}, \ \ldots, \ \bar{x}_k = e_{n+k-2}e_{n+k-1},$$
respectively. Then, since the elements $\bar{x}_2, \bar{x}_3, \ldots, \bar{x}_k$ are central in G, we obtain
$$w_k = \left(\sum_{\sigma \in S_k} \sigma\right) \bar{x}_2 \cdots \bar{x}_k St_{n-k+1}(\bar{x}_1, \bar{x}_{k+1}, \ldots, \bar{x}_n)$$
where
$$St_r(x_1, \ldots, x_r) = \sum_{\tau \in S_r} (\text{sgn } \tau) x_{\tau(1)} \cdots x_{\tau(r)}$$
is the standard polynomial of degree r. Then, since
$$St_{n-k+1}(\bar{x}_1, \bar{x}_{k+1}, \ldots, \bar{x}_n)$$
$$= (n-k+1)! \bar{x}_1 \bar{x}_{k+1} \cdots \bar{x}_n = (n-k+1)! e_1 \cdots e_{n-k+1} \neq 0,$$
we obtain
$$w_k(\bar{x}_1, \ldots, \bar{x}_n) = (n-k+1)! \left(\sum_{\sigma \in S_k} \sigma\right) \bar{x}_2 \cdots \bar{x}_k \bar{x}_1 \bar{x}_{k+1} \cdots \bar{x}_n$$
$$= (n-k+1)! k! \bar{x}_1 \cdots \bar{x}_k \bar{x}_{k+1} \cdots \bar{x}_n = (n-k+1)! k! e_1 e_2 \cdots e_{n+k-1} \neq 0.$$

We have proved that w_k is not an identity of G. This implies that in the decomposition
$$(4.9) \qquad \chi_n(G) = \sum_{\mu \vdash n} m_\mu \chi_\mu$$
all multiplicities m_λ for $\lambda = \lambda^{(k)} = (k, 1^{n-k}), k = 1, \ldots, n$, are greater or equal to one and
$$c_n(G) \geq \sum_{k=1}^{n} \deg \chi_{\lambda^{(k)}}.$$
By the hook formula (Proposition 2.2.8)
$$\deg \chi_{\lambda(k)} = \frac{n!}{(k-1)!(n-k)!n} = \frac{(n-1)!}{(k-1)!((n-1)-(k-1))!} = \binom{n-1}{k-1}$$

and we get the lower bound

$$c_n(G) \geq \sum_{k=1}^{n} \binom{n-1}{k-1} = 2^{n-1}.$$

Since $\mathrm{var}(G)$ is a subvariety of $\mathcal{V}$ we conclude that $c_n(G) = 2^{n-1}$ and $\mathrm{var}(G) = \mathcal{V}$, i.e., all identities of G follow from (4.4). Thus, all characters $\chi_{\lambda(k)}$, $k = 1, 2, \ldots, n$, have non-zero multiplicity. Clearly, 4) follows from 3) and the proof is complete. $\square$

It is worth noting that from the proof of the previous theorem it follows that all polynomials (4.7) form a basis of P_n (mod $P_n \cap \mathrm{Id}(G)$).

We conclude this section by showing that for any PI-algebra A over F, the codimensions do not change upon extension of the base field F provided F is infinite. In the case of characteristic zero, the same conclusion can be drawn about the multiplicities in the cocharacters and the colengths. Let K be a field extension of F and let us consider the K-algebra $\bar{A} = A \otimes_F K$. By Lemma 1.4.2, if F is infinite, the algebra $\bar{A}$, viewed as an F-algebra, satisfies the same multilinear identities as A. On the other hand, $\bar{A}$ is a K-algebra and one can consider identities with coefficients in K. Denote by $c_n^K(\bar{A})$ and $l_n^K(\bar{A})$ the nth codimension and the nth colength of $\bar{A}$ as a K-algebra, respectively. Let also $\chi_\lambda(A) = \sum_{\lambda \vdash n} m_\lambda \chi_\lambda$ be the nth cocharacter of A and $\chi_\lambda(\bar{A}) = \sum_{\lambda \vdash n} m_\lambda^K \chi_\lambda$ the nth cocharacter of $\bar{A}$ as a K-algebra.

THEOREM 4.1.9. *If A is an algebra over the infinite field F and K is an extension field of F, then*

$$c_n^K(\bar{A}) = c_n(A), \; n = 1, 2, \ldots .$$

If char $F = 0$, *then* $m_\lambda = m_\lambda^K$ *also for all* $\lambda \vdash n$, *and* $l_n^K(\bar{A}) = l_n(A)$, $n = 1, 2, \ldots .$

PROOF. Let $c_n(A) = m$ and fix a basis $f_1, \ldots, f_m$ of P_n (mod $P_n \cap \mathrm{Id}(A)$). Then any non-trivial linear combination $\alpha_1 f_1 + \cdots + \alpha_m f_m$ is still not an identity of A. On the other hand, for any $f \in P_n$ one can find $\beta_1, \ldots, \beta_m \in F$ such that

$$h = h(x_1, \ldots, x_n) = f - \sum_i \beta_i f_i \equiv 0$$

is an identity of A. Hence $h(x_1, \ldots, x_n)$ is an identity of $\bar{A}$ with coefficients in K. It follows that $c_n^K(\bar{A}) \leq c_n(A)$.

Let us verify now that $f_1, \ldots, f_m$ are linearly independent (mod $\mathrm{Id}(\bar{A})$) over K. Assume that

$$f = \beta_1 f_1 + \cdots + \beta_m f_m \equiv 0$$

is an identity of $\bar{A}$ with $\beta_1, \ldots, \beta_m \in K$. Choose a basis $\mathcal{B}$ of the vector space K over F. Then there exist scalars $\gamma_{ij} \in F$ such that

$$\beta_i = \sum_j \gamma_{ij} t_j, \quad i = 1, \ldots, m, \quad t_1, t_2, \ldots \in \mathcal{B}.$$

Now we evaluate $f = f(x_1, \ldots, x_n)$ on the tensors $a_1 \otimes 1, \ldots, a_n \otimes 1$ with $a_1, \ldots, a_n \in A$. We obtain

$$0 = f(a_1 \otimes 1, \ldots, a_n \otimes 1) = \sum_{i,j} \gamma_{ij} t_j f_i(a_1 \otimes 1, \ldots, a_n \otimes 1)$$

$$= \sum_j \left(\sum_i \gamma_{ij} f_i(a_1, \ldots, a_n) \right) \otimes t_j.$$

It follows that $\sum_i \gamma_{ij} f_i(a_1, \ldots, a_n) = 0$ for all $a_1, \ldots, a_n \in A$ and $\sum_i \gamma_{ij} f_i \equiv 0$ is an identity of $\bar{A}$, for all $j = 1, 2, \ldots$. But then by the choice of $f_1, \ldots, f_m$ all coefficients γ_{ij} should be zero, so $\beta_1 = \ldots = \beta_m = 0$. We have proved that $f_1, \ldots, f_m$ are linearly independent modulo $\mathrm{Id}(\bar{A})$. Thus $c_n^K(\bar{A}) = c_n(A)$, and the first part of the theorem is proved.

The second part of the theorem concerning the equalities $m_\lambda = m_\lambda^K$ and $l_n^K(\bar{A}) = l_n(A)$ follows immediately from the fact that in characteristic zero any irreducible S_n-representation is absolutely irreducible. □

4.2. An exponential upper bound for the codimensions

In this section we present one of the key results of this book, i.e., that the codimension sequence of a PI-algebra A over a field of any characteristic is exponentially bounded. This was proved by Regev in [**Re1**] as a tool in order to show that the tensor product of two PI-algebras is still a PI-algebra. As we shall see later, this can be used as a tool in order to prove many other existence theorems in PI-theory. The proof we present here, based on a paper of Latyshev ([**L1**]), is a modification of the original one and provides the best known upper bound to the sequence of codimensions. This result will be the starting point of the investigation of the next chapters where we shall try to estimate the exponential growth of the sequence of codimensions.

We start with some combinatorial properties of the permutation group S_n. Let $2 \leq d \leq n$ be an integer.

DEFINITION 4.2.1. A permutation $\sigma \in S_n$ is d-bad if there exist $1 \leq i_1 < \cdots < i_d \leq n$ such that $\sigma(i_1) > \cdots > \sigma(i_d)$. If σ is not d-bad, we say that σ is d-good.

In other words, σ is d-good if for any $1 \leq i_1 < \ldots < i_d \leq n$, there exists a pair of indices $1 \leq k < l \leq d$ such that $\sigma(i_k) < \sigma(i_l)$. Our next goal is to find an appropriate upper bound to the number of d-good permutations in the symmetric group S_n. We do this by using a well-known result of Dilworth. Denote for short $\mathbf{n} = \{1, \ldots, n\}$.

LEMMA 4.2.2. Let $\sigma \in S_n$ be a d-good permutation. Then there exists $k \leq d-1$ such that $\mathbf{n}$ can be decomposed into a disjoint union

$$\mathbf{n} = I_1 \cup \ldots \cup I_k$$

with the following property: for any $j = 1, \ldots, k$ and for any $a, b \in I_j$ with $a < b$ we have $\sigma(a) < \sigma(b)$.

PROOF. Define a partial ordering $<<$ on $\mathbf{n}$ by setting $x << y$ if $x < y$ and $\sigma(x) < \sigma(y)$. Since σ is d-good, any subset $\{x_1, \ldots, x_d\}$ of $\mathbf{n}$ with d-elements contains at least one pair x_i, x_j of comparable elements, i.e., $x_i << x_j$. By [**Ha**, Lemma 7.2.1], the partially ordered set $\mathbf{n}$ can be decomposed into the disjoint union of at most $d-1$ chains, i.e., $\mathbf{n} = I_1 \cup \ldots \cup I_k$ with $k \leq d-1$, with the following property: for any $j = 1, \ldots, k$, all elements of I_j are linearly ordered with respect to the order $<<$. From the definition of this order it follows that for any $a, b \in I_j$ with $a < b$, we must have $\sigma(a) < \sigma(b)$. □

As a consequence we immediately get an upper bound of the number of d-good permutations of S_n.

LEMMA 4.2.3. *The total number of d-good permutations in S_n does not exceed $(d-1)^{2n}$.*

PROOF. The number of decompositions of $\mathbf{n} = I_1 \cup \ldots \cup I_k$ into $k \leq d-1$ disjoint subsets is bounded by $(d-1)^n$. Now we fix a decomposition $\mathbf{n} = I_1 \cup \ldots \cup I_k$ and we estimate the number of permutations σ which preserve the natural order of the integers in each set I_j, $j = 1, \ldots, k$. Let $L_1 = \sigma(I_1), \ldots, L_k = \sigma(I_k)$. Clearly $\mathbf{n} = L_1 \cup \ldots \cup L_k$ and $\sigma(1) \in \{x_1, \ldots, x_k\}$ where $x_j = \min\{a \mid a \in L_j\}$. If we denote $L'_t = L_t \setminus \{\sigma(1)\}, t = 1, \ldots, k$, then $\sigma(2) \in \{y_1, \ldots, y_k\}$ where $y_j = \min\{a \mid a \in L'_j\}$, and so on. From this it follows that the number of permutations preserving the order in each I_j is bounded by $k^n \leq (d-1)^n$. Applying Lemma 4.2.2 we get the required upper bound for the d-good permutations, and the proof of the lemma is complete. □

The notion of d-good permutation will be exploited in order to find an upper bound to the codimension growth of a PI-algebra. Let us say that a monomial $x_{\sigma(1)} \cdots x_{\sigma(n)} \in P_n$ is d-good (d-bad) if the corresponding permutation $\sigma \in S_n$ is d-good (d-bad resp.).

THEOREM 4.2.4 (Regev [**Re1**]). *If the algebra A satisfies an identity of degree $d \geq 1$, then $c_n(A) \leq (d-1)^{2n}$.*

PROOF. We may clearly assume that A satisfies a multilinear identity of degree d of the form
$$x_1 \cdots x_d - \sum_{\substack{\sigma \in S_d \\ \sigma \neq 1}} \alpha_\sigma x_{\sigma(1)} \cdots x_{\sigma(d)} \equiv 0.$$
We claim that the space P_n is spanned (mod $P_n \cap \mathrm{Id}(A)$) by the d-good monomials $x_{\pi(1)} \cdots x_{\pi(n)}$. Suppose not and, among all monomials that cannot be written as a linear combination of d-good monomials (mod $P_n \cap \mathrm{Id}(A)$), pick one $x_{i_1} \cdots x_{i_n} = f$ which is minimal in the left lexicographic order. Then, in particular, the permutation
$$\rho = \begin{pmatrix} 1 & 2 & \ldots & n \\ i_1 & i_2 & \ldots & i_n \end{pmatrix}$$
is d-bad. Hence there exist $1 \leq j_1 < \cdots < j_d \leq n$ such that $\rho(j_1) > \cdots > \rho(j_d)$. Denote
$$a_0 = x_{i_1} \cdots x_{\rho(j_1 - 1)}, \ a_1 = x_{\rho(j_1)} \cdots x_{\rho(j_2 - 1)}, \ \ldots, \ a_d = x_{\rho(j_d)} \cdots x_{i_n}.$$
Then $a_1 > \cdots > a_d$ in the left lexicographic order of monomials, so
$$a_0 a_{\sigma(1)} \cdots a_{\sigma(d)} < a_0 a_1 \cdots a_d = f$$
for any $\sigma \in S_d, \sigma \neq 1$. By the minimality of f it follows that any monomial $a_0 a_{\sigma(1)} \cdots a_{\sigma(d)}$ with $\sigma \neq 1$ is a linear combination (mod $P_n \cap \mathrm{Id}(A)$) of d-good monomials $x_{\pi(1)} \cdots x_{\pi(n)}$. But then, since
$$f = \sum_{\substack{\sigma \in S_d \\ \sigma \neq 1}} \alpha_\sigma a_0 a_{\sigma(1)} \cdots a_{\sigma(d)} \ (\mathrm{mod}\ P_n \cap \mathrm{Id}(A)),$$
we obtain that f is a linear combination (mod $P_n \cap \mathrm{Id}(A)$) of d-good monomials, a contradiction.

We have proved that P_n is spanned modulo $P_n \cap \text{Id}(A)$ by d-good monomials. Hence, by Lemma 4.2.3, $c_n(A)$ is bounded by $(d-1)^{2n}$. □

As an application of the previous theorem we present two results. The first is concerned with the identities of the tensor product of two PI-algebras and we shall prove it in this section. The second result deals with group graded algebras and will be presented in the next section. The next theorem was proved by Regev in [**Re1**].

THEOREM 4.2.5. *Let A and B be two PI-algebras over an arbitrary field F. Then $c_n(A \otimes B) \leq c_n(A) c_n(B)$, for all $n \geq 1$.*

PROOF. Let $\bar{A}$ and $\bar{B}$ be the relatively free algebras of countable rank of the varieties $\text{var}(A)$ and $\text{var}(B)$, respectively. Denote by $\{u_1, u_2, \ldots\}$ a set of free generators of $\bar{A}$ and $\{v_1, v_2, \ldots\}$ a set of free generators of $\bar{B}$. Write $p = c_n(A), q = c_n(B)$. Let $m_1, \ldots, m_p \in \bar{A}$ be linearly independent multilinear monomials in the elements $u_1, \ldots, u_n$ and let $r_1, \ldots, r_q \in \bar{B}$ be linearly independent multilinear monomials in $v_1, \ldots, v_n$. Then in $\bar{A} \otimes \bar{B}$ any multilinear polynomial

$$f(u_1 \otimes v_1, \ldots, u_n \otimes v_n)$$

can be written as a linear combination of $m_i \otimes r_j$, $1 \leq i \leq p, 1 \leq j \leq q$. Hence, for any multilinear elements $f_1, \ldots, f_{pq+1}$ one can find non-trivial scalars $\alpha_1, \ldots, \alpha_{pq+1} \in F$ such that

$$(\alpha_1 f_1 + \cdots + \alpha_{pq+1} f_{pq+1})(u_1 \otimes v_1, \ldots, u_n \otimes v_n) = 0$$

in $\bar{A} \otimes \bar{B}$. Write $g = \alpha_1 f_1 + \cdots + \alpha_{pq+1} f_{pq+1}$. By the definition of relatively free algebra and by the properties of the tensor product, any map $u_i \to a_i \in A$, $v_i \to b_i \in B, i = 1, \ldots, n$, can be extended to a homomorphism $\varphi : \bar{A} \otimes \bar{B} \to A \otimes B$. It follows that

$$g(a_1 \otimes b_1, \ldots, a_n \otimes b_n) = 0$$

for all $a_1, \ldots, a_n \in A, b_1, \ldots, b_n \in B$. Since g is multilinear, g is an identity of $A \otimes B$. Hence the dimension of P_n modulo $P_n \cap \text{Id}(A \otimes B)$ does not exceed $pq = c_n(A) c_n(B)$ and the proof of the theorem is complete. □

As an immediate consequence we obtain.

THEOREM 4.2.6 (Regev [**Re1**]). *If A and B are two PI-algebras, then $A \otimes B$ is also a PI-algebra.*

PROOF. By Theorem 4.2.4 there exist integers d and l such that $c_n(A) \leq d^n$ and $c_n(B) \leq l^n$ for all $n \geq 1$. Then

$$c_n(A \otimes B) \leq (dl)^n,$$

for all n. Since for any k, $n! > k^n$ for n large enough, there exists m such that $c_m(A \otimes B) < m!$, i.e., $A \otimes B$ satisfies a non-trivial multilinear identity of degree m. □

4.3. Identities of graded algebras

As a second application of Theorem 4.2.4, in this section we prove a theorem about the ordinary identities of a group graded algebra over a field of arbitrary characteristic. Let H be an arbitrary finite group and let $A = \oplus_{h \in H} A^{(h)}$ be an H-graded algebra over the field F. Recall (see Section 3.3) that if $F\langle X \rangle$ is the free H-graded algebra of countable rank over F, then the set X decomposes into the disjoint union of $m = |H|$ countable sets

$$X = \cup_{h \in H} X^{(h)}$$

where $X^{(h)} = \{x_1^{(h)}, x_2^{(h)}, \ldots\}$. Then $F\langle X \rangle$ is endowed with the H-grading

$$F\langle X \rangle = \bigoplus_{h \in H} F\langle X \rangle^{(h)}$$

where $F\langle X \rangle^{(h)}$ is spanned by all monomials of the type $x_{i_1}^{(g_1)} x_{i_2}^{(g_2)} \cdots x_{i_t}^{(g_t)}$ such that $g_1 g_2 \cdots g_t = h$. Recall that a polynomial $f = f(x_1, \ldots, x_n)$ is a graded polynomial identity of A if $f(a_1, \ldots, a_n) = 0$ for any homogeneous $a_1 \in A^{(h_1)}, \ldots, a_n \in A^{(h_n)}$ provided that $x_1 \in X^{(h_1)}, \ldots, x_n \in X^{(h_n)}$. Also $\mathrm{Id}^{gr}(A)$ denotes the ideal of $F\langle X \rangle$ of all graded identities of A and $\mathrm{Id}^{gr}(A)$ is stable under all graded endomorphisms of $F\langle X \rangle$.

As in the ordinary case it can be easily proved that any non-trivial graded identity has non-trivial graded multilinear consequences and if $\mathrm{char}\, F = 0$, $\mathrm{Id}^{gr}(A)$ is uniquely determined by all the multilinear polynomials it contains. Notice that for an algebra A to satisfy a graded identity is a much weaker condition than satisfying an ordinary polynomial identity. For instance, any algebra B can be considered as a graded algebra with trivial grading ($B = B_1$, where $1 \in H$ is the unit element of H) and it satisfies any identity of the type $x \equiv 0$, $x \in X^{(h)}$, $h \neq 1$. Nevertheless, as we shall prove below, any non-trivial identity of the component A_e implies the existence of an ordinary polynomial identity on $A = \oplus_{h \in H} A^{(h)}$ provided H is a finite group. The main tool for proving this result is Theorem 4.2.4.

Recalling that the indeterminates from $X^{(h)}$ are denoted $x_i^{(h)}$, then any multilinear graded polynomial can be written as

$$f = f(x_1^{(h_1)}, \ldots, x_n^{(h_n)}) = \sum_{\sigma \in S_n} \alpha_\sigma x_{\sigma(1)}^{(h_{\sigma(1)})} \cdots x_{\sigma(n)}^{(h_{\sigma(n)})}.$$

For a fixed n-tuple $(h_1, \ldots, h_n) \in H^n$, all multilinear polynomials in the variables $x_1^{(h_1)}, \ldots, x_n^{(h_n)}$ form the subspace

$$P_n^{(h_1, \ldots, h_n)} = \mathrm{span}_F \{x_{\sigma(1)}^{(h_{\sigma(1)})} \cdots x_{\sigma(n)}^{(h_{\sigma(n)})} \mid \sigma \in S_n\}$$

and $\dim P_n^{(h_1, \ldots, h_n)} = n!$. The intersection

$$P_n^{(h_1, \ldots, h_n)} \cap \mathrm{Id}^{gr}(A)$$

consists of all multilinear graded identities of A in the variables $x_1^{(h_1)}, \ldots, x_n^{(h_n)}$. There is an obvious way of relating ordinary identities and graded identities of the algebra A. In fact, for every $i = 1, 2, \ldots$, define

$$\bar{x}_i = \sum_{h \in H} x_i^{(h)}.$$

It is clear that the set $\{\bar{x}_1, \bar{x}_2, \ldots\}$ generates the free associative algebra of countable rank. Moreover, given a polynomial $f(x_1, \ldots, x_n) \in F\langle X \rangle$, f is an ordinary polynomial identity of A if and only if

$$f(\bar{x}_1, \ldots, \bar{x}_n) \in \mathrm{Id}^{gr}(A).$$

This basic observation implies the following technical result. For $(h_1, \ldots, h_n) \in H^n$,

$$c_n^{(h_1,\ldots,h_n)}(A) = \dim \frac{P_n^{(h_1,\ldots,h_n)}}{P_n^{(h_1,\ldots,h_n)} \cap \mathrm{Id}^{gr}(A)}$$

is called the homogeneous nth codimension associated to $(h_1, \ldots, h_n)$.

LEMMA 4.3.1. *The homogeneous codimensions $c_n^{(h_1,\ldots,h_n)}(A)$ and the ordinary codimensions $c_n(A)$ satisfy the inequality*

$$c_n(A) \leq \sum_{(h_1,\ldots,h_n) \in H^n} c_n^{(h_1,\ldots,h_n)}(A).$$

PROOF. For any n-tuple $(h_1, \ldots, h_n) \in H^n$, fix a basis $B^{(h_1,\ldots,h_n)}$ of $P_n^{(h_1,\ldots,h_n)}$ (mod $\mathrm{Id}^{gr}(A)$) and let

$$B = \bigcup_{(h_1,\ldots,h_n) \in H^n} B^{(h_1,\ldots,h_n)}.$$

Then clearly $|B^{(h_1,\ldots,h_n)}| = c_n^{(h_1,\ldots,h_n)}(A)$ and

$$k_n = |B| = \sum_{(h_1,\ldots,h_n) \in H^n} c_n^{(h_1,\ldots,h_n)}(A).$$

Hence if W_n is the subspace spanned by B, $\dim W_n \leq k_n$. On the other hand, any multilinear polynomial in $\bar{x}_1, \ldots, \bar{x}_n$ lies in $W_n + \mathrm{Id}^{gr}(A)$. It follows that any $k_n + 1$ multilinear polynomials in $\bar{x}_1, \ldots, \bar{x}_n$ are linearly dependent modulo $\mathrm{Id}(A)$ and $c_n(A) \leq k_n$. □

Note that the number of summands on the right-hand side is $|H|^n$, i.e., it is exponentially bounded. If we show that for all $(h_1, \ldots, h_n)$, $c_n^{(h_1,\ldots,h_n)}(A)$ is exponentially bounded, then by the above inequality, we will get that $c_n(A)$ is also exponentially bounded; hence $c_n(A)$ will be strictly smaller than $n!$, for n sufficiently large and this implies $\mathrm{Id}(A) \neq 0$. In order to prove these inequalities we shall make use of the results on d-good permutations as in the previous section.

LEMMA 4.3.2. *Let H be a finite group and let $A = \oplus_{h \in H} A^{(h)}$ be an H-graded algebra over a field F. Suppose that $A^{(1)}$ satisfies a multilinear identity of degree d and let $t = |H|d$. For $(h_1, \ldots, h_n) \in H^n$, denote $z_1 = x_1^{(h_1)}, \ldots, z_n = x_n^{(h_n)}$. Then, if $n \geq t$, any multilinear monomial in $z_1, \ldots, z_n$ can be written (mod $\mathrm{Id}^{gr}(A)$) as a linear combination of monomials of the type*

$$z_{\sigma(1)} \cdots z_{\sigma(n)}$$

where $\sigma \in S_n$ is a t-good permutation.

PROOF. Suppose that the conclusion of the lemma is false and among all multilinear monomials in $z_1, \ldots, z_n$ which cannot be expressed (mod $\mathrm{Id}^{gr}(A)$) as a linear

4.3. IDENTITIES OF GRADED ALGEBRAS

combination of t-good monomials, choose one $f = z_{i_1} \cdots z_{i_n}$ which is minimal in the left lexicographic order of the monomials. In particular, the permutation

$$\rho = \begin{pmatrix} 1 & 2 & \cdots & n \\ i_1 & i_2 & \cdots & i_n \end{pmatrix}$$

is not t-good; hence there exist $1 \leq j_1 < \cdots < j_t \leq n$ such that $\rho(j_1) > \cdots > \rho(j_t)$. As in the proof of Theorem 4.2.4 let $a_0 = z_{i_1} \cdots z_{\rho(j_1-1)}$, $a_1 = z_{\rho(j_1)} \cdots z_{\rho(j_2-1)}, \ldots, a_t = z_{\rho(j_t)} \cdots z_{i_n}$. Then $a_0, a_1, \ldots, a_n$ are homogeneous in the H-grading and $a_1 > \cdots > a_n$ in the left lexicographic order of monomials. Hence $a_0 a_{\sigma(1)} \cdots a_{\sigma(t)} < f = a_0 a_1 \cdots a_t$ for any $\sigma \in S_t$, $\sigma \neq 1$.

Suppose that $a_1 \in F\langle X \rangle^{(g_1)}, \ldots, a_t \in F\langle X \rangle^{(g_t)}$ and compute the products

$$w_1 = g_1, w_2 = g_1 g_2, \ldots, w_t = g_1 \cdots g_t$$

in the group H. Since $t = |H|d$, then either the unit $1 \in H$ appears at least d times among $w_1, w_2, \ldots, w_t$ or there exists $h \in H$ appearing at least $d+1$ times. In the first case $1 = w_{k_1} = w_{k_2} = \cdots = w_{k_d}$ with $1 \leq k_1 < \cdots < k_d$ and

$$b_1 = a_1 \cdots a_{k_1}, \; b_2 = a_{k_1+1} \cdots a_{k_2}, \ldots, b_d = a_{k_{d-1}+1} \cdots a_{k_d}$$

are homogeneous elements of $F\langle X \rangle^{(1)}$. In the second case $h = w_{k_0} = w_{k_1} = \cdots = w_{k_d}$ with $1 \leq k_0 < k_1 < \cdots < k_d$ and again

$$b_1 = a_{k_0+1} \cdots a_{k_1}, \ldots, b_d = a_{k_{d-1}+1} \cdots a_{k_d}$$

are homogeneous elements of $F\langle X \rangle^{(1)}$. In any case we have

(4.10) $$f = a_0 \cdots a_t = b_0 b_1 \cdots b_d b_{d+1}.$$

By hypothesis, $A^{(1)}$ satisfies a multilinear polynomial identity of degree d. Hence for some $\alpha_\sigma \in F$, we can write

$$x_1^{(1)} \cdots x_d^{(1)} \equiv \sum_{\substack{\sigma \in S_d \\ \sigma \neq 1}} \alpha_\sigma x_{\sigma(1)}^{(1)} \cdots x_{\sigma(d)}^{(1)} \pmod{\mathrm{Id}^{gr}(A)}.$$

By applying this to (4.10) we get

$$b_1 \cdots b_d \equiv \sum_{\substack{\sigma \in S_d \\ \sigma \neq 1}} \alpha_\sigma b_{\sigma(1)} \cdots b_{\sigma(d)} \pmod{\mathrm{Id}^{gr}(A)}.$$

It follows that $f = a_0 a_1 \cdots a_t$ can be written (mod $\mathrm{Id}^{gr}(A)$) as a linear combination of monomials of the type

$$a_0 a_{\sigma(1)} \cdots a_{\sigma(t)}$$

with $\sigma \neq 1$. But any such monomial is lexicographically smaller than f; hence, due to the minimality of f, it is a linear combination (mod $\mathrm{Id}^{gr}(A)$) of t-good monomials. It follows that f itself can be expressed as a linear combination (mod $\mathrm{Id}^{gr}(A)$) of t-good monomials. This contradiction completes the proof of the lemma. □

The following is the main result of this section. Let $e = 2.71\ldots$ denote the base of the natural logarithms.

THEOREM 4.3.3. *Let $A = \oplus_{h \in H} A^{(h)}$ be an algebra graded by a finite group H. If $A^{(1)}$ satisfies an identity of degree d, then, for any integer n satisfying the inequality*
$$n > e|H|(d|H| - 1)^2,$$
A satisfies a polynomial identity of degree n.

PROOF. By Lemma 4.3.2, for any $(h_1, \ldots, h_n) \in H^n$,
$$c_n^{(h_1,\ldots,h_n)}(A) = \dim \frac{P_n^{(h_1,\ldots,h_n)}}{P_n^{(h_1,\ldots,h_n)} \cap \operatorname{Id}^{gr}(A)}$$
is bounded above by the number of t-good permutations where $t = d|H|$. Hence, by Lemma 4.2.3,
$$c_n^{(h_1,\ldots,h_n)}(A) \leq (t-1)^{2n} = (d|H| - 1)^{2n}.$$
But then, by Lemma 4.3.1,
$$c_n(A) \leq |H|^n (d|H| - 1)^{2n}.$$
It is well known that
$$n! > \left(\frac{n}{e}\right)^n$$
(see for instance [**FR**, page 105]). Hence, for $n > e|H|(d|H| - 1)^2$ one has
$$n! > \left(\frac{n}{e}\right)^n > |H|^n (d|H| - 1)^{2n} \geq c_n(A).$$
Since $c_n(A)$ is strictly less that $n! = \dim P_n$, we conclude that $\operatorname{Id}(A) \cap P_n \neq 0$, and A satisfies a multilinear polynomial identity of degree n. □

The above result was proved by Bahturin, Giambruno and Riley in [**BGR**]. The existence of an ordinary identity for the algebra A was first proved by Bergen and Cohen in [**BC**]. Nevertheless, no information on the degree of the identity satisfied by A was given. This occurred since the theorem was first proved for semiprime algebras through structure theory. Then the result was extended to arbitrary algebras by making use of the so-called Amitsur trick (see [**Ro2**]). This is a powerful technique that allows passing from semiprime algebras to arbitrary algebras but that gives no information on the degree of the identities satisfied by the algebra.

As an immediate cosequence of the above theorem we get.

COROLLARY 4.3.4. *Let $A = A^{(0)} \oplus A^{(1)}$ be a superalgebra and suppose that $A^{(0)}$ satisfies a non-trivial polynomial identity. Then the whole of A is also a PI-algebra.*

Another consequence concerns the algebra of invariants under a group action (compare with [**Ka**]).

COROLLARY 4.3.5. *Let A be an algebra over a field of characteristic zero and let H be a finite solvable group acting on A by automorphisms. If $A^H = \{a \in A \mid h(a) = a \text{ for all } h \in H\}$ is a PI-algebra, then A is also a PI-algebra.*

PROOF. First note that if $K \supseteq F$ is an extension field of F, then H acts naturally on $\bar{A} = A \otimes_F K$ by $h(a \otimes k) = h(a) \otimes k$ and $A^H \subseteq \bar{A}^H = A^H \otimes K$. Hence we may assume F to be algebraically closed. We will prove the corollary by induction on the length of solubility of H.

If H is abelian then, by Theorem 3.2.1, the H-action induces an H-grading on A. Hence $A = \oplus_{h \in H} A^{(h)}$, and $A^{(1)} = A^H$, in this case. But then, by Theorem 4.3.3 A is a PI-algebra and we are done. Suppose now that $H^{(1)} = H' \neq \{1\}$ and consider the chain of subalgebras

$$A^H \subseteq A^{H^{(1)}} \subseteq \ldots \subseteq A^{H^{(i)}} \subseteq \ldots \subseteq A^{H^{(k)}} = A$$

where $H^{(i)}$ is the ith commutator subgroup of H, $H^{(k)} = 1$ and $A^{H^{(i)}}$ consists of all $H^{(i)}$-stable elements of A. Notice that if T is any normal subgroup of H, the subalgebra A^T is invariant under the H-action since $t(h(a)) = h(h^{-1}th)(a) = h(a)$ as soon as $a \in A^T$, $t \in T$. In particular, each $A_i = A^{H^{(i)}}$ has an induced H-action. Since A_i is $H^{(i)}$-stable, the restriction of this action to $H^{(i-1)}$ gives an $H_i = H^{(i-1)}/H^{(i)}$-action on A_i. But all groups $H_1, \ldots, H_{k-1}$ are abelian. Hence, by the first part of the proof, we obtain that $A_1 = A^{H^{(1)}}$ is a PI-algebra. Hence $A_2 = A^{H^{(2)}}$ is a PI-algebra, and so on. Thus at the end, $A = A_k$ is a PI-algebra and the proof is complete. $\square$

4.4. Robinson-Schensted correspondence

There is a way to relate the number of d-good permutations of S_n to the degrees of the irreducible S_n-characters lying in a strip. This is done through the Robinson-Schensted correspondence (see [**Kn**], [**St**]) that we shall describe below. Besides the application mentioned above, such correspondence and its consequences have interesting connections to the theory of Schur functions (see for instance [**St**]).

First we generalize in an obvious way the notion of standard tableau given before: a Young tableau whose elements are positive integers is called standard if the integers are increasing along the rows from left to right and along the columns from top to bottom. We write a tableau $T = (a_{ij})$ where a_{ij} is the element in the (i,j) position.

We now describe an algorithm called "row insertion" which is the basic operation of the Robinson-Schensted correspondence and allows us to insert a positive integer k into a standard tableau T obtaining a new standard tableau with an extra box, denoted $T \leftarrow k$.

Row insertion algorithm ($T \leftarrow k$). Let T be a standard tableau and let k be an integer distinct from the entries of T. Let r be the largest integer such that $a_{1,r-1} \leq k$ (if $a_{11} > k$, then we let $r = 1$). If a_{1r} does not exist (e.g., T has $r-1$ columns), then simply place k at the end of the first row. The insertion process stops and the resulting standard tableau is $T \leftarrow k$. If, on the other hand, a_{1r} exists, then replace a_{1r} by k. The element then bumps $a_{1r} = k'$ into the second row, i.e., we insert k' into the second row of T by the insertion rule just described. Continue until an element is inserted at the end of a row (eventually as the first element of a new row). The resulting tableau is $T \leftarrow k$.

EXAMPLE 4.4.1.

If $T = \begin{array}{|c|c|c|c|} \hline 1 & 3 & 6 & 8 \\ \hline 2 & 5 & 9 \\ \cline{1-3} 7 \\ \cline{1-1} \end{array}$, then $T \leftarrow 4$ is the tableau $\begin{array}{|c|c|c|c|} \hline 1 & 3 & 4 & 8 \\ \hline 2 & 5 & 6 \\ \cline{1-3} 7 & 9 \\ \cline{1-2} \end{array}$.

The above algorithm is the basis for the proof of the following theorem.

THEOREM 4.4.2 (Robinson-Schensted Correspondence). *There is a one-to-one correspondence between the set of n-tuples of distinct positive integers $(\alpha_1, \ldots, \alpha_n)$ and the set of pairs (P, Q) of standard tableaux, where P and Q are of the same shape, P is on $\{\alpha_1, \ldots, \alpha_n\}$ and Q is on $\{1, \ldots, n\}$.*

PROOF. Let $\sigma = (\alpha_1, \ldots, \alpha_n)$ with the α_i's all distinct. As first step we construct two one-box tableaux $(P(1), Q(1))$ with entries α_1 and 1, respectively. If $t < n$ and $(P(t), Q(t))$ have been constructed, let $P(t+1) = P(t) \leftarrow \alpha_{i_{t+1}}$ and let $Q(t+1)$ be the tableau obtained from $Q(t)$ by inserting the integer $t+1$ into a new box (leaving all parts of $Q(t)$ unchanged), so that $P(t+1)$ and $Q(t+1)$ have the same shape. The process ends at $(P(n), Q(n))$ and we define $(P, Q) = (P(n), Q(n))$.

It is clear from the construction that P and Q are standard and of the same shape.

Conversely, given an ordered pair (P, Q) of standard tableaux of the same shape, P on $\{\alpha_1, \ldots, \alpha_n\}$ and Q on $\{1, \ldots, n\}$, we can determine an n-tuple of positive integers (actually a permutation σ of $\{\alpha_1, \ldots, \alpha_n\}$) as follows.

Suppose $Q = (q_{ij})$ and $P = (p_{ij})$. Let q_{rs} be the largest entry of Q. Clearly $q_{rs} = n$; then set $Q(n-1) = Q \setminus q_{rs}$ (i.e., the tableau on $n-1$ boxes obtained from Q by deleting the box with the element q_{rs}). Consider the element p_{rs} and let $p_{r-1,t}$ be the rightmost element of row $r-1$ of P that is smaller than p_{rs}. Then exclude the box with p_{rs} from P, replace $p_{r-1,t}$ with p_{rs} and continue by replacing the rightmost element of row $r-2$ of P that is smaller than $p_{r-1,t}$ with $p_{r-1,t}$, etc. Eventually some element α_{i_n} is removed from the first row of P. Let $P(n-1)$ be the resulting tableau. Hence we have uniquely recovered α_{i_n} and $(P(n-1), Q(n-1))$. By iterating this procedure we determine a permutation σ of $\{\alpha_1, \ldots, \alpha_n\}$, i.e., the n-tuple $(\alpha_{i_1}, \ldots, \alpha_{i_n})$. It is clear that the algorithm described above is the inverse of the row insertion algorithm and the stated bijection follows. □

EXAMPLE 4.4.3. For $\sigma = (5, 2, 7, 9, 3, 4, 1, 6) \in \mathbb{Z}^8$ we find the corresponding ordered pair of standard tableaux (P, Q) by a repeated application of the row insertion algorithm.

Insert 5 : $\boxed{5}$ $\boxed{1}$

Insert 2 : $\begin{array}{|c|}\hline 2 \\\hline 5 \\\hline\end{array}$ $\begin{array}{|c|}\hline 1 \\\hline 2 \\\hline\end{array}$

Insert 7 : $\begin{array}{|c|c|}\hline 2 & 7 \\\hline 5 \\\cline{1-1}\end{array}$ $\begin{array}{|c|c|}\hline 1 & 3 \\\hline 2 \\\cline{1-1}\end{array}$

Insert 9 : $\begin{array}{|c|c|c|}\hline 2 & 7 & 9 \\\hline 5 \\\cline{1-1}\end{array}$ $\begin{array}{|c|c|c|}\hline 1 & 3 & 4 \\\hline 2 \\\cline{1-1}\end{array}$

Insert 3 : $\begin{array}{|c|c|c|}\hline 2 & 3 & 9 \\\hline 5 & 7 \\\cline{1-2}\end{array}$ $\begin{array}{|c|c|c|}\hline 1 & 3 & 4 \\\hline 2 & 5 \\\cline{1-2}\end{array}$

Insert 4 :
$\begin{array}{|c|c|c|} \hline 2 & 3 & 4 \\ \hline 5 & 7 & 9 \\ \hline \end{array}$
$\begin{array}{|c|c|c|} \hline 1 & 3 & 4 \\ \hline 2 & 5 & 6 \\ \hline \end{array}$

Insert 1 :
$\begin{array}{|c|c|c|} \hline 1 & 3 & 4 \\ \hline 2 & 7 & 9 \\ \hline 5 \\ \hline \end{array}$
$\begin{array}{|c|c|c|} \hline 1 & 3 & 4 \\ \hline 2 & 5 & 6 \\ \hline 7 \\ \hline \end{array}$

Insert 6 : $P = \begin{array}{|c|c|c|c|} \hline 1 & 3 & 4 & 6 \\ \hline 2 & 7 & 9 \\ \cline{1-3} 5 \\ \cline{1-1} \end{array}$ $Q = \begin{array}{|c|c|c|c|} \hline 1 & 3 & 4 & 8 \\ \hline 2 & 5 & 6 \\ \cline{1-3} 7 \\ \cline{1-1} \end{array}$

Let $\sigma = (\alpha_1, \ldots, \alpha_n) \in \mathbb{Z}^n$ be an n-tuple of distinct positive integers. A subsequence $\tau = (\alpha_{i_1}, \ldots, \alpha_{i_k})$ is called a chain in σ if $1 \leq i_1 < \cdots < i_k \leq n$ and $\alpha_{i_1} < \cdots < \alpha_{i_k}$; τ is an antichain if $1 \leq i_1 < \cdots < i_k \leq n$ and $\alpha_{i_1} > \cdots > \alpha_{i_k}$. It is clear that in this terminology σ is d-good if σ does not contain any antichain of length d.

Recall that given a Young tableau T, $h(T)$ is the height of T, i.e., the number of rows of T.

LEMMA 4.4.4. *Let $\sigma = (\alpha_1, \ldots, \alpha_n)$ be an n-tuple of distinct positive integers. If σ is d-bad, then $h(P) \geq d$, where (P, Q) is the pair of tableaux corresponding to σ by the previous theorem.*

PROOF. Consider the tableau P' obtained from P by erasing the first row of P and define a new sequence $(\beta_1, \ldots, \beta_s)$ as follows: β_1 is the first integer that was pushed down into P' from the first row of P in the row insertion algorithm applied to $(\alpha_1, \ldots, \alpha_n)$; β_2 is the second integer with the same property, and so on. Notice that P' can be constructed by applying the Robinson-Schensted algorithm to the sequence $(\beta_1, \ldots, \beta_s)$.

Denote by $(\beta_1, \ldots, \beta_{s(i)})$ the sequence obtained after the first i steps of the algorithm applied to $(\alpha_1, \ldots, \alpha_i)$. We shall prove by induction on r that as soon as $(\alpha_1, \ldots, \alpha_i)$ contains an antichain $j_1 > \cdots > j_r$, then $(\beta_1, \ldots, \beta_{s(i)})$ contains an antichain $k_1 > \cdots > k_{r-1}$ with $k_{r-1} > j_r$.

This statement is obvious for $r = 2$. Suppose that the conclusion is valid for all integers strictly smaller than r. If $\alpha_q = j_{r-1} (q < i)$, then by induction $(\beta_1, \ldots, \beta_{s(q)})$ contains an antichain $k_1 > \cdots > k_{r-2}$ with $k_{r-2} > j_{r-1}$. Let $\alpha_l = j_r$ for some $i \geq l > q$. If after $(l-1)$ steps of the algorithm the integer j_{r-1} is still in the first row of P, then in the lth step j_r will push down some $x \leq j_{r-1}$. In this case $k_{r-1} = x$ satisfies $k_{r-1} < k_{r-2}$ and $k_{r-1} > j_r$. If j_{r-1} is not in the first row of P, after the first $(l-1)$ steps, then some α_t with $q < t \leq l-1$ pushed down j_{r-1} and we can set $k_{r-1} = j_{r-1}$. So, we obtain the required antichain $k_1 > \cdots > k_{r-1}$ with $k_{r-1} > j_r$ even before the lth step.

In particular, for $i = n$, we find a sequence $(\beta_1, \ldots, \beta_s)$ which contains an antichain of length at least $d - 1$. The proof is completed by induction on d. □

Now fix $\{\alpha_1, \ldots, \alpha_n\} = \{1, \ldots, n\}$. Then, the Robinson-Schensted correspondence is a bijection between the group of permutations S_n and all pairs (P, Q) of standard Young tableaux with n boxes, P and Q of the same shape. We can now state the precise relation between the badness of permutations in S_n and the height of the corresponding tableaux.

PROPOSITION 4.4.5. *Let $\sigma \in S_n$ correspond to the pair (P, Q) in the Robinson-Schensted correspondence. Then $h(P) = d$ if and only if σ is d-bad and $(d+1)$-good.*

PROOF. If σ is d-bad, then by Lemma 4.4.4, $h(P) \geq d$. On the other hand, if $h(P) = r$, any integer in the last row of P belongs to an antichain of length $\geq r$. Thus r is the maximal length of an antichain of σ. □

Recalling that for any partition λ of n, the degree d_λ of the irreducible S_n-character associated to λ is equal to the number of standard Young tableaux on $\{1, \ldots, n\}$, we easily get the following:

COROLLARY 4.4.6. *The number of t-good permutations of S_n equals*

$$\sum_{h(\lambda) \leq t-1} d_\lambda^2.$$

Hence $\sum_{h(\lambda) \leq t-1} d_\lambda^2 \leq (t-1)^{2n}$.

PROOF. The proof follows from the results of this section and Lemma 4.2.3 on the computation of an upper bound on the number of d-good permutations. □

We shall rediscover the inequality $\sum_{h(\lambda) \leq t-1} d_\lambda^2 \leq (t-1)^{2n}$ in Chapter 5 as a consequence of celebrated results of Schur and Weyl. Here we record another consequence that we shall need in the following sections.

COROLLARY 4.4.7. *Let $\lambda \vdash n$ be such that $h(\lambda) \leq t$, then $d_\lambda \leq t^n$.*

4.5. Cocharacters of PI-algebras

In this and the following sections until the end of the chapter we shall assume that A is an algebra over a field of characteristic zero. In this section we shall get some restrictions on the shape of the diagrams whose corresponding S_n-character appears with non-zero multiplicity in the nth cocharacter of a PI-algebra. Namely we shall prove that for any PI-algebra A, there exists an infinite hook $H(d, l)$ containing all partitions λ such that χ_λ has non-zero multiplicity in $\chi_n(A), n = 1, 2, \ldots$.

Recall from Section 2.5, that given any integers $d, l \geq 0$,

$$H(d, l) = \bigcup_{n \geq 1} \{\lambda = (\lambda_1, \lambda_2, \ldots) \vdash n \mid \lambda_{d+1} \leq l\}.$$

Also recall that given a PI-algebra A over a field F of characteristic zero we say that its nth cocharacter $\chi_n(A)$ lies in the hook $H(d, l)$ if in the decomposition

$$\chi_n = \sum_{\lambda \vdash n} m_\lambda \chi_\lambda$$

of $\chi_n(A)$ into irreducible S_n-characters, all multiplicities m_λ are equal to zero for all $\lambda \notin H(d, l)$.

We now state the main result of this section. This theorem is due to Amitsur and Regev.

THEOREM 4.5.1 (The Hook Theorem, [**AR**]). *For any PI-algebra A, there exist integers $d, l \geq 0$ such that*

$$\chi_n = \sum_{\substack{\lambda \vdash n \\ \lambda \in H(d,l)}} m_\lambda \chi_\lambda$$

for all $n = 1, 2, \ldots$.

In order to prove the above theorem we need some technical results concerning the permutation action of the symmetric group on multilinear polynomials. We first derive a lower bound for the dimensions of square partitions, i.e., partitions whose Young diagram is a square.

LEMMA 4.5.2. *Let d be a positive real number and let $n = m^2 > e^4 d^2$ where e is the base of the natural logarithms. If $\lambda = (m^m) \vdash n$ is a square partition of n, then $\deg \chi_\lambda > d^n$.*

PROOF. Recall that by the hook formula (Proposition 2.2.8)

$$\deg \chi_\lambda = \frac{n!}{\prod h_{ij}}$$

where the h_{ij}'s are the hook numbers of the Young diagram D_λ. In our case D_λ is a $m \times m$-square, hence the hook h_{ij} corresponding to the (i,j)-box is

$$h_{ij} = m - (i-1) + m - (j-1) - 1 = 2m - i - j + 1 < 2m.$$

Therefore $p < (2m)^n = 2^n n^{\frac{1}{2}n}$. On the other hand, by the Stirling formula

$$n! \simeq \sqrt{2\pi n} \left(\frac{n}{e}\right)^n > \left(\frac{n}{e}\right)^n$$

hence

$$\deg \chi_\lambda = \frac{n!}{p} > \frac{n^n}{e^n 2^n n^{\frac{n}{2}}} = \frac{n^{\frac{1}{2}n}}{e^n 2^n}.$$

Since by hypothesis $n > e^4 d^2$, we have $n^{\frac{1}{2}n} > e^{2n} d^n$ and $\deg \chi_\lambda > d^n$. □

Now let $n_1 \geq n_2$ be two positive integers and let $\lambda = (\lambda_1, \ldots, \lambda_p) \vdash n_1$, $\mu = (\mu_1, \ldots, \mu_q) \vdash n_2$. Recall that we write $\lambda \geq \mu$ if $p \geq q$ and $\lambda_1 \geq \mu_1, \ldots, \lambda_q \geq \mu_q$. In the language of Young diagrams this means that D_μ is a subdiagram of D_λ.

LEMMA 4.5.3. *Let $\lambda \vdash n, \mu \vdash n' \leq n$ and suppose that $\lambda \geq \mu$. Consider a Young tableau T_λ such that the integers $1, \ldots, n'$ are arranged into the subtableau T_μ. Then in the group algebra FS_n we can write*

$$e_{T_\lambda} = a e_{T_\mu} b$$

for some $a, b \in FS_n$.

PROOF. Recall that $R_{T_\lambda}(C_{T_\lambda})$ is the subgroup of row (column) permutations of T_λ in S_n. Starting with T_μ we construct the analogous subgroups R_{T_μ} and C_{T_μ} of $S_{n'}$. Clearly by the hypotheses, R_{T_μ} is a subgroup of R_{T_λ} and C_{T_μ} is a subgroup of C_{T_λ}. Consider the decompositions of $R_{T_\lambda}, C_{T_\lambda}$ into left and right cosets of R_{T_μ}, C_{T_μ}, respectively,

$$R_{T_\lambda} = \cup_i \tau_i R_{T_\mu}, \quad C_{T_\lambda} = \cup_j C_{T_\mu} \sigma_j.$$

Then

$$e_{T_\lambda} = \left(\sum_{\tau \in R_{T_\lambda}} \tau\right) \left(\sum_{\sigma \in C_{T_\lambda}} (\operatorname{sgn} \sigma) \sigma\right)$$

$$= \left(\sum_i \tau_i\right)\left(\sum_{\tau \in R_{T_\mu}} \tau\right)\left(\sum_{\sigma \in C_{T_\mu}} (\text{sgn}\,\sigma)\sigma\right)\left(\sum_j (\text{sgn}\,\sigma_j)\sigma_j\right) = ae_{T_\mu}b.$$

□

PROOF OF THEOREM 4.5.1. Given the PI-algebra A, by Theorem 4.2.4 there exists an integer d such that $c_n(A) \leq d^n$ for all $n = 1, 2, \ldots$. We shall prove that for any n, the nth cocharacter of A lies in the hook $H(m,m)$ where m is such that $e^2q + 2 \geq m \geq e^2q + 1$ and $q = d^3$.

Suppose that there exists $n \geq 1$ such that

(4.11) $$\chi_n(A) = \sum_{\lambda \vdash n} m_\lambda \chi_\lambda$$

and $m_\mu \neq 0$ for some $\mu \notin H(m,m)$. Since μ does not belong to $H(m,m)$, the Young diagram D_μ contains the $m \times m$ square diagram D_ν, i.e.,

$$\mu \geq \nu = (m^m) \vdash m^2.$$

By Theorem 2.4.5, there exist a tableau T_μ and a monomial $f = x_{i_1} \cdots x_{i_n}$ such that the polynomial $e_{T_\mu} f$ is not an identity of A. Since we can renumerate the variables $x_1, \ldots, x_n$, without loss of generality, we may assume that T_μ contains the $m \times m$ square tableau T_ν and the first m^2 integers in T_μ are placed into T_ν. Then by Lemma 4.5.3 $e_{T_\mu} f = a e_{T_\nu} bf$. Hence $e_{T_\nu} bf$ is not an identity of A. Therefore there exists a monomial $g = g(x_1, \ldots, x_n) = x_{j_1} \cdots x_{j_n}$ such that the multilinear polynomial $e_{T_\nu} g$ is not equal to zero on A. We can write g as

$$g = u_0 x_{i_1} u_1 \cdots x_{i_k} u_k$$

where $k = m^2, \{i_1, i_2, \ldots, i_k\} = \{1, 2, \ldots, m^2\}$ and $u_0, u_1, \ldots, u_k$ depend on the variables x_j with $j > m^2$. Now we can replace any non-empty word $u_j, 0 \leq j \leq k$, in g with one variable and obtain a new monomial of degree not less than m^2 and not greater than $2m^2 + 1$. Since $e_{T_\nu} g \not\equiv 0$ on A, it follows that there exist an integer $m^2 \leq t \leq 2m^2 + 1$ and a multilinear monomial $w = w(x_1, \ldots, x_t)$, such that $e_{T_\nu} w \not\equiv 0$ on A.

We consider P_t as an S_{m^2}-module by letting S_{m^2} act on the first m^2 variables $x_1, \ldots, x_{m^2}$. Since $e_{T_\nu} w \notin \text{Id}(A)$, if M is the S_{m^2}-module generated by $e_{T_\nu} w$, then $M \cap (\text{Id}(A) \cap P_t) = 0$. This says that $c_t(A) \geq \dim M = \deg \chi_\nu$.

Since $\nu \vdash m^2$ and $m^2 \geq (e^2q+1)^2 > e^4q^2$, by Lemma 4.5.2 one has

$$c_t(A) \geq \deg \chi_\nu > q^{m^2}.$$

Recall that $q = d^3$ and $2m^2 \geq t - 1$. Hence

$$c_t(A) > (d^3)^{\frac{t-1}{2}} > d^t.$$

The latter inequality contradicts the condition $c_n(A) \leq d^n$. It follows that all multiplicities m_μ in (4.11) should be equal to zero if $\mu \notin H(m,m)$, and the proof of Theorem 4.5.1 is complete. □

As a consequence of Theorem 4.5.1 we can prove a well-known result of Amitsur stating that any PI-algebra satisfies some power of a standard identity.

THEOREM 4.5.4 (Amitsur [**A1**]). *For any PI-algebra A there exist integers r and m such that A satisfies the identity $St_r^m \equiv 0$.*

PROOF. By Theorem 4.5.1, for all $n \geq 1$, the nth cocharacter $\chi_n(A)$ lies in an infinite hook $H(k,l)$. Hence χ_λ appears with zero multiplicity in $\chi_n(A)$ as soon as $\lambda \notin H(k,l)$. Consider the partition $\lambda = ((l+1)^{k+1})$ of $n = (l+1)(k+1)$ and the Young tableau

$$T_\lambda = \begin{array}{|c|c|c|c|} \hline 1 & k+2 & \cdots & (k+1)l+1 \\ \hline \vdots & \vdots & & \vdots \\ \hline k+1 & 2k+2 & \cdots & (k+1)(l+1) \\ \hline \end{array}$$

The polynomial $e_{T_\lambda}(x_1,\ldots x_n)$ is symmetric in each set of variables
$$X_1 = \{x_1, x_{k+2}, \ldots, x_{(k+1)l+1}\}, \ldots, X_{k+1} = \{x_{k+1}, x_{2k+2}, \ldots, x_{(k+1)(l+1)}\}.$$
Since $\lambda \notin H(k,l)$, the polynomial f is an identity of A. If we identify all the indeterminates of each set X_i with y_i, $i = 1, \ldots, k+1$, we obtain that
$$f(y_1, \ldots, y_{k+1}, \ldots, y_1, \ldots, y_{k+1}) = St_{k+1}(y_1, \ldots, y_{k+1})^{l+1}$$
is an identity of A. $\square$

4.6. Capelli polynomials and the strip theorem

In this section we shall characterize the cocharacter sequence of those PI-algebras satisfying a Capelli polynomial Cap_k by proving the so-called strip theorem. Recall that
$$Cap_k(x_1, \ldots, x_k; y_1, \ldots, y_{k+1}) = \sum_{\sigma \in S_k} (\operatorname{sgn} \sigma) y_1 x_{\sigma(1)} y_2 \cdots x_{\sigma(k)} y_{k+1}$$
and, according to Definition 1.5.4, an algebra A satisfies the Capelli identity of rank k if A satisfies all polynomials obtained from Cap_k by eventually setting the variables y_i equal to 1, in all possible ways.

We shall prove that an algebra A satisfies the Capelli identity of rank k if and only if its cocharacter is contained in a strip of height $k-1$, i.e., for such algebras the hook becomes a strip of hand k and empty foot. Historically this result was proved before the strip theorem by Regev.

THEOREM 4.6.1 (The Strip Theorem, [**Re2**]). *Let A be a PI-algebra and let $\chi_n = \sum_{\lambda \vdash n} m_\lambda \chi_\lambda$ be its nth cocharacter. Then A satisfies Cap_k, the Capelli identity of rank k, if and only if $m_\lambda = 0$ whenever $h(\lambda) \geq k$.*

PROOF. Suppose first that $Cap_k \subseteq \operatorname{Id}(A)$ and let $\lambda \vdash n$ be such that $h(\lambda) \geq k$. By Theorem 2.4.5, $m_\lambda = 0$ if and only if for any multilinear polynomial f of degree n, $e_{T_\lambda} f \in \operatorname{Id}(A)$ for all tableaux T_λ of shape λ.

Let T_0 be a fixed tableau of shape λ. If $e_{T_0} f = 0$, then $e_{T_0} f \in \operatorname{Id}(A)$. Suppose that $e_{T_0} f \neq 0$ for some f. Since e_{T_0} generates an irreducible left S_n-module, it is clear that it is enough to show that $a e_{T_0} f \in \operatorname{Id}(A)$ for some $a \in FS_n$ such that $a e_{T_0} f \neq 0$. Let
$$a = \sum_{\tau \in C_{T_0}} (\operatorname{sgn} \tau) \tau$$
where C_{T_0} is the subgroup of S_n stabilizing the columns of e_{T_0} respectively. By Lemma 2.5.1, $a e_{T_0} f \neq 0$. On the other hand, if $h = h(\lambda)$ and $i_1, \ldots, i_h$ are the entries appearing in the first column of T_0, then $a e_{T_0} f$ is a polynomial alternating

on $x_{i_1}, \ldots, x_{i_h}$. But since $Cap_k \subseteq \mathrm{Id}(A)$, any such polynomial is an identity of A; hence $ae_{T_0}f \in \mathrm{Id}(A)$, and this proves the first part of the theorem.

Conversely, suppose $m_\lambda = 0$ whenever $h(\lambda) \geq k$ and consider the polynomial
$$f = f(x_1, \ldots, x_{2k-1}) = \sum_{\sigma \in S_k} (\mathrm{sgn}\,\sigma) x_{\sigma(1)} x_{k+1} x_{\sigma(2)} x_{k+2} \cdots x_{2k-1} x_{\sigma(k)}.$$
We claim that f is an identity of A, i.e., $FS_{2k-1}f \subseteq \mathrm{Id}(A)$. To see this, clearly it is enough to show that $e_{T_\lambda}f \in \mathrm{Id}(A)$ for all $\lambda \vdash 2k-1$ and for all tableaux T_λ of shape λ. If $h(\lambda) \geq k$, then $e_{T_\lambda}f \in \mathrm{Id}(A)$ by definition of m_λ.

Suppose now that $h(\lambda) < k$. Since f is alternating on $x_1, \ldots, x_k$, for all $\tau \in S_{2k-1}$, the polynomial
$$\tau f(x_1, \ldots, x_{2k-1})$$
is alternating on $x_{\tau(1)}, \ldots, x_{\tau(k)}$. Notice that since $h(\lambda) = t < k$, the group R_{T_λ} is the direct product of $t < k$ factors, $R_{T_\lambda} = H_1 \times \cdots \times H_t$ where H_i is the stabilizer of the ith row of T_λ. Hence $\sum_{\sigma \in R_{T_\lambda}} \sigma$ is the product of $t < k$ commuting factors
$$\sum_{\sigma \in R_{T_\lambda}} \sigma = \left(\sum_{\sigma \in H_1} \sigma\right) \cdots \left(\sum_{\sigma \in H_t} \sigma\right).$$
It follows that at least two integers among $\tau(1), \ldots, \tau(k)$ belong to the same row of T_λ, say, the ith row. Then
$$\left(\sum_{\sigma \in H_i} \sigma\right) \tau f = 0,$$
i.e., it is the zero polynomial in $F\langle X\rangle$ since τf is alternating on $x_{\tau(1)}, \ldots, x_{\tau(k)}$. Thus $\left(\sum_{\sigma \in R_{T_\lambda}} \sigma\right) \tau f = 0$ and $e_{T_\lambda} f = 0$. We have proved that for all $\lambda \vdash n$ with $h(\lambda) < k$, $e_{T_\lambda} f \in \mathrm{Id}(A)$, so $f \in \mathrm{Id}(A)$.

Finally, if in the polynomial f we replace some of the variables $x_{k+1}, \ldots, x_{2k-1}$ with 1, then the same proof as above will lead to the conclusion that this polynomial is also an identity for A. $\square$

Combining the previous result with Theorem 1.5.8, we get

THEOREM 4.6.2. *Let A be a finite dimensional algebra, $\dim_F A = k$. Then for any $n \geq 1$,*
$$\chi_n(A) = \sum_{\substack{\lambda \vdash n \\ h(\lambda) \leq k}} m_\lambda \chi_\lambda,$$
i.e., the nth cocharacter of A lies in a strip of height k.

4.7. Amitsur polynomials and hooks

As we saw in the previous section the Capelli identities can be used as a test for a PI-algebra to have its cocharacter sequence lying in a strip. Generalizing this approach Amitsur and Regev in [**AR**] defined new polynomials characterizing algebras whose cocharacter sequence lies in a hook.

Let $n = (d+1)(l+1)$ and let $\lambda = ((l+1)^{d+1})$ be a partition of n with rectangular diagram. Given the group algebra element
$$e_\lambda = \sum_{\sigma \in S_n} \chi_\lambda(\sigma)\sigma,$$

define the polynomial

$$E_{d,l}^*(x_1,\ldots,x_n;y_1,\ldots,y_{n-1}) = \sum_{\sigma \in S_n} \chi_\lambda(\sigma) x_{\sigma(1)} y_1 x_{\sigma(2)} y_2 \cdots y_{n-1} x_{\sigma(n)}.$$

As in the case of Capelli polynomials we can make the following definition.

DEFINITION 4.7.1. An algebra A satisfies the Amitsur identity $E_{d,l}^* \equiv 0$ if A satisfies all polynomials obtained from $E_{d,l}^* = (x_1,\ldots,x_n;y_1,\ldots,y_{n-1})$ by setting the variables y_i equal to 1 in all possible ways.

The following result is the analogue of Theorem 4.6.1 for the hook $H(d,l)$.

THEOREM 4.7.2 ([**AR**]). Let A be a PI-algebra and let $\chi_n(A) = \sum_{\lambda \vdash n} m_\lambda \chi_\lambda$ be its nth cocharacter. Then A satisfies the Amitsur identity $E_{d,l}^* \equiv 0$ if and only if $m_\lambda = 0$ whenever $\lambda \notin H(d,l)$.

PROOF. Suppose first that $\chi_m(A) \not\subseteq H(d,l)$ for some $m \geq n = (d+1)(l+1)$. Hence there exists a partition $\mu \vdash m$ such that $\mu \notin H(d,l)$ and χ_μ has multiplicity $m_\mu \neq 0$ in $\chi_m(A)$. By Theorem 2.4.5, this says that there exists a Young tableau T_μ such that

$$e_{T_\mu} \cdot x_{i_1} \cdots x_{i_m} \not\equiv 0,$$

for some multilinear monomial $x_{i_1} \cdots x_{i_m}$. Since $\mu \notin H(d,l)$, μ contains the rectangle $((l+1)^{d+1})$ and let T_λ be the corresponding subtableau of T_μ. After renaming the variables we may clearly assume that T_λ is filled up with the integers $1,\ldots,n$.

We then decompose C_{T_μ} into the union of the right cosets of its subgroup C_{T_λ} and R_{T_μ} into the union of the left cosets of R_{T_λ}. Then $e_{T_\mu} x_{i_1} \cdots x_{i_m}$ is a linear combination of polynomials of the type

$$\rho \sum_{\substack{\sigma \in R_{T_\lambda} \\ \tau \in C_{T_\lambda}}} (\operatorname{sgn} \tau) y_1 x_{\sigma\tau\pi(1)} y_2 \cdots y_n x_{\sigma\tau\pi(n)} y_{n+1}$$

where $\rho, \pi \in S_m$ and $y_1,\ldots,y_{n+1}$ are (eventually trivial) monomials in the remaining variables $x_{\pi(n+1)},\ldots,x_{\pi(m)}$. Moreover, at least one of these polynomials is not an identity of A. Since ρ is invertible, by multiplying on the left by $\pi^{-1}\rho^{-1}$, we deduce that

$$\pi^{-1} \sum_{\substack{\sigma \in R_{T_\lambda} \\ \tau \in C_{T_\lambda}}} (\operatorname{sgn} \tau) y_1 x_{\sigma\tau\pi(1)} y_2 \cdots y_n x_{\sigma\tau\pi(n)} y_{n+1}$$

$$(4.12) \qquad = (\pi^{-1} e_{T_\lambda} \pi)(y_1 x_1 y_2 \cdots y_{n-1} x_n y_n) \not\equiv 0.$$

By Proposition 2.2.13, $\pi^{-1} e_{T_\lambda} \pi = e_{\pi^{-1} T_\lambda} = b$ is an essential idempotent corresponding to λ, so $b \in I_\lambda$.

Recall that the element $e_\lambda \in FS_n$ generates I_λ as a left ideal (see Proposition 2.2.2). Hence, since $b \in I_\lambda$, there exists an element $a = \sum_{\rho \in S_n} \alpha_\rho \rho \in FS_n$ such that $ae_\lambda = b$. Recalling that $e_\lambda = \sum_{\sigma \in S_n} \chi_\lambda(\sigma) \sigma$, and $e_\lambda(x_1 y_1 x_2 \cdots y_{n-1} x_n) = E_{d,l}^*(x_1,\ldots,x_n;y_1,\ldots,y_{n-1})$, we obtain

$$aE_{d,l}^*(x_1,\ldots,x_n;y_1,\ldots,y_{n-1}) = \sum_{\rho \in S_n} \sum_{\sigma \in S_n} \alpha_\rho \chi_\lambda(\sigma) x_{\rho\sigma(1)} y_1 \cdots y_{n-1} x_{\rho\sigma(n)}$$

$$= b(x_1 y_1 x_2 \cdots y_{n-1} x_n).$$

Notice that the above equalities hold in $F\langle X\rangle$ also if we set the variables y_i equal to 1 in all possible ways. From (4.12) we obtain that also the polynomial $aE^*_{d,l}$ does not vanish in A. It follows that the polynomial $E^*_{d,l}$ is not an identity of A. This proves the first part of the theorem.

Now assume that $\chi_m(A) \subseteq H(d,l)$ for any $m \geq n = (d+1)(l+1)$. Set some of the variables equal to 1 and denote by $m = \deg E^*_{d,l}$, the degree of the resulting polynomial. Also set $\lambda = ((l+1)^{d+1}) \vdash n$. Since $FS_n e_\lambda = I_\lambda$, a two-sided ideal of FS_n with character $d_\lambda \chi_\lambda$, by the branching rule of the symmetric group (Theorem 2.3.1), it follows that $FS_m E^*_{d,l}$ is an S_m-submodule of $P_m = P_m(x_1,\ldots,x_m)$ with character

$$\chi(FS_m E^*_{d,l}) = \sum_{\substack{\mu \vdash m \\ \mu \geq \lambda}} m_\mu \chi_\mu$$

for some multiplicities m_λ. By hypothesis $\chi_m(A)$ lies in the hook $H(d,l)$, hence $FS_m E^*_{d,l} \subseteq P_m \cap \mathrm{Id}(A)$ and $E^*_{d,l} \equiv 0$ is an identity of A. □

4.8. Finitely generated superalgebras

Let $\mathcal{V}$ be a non-trivial variety. In the previous sections we proved the existence of an infinite hook $H(d,l)$ such that all cocharacters $\chi_n(\mathcal{V}), n=1,2,\ldots$ lie in this hook. Using this phenomenon we will prove in this section that there exists a finitely generated superalgebra $A = A^{(0)} \oplus A^{(1)}$ such that $\mathcal{V} = \mathrm{var}(G(A))$, i.e., $\mathcal{V}$ is generated by the Grassmann envelope of A. This result was proved by Kemer in [**Ke3**]. Nevertheless, we should warn the reader that Kemer later proved a much stronger result: $\mathcal{V}$ can be generated by the Grassmann envelope of a finite dimensional superalgebra ([**Ke6**]).

In Section 3.7, we introduced a linear isomorphism $\sim : P_{k,m} \to P_{k,m}$ where $P_{k,m}$ is the space of multilinear polynomials in the free superalgebra in k even and m odd variables. Recall that if $f(y_1,\ldots,y_k,z_1,\ldots,z_m) \in P_{k,m}$ is written in the form

$$f = \sum_{\substack{\sigma \in S_m \\ W=(w_0,w_1,\ldots,w_m)}} \alpha_{\sigma,W}\, w_0 z_{\sigma(1)} w_1 \cdots w_{m-1} z_{\sigma(m)} w_m$$

where $w_0, w_1, \ldots, w_m$ are monomials in $y_1, \ldots, y_k$ and $\alpha_{\sigma,W} \in F$, then

$$\widetilde{f} = \sum_{\substack{\sigma \in S_m \\ W=(w_0,w_1,\ldots,w_m)}} (\mathrm{sgn}\,\sigma)\alpha_{\sigma,W}\, w_0 z_{\sigma(1)} w_1 \cdots w_{m-1} z_{\sigma(m)} w_m.$$

Recall also that $\mathcal{V}^*$ is the class of all superalgebras B such that $G(B) \in \mathcal{V}$ (see Section 3.7). By Theorem 3.7.5, $\mathcal{V}^*$ is a supervariety, hence we can speak of relatively free superalgebras of $\mathcal{V}^*$.

LEMMA 4.8.1. *Let*

$$f = f(y_1^1,\ldots,y_1^{m_1},\ldots,y_k^1,\ldots,y_k^{m_k}, z_1^1,\ldots,z_1^{r_1},\ldots,z_l^1,\ldots,z_l^{r_l})$$

be a multilinear polynomial in the variables $y_1^1,\ldots,y_k^{m_k},z_1^1,\ldots,z_l^{r_l}$ such that f is symmetric on each set $\{y_i^1,\ldots,y_i^{m_i}, 1 \leq i \leq k\}$, and is alternating on each set $\{z_j^1,\ldots,z_j^{r_j}, 1 \leq j \leq l\}$. Given a variety $\mathcal{V}$, denote by $L = L(u_1,\ldots,u_k,w_1,\ldots,w_l)$ the relatively free superalgebra of the supervariety $\mathcal{V}^*$ on k even free generators

4.8. FINITELY GENERATED SUPERALGEBRAS

$u_1, \ldots, u_k$ and l odd free generators $w_1, \ldots, w_l$. Considering all y_i^j as even variables and all z_i^j as odd variables construct the multilinear polynomial $\widetilde{f}$. Then, if

$$(4.13) \qquad \widetilde{f}(\underbrace{u_1, \ldots, u_1}_{m_1}, \ldots, \underbrace{u_k, \ldots, u_k}_{m_k}, \underbrace{w_1, \ldots, w_1}_{r_1}, \ldots, \underbrace{w_l, \ldots, w_l}_{r_l}) = 0$$

in L, $f \equiv 0$ is an ordinary identity of $\mathcal{V}$.

PROOF. Since $u_1, \ldots, u_k, w_1, \ldots, w_l$ are free generators of L, the equality (4.13) says that

$$\widetilde{f}(y_1, \ldots, y_1, \ldots, y_k, \ldots, y_k, z_1, \ldots, z_1, \ldots, z_l, \ldots, z_l) \equiv 0$$

is a graded identity of $\mathcal{V}^*$. Also, from the definition of $*$ it follows that the polynomial

$$\widetilde{f}(y_1^1, \ldots, y_1^{m_1}, \ldots, y_k^1, \ldots, y_k^{m_k}, z_1^1, \ldots, z_1^{r_1}, \ldots, z_l^1, \ldots, z_l^{r_l})$$

is symmetric in each set of variables $\{y_i^1, \ldots, y_i^{m_i}\}$ and $\{z_j^1, \ldots, z_j^{r_j}\}$.

Now denote by $S = S\langle \bar{y}_i^j, \bar{z}_i^j \mid i, j = 1, 2, \ldots \rangle$ the relatively free algebra of the variety $\mathcal{V}$ in the countable set of free generators $\{\bar{y}_i^j, \bar{z}_i^j \mid i, j = 1, 2, \ldots\}$ and construct the superalgebra $Q^{(0)} \oplus Q^{(1)}$ where $Q^{(0)} = S \otimes G^{(0)}, Q^{(1)} = S \otimes G^{(1)}$. If $G(Q)$ is the Grassmann envelope of Q, then $G(Q) = (Q^{(0)} \otimes G^{(0)}) \oplus (Q^{(1)} \otimes G^{(1)}) \subseteq S \otimes R$ where $R = (G^{(0)} \otimes G^{(0)}) \oplus (G^{(1)} \otimes G^{(1)})$ is a commutative algebra. It follows that $G(Q)$ satisfies all multilinear identities of S, so $G(Q) \in \mathcal{V}$. Thus $Q \in \mathcal{V}^*$. In particular,

$$(4.14) \qquad \widetilde{f}(\underbrace{q_1, \ldots, q_1}_{m_1}, \ldots, \underbrace{q_k, \ldots, q_k}_{m_k}, \underbrace{p_1, \ldots, p_1}_{r_1}, \ldots, \underbrace{p_l, \ldots, p_l}_{r_l}) = 0$$

for any $q_1, \ldots, q_k \in Q^{(0)}$, $p_1, \ldots, p_l \in Q^{(1)}$. Now take

$$q_i = \bar{y}_i^1 \otimes a_i^1 + \cdots + \bar{y}_i^{m_i} \otimes a_i^{m_i}, \quad i = 1, \ldots, k,$$

$$p_j = \bar{z}_j^1 \otimes b_j^1 + \cdots + \bar{z}_j^{r_j} \otimes b_j^{r_j}, \quad j = 1, \ldots, l,$$

in Q where $a_1^1, \ldots, a_1^{m_1}, \ldots, a_k^1, \ldots, a_k^{m_k}, b_1^1, \ldots, b_1^{r_1}, \ldots, b_l^1, \ldots, b_l^{r_l}$ are monomials of the Grassmann algebra G written on distinct generators of G, $a_i^j \in G^{(0)}, b_i^j \in G^{(1)}$ (hence $a_1^1 \cdots a_k^{m_k} b_1^1 \cdots b_l^{r_l} \neq 0$). We now compute the left-hand side of (4.14). Since $(a_i^j)^2 = (b_i^j)^2 = 0$ for all i, j, using the symmetry of $\widetilde{f}$ on any set $\{y_i^1, \ldots, y_i^{m_i}\}$ and on any set $\{z_j^1, \ldots, z_j^{r_j}\}$ we obtain

$$\widetilde{f}(q_1, \ldots, p_l)$$

$$= m_1! \cdots m_k! r_1! \cdots r_l! \widetilde{f}(\bar{y}_1^1, \ldots, \bar{y}_1^{m_1}, \ldots, \bar{z}_l^1, \ldots, \bar{z}_l^{r_l}) \otimes a_1^1 \cdots a_1^{m_1} \cdots b_l^1 \cdots b_l^{r_l}.$$

By Lemma 3.7.4, $\widetilde{\widetilde{f}} = f$. Hence

$$f(\bar{y}_1^1, \ldots, \bar{z}_l^{r_l}) = 0$$

in L. Recall now that the elements $\bar{y}_i^j, \bar{z}_i^j$ are free generators of S, and S is a relatively free algebra of $\mathcal{V}$. It follows that $f \equiv 0$ on $\mathcal{V}$. □

We can now prove the main result of this section ([**Ke3**]).

THEOREM 4.8.2. *For any non-trivial variety $\mathcal{V}$ there exists a finitely generated superalgebra $A = A^{(0)} \oplus A^{(1)}$ such that $\mathcal{V} = \mathrm{var}(G(A))$. Moreover, as an ordinary algebra A satisfies a non-trivial polynomial identity.*

PROOF. By Theorem 4.5.1 there exist integers k, l such that the nth cocharacter $\chi_n(\mathcal{V})$ of $\mathcal{V}$ lies in the hook $H(k, l)$ for all $n = 1, 2, \ldots$. As in Lemma 4.8.1 consider the corresponding supervariety $\mathcal{V}^*$ and its relatively free superalgebra $L = L(u_1, \ldots, u_k, w_1, \ldots, w_l)$ on k even free generators $u_1, \ldots, u_k$ and l odd free generators $w_1, \ldots, w_l$. We shall prove that $\mathcal{V}$ is generated by $G(L)$.

Clearly $G(L) \in \mathcal{V}$ by the definition of $\mathcal{V}^*$. We next verify that $\mathcal{V}$ satisfies all identities of $G(L)$.

Let $f \equiv 0$ be a multilinear identity of $G(L)$ of degree n. By Theorem 2.4.7, f is equivalent to a system of identities of the type

$$(4.15) \qquad e_{T_\lambda} g \equiv 0$$

where $g = g(x_1, \ldots, x_n)$ is a multilinear polynomial in $x_1, \ldots, x_n$, λ is a partition of n and T_λ is some Young tableau corresponding to λ. Hence it is enough to prove our statement only in case f is of type (4.15).

If $\lambda \notin H(k, l)$, then $f \equiv 0$ is an identity of $\mathcal{V}$ since $\chi_n(\mathcal{V})$ lies in the hook $H(k, l)$. Now let $\lambda \in H(k, l)$. By applying Lemma 2.5.6, we may assume that f depends on k sets of symmetric variables $\{y_i^1, \ldots, y_i^{m_i}\}$, $1 \le i \le k$ and on l sets of alternating variables $z_j^1, \ldots, z_j^{r_j}$ $1 \le j \le l$ (without loss of generality, we are assuming that $l' = l, k' = k$). But then the polynomial

$$f = f(y_1^1, \ldots, y_1^{m_1}, \ldots, y_k^1, \ldots, y_k^{m_k}, z_1^1, \ldots, z_1^{r_1}, \ldots, z_l^1, \ldots, z_l^{r_l})$$

satisfies the hypotheses of Lemma 4.8.1. In fact, if we consider all y_i^j as even variables and all z_i^j as odd variables, then the relation $f \equiv 0$ can be considered as a graded identity of the superalgebra $G(L)$. But then, by Lemma 3.7.4, the superalgebra L satisfies the graded identity $\widetilde{f} \equiv 0$, so in particular,

$$\widetilde{f}(\underbrace{u_1, \ldots, u_1}_{m_1}, \ldots, \underbrace{u_k, \ldots, u_k}_{m_k}, \underbrace{w_1, \ldots, w_1}_{r_1}, \ldots, \underbrace{w_l, \ldots, w_l}_{r_l}) = 0.$$

We can now apply Lemma 4.8.1 and deduce that $f \equiv 0$ is an identity of $\mathcal{V}$. We have proved that $\mathcal{V} = \operatorname{var}(G(L))$ and the first part of the proof is completed.

Since $G(A) = (A^{(0)} \otimes G^{(0)}) \oplus (A^{(1)} \otimes G^{(1)})$ satisfies all identities of $\mathcal{V}$, $A^{(0)} \otimes G^{(0)}$ is a PI-algebra. Hence $A^{(0)}$ is also a PI-algebra. By Corollary 4.3.4, A satisfies a polynomial identity. $\square$

As we mentioned at the beginning of the section, the previous theorem can be drastically improved since it can be shown that $\mathcal{V}$ is generated by the Grassmann envelope of a finite dimensional superalgebra. Next we deduce this result from the following theorem of Kemer.

THEOREM 4.8.3 (Kemer). *For any finitely generated superalgebra $A = A^{(0)} \oplus A^{(1)}$ there exists a finite dimensional superalgebra $B = B^{(0)} \oplus B^{(1)}$ such that A and B have the same superidentities.*

Here we shall not prove this theorem since it would require a long digression away from the main objective of the book. A complete proof is given in Kemer's monograph [**Ke7**, Theorem 2.2]. Another proof is also available in [**BR**, Theorem 6.33].

Applying the above result together with Theorem 4.8.2 and Remark 3.7.6 we immediately obtain the following improvement of Theorem 4.8.2.

COROLLARY 4.8.4. *For any non-trivial variety $\mathcal{V}$ there exists a finite dimensional superalgebra $A = A^{(0)} \oplus A^{(1)}$ such that $\mathcal{V} = \mathrm{var}(G(A))$.*

In Section 1.12 we showed that the nilpotency of the Jacobson radical of a finitely generated PI-algebra A, forces A to satisfy the identities of some finite dimensional algebra (see Theorem 1.12.2). The next corollary is a far reaching generalization of Theorem 1.12.2 in characteristic zero and is based on Theorem 4.8.3 above.

COROLLARY 4.8.5. *For any finitely generated PI-algebra A there exists a finite dimensional algebra B such that $\mathrm{var}(A) = \mathrm{var}(B)$.*

PROOF. If we set $A^{(0)} = A$ and $A^{(1)} = 0$, the algebra A becomes a superalgebra with trivial grading. Hence A satisfies the graded identity $z \equiv 0$, where z is an odd variable. By Theorem 4.8.3 there exists a finite dimensional superalgebra B with the same superidentities. In particular, B satisfies the identity $z \equiv 0$ and $B^{(1)} = 0$ follows. Thus $B = B^{(0)}$ and the supervariety generated by B coincides with the ordinary variety generated by B. Thus $\mathrm{var}(A) = \mathrm{var}(B)$. □

There is a close relation between finite dimensional superalgebras and infinite hooks in the following sense.

Recall that given a partition $\lambda \vdash n$, the conjugate partition of λ is $\lambda' = (\lambda'_1, \ldots, \lambda'_r)$ where for every $i = 1, \ldots, r$, λ'_i is the length of the ith column of the diagram D_λ. Accordingly, given a tableau T_λ, we say that a tableau $T_{\lambda'}$ of shape λ' is conjugated to the tableau T_λ if for every j, the jth column of $T_{\lambda'}$ equals the jth row of T_λ. There is a close connection between conjugate tableaux and the map $\tilde{\ }$ defined in Section 3.7. For any tableau T_λ, let us write

$$e^*_{T_\lambda} = \sum_{\sigma \in C_{T_\lambda}} (\mathrm{sgn}\, \sigma) \sigma \sum_{\tau \in R_{T_\lambda}} \tau.$$

It is clear that $e^*_{T_\lambda}$ is an essential idempotent of FS_n and we have

LEMMA 4.8.6. *Suppose that in the $\mathbb{Z}_2$-grading of $F\langle X \rangle$, $x_1, \ldots, x_n$ are odd variables and $x_{n+1}, \ldots, x_{n+m}$ are even variables and let $g = g(x_1, \ldots, x_{n+m})$ be a multilinear monomial. If P and Q are two subsets of S_n and*

$$f(x_1, \ldots, x_{n+m}) = \sum_{\sigma \in P} \sum_{\tau \in Q} (\mathrm{sgn}\, \tau) \sigma \tau g(x_1, \ldots, x_{n+m}),$$

then

$$\tilde{f}(x_1, \ldots, x_{n+m}) = \pm \sum_{\sigma \in P} \sum_{\tau \in Q} (\mathrm{sgn}\, \sigma) \sigma \tau g(x_1, \ldots, x_{n+m}).$$

In particular, if T_λ is a tableau of shape $\lambda \vdash n$ and

$$f(x_1, \ldots, x_{n+m}) = e_{T_\lambda} g(x_1, \ldots, x_{n+m}),$$

where the symmetric group S_n acts on $x_1, \ldots, x_n$, then

$$\tilde{f}(x_1, \ldots, x_{n+m}) = \pm e^*_{T_{\lambda'}} g(x_1, \ldots, x_{n+m}),$$

where $\lambda' \vdash n$ is the conjugate partition of λ.

PROOF. Let $g = a_0 x_{i_1} a_1 \cdots a_{n-1} x_{i_n} a_n$ where $\{i_1, \ldots, i_n\} = \{1, \ldots, n\}$ and $a_0, \ldots, a_n$ are (possibly trivial) monomials in $x_{n+1}, \ldots, x_{n+m}$. Then

$$f(x_1, \ldots, x_{n+m}) = \sum_{\sigma \in P} \sum_{\tau \in Q} (\operatorname{sgn} \tau) a_0 x_{\sigma\tau(i_1)} a_1 \cdots a_{n-1} x_{\sigma\tau(i_n)} a_n$$

and, by the definition of $\tilde{f}$,

$$\tilde{f}(x_1, \ldots, x_{n+m}) = \sum_{\sigma \in P} \sum_{\tau \in Q} (\operatorname{sgn} \tau)(\operatorname{sgn} \sigma\tau\rho) a_0 x_{\sigma\tau(i_1)} a_1 \cdots a_{n-1} x_{\sigma\tau(i_n)} a_n,$$

where $\rho = \begin{pmatrix} 1 & 2 & \cdots & n \\ i_1 & i_2 & \cdots & i_n \end{pmatrix}$.

Since $(\operatorname{sgn} \tau)(\operatorname{sgn} \sigma\tau\rho) = (\operatorname{sgn} \sigma\rho)$, we have

$$\tilde{f}(x_1, \ldots, x_{n+m}) = (\operatorname{sgn} \rho) \sum_{\sigma \in P} \sum_{\tau \in Q} (\operatorname{sgn} \sigma) \sigma\tau g(x_1, \ldots, x_{n+m}),$$

as required. If $f = e_{T_\lambda} g$, then $P = R_{T_\lambda}, Q = C_{T_\lambda}$. Denote by λ' the conjugate partition of λ. Since the columns of $T_{\lambda'}$ coincide with the rows of T_λ and the rows of $T_{\lambda'}$ are the columns of T_λ, we have $R_{T_{\lambda'}} = C_{T_\lambda}$ and $C_{T_{\lambda'}} = R_{T_\lambda}$. Hence

$$\tilde{f} = (\operatorname{sgn} \rho) \sum_{\sigma \in C_{T_{\lambda'}}} \sum_{\tau \in R_{T_{\lambda'}}} (\operatorname{sgn} \sigma) \sigma\tau g(x_1, \ldots, x_{n+m}),$$

and the proof of the lemma is complete. $\square$

THEOREM 4.8.7. *Let $A = A^{(0)} \oplus A^{(1)}$ be a finite dimensional superalgebra. Then, for all $n \geq 1$, $\chi_n(G(A)) \subseteq H(k, l)$ where $k = \dim A^{(0)}$ and $l = \dim A^{(1)}$.*

PROOF. Let $\lambda \vdash n$ be a partition such that χ_λ has non-zero multiplicity in $\chi_n(G(A))$. Then there exists a tableau T_λ and a multilinear monomial $g(x_1, \ldots, x_n)$ such that

$$f(x_1, \ldots, x_n) = e_{T_\lambda} g(x_1, \ldots, x_n)$$

is not an identity of $G(A)$. Consider a non-zero evaluation of f on homogeneous elements of $G(A)$. Suppose that the variables $x_1, \ldots, x_p$ are evaluated into odd elements and $x_{p+1}, \ldots, x_n$ into even elements. By Lemma 3.7.4, the polynomial $\tilde{f}(x_1, \ldots, x_n)$ is not a graded identity of A.

Consider the $S_p \times S_{n-p}$ permutation action on P_n, where S_p acts on the variables $x_1, \ldots, x_p$ and S_{n-p} on the variables $x_{p+1}, \ldots, x_n$ and let M be the $S_p \times S_{n-p}$-submodule generated by $f(x_1, \ldots, x_n)$. Decompose M into $S_p \times S_{n-p}$-irreducible submodules $M_{\mu,\nu}$ with character $\chi_\mu \otimes \chi_\nu$ where $\mu \vdash p$ and $\nu \vdash n-p$. We shall next examine the shape of the Young diagrams D_μ and D_ν.

Let T_μ and T_ν be two Young tableaux and let $e_{T_\mu} \in FS_p$ and $e_{T_\nu} \in FS_{n-p}$ be the corresponding essential idempotents. We claim that for any multilinear monomial $h(x_1, \ldots, x_n)$, the polynomial $e_{T_\mu} e_{T_\nu} h(x_1, \ldots, x_n)$ is a graded identity of $G(A)$ as soon as $\mu_1 \geq l+1$ or $h(\nu) \geq k+1$.

Suppose first that $h(\nu) \geq k+1$. Then, since $\dim A^{(0)} = k$, any polynomial alternating on $k+1$ even variables vanishes on A. It follows that $e_{T_\nu} h(x_1, \ldots, x_n)$, so $e_{T_\mu} e_{T_\nu} h(x_1, \ldots, x_n)$ is a graded identity of A.

Similarly, suppose that $\mu_1 \geq l+1$ and let $w = e_{T_\mu} e_{T_\nu} h(x_1, \ldots, x_n)$. By Lemma 3.7.4, w is a graded identity of $G(A)$ if $\tilde{w}$ is a graded identity of A. Now, by Lemma 4.8.6, we have

$$\tilde{w}(x_1, \ldots, x_n) = \pm e^*_{T_{\mu'}} e_{T_\nu} h(x_1, \ldots, x_n)$$

where
$$e^*_{T_{\mu'}} = \sum_{\pi \in R_{T_\mu}} (\operatorname{sgn} \pi)\pi \sum_{\rho \in C_{T_\mu}} \rho.$$

In particular, $\tilde{w}(x_1, \ldots, x_n)$ is alternating on some odd variables $x_{i_1}, \ldots, x_{i_{\mu_1}}$, $1 \le i_1, \ldots, i_{\mu_1} \le p$. Since $\mu_1 > l = \dim A^{(1)}$, it follows that $\tilde{w}$ is a graded identity of A and, so, w is a graded identity of $G(A)$. This proves the claim.

Now recall that our original polynomial $f(x_1, \ldots, x_n)$ generates an irreducible S_n-module M_λ whose character is χ_λ. This module can be induced from an irreducible $S_p \times S_{n-p}$-module $M_{\mu,\nu}$ with character $\chi_\mu \otimes \chi_\nu$ where $\mu \vdash p, \nu \vdash n - p$. If $e_{T_\mu} e_{T_\nu} h(x_1, \ldots, x_n)$ is a generator of $M_{\mu,\nu}$, then $e_{T_\mu} e_{T_\nu} h(x_1, \ldots, x_n)$ is not a graded identity of $G(A)$ since f is not. By the above this implies that $\mu_1 \le l$ and $h(\nu) \le k$. Now, by the Littlewood-Richardson rule (Theorem 2.3.9), since $\mu_1 \le l$ and $h(\nu) \le k$, it follows that $\lambda \in H(k, l)$ and the proof of the theorem is complete. □

4.9. Colength growth: a polynomial upper bound

Recall that if

(4.16) $$\chi_n(\mathcal{V}) = \sum_{\lambda \vdash n} m_\lambda \chi_\lambda$$

is the nth cocharacter of a variety $\mathcal{V}$, then

$$l_n(\mathcal{V}) = \sum_{\lambda \vdash n} m_\lambda$$

is the nth colength.

In this section we shall prove that for a non-trivial variety $\mathcal{V}$, the sequence of colengths $l_n(\mathcal{V})$, $n = 1, 2, \ldots$, is polynomially bounded. This result was proved by Berele and Regev. Our proof differs from the original one and will be based on two main results of the previous sections: the existence of an infinite hook $H(k, l)$ containing all cocharacters of $\mathcal{V}$ and the existence of a finitely generated superalgebra A whose Grassmann envelope $G(A)$ generates $\mathcal{V}$.

We start with an easy remark.

LEMMA 4.9.1. *Let $\mathcal{V}$ be a non-trivial variety and let $H(k, l)$ be an infinite hook such that $\chi_n(\mathcal{V}) \subseteq H(l, k)$ for all $n = 1, 2, \ldots$. Then the number of partitions $\lambda \vdash n$ for which $m_\lambda \ne 0$, is bounded by some polynomial in n of degree $k + l$.*

PROOF. In the language of Young diagrams the inclusion $\lambda \in H(k, l)$ means that all boxes of D_λ lie in the first k rows and l columns. The lengths of the first k rows of D_λ, i.e., the integers $\lambda_1, \ldots, \lambda_k$, satisfy the inequalities $0 \le \lambda_i \le n$ and for the choice of $\lambda_1, \ldots, \lambda_k$ we have at most $(n+1)^k$ options. Similarly, we have at most $(n+1)^l$ options for the choice of the heights of the first l columns. Hence, the total number of Young diagrams D_λ with $\lambda \in H(k, l)$ is bounded by $(n+1)^{k+l}$. □

Let us show now that for all $\lambda \vdash n$ the corresponding multiplicity m_λ in (4.16) is polynomially bounded. Fix the infinite hook $H(k, l)$ such that $\chi_n(\mathcal{V}) \subseteq H(k, l)$ for all $n = 1, 2, \ldots$, and consider the supervariety $\mathcal{V}^*$ corresponding to $\mathcal{V}$ (see Section 3.7). Let $L = L(u_1, \ldots, u_k, w_1, \ldots, w_l)$ be the relatively free superalgebra of $\mathcal{V}^*$ with even generators $u_1, \ldots, u_k$ and odd generators $w_1, \ldots, w_l$. By Theorem 4.8.2

we know that $\mathcal{V} = \mathrm{var}(G(L))$ and L is a finitely generated algebra satisfying an ordinary polynomial identity.

Denote by L_n the subspace of L spanned by all products $v_1 \cdots v_i$ where $v_1, \ldots, v_i \in \{u_1, \ldots, u_k, w_1, \ldots, w_l\}$ and $1 \le i \le n$. By Theorem 1.13.5 or its proof, there exist constants a, T such that

(4.17) $$\dim L_n \le an^T,$$

for all $n = 1, 2, \ldots$. We shall use the constants a and T in the next lemma.

LEMMA 4.9.2. *For every $n \ge 1$ and every $\lambda \vdash n$ the multiplicity m_λ in (4.16) satisfies the inequality*
$$m_\lambda \le an^T$$
where a and T are constants not depending on n.

PROOF. Suppose that there exist n and $\lambda \vdash n$ such that $m_\lambda > an^T \ge \dim L_n$ where a, T and L_n are as in (4.17). Then the S_n-module
$$P_n(\mathcal{V}) = \frac{P_n}{P_n \cap \mathrm{Id}(\mathcal{V})}$$
contains $m_\lambda > an^T$ copies of a module whose character is χ_λ. In other words, the space P_n as an S_n-module contains a submodule of the type

(4.18) $$M_1 \oplus \cdots \oplus M_q \oplus (P_n \cap \mathrm{Id}(\mathcal{V}))$$

where $M_1, \ldots, M_q$ are isomorphic irreducible S_n-modules with character χ_λ and $q > an^T$. Now by Lemma 2.4.1, we can choose a tableau T_λ such that each summand $M_i, i = 1, \ldots, q$, can be generated by a non-zero polynomial of the type $f_i = e_{T_\lambda} g_i$. Moreover, by Lemma 2.5.6, each $f_i \equiv 0$ is equivalent, as an identity, to a polynomial $h_i \equiv 0$ where h_i has the following properties: the set of variables $\{x_1, \ldots, x_n\}$ can be decomposed into a disjoint union
$$X_1 \cup \ldots \cup X_{k'} \cup Y_1 \cup \ldots \cup Y_{l'}$$
where $k' \le k, l' \le l$, and all $h_i, i = 1, \ldots, q$, are symmetric in the variables of each set $X_j, 1 \le j \le k'$, and alternating in the variables of each set $Y_j, 1 \le j \le l'$. By eventually adding some empty sets, we may assume that $k' = k$ and $l' = l$.

Since $h_1 \in M_1, \ldots, h_q \in M_q$, by (4.18) for every choice of not all zero scalars $\alpha_1, \ldots, \alpha_q \in F$, we must have $\alpha_1 h_1 + \cdots + \alpha_q h_q \notin \mathrm{Id}(\mathcal{V})$.

Now, the multilinear polynomial $\alpha_1 h_1 + \cdots + \alpha_q h_q$ is also symmetric on each of the sets $X_1, \ldots, X_k$ and alternating on each of the sets $Y_1, \ldots, Y_l$. Denote
$$s_1 = |X_1|, \ldots, s_k = |X_k|, r_1 = |Y_1|, \ldots, r_l = |Y_l|$$
and let
$$g_i = g_i(x_1, \ldots, x_k, y_1, \ldots, y_l) =$$
$$\widetilde{h}_i(\underbrace{x_1, \ldots, x_1}_{s_1}, \ldots, \underbrace{x_k, \ldots, x_k}_{s_k}, \underbrace{y_1, \ldots, y_1}_{r_1}, \ldots, \underbrace{y_l, \ldots, y_l}_{r_l}), \quad i = 1, \ldots, q,$$
where $\widetilde{h}_i$ is constructed from h_i using the procedure described in Section 3.7 under the assumption that the variables of $X_1 \cup \ldots \cup X_k$ are even and the variables of $Y_1 \cup \ldots \cup Y_l$ are odd.

Recall that $L = L(u_1, \ldots, u_k, w_1, \ldots, w_l)$ is a relatively free superalgebra of the supervariety $\mathcal{V}^*$ and

(4.19) $$\dim L_n \le an^T < q.$$

Since the total degree of g_i is equal to n, its valuation $g_i(u_1, \ldots, u_k, w_1, \ldots, w_l) = \bar{g}_i$ belongs to L_n. From the inequality (4.19) since $q > an^T$, it follows that there exist scalars $\alpha_1, \ldots, \alpha_q$, not all zero, such that

$$\alpha_1 \bar{g}_1 + \cdots + \alpha_q \bar{g}_q = 0$$

in L. But then $f = \alpha_1 h_1 + \cdots + \alpha_q h_q$ satisfies the hypothesis of Lemma 4.8.1, hence $f \equiv 0$ is an identity of $\mathcal{V}$ and $\alpha_1 h_1 + \cdots + \alpha_q h_q \in \mathrm{Id}(\mathcal{V})$. This contradiction completes the proof of the lemma. $\square$

From Lemmas 4.9.1 and 4.9.2 it immediately follows

THEOREM 4.9.3 ([**BR1**]). *If $\mathcal{V}$ is a non-trivial variety, the sequence of colengths of $\mathcal{V}$ is polynomially bounded, i.e., there exist constants C and k such that*

$$l_n(\mathcal{V}) = \sum_{\lambda \vdash n} m_\lambda \leq Cn^k.$$

for all $n \geq 1$.

CHAPTER 5

Matrix Invariants and Central Polynomials

In this chapter we present a brief introduction to the theory of matrix invariants over a field of characteristic zero as developed by Procesi in [**Pr2**] and we apply the theory in order to draw some important consequences to the theory of polynomial identities.

As in general invariant theory, we present a first fundamental theorem giving the generators of the ring of invariants and a second fundamental theorem giving the relations among the (multilinear) generators. This leads to the notion of trace polynomial identity for the algebra of $n \times n$ matrices and a thorough study of these identities allows us to prove a theorem of Formanek (Theorem 5.7.4) on the existence of certain multilinear central polynomials for $M_k(F)$, which are alternating on two distinct sets of variables each of order k^2. This polynomial will be a basic tool in the proof of the existence of the PI-exponent of an algebra defined in the next chapter.

There is a natural action of the symmetric group on the space of multilinear trace identities in a fixed number of variables. Accordingly, one can also define, as in the ordinary case, the sequence of trace cocharacters and the sequence of trace codimensions. These sequences are easier to handle since the action of the symmetric group is better understood. The last part of the chapter is devoted to the proof that the trace codimensions and the ordinary codimensions are asymptotically equal and this leads to an explicit computation of the asymptotics for the codimensions of $n \times n$ matrices.

Throughout this chapter we assume that the ground field F is of characteristic zero.

5.1. S_n-action on tensor space

Let V be a vector space, $\dim_F V = k < \infty$, and let

$$V^{\otimes n} = \underbrace{V \otimes \cdots \otimes V}_{n}$$

be the tensor product of n copies of V. The symmetric group S_n acts on $V^{\otimes n}$ from the left by permuting coordinates and the corresponding representation is $\psi : FS_n \to \mathrm{End}(V^{\otimes n})$ where

(5.1) $$\psi(\sigma)(v_1 \otimes \cdots \otimes v_n) = v_{\sigma^{-1}(1)} \otimes \cdots \otimes v_{\sigma^{-1}(n)}$$

for $\sigma \in S_n$, $v_1, \ldots, v_n \in V$. Also, the general linear group $GL(V)$ acts diagonally on $V^{\otimes n}$: if $A \in GL(V)$,

$$A(v_1 \otimes \cdots \otimes v_n) = Av_1 \otimes \cdots \otimes Av_n,$$

for $v_1, \ldots, v_n \in V$.

A celebrated theorem of Schur ([**S**]) asserts that these two actions have a double centralizing property in $\text{End}(V^{\otimes n})$. As a consequence one can deduce the finite dimensional polynomial representations of $GL(V)$ from those of S_n. Recall that the group algebra FS_n has the decomposition $FS_n = \oplus_{\lambda \vdash n} I_\lambda$ where I_λ is the minimal two-sided ideal associated to λ.

Continuing the work of Schur, Weyl ([**W**, p. 127ff]) determined $\psi(FS_n)$ as follows.

THEOREM 5.1.1 ([**S**], [**W**]).
$$\ker \psi = \bigoplus_{\substack{\lambda \vdash n \\ h(\lambda) > k}} I_\lambda. \quad \text{Hence} \quad \psi(FS_n) \simeq \bigoplus_{\substack{\lambda \vdash n \\ h(\lambda) \leq k}} I_\lambda.$$

An immediate consequence of Theorem 5.1.1 is the following result that we shall use in the sequel. Recall that for $\lambda \vdash n$, $d_\lambda = \chi_\lambda(1)$ is the degree of the irreducible S_n-character associated to λ.

COROLLARY 5.1.2. *For any* $n \geq k$, $\sum_{\substack{\lambda \vdash n \\ h(\lambda) \leq k}} d_\lambda^2 \leq k^{2n}$.

PROOF. Since $k^{2n} = \dim_F \text{End}(V^{\otimes n})$, we obtain $k^2 \geq \dim \psi(FS_n) = \sum_{\substack{\lambda \vdash n \\ h(\lambda) \leq k}} d_\lambda^2$. □

Let V^* be the dual space of V. For $\omega \in V^*$ and $v \in V$ we use the notation $\omega(v) = \langle v, \omega \rangle$. Then there is a natural identification of $V \otimes V^*$ and $\text{End}(V)$ if we set
$$(v \otimes \omega)(v') = \langle v', \omega \rangle v,$$
for $v, v' \in V, w \in V^*$.

Note that if $x_1 \otimes \omega_1, x_2 \otimes \omega_2 \in V \otimes V^*$, then
$$(x_1 \otimes \omega_1) \cdot (x_2 \otimes \omega_2) = \langle x_2, \omega_1 \rangle x_1 \otimes \omega_2$$
and $\text{tr}(x \otimes \omega) = \langle x, \omega \rangle$, for $x \otimes \omega \in V \otimes V^*$. Also, $(V^*)^{\otimes n}$ is naturally identified with $(V^{\otimes n})^*$ where
$$\langle x_1 \otimes \cdots \otimes x_n, \omega_1 \otimes \cdots \otimes \omega_n \rangle = \prod_{i=1}^n \langle x_i, \omega_i \rangle.$$

Next we prove a result due to Kostant ([**Ks**]) giving properties of the trace of the endomorphism $\psi(\sigma)$ defined in (5.1).

LEMMA 5.1.3. *Let* $\sigma \in S_n$ *and suppose that*
$$\sigma^{-1} = (i_1 \ldots i_{r_1}) \cdots (l_1 \ldots l_{r_t})$$
is the decomposition of σ^{-1} *into disjoint cycles. Then, for all* $A_1, \ldots, A_n \in \text{End}(V)$, *we have that*
$$\text{tr}(\psi(\sigma) \cdot (A_1 \otimes \cdots \otimes A_n)) = \text{tr}(A_{i_1} \cdots A_{i_{r_1}}) \cdots \text{tr}(A_{l_1} \cdots A_{l_{r_t}}).$$

PROOF. Let $\{e_1, \ldots, e_n\}$ be a basis of V and $\{\theta_1, \ldots, \theta_n\}$ the corresponding dual basis of V^*. Then $\{e_{j_1} \otimes \cdots \otimes e_{j_n}\}_{j_1, \ldots, j_n}$ and $\{\theta_{j_1} \otimes \cdots \otimes \theta_{j_n}\}_{j_1, \ldots, j_n}$ are dual bases of $V^{\otimes n}$ and $(V^*)^{\otimes n}$ respectively, and for $C \in \text{End}(V^{\otimes n})$, we have
$$C(e_{j_1} \otimes \cdots \otimes e_{j_n}) = \sum_{1 \leq s_i \leq n} \langle C(e_{j_1} \otimes \cdots \otimes e_{j_n}), \theta_{s_1} \otimes \cdots \otimes \theta_{s_n} \rangle e_{s_1} \otimes \cdots \otimes e_{s_n}.$$

It follows that
$$\operatorname{tr} C = \sum_{1 \leq j_i \leq n} \langle C(e_{j_1} \otimes \cdots \otimes e_{j_n}), \theta_{j_1} \otimes \cdots \otimes \theta_{j_n} \rangle.$$

Now, for $i = 1, \ldots, n$, take $A_i = x_i \otimes \omega_i \in V \otimes V^*$, and we wish to compute
$$\operatorname{tr}(A_1 \otimes \cdots \otimes A_n \cdot \psi(\sigma)).$$

We first compute
$$\langle A_1 \otimes \cdots \otimes A_n \cdot \psi(\sigma)(e_{j_1} \otimes \cdots \otimes e_{j_n}), \theta_{j_1} \otimes \cdots \otimes \theta_{j_n} \rangle$$
$$= \langle A_1 \otimes \cdots \otimes A_n(e_{j_{\sigma^{-1}(1)}} \otimes \cdots \otimes e_{j_{\sigma^{-1}(n)}}), \theta_{j_1} \otimes \cdots \otimes \theta_{j_n} \rangle$$
$$= \langle (x_1 \otimes \omega_1)(e_{j_{\sigma^{-1}(1)}}) \otimes \cdots \otimes (x_n \otimes \omega_n)(e_{j_{\sigma^{-1}(n)}}), \theta_{j_1} \otimes \cdots \otimes \theta_{j_n} \rangle$$
$$= \langle \langle e_{j_{\sigma^{-1}(1)}}, \omega_1 \rangle x_1 \otimes \cdots \otimes \langle e_{j_{\sigma^{-1}(n)}}, \omega_n \rangle x_n, \theta_{j_1} \otimes \cdots \otimes \theta_{j_n} \rangle$$
$$= \prod_{i=1}^{n} \langle e_{j_{\sigma^{-1}(i)}}, \omega_i \rangle \langle x_1 \otimes \cdots \otimes x_n, \theta_{j_1} \otimes \cdots \otimes \theta_{j_n} \rangle$$
$$= \prod_{i=1}^{n} \langle x_i, \theta_{j_i} \rangle \langle e_{j_i}, \omega_{\sigma(i)} \rangle.$$

Thus we obtain
$$\operatorname{tr}(A_1 \otimes \cdots \otimes A_n \cdot \psi(\sigma)) = \sum_{1 \leq j_i \leq n} \prod_{i=1}^{n} \langle \langle x_i, \theta_{j_i} \rangle e_{j_i}, \omega_{\sigma(i)} \rangle$$
$$= \prod_{i=1}^{n} \sum_{j_i} \langle \langle x_i, \theta_{j_i} \rangle e_{j_i}, \omega_{\sigma(i)} \rangle = \prod_{i=1}^{n} \langle x_i, \omega_{\sigma(i)} \rangle.$$

We now rearrange the factors in the above product as follows
$$\prod_{i=1}^{n} \langle x_i, \omega_{\sigma(i)} \rangle = \left(\langle x_{i_1}, \omega_{i_{\delta_1}} \rangle \prod_{p=2}^{\delta_1} \langle x_{i_p}, \omega_{i_{p-1}} \rangle \right)$$
$$\cdot \left(\langle x_{i_{\delta_1+1}}, \omega_{i_{\delta_1+\delta_2}} \rangle \prod_{p=2}^{\delta_2} \langle x_{i_{\delta_1+p}}, \omega_{i_{\delta_1+p-1}} \rangle \right) \cdot \ldots$$
$$= \operatorname{tr}(A_{i_1} A_{i_2} \cdots A_{i_{\delta_1}}) \operatorname{tr}(A_{i_{\delta_1+1}} \cdots A_{i_{\delta_1+\delta_2}}) \cdots \quad .$$

In order to check the last equality, we have, for instance,
$$\operatorname{tr}(A_{i_1} A_{i_2} \cdots A_{i_{\delta_1}}) = \operatorname{tr}((x_{i_1} \otimes \omega_{i_1}) \cdots (x_{i_{\delta_1}} \otimes \omega_{i_{\delta_1}}))$$
$$= \langle x_{i_2}, \omega_{i_1} \rangle \operatorname{tr}((x_{i_1} \otimes \omega_{i_3}) \cdots) = \ldots$$
$$= \langle x_{i_2}, \omega_{i_1} \rangle \langle x_{i_3}, \omega_{i_2} \rangle \cdots \langle x_{i_{\delta_1}}, \omega_{i_{\delta_1-1}} \rangle \langle x_{i_1}, \omega_{i_{\delta_1}} \rangle.$$

$\square$

5.2. Trace identities

In this section we apply the previous lemma to the study of the so-called trace identities of the algebra $M_k(F)$.

In general, an algebra with trace is an associative algebra A endowed with a linear map tr $: A \to C$, for some commutative F-algebra C, such that for all $a, b \in A$,

1) $\text{tr}(a)b = b\text{tr}(a)$,
2) $\text{tr}(\text{tr}(a)b) = \text{tr}(a)\text{tr}(b)$,
3) $\text{tr}(ab) = \text{tr}(ba)$.

A typical example is the matrix algebra $M_k(F)$ with tr being the usual trace.

Accordingly, one can construct $F\langle X, \text{Tr}\rangle$, the free algebra with trace on the set X where Tr is a formal trace. Let $\mathcal{M}$ denote the set of all monomials in the elements of X. Then $F\langle X, \text{Tr}\rangle$ is the algebra generated by the free algebra $F\langle X\rangle$ together with the set of central (commuting) indeterminates $\text{Tr}(M), M \in \mathcal{M}$, subject to the condition that $\text{Tr}(MN) = \text{Tr}(NM)$, for all $M, N \in \mathcal{M}$. In other words, $F\langle X, \text{Tr}\rangle \cong F\langle X\rangle \otimes F[\text{Tr}(M) \mid M \in \mathcal{M}]$.

The elements of the free algebra with trace are called trace polynomials.

DEFINITION 5.2.1. If A is an algebra with trace tr, then a trace polynomial $f(x_1, \ldots, x_n, \text{Tr}) \in F\langle X, \text{Tr}\rangle$ is said to be a trace identity for A if after substituting the variables x_i with arbitrary elements $a_i \in A$ and Tr with the trace tr, we obtain 0.

It is readily seen that the process of multilinearization described in Chapter 1 can be extended to the case of trace polynomials. We make the following.

DEFINITION 5.2.2. The space of multilinear elements of the free algebra with trace in the first n variables is called the space of multilinear mixed trace polynomials in $x_1, \ldots, x_n$ and is denoted by MT_n.

Hence the elements of MT_n are linear combinations of expressions of the type

$$\text{Tr}(x_{i_1} \cdots x_{i_a}) \cdots \text{Tr}(x_{j_1} \cdots x_{j_b}) x_{l_1} \cdots x_{l_c}$$

where $\{i_1, \ldots, i_a, \ldots, j_1, \ldots, j_b, l_1, \ldots, l_c\} = \{1, \ldots, n\}$.

A prominent role among the elements of MT_n is played by the so-called pure trace polynomials, i.e., polynomials such that all the variables $x_1, \ldots, x_n$ appear inside a trace.

DEFINITION 5.2.3. The space of multilinear pure trace polynomials in $x_1, \ldots, x_n$ is the space

$$PT_n = \text{span}\{\text{Tr}(x_{i_1} \cdots x_{i_a}) \cdots \text{Tr}(x_{j_1} \cdots x_{j_b}) \mid \{i_1, \ldots, i_a,$$
$$\ldots, j_1, \ldots, j_b\} = \{1, \ldots, n\}\}.$$

As in the case of the free associative algebra we shall now introduce an action of the symmetric group on the space of multilinear trace polynomials.

For a permutation $\sigma \in S_n$ write

$$\sigma^{-1} = (i_1 \ldots i_{r_1})(j_1 \ldots j_{r_2}) \cdots (l_1 \ldots l_{r_t})$$

as a product of disjoint cycles, including one-cycles and let us assume that $r_1 \geq r_2 \geq \cdots \geq r_t$. In this case we say that σ has cycle type $\lambda = (r_1, \ldots, r_t)$.

5.2. TRACE IDENTITIES

We then define the pure trace monomial

$$T_\sigma(x_1,\ldots,x_n) = \text{Tr}(x_{i_1}\cdots x_{i_{r_1}})\text{Tr}(x_{j_1}\cdots x_{j_{r_2}})\cdots \text{Tr}(x_{l_1}\cdots x_{l_{r_t}})$$

and, if $a = \sum_{\sigma \in S_n} \alpha_\sigma \sigma \in FS_n$, we also define

$$T_a(x_1,\ldots,x_n) = \sum_{\sigma \in S_n} \alpha_\sigma T_\sigma(x_1,\ldots,x_n).$$

Let

(5.2) $$\varphi : FS_n \to PT_n$$

be the map defined by $a \mapsto T_a(x_1,\ldots,x_n)$. Clearly, φ is a linear isomorphism. There is a natural left action (permutation action) of the symmetric group on the space PT_n defined by

$$\tau T_a(x_1,\ldots,x_n) = T_a(x_{\tau(1)},\ldots,x_{\tau(n)})$$

where $\tau \in S_n$. Since

$$T_{\tau\sigma\tau^{-1}}(x_1,\ldots,x_n) = T_\sigma(x_{\tau(1)},\ldots,x_{\tau(n)}),$$

we see that φ is actually an S_n-module isomorphism and the permutation action of S_n on PT_n corresponds to the conjugation action on FS_n.

Let $PT_n(k)$ be the subspace of PT_n of trace identities of $M_k(F)$. Hence $PT_n(k)$ is the space of multilinear pure trace identities of $M_k(F)$ in $x_1,\ldots,x_n$. It is clear that $PT_n(k)$ is an S_n-submodule of PT_n. The character of the module $PT_n/PT_n(k)$ is called the nth pure trace cocharacter of $k \times k$ matrices and is denoted by

$$\chi\left(\frac{PT_n}{PT_n(k)}\right) = \chi_n^{tr}(M_k(F)).$$

Moreover $t_n(M_k(F)) = \dim_F PT_n/PT_n(k)$ is called the nth trace codimension of $M_k(F)$. We next compute the preimage of $PT_n(k)$ under the map φ given in (5.2). Our proof will be based on a reduction to Theorem 5.1.1 through Lemma 5.1.3.

THEOREM 5.2.4 (Procesi [**Pr2**], Razmyslov [**R2**]).

$$\bigoplus_{\substack{\lambda \vdash n \\ h(\lambda) > k}} I_\lambda \stackrel{\varphi}{\cong} PT_n(k).$$

PROOF. Let $T_a(x_1,\ldots,x_n) \in PT_n(k)$ where $a = \sum_{\sigma \in S_n} \alpha_\sigma \sigma$. Then

$$\sum_{\sigma \in S_n} \alpha_\sigma T_\sigma(A_1,\ldots,A_n) = 0,$$

for all $A_1,\ldots,A_n \in \text{End}(V) \equiv M_k(F)$. Recall the map $\psi : FS_n \to \text{End}(V^{\otimes n})$ defined by (5.1). Since by Lemma 5.1.3, $T_\sigma(A_1,\ldots,A_n) = \text{tr}(\psi(\sigma)\cdot(A_1\otimes\cdots\otimes A_n))$, we obtain that $T_a(x_1,\ldots,x_n) \in PT_n(k)$ if and only if

$$\text{tr}(\psi(a)\cdot(A_1\otimes\cdots\otimes A_n)) = \sum_{\sigma \in S_n} \alpha_\sigma \text{tr}(\psi(\sigma)\cdot(A_1\otimes\cdots\otimes A_n)) = 0,$$

for all $A_1,\ldots,A_n \in \text{End}(V)$.

It follows that $T_a \in PT_n(k)$ if and only if $\text{tr}(\psi(a)\cdot B) = 0$, for all $B \in \text{End}(V^{\otimes n})$. By the non-degeneracy of the trace, this is equivalent to $\psi(a) = 0$. Therefore $PT_n(k) = \ker \psi$ and, by applying Theorem 5.1.1 the proof is complete. □

As an immediate consequence we can get some information on the nth trace cocharacter of $M_k(F)$. In fact, since the S_n-character of I_λ under the conjugation action is $\chi_\lambda \otimes \chi_\lambda$, we obtain the following corollary.

COROLLARY 5.2.5. *The nth pure trace cocharacter of $M_k(F)$ is*
$$\chi_n^{tr}(M_k(F)) = \sum_{\substack{\lambda \vdash n \\ h(\lambda) \leq k}} \chi_\lambda \otimes \chi_\lambda$$

and the nth trace codimension satisfies $t_n(M_k(F)) = \sum_{\substack{\lambda \vdash n \\ h(\lambda) \leq k}} d_\lambda^2$.

5.3. A primer of matrix invariants

Let U be a finite dimensional vector space with basis $\{u_1, \ldots, u_m\}$ over the field F of characteristic zero. Let $F[U]$ denote the symmetric algebra of U over F, i.e., the polynomial algebra $F[U] = F[u_1, \ldots, u_m]$ in the m variables $u_1, \ldots, u_m$. $F[U]$ is naturally graded by the degree

$$F[U] = F \oplus U \oplus S^2(U) \oplus \cdots,$$

where $S^k(U)$ is the kth symmetric power of U, that is, the vector space spanned by all monomials in $u_1, \ldots, u_m$ of total degree k.

The general linear group $GL(U) \simeq GL(m, F)$ acts naturally on U, and this action extends diagonally to all of $F[U]$, inducing the group of homogeneous automorphisms of $F[U]$: $g(f(u_1, \ldots, u_m)) = f(g(u_1), \ldots, g(u_m))$ where $g \in GL(U)$ and $f(u_1, \ldots, u_m) \in F[U]$.

If G is a subgroup of $GL(U)$, G acts on $F[U]$ and we denote by

$$F[U]^G = \{f \in F[U] \mid g(f) = f, \text{ for all } g \in G\}$$

the algebra of invariants of this action. Notice that since G acts as a group of homogeneous automorphisms, $F[U]^G$ inherits the grading of $F[U]$.

Now, in general a theorem giving an explicit set of generators for the ring of invariants of a group $G \subseteq GL(U)$ is called a first fundamental theorem of the invariant theory of G. Whereas a theorem describing the relations among a set of generators is called a second fundamental theorem. We shall next describe the first and the second fundamental theorem for a special subgroup of $GL(U)$ which is related to the study of the trace identities of matrices.

Fix $k \geq 1$ and let $U = M_k(F) \oplus \cdots \oplus M_k(F) = M_k(F)^n$ be the vector space direct sum of n copies of $M_k(F)$. We let the group $GL(k, F)$ act on $M_k(F)^n$ via the adjoint action:

$$(A_1, \ldots, A_n) \to (PA_1 P^{-1}, \ldots, PA_n P^{-1})$$

where $A_1, \ldots, A_n \in M_k(F)$ and $P \in GL(k, F)$. Note that actually the acting group is $PGL(k, F) = GL(k, F)/Z$, where Z is the center of $GL(k, F)$.

DEFINITION 5.3.1. *The ring of invariants* $F[M_k(F)^n]^{GL(k,F)}$ *is called the ring of invariants of n $k \times k$ matrices and is denoted by* $C(k, n)$.

The ring of invariants can also be defined in terms of generic $k \times k$ matrices. If $\xi_t = (\xi_{ij}^{(t)}), t = 1, \ldots, n$, are n generic matrices, then the group $GL(k, F)$ acts on the polynomial ring $F[\xi_{ij}^{(t)} \mid 1 \le i,j \le k, t = 1,\ldots,n]$ as follows: if $P \in GL(k,F)$ and $P\xi_t P^{-1} = \sum \bar{\xi}_{ij}^{(t)} e_{ij}$, then $P \cdot \xi_{ij}^{(t)} = \bar{\xi}_{ij}^{(t)}$.

For this action $C(k,n) = F[\xi_{ij}^{(t)} \mid 1 \le i,j \le k, t = 1,\ldots,n]^{GL(k,F)}$ is the ring of invariants.

In case $n = 1$, i.e., there is only one generic matrix ξ_1, a classical result (see [**Sr**], p. 10) states that the ring of invariants $C(k,1)$ is a polynomial ring in k variables. Moreover, these can be choosen to be the coefficients of the characteristic polynomial of ξ_1. In characteristic zero, by Newton's formulas (Proposition 1.6.3) these coefficients can be expressed in terms of $\mathrm{tr}(\xi_1), \mathrm{tr}(\xi_1^2), \ldots, \mathrm{tr}(\xi_1^k)$. Thus $C(k,1)$ is the polynomial ring in the variables $\mathrm{tr}(\xi_1^i)$, $i = 1, \ldots, k$. This fact can be generalized to $n \ge 2$. First note that if $\xi_{i_1}, \ldots, \xi_{i_s}$ are not necessarily distinct generic matrices, then for $P \in GL(k,F)$,

$$P \cdot \mathrm{tr}(\xi_{i_1} \cdots \xi_{i_s}) = \mathrm{tr}(P\xi_{i_1} P^{-1} \cdots P\xi_{i_s} P^{-1}) = \mathrm{tr}(\xi_{i_1} \cdots \xi_{i_s}).$$

Hence $\mathrm{tr}(\xi_{i_1} \cdots \xi_{i_s}) \in C(k,n)$. Conversely, in characteristic zero all the invariants can be expressed in terms of traces of products of generic matrices. This is the content of the following.

THEOREM 5.3.2 (First Fundamental Theorem of Matrix Invariants; Gurevich [**Gu**], Sibirskii [**Si**], Procesi [**Pr2**]). *The ring of invariants $C(k,n)$ is generated, as an F-algebra by all traces $\mathrm{tr}(\xi_{i_1} \cdots \xi_{i_s})$, where $\xi_{i_1} \cdots \xi_{i_s}$ is a monomial in generic matrices.*

Concerning a second fundamental theorem of matrix invariants, in characteristic zero the usual multilinearization process shows that all relations are determined by the multilinear ones. In this case a description of the multilinear relations among the matrix invariants is given as follows.

Let $C(k,n)^{mult}$ denote the subspace of $C(k,n)$ consisting of all multilinear elements in $\xi_1, \ldots, \xi_n$. Let $\varphi : FS_n \to PT_n$ be the map defined in (5.2) of the previous section by $a \mapsto T_a(x_1, \ldots, x_n)$, and let $\nu : PT_n \to C(k,n)^{mult}$ be the evaluation map obtained by mapping $x_i \mapsto \xi_i$ for $i = 1, \ldots, n$. Then the kernel of the composition map

(5.3) $$\theta = \nu \circ \varphi : FS_n \to C(k,n)^{mult}$$

such that $a \mapsto T_a(\xi_1, \ldots, \xi_n)$ gives all relations among the multilinear invariants. Now, $\ker \theta$ has been describe in Theorem 5.2.4 above:

$$\ker \theta = \bigoplus_{\substack{\lambda \vdash n \\ h(\lambda) > k}} I_\lambda.$$

This is the second fundamental theorem of matrix invariants.

5.4. The discriminant

As an application of the first (and second) fundamental theorem of matrix invariants we shall deduce a formula for the discriminant of $n \times n$ matrices.

DEFINITION 5.4.1. The discriminant $\Delta = \Delta(\xi_1, \ldots, \xi_{k^2})$ of the $k \times k$ matrices $\xi_1, \ldots, \xi_{k^2}$ is the determinant of the $k^2 \times k^2$ matrix U whose ith row is

$$(\xi_{11}^{(i)}, \ldots, \xi_{1k}^{(i)}, \ldots, \xi_{k1}^{(i)}, \ldots, \xi_{kk}^{(i)}),$$

i.e., the matrix ξ_i written as a $1 \times k^2$ row vector.

Clearly $\Delta(\xi_1, \ldots, \xi_{k^2})$ is non-zero as a function of $\xi_1, \ldots, \xi_{k^2}$. Also, since, for any $\pi \in S_{k^2}$, $\Delta(\xi_{\pi(1)}, \ldots, \xi_{\pi(k^2)}) = (\operatorname{sgn} \pi)\Delta(\xi_1, \ldots, \xi_{k^2})$, then Δ is a multilinear alternating function of $\xi_1, \ldots, \xi_{k^2}$.

In order to describe some properties of the discriminant we first prove two technical lemmas.

Recall that in Example 3.1.7, we proved that if F contains a primitive kth root of 1, then $M_k(F)$ can be graded by the group $H = \langle a \rangle_k \times \langle b \rangle_k$, the direct product of two cyclic groups of order k,

$$M_k(F) = \bigoplus_{h \in H} M_k(F)^{(h)}.$$

Moreover, for all $h \in H$, $M_k(F)^{(h)} = \operatorname{span}\{r_h\}$ is one-dimensional and $M_k(F)^{(1)}$ consists of scalar matrices. We now apply this to the next lemma. In fact, by using the basis given by the r_h's we can easily evaluate in the matrix algebra $M_k(F)$ any multilinear alternating polynomial.

LEMMA 5.4.2. *Let $f = f(x_1^{(1)}, \ldots, x_{k^2}^{(1)}, \ldots, x_1^{(r)}, \ldots, x_{k^2}^{(r)})$ be a multilinear polynomial in rk^2 indeterminates, alternating on each set $x_1^{(i)}, \ldots, x_{k^2}^{(i)}$, $i = 1, \ldots, r$. Then all values of f on $M_k(F)$ lie in the center of $M_k(F)$.*

PROOF. Clearly, it is enough to prove the lemma only when F is algebraically closed. Let $M_k(F) = \bigoplus_{h \in H} M_k(F)^{(h)}$ be the H-grading described above. Since f is multilinear, we can evaluate it only on of the basis $\{r_h | h \in H = \langle a \rangle_k \times \langle b \rangle_k\}$. Any alternating set $\{x_1^{(i)}, \ldots, x_{k^2}^{(i)}\}$ must be evaluated into all the k^2 basic vectors r_h, otherwise f would have zero value. But in this case f takes values in the H-homogeneous component $M_k(F)^{(g)}$ where

$$g = \left(\prod_{h \in H} h\right)^r = 1.$$

Since $M_k(F)^{(1)}$ consists of scalar matrices, our lemma is proven. $\square$

The next technical lemma is a standard result of linear algebra.

LEMMA 5.4.3. *Let W be a left module over the polynomial ring $R = F[\xi_{\alpha\beta}^{(\gamma)}]$ and let*

$$f : (M_k(R))^{k^2} \to W$$

be an R-multilinear alternating function in k^2 variables. Then for any generic matrices $\xi_1, \ldots, \xi_{k^2} \in M_k(R)$ we have

$$f(\xi_1, \ldots, \xi_{k^2}) = \Delta(\xi_1, \ldots, \xi_{k^2}) f(e_{11}, \ldots, e_{1k}, e_{21}, \ldots, e_{2k}, \ldots, e_{k1}, \ldots, e_{kk}).$$

PROOF. Let $\xi_n = (\xi_{ij}^{(n)})$, $n = 1, \ldots, k^2$. For convenience rename $k^2 = N$ and $(e_{11}, e_{12}, \ldots, e_{kk}) = (e_1, e_2, \ldots, e_N)$. Then

$$\xi_n = \sum_j w_{nj} e_j,$$

where $(w_{n1}, \ldots, w_{nN}) = (\xi_{11}^{(n)}, \ldots, \xi_{kk}^{(n)})$. Now

$$f(\xi_1, \ldots, \xi_N) = f(\sum_{j_1} w_{1j_1} e_{j_1}, \ldots, \sum_{j_N} w_{Nj_N} e_{j_N})$$

$$= \sum_{j_1} \cdots \sum_{j_N} w_{1j_1} \cdots w_{Nj_N} f(e_{j_1}, \ldots, e_{j_N}).$$

Since f is alternating, all summands with $j_\alpha = j_\beta$, $\alpha \neq \beta$, are zero. Hence

$$f(\xi_1, \ldots, \xi_N) = \sum_{\sigma \in S_N} w_{1\sigma(1)} \cdots w_{N\sigma(N)} f(e_{\sigma(1)}, \ldots, e_{\sigma(N)})$$

$$= \left(\sum_{\sigma \in S_N} (\operatorname{sgn} \sigma) w_{1\sigma(1)} \cdots w_{N\sigma(N)} \right) f(e_1, \ldots, e_N)$$

$$= \Delta(\xi_1, \ldots, \xi_{k^2}) f(e_{11}, \ldots, e_{1k}, \ldots, e_{k1}, \ldots, e_{kk}).$$

$\square$

In particular, in case $W = F[\xi_{\alpha\beta}^{(\gamma)}]$ we obtain that $\Delta(\xi_1, \ldots, \xi_{k^2})$ is, up to a multiplicative constant, the unique multilinear alternating function of $\xi_1, \ldots, \xi_{k^2}$.

Next we claim that if $P \in GL(k, n)$, then

$$\Delta(P\xi_1 P^{-1}, \ldots, P\xi_{k^2} P^{-1}) = \Delta(\xi_1, \ldots, \xi_{k^2}),$$

i.e., Δ is a multilinear invariant of $k \times k$ matrices. In fact, let Φ^P be the linear transformation of $M_k(F)$ defined by $\Phi^P(A) = PAP^{-1}$ and let (Φ_{ij}^P) be the matrix of Φ^P relative to the ordered basis $\{e_{11}, e_{12}, \ldots, e_{kk}\}$. Then it is not difficult to compute

$$\Delta(P\xi_1 P^{-1}, \ldots, P\xi_{k^2} P^{-1}) = \det \begin{pmatrix} \Phi^P(\xi^{(1)}) \\ \vdots \\ \Phi^P(\xi^{(k^2)}) \end{pmatrix} = \det((\Phi_{ij}^P) \cdot U)$$

$$= (\det(\Phi_{ij}^P))(\det U) = \det U = \Delta(\xi_1, \ldots, \xi_{k^2}).$$

It follows that $\Delta(\xi_1, \ldots, \xi_{k^2}) \in C(k, k^2)$. Hence the discriminant can be written as $\Delta = \sum_{\sigma \in S_{k^2}} a_\sigma T_\sigma(\xi_1, \ldots, \xi_{k^2})$. We now determine which coefficients a_σ are non-zero. For a partition $\lambda \vdash k^2$, let σ_λ be any permutation of S_{k^2} of cycle type λ. Then define

$$A_\lambda(\xi_1, \ldots, \xi_{k^2}) = \sum_{\pi \in S_{k^2}} (\operatorname{sgn} \pi) T_{\pi \sigma \pi^{-1}}(\xi_1, \ldots, \xi_{k^2}).$$

Since Δ is alternating as a function of $\xi_1, \ldots, \xi_{k^2}$, it can be written as a linear combination of the A_λ, i.e.,

$$\Delta(\xi_1, \ldots, \xi_{k^2}) = \sum_{\lambda \vdash k} c_\lambda A_\lambda(\xi_1, \ldots, \xi_{k^2})$$

for some $c_\lambda \in F$.

Next we shall prove that $A_\lambda = 0$ except when $\lambda = (2k-1, 2k-3, \ldots, 5, 3, 1)$.

In fact, recall that the conjugacy class of the permutation σ_λ in S_{k^2} either coincides with a conjugacy class of the alternating group A_{k^2} or splits into two classes of A_{k^2} of the same order. This last case occurs if and only if λ is a partition of k^2 in odd distinct parts. It follows that the element $\sum_{\pi \in S_{k^2}} (\operatorname{sgn} \pi) \pi \sigma_\lambda \pi^{-1} = 0$ in FS_{k^2} unless $\lambda = (\lambda_1, \ldots, \lambda_t)$ with all λ_i odd and distinct. But, if $\lambda_1 \geq 2k$, we can rearrange A_λ so that it can be written as a linear combination of products, each containing the trace of a standard polynomial St_{λ_i} as the first factor. By the Amitsur-Levitzki theorem $A_\lambda = 0$ in this case. Thus $A_\lambda(\xi_1, \ldots, \xi_{k^2}) = 0$, unless $\lambda = (2k-1, 2k-3, \ldots, 5, 3, 1)$.

We record this fact in the following.

THEOREM 5.4.4. *Let $\delta = (2k-1, 2k-3, \ldots, 5, 3, 1) \vdash k^2$ and let σ_δ be a permutation of S_{k^2} of cycle type δ. Then*

$$\Delta(\xi_1, \ldots, \xi_{k^2}) = C_k \sum_{\pi \in S_{k^2}} (sgn\,\pi) T_{\pi \sigma_\delta \pi^{-1}}(\xi_1, \ldots, \xi_{k^2})$$

for some non-zero constant C_k.

In [**F3**] Formanek computed the constant C_k. It turns out that

$$C_k = \pm \frac{1! 2! \cdots (k-1)!}{1! 3! \cdots (2k-1)!}.$$

5.5. Invariants and central polynomials

As an application of the first and second fundamental theorem of matrix invariants we shall next prove the existence of central polynomials for $M_k(F)$ of degree $2k^2$ which are multilinear and alternating in two distinct sets of k^2 variables.

Recall that a sequence of positive integers $\lambda = (\lambda_1, \ldots, \lambda_t)$ is an unordered partition of k if $\sum_{i=1}^t \lambda_i = k$ and we write $\lambda \models k$. Then, in general if $\lambda = (\lambda_1, \ldots, \lambda_m)$ and $\mu = (\mu_1, \ldots, \mu_m)$ are two unordered partitions of k^2 with the same number of parts, we can define the polynomial

$$L_k^{\lambda, \mu}(x; y) = L_k^{\lambda, \mu}(x_1, \ldots, x_{k^2}, y_1, \ldots, y_{k^2})$$

as

$$L_k^{\lambda, \mu}(x; y) = \sum_{\sigma, \tau \in S_{k^2}} (\text{sgn}\,\sigma\tau) x_{\sigma(1)} \cdots x_{\sigma(\lambda_1)} y_{\tau(1)} \cdots y_{\tau(\mu_1)} \cdots x_{\sigma(\lambda_1 + \cdots + \lambda_{m-1} + 1)}$$

$$\cdots x_{\sigma(k^2)} y_{\tau(\mu_1 + \cdots + \mu_{m-1} + 1)} \cdots y_{\tau(k^2)}.$$

Notice that $L_k^{\lambda, \mu}(x; y)$ is multilinear and alternating as a function of $x_1, \ldots, x_{k^2}$ and of $y_1, \ldots, y_{k^2}$. Hence, if $\xi_1, \ldots, \xi_{k^2}, \eta_1, \ldots, \eta_{k^2}$ are distinct $k \times k$ generic matrices, by Lemma 5.4.3,

$$L_k^{\lambda, \mu}(\xi_1, \ldots, \xi_{k^2}, \eta_1, \ldots, \eta_{k^2})$$

$$= \Delta(\xi_1, \ldots, \xi_{k^2}) \Delta(\eta_1, \ldots, \eta_{k^2}) L_k^{\lambda, \mu}(e_{11}, \ldots, e_{kk}; e_{11}, \ldots, e_{kk})$$

and by Lemma 5.4.2 the factor $L_k^{\lambda, \mu}(e_{11}, \ldots, e_{kk}; e_{11}, \ldots, e_{kk})$ is a scalar matrix. Thus the general problem arises: classify all partitions $\lambda, \mu \models k^2$ for which $L_k^{\lambda, \mu}(x; y)$ is a central polynomial for $M_k(F)$.

For the partition $\lambda = \mu = (1, 3, 5, \ldots, 2k-1) \models k^2$ Regev conjectured that $L_k^{\lambda, \lambda}(x; y)$ is central for $M_k(F)$ and this conjecture was verified by Formanek in [**F3**]. Here we shall present his proof. We should remark that, by using Formanek approach, Giambruno and Valenti in [**GV**] studied the above problem in its generality and showed that actually Regev's polynomial plays a prominent role in the classification of central polynomials.

We start by making a formal definition of Regev's polynomial $L_k(x; y) = L_k^{\delta, \delta}(x_1, \ldots, x_{k^2}, y_1, \ldots, y_{k^2})$ when $\delta = (1, 3, 5, \ldots, 2k-1) \models k^2$.

DEFINITION 5.5.1.
$$L_k(x;y) = \sum_{\sigma,\tau \in S_{k^2}} (\operatorname{sgn} \sigma\tau) x_{\sigma(1)} y_{\tau(1)} x_{\sigma(2)} x_{\sigma(3)} x_{\sigma(4)} y_{\tau(2)} y_{\tau(3)} y_{\tau(4)}$$
$$\cdots x_{\sigma(k^2-2k+2)} \cdots x_{\sigma(k^2)} y_{\tau(k^2-2k+2)} \cdots y_{\tau(k^2)}.$$

Our aim in the next section will be to prove that $L_k(x;y)$ is a central polynomial for $M_k(F)$.

We first recall that the linear map $\varphi : FS_n \to C(k,n)^{mult}$ given by $\varphi(\Sigma \alpha_\sigma \sigma) = \Sigma \alpha_\sigma T_\sigma(\xi_1, \ldots, \xi_n)$ is not in general injective and we let
$$\theta_n : FS_n / \ker \varphi \to C(k,n)^{mult}$$
be the induced linear isomorphism. We write $FS_n / \ker \varphi = \overline{FS_n}$ and we denote the elements of $\overline{FS_n}$ as linear combinations $\sum_{\sigma \in S_n} \alpha_\sigma \sigma$, $\alpha_\sigma \in F$, even if this representation is not unique when $\ker \varphi \neq 0$. Also, the space $C(k,n)^{mult}$ becomes a left and right S_n-module through the map θ_n. Recall also that the permutation action on $C(k,n)^{mult}$ is induced by conjugation in S_n.

We now fix some notation:
- S_k will be the symmetric group acting on $\{1, \ldots, k\}$, S_{k^2} the symmetric group acting on $\{1^*, \ldots, (k^2)^*\}$ and S_{k+k^2} the symmetric group acting on $\{1, \ldots, k, 1^*, \ldots, (k^2)^*\}$;
- $\xi_1, \ldots, \xi_k, \eta_1, \ldots, \eta_{k^2}$ will be $k + k^2$ generic matrices;
- $C(k) = C(k,k)^{mult}$ will be the multilinear invariants of $\xi_1, \ldots, \xi_k$, $C(k^2) = C(k,k^2)^{mult}$, the multilinear invariants of $\eta_1, \ldots, \eta_{k^2}$ and $C(k+k^2) = C(k, k+k^2)^{mult}$, the multilinear invariants of $\xi_1, \ldots, \xi_k, \eta_1, \ldots, \eta_{k^2}$.

We denote by
$$\theta_k : \overline{FS_k} \to C(k), \quad \theta_{k^2} : \overline{FS_{k^2}} \to C(k^2)$$
and
$$\theta_{k+k^2} : \overline{FS_{k+k^2}} \to C(k+k^2)$$
the corresponding isomorphisms. Note that $\overline{FS_k} = FS_k$, since $\ker \varphi = 0$ in this case. Also, $C(k)C(k^2) \subseteq C(k+k^2)$ and the maps θ_k, θ_{k^2}, θ_{k+k^2} have the following property.

LEMMA 5.5.2. If $f(\xi_1, \ldots, \xi_k) \in C(k)$ and $g(\eta_1, \ldots, \eta_{k^2}) \in C(k^2)$, then
$$(\theta_{k+k^2})^{-1}(fg) = (\theta_k)^{-1}(f) \cdot (\theta_{k^2})^{-1}(g).$$

We now define a function $\Gamma : C(k) \to C(k+k^2)$ by
$$\Gamma(f(\xi_1, \ldots, \xi_k)) =$$
$$\sum (\operatorname{sgn} \pi) f(\xi_1 \eta_{\pi(1^*)}, \xi_2 \eta_{\pi(2^*)} \eta_{\pi(3^*)} \eta_{\pi(4^*)}, \ldots, \xi_k \eta_{\pi((k^2-2k+2)^*)} \cdots \eta_{\pi((k^2)^*)}).$$

Through the map Γ we shall define a linear map $C(k) \to C(k)$ that will be the main tool for proving the theorem. First we need the following basic lemma.

LEMMA 5.5.3. Let $f(\xi_1, \ldots, \xi_k) \in C(k)$. Then there exists a unique
$$\hat{f} = \hat{f}(\xi_1, \ldots, \xi_k) \in C(k)$$
such that $\Gamma(f) = \hat{f} \Delta(\eta)$ where $\Delta(\eta) = \Delta(\eta_1, \ldots, \eta_{k^2})$.

PROOF. Recall that $\Gamma(f) \in F[\xi_{ij}^{(t)}, \eta_{ql}^{(s)} | 1 \leq t \leq k, 1 \leq s \leq k^2]^{GL(k,F)}$ where $\eta_i = (\eta_{ql}^{(i)})$, $i = 1, \ldots, k^2$. Hence we may regard the entries of $\xi_1, \ldots, \xi_k$ as constants and $\Gamma(f)$ as an invariant of $\eta_1, \ldots, \eta_{k^2}$ over the field $F(\xi_{ij}^{(t)})$. Since $\Gamma(f)$ is multilinear and alternating as a function on $\eta_1, \ldots, \eta_{k^2}$, it follows that $\Gamma(f) = \hat{f}\Delta(\eta)$ for a suitable $\hat{f} \in F(\xi_{ij}^{(t)})$. Since both $\Gamma(f)$ and $\Delta(\eta)$ belong to $F[\xi_{ij}^{(t)}, \eta_{ql}^{(s)}]$, then actually $\hat{f} \in F[\xi_{ij}^{(t)}]$. Moreover, $\hat{f}$ is multilinear in $\xi_1, \ldots, \xi_k$ and is an invariant of $\xi_1, \ldots, \xi_k$. Thus $\hat{f} \in C(k)$. Since $F[\xi_{ij}^{(t)}, \eta_{ql}^{(s)}]$ is a domain, $\hat{f}$ is unique. □

We now translate the previous lemma into a property of FS_k. Recall that the isomorphisms $\theta_k : \overline{FS_k} \to C(k)$ and $\theta_{k+k^2} : \overline{FS_{k+k^2}} \to C(k+k^2)$ were defined just before Lemma 5.5.2. Then define
$$\Gamma_0 = (\theta_{k+k^2})^{-1}\Gamma\theta_k,$$
thus Γ_0 is defined by the following diagram:

$$\begin{array}{ccc} FS_k & \xrightarrow{\Gamma_0} & \overline{FS_{k+k^2}} \\ \theta_k \downarrow & & \downarrow \theta_{k+k^2} \\ C(k) & \xrightarrow{\Gamma} & C(k+k^2) \end{array}$$

The analogue of the previous lemma is

LEMMA 5.5.4. *Let* $\alpha = (1 1^*)(2 2^* 3^* 4^*) \cdots (k(k^2 - 2k - 2)^* \cdots (k^2)^*)$ *and* $\beta = \sum_{\pi \in S_{k^2}} (\text{sgn } \pi) \pi \alpha \pi^{-1}$. *If* $f_0 \in FS_k$, *then there exists a unique* $\hat{f}_0 \in FS_k$ *such that*
$$\Gamma_0(f_0) = f_0\beta = \hat{f}_0\Delta_0$$
where $\Delta_0 = (\theta_{k^2})^{-1}(\Delta(\eta)) \in \overline{FS_{k^2}}$.

PROOF. Let σ be a product of disjoint t-cycles of the type $(i \, l \cdots j), t \geq 1$. Then $\sigma\alpha$ is the product of $(2i-1) + (2l-1) + \cdots + (2j-1) + t = (2i + 2l + \cdots + 2j)$-cycles of the type
$$(i(i^2 - 2i + 2)^* \cdots (i^2)^* l(l^2 - 2l + 2)^* \cdots (l^2)^* \cdots j(j^2 - 2j + 2)^* \cdots (j^2)^*).$$
Hence, for any $\pi \in S_{k^2}$, $\sigma\pi\alpha\pi^{-1}$ is the product of the cycles
$$(i\pi((i^2 - 2i + 2)^*) \cdots \pi((i^2)^*) \cdots j\pi((j^2 - 2j + 2)^*) \cdots \pi((j^2)^*)).$$
It follows that
$$\sigma \xrightarrow{\theta_k} T_\sigma(\xi_1, \ldots, \xi_k) = \prod \text{tr}(\xi_i \cdots \xi_j) \xrightarrow{\Gamma}$$
$$\sum_{\pi \in S_{k^2}} (\text{sgn } \pi) \prod \text{tr}(\xi_i \eta_{\pi((i^2-2i+2)^*)} \cdots \eta_{\pi((i^2)^*)} \cdots \xi_j \eta_{\pi((j^2-2j+2)^*)} \cdots \eta_{\pi((j^2)^*)})$$
$$\xrightarrow{(\theta_{k+k^2})^{-1}} \sum_{\pi \in S_{k^2}} (\text{sgn } \pi)\sigma\pi\alpha\pi^{-1} = \sigma\beta.$$

Hence $\Gamma_0(\sigma) = \sigma\beta$ and this is easily extended to show that $\Gamma_0(f_0) = f_0\beta$, for all $f_0 \in FS_k$. On the other hand, by applying the previous two lemmas we get
$$\Gamma_0(f_0) = (\theta_{k+k^2})^{-1}\Gamma\theta_k(f_0) = (\theta_{k+k^2})^{-1}\widehat{\theta_k(f_0)}\Delta(\eta)$$

$$= \theta_k^{-1}(\widehat{\theta_k(f_0)})\theta_{k^2}^{-1}(\Delta(\eta)) = \theta_k^{-1}(\widehat{\theta_k(f_0)})\Delta_0 = \widehat{f_0}\Delta_0,$$

where $\widehat{f_0} = \theta_k^{-1}(\widehat{\theta_k(f_0)})$. □

5.6. Constructing S_k-maps

Through the maps Γ and Γ_0 we can construct S_k-endomorphisms of $C(k)$ and FS_k. We define maps $\varphi : C(k) \to C(k)$ by $\varphi(f) = \widehat{f}$ and $\varphi_0 : FS_k \to FS_k$ by $\varphi_0(f_0) = \widehat{f_0}$. It is clear that $\varphi_0 = (\theta_k)^{-1}\varphi\theta_k$. We have

LEMMA 5.6.1. *The maps $\varphi : C(k) \to C(k)$ and $\varphi_0 : FS_k \to FS_k$ are left S_k-module homomorphisms.*

PROOF. Since $\varphi = \theta_k \varphi_0 (\theta_k)^{-1}$ and θ_k is a left S_k-homomorphism, it is enough to show that φ_0 is an S_k-homomorphism. Let $\sigma \in S_k$ and $f_0 \in FS_k$. Then $(\sigma f_0)\beta = \sigma(f_0\beta) = \sigma(\widehat{f_0}\Delta_0) = (\sigma\widehat{f_0})\Delta_0$. On the other hand, $(\sigma f_0)\beta = \widehat{\sigma f_0}\Delta_0$ and by the uniqueness of $\widehat{\sigma f_0}$ we get that $\sigma\widehat{f_0} = \widehat{\sigma f_0}$, i.e., φ_0 is a left S_k-module homomorphism. □

Since by the previous lemma $\varphi_0 : FS_k \to FS_k$ is a left S_k-module homomorphism, it is completely determined by $\varphi_0(1)$. The next step of our investigation will be the computation of $\varphi_0(1)$.

First we introduce some notation. If $\sigma \in S_k$, let us denote by $z(\sigma)$ the number of cycles appearing in the decomposition of σ into disjoint cycles.

Let I denote the $k \times k$ identity matrix and let C_k be the non-zero constant of Theorem 5.4.4. We start with the following.

LEMMA 5.6.2. *For any $\sigma \in S_k$, we have*

$$\varphi(T_\sigma)(I, \ldots, I) = \begin{cases} C_k, & \text{if } \sigma = 1, \\ 0, & \text{if } \sigma \neq 1. \end{cases}$$

PROOF. Recall that $\Gamma(T_\sigma(\xi_1, \ldots, \xi_k)) = \varphi(T_\sigma)\Delta(\eta)$, for every $\sigma \in S_k$. Hence

$$\varphi(T_\sigma)(I, \ldots, I)\Delta(\eta) = \Gamma(T_\sigma)(I, \ldots, I, \eta_1, \ldots, \eta_{k^2}) = A_{\lambda_\sigma}(\eta_1, \ldots, \eta_{k^2}),$$

where $\lambda_\sigma \vdash k^2$ is a suitable partition whose number of parts equals $z(\sigma)$. Now, if $\sigma \neq 1$, such a number is less than k; in particular, $\lambda_\sigma \neq \delta = (2k-1, 2k-3, \ldots, 3, 1)$. Then, by the discussion before Theorem 5.4.4, $A_{\lambda_\sigma} = 0$ whenever $\sigma \neq 1$ and, so, $\varphi(T_\sigma)(I, \ldots, I) = 0$. If $\sigma = 1$, then $\varphi(T_1)(I, \ldots, I)\Delta(\eta) = A_\delta(\eta_1, \ldots, \eta_{k^2}) = C_k\Delta(\eta)$, by Theorem 5.4.4. Hence $\varphi(T_1)(I, \ldots, I) = C_k$ and the proof is complete. □

We now can compute $\varphi_0(1)$.

THEOREM 5.6.3. $\varphi_0(1) = C_k(\sum_{\sigma \in S_k} k^{z(\sigma)}\sigma)^{-1}$.

PROOF. Let $\varphi_0(1) = \sum_{\sigma \in S_k} \alpha_\sigma \sigma$. Since $\varphi = \theta_k\varphi_0\theta_k^{-1}$, it follows that $\varphi(T_1) = \sum_{\sigma \in S_k} \alpha_\sigma T_\sigma$. Recalling that, by Lemma 5.6.2, φ is a left S_k-module homomorphism, we get for every $\rho \in S_k$,

$$\varphi(T_\rho)(\xi_1, \ldots, \xi_k) = \sum_{\sigma \in S_k} \alpha_\sigma T_{\rho\sigma}(\xi_1, \ldots, \xi_k).$$

Note that $T_\rho(I,\ldots,I) = k^{z(\sigma)}$. Hence, if in the equality above we specialize $\xi_1 = \cdots = \xi_k = I$, from the previous lemma we obtain

$$\sum_{\sigma \in S_k} \alpha_\sigma k^{z(\rho\sigma)} = \begin{cases} C_k, & \text{if } \rho = 1, \\ 0, & \text{if } \rho \neq 1. \end{cases}$$

Since every permutation ρ and its inverse ρ^{-1} have the same structure as products of disjoint cycles, $z(\rho) = z(\rho^{-1})$. Hence we can compute

$$\left(\sum_{\sigma \in S_k} \alpha_\sigma \sigma\right)\left(\sum_{\rho \in S_k} k^{z(\rho)} \rho\right) = \sum_{\sigma,\rho \in S_k} \alpha_\sigma k^{z(\rho)} \sigma\rho$$

$$= \sum_{\sigma,\rho \in S_k} \alpha_\sigma k^{z(\rho\sigma)} \rho = \sum_{\rho \in S_k}(\sum_{\sigma \in S_k} \alpha_\sigma k^{z(\rho\sigma)}) \rho = C_k$$

and, $\varphi_0(1) = C_k (\sum_{\rho \in S_k} k^{z(\rho)} \rho)^{-1}$ follows. □

COROLLARY 5.6.4. φ and φ_0 are S_k-bimodule isomorphisms.

PROOF. Since $\varphi_0(1)$ is a unit, φ_0 is an isomorphism. Also, since $\varphi_0(1)$ is central, φ_0 is an S_k-bimodule isomorphism. Recalling that $\varphi = \theta_k \varphi_0 \theta_k^{-1}$, the same conclusion holds for φ. □

5.7. Computing central polynomials

In this section we shall prove that Regev's polynomial $L_k(x;y)$ of Definition 5.5.1 is a central polynomial for $M_k(F)$.

We first derive a different formula for $\varphi_0(1)$. Let V be a k-dimensional vector space over F and let $\psi : FS_k \to \text{End}(V^{\otimes k})$ be the S_k-representation induced by $\psi(\sigma)(v_1 \otimes \cdots \otimes v_k) = v_{\sigma^{-1}(1)} \otimes \cdots \otimes v_{\sigma^{-1}(k)}$, where $\sigma \in S_k$, $v_1,\ldots,v_k \in V$ (see Section 5.1). Let χ_ψ denote the S_k-character of ϕ. Also, recall that $\{\chi_\lambda \mid \lambda \vdash k\}$ denotes the set of irreducible S_k-characters. If χ_1, χ_2 are two characters of a group G, let $\langle \chi_1, \chi_2 \rangle$ denote the usual inner product of χ_1 and χ_2 (recall that $\langle \chi_1, \chi_2 \rangle = \frac{1}{|G|} \sum_\sigma \chi_1(\sigma) \chi_2(\sigma^{-1})$).

We can now restate Theorem 5.6.3 as follows.

LEMMA 5.7.1.
$$\varphi_0(1) = \frac{C_k}{(k!)^2} \sum_{\sigma \in S_k} (\sum_{\lambda \vdash k} \frac{\chi_\lambda(1)^2 \chi_\lambda(\sigma)}{\langle \chi_\psi, \chi_\lambda \rangle}) \sigma.$$

PROOF. Let $\{e_\lambda \mid \lambda \vdash k\}$ be the set of minimal central idempotents of the group algebra FS_k. Recall that $e_\lambda = \frac{1}{k!} \chi_\lambda(1) \sum_{\sigma \in S_k} \chi_\lambda(\sigma) \sigma$, for every $\lambda \vdash k$ (see Remark 2.1.11).

Since $\sum_{\sigma \in S_k} k^{z(\sigma)} \sigma \in Z(FS_k)$, the center of FS_k, we can write

$$\sum_{\sigma \in S_k} k^{z(\sigma)} \sigma = \sum_{\lambda \vdash k} \alpha_\lambda e_\lambda$$

for suitable non-zero $\alpha_\lambda \in F$. Hence $(\sum_{\sigma \in S_k} k^{z(\sigma)}\sigma)^{-1} = \sum_{\lambda \vdash k} \alpha_\lambda^{-1} e_\lambda$. The α_λ's are easily computed:

$$\sum_{\sigma \in S_k} k^{z(\sigma)} \chi_\lambda(\sigma) = \chi_\lambda(\sum_{\sigma \in S_k} k^{z(\sigma)}\sigma) = \chi_\lambda(\sum_{\mu \vdash k} \alpha_\mu e_\mu) = \alpha_\lambda \chi_\lambda(e_\lambda) = \alpha_\lambda \chi_\lambda(1).$$

Hence
$$\alpha_\lambda = \frac{\sum_{\sigma \in S_k} k^{z(\sigma)} \chi_\lambda(\sigma)}{\chi_\lambda(1)}.$$

To compute α_λ^{-1}, we proceed as follows: first notice that the character χ_ψ of the representation $\psi : FS_k \to End(V^{\otimes k})$ satisfies $\chi_\psi(\sigma) = k^{z(\sigma)}$ for all $\sigma \in S_k$. Hence, for all $\lambda \vdash k$, we have that

$$\langle \chi_\psi, \chi_\lambda \rangle = \frac{1}{k!} \sum_{\sigma \in S_k} \chi_\psi(\sigma)\chi_\lambda(\sigma) = \frac{1}{k!} \sum_{\sigma \in S_k} k^{z(\sigma)}\chi_\lambda(\sigma) = \frac{\chi_\lambda(1)\alpha_\lambda}{k!}.$$

It follows that
$$\alpha_\lambda^{-1} = \frac{\chi_\lambda(1)}{k!\langle \chi_\psi, \chi_\lambda \rangle}$$

and we finally obtain

$$\varphi_0(1) = C_k (\sum_{\sigma \in S_k} k^{z(\sigma)}\sigma)^{-1} = C_k \sum_{\lambda \vdash k} \alpha_\lambda^{-1} e_\lambda$$
$$= C_k \sum_{\lambda \vdash k} (\frac{\chi_\lambda(1)}{k!\langle \chi_\psi, \chi_\lambda \rangle})(\frac{1}{k!}\chi_\lambda(1) \sum_{\sigma \in S_k} \chi_\lambda(\sigma)\sigma) = \frac{1}{(k!)^2} \sum_{\sigma \in S_k} \sum_{\lambda \vdash k} \frac{\chi_\lambda(1)^2 \chi_\lambda(\sigma)}{\langle \chi_\psi, \chi_\lambda \rangle} \sigma$$

as claimed. □

Next we compute the coefficient of k-cycle in the expression of $\varphi_0(1)$.

First we need to recall the following well-known facts of the irreducible S_k-characters.

LEMMA 5.7.2. *Let μ be an k-cycle in S_k. Then, for all $\lambda \vdash k$,*
$$\chi_\lambda(\mu) = \begin{cases} (-1)^i & \text{if } \lambda = (k-i, 1^i),\ i = 0, \ldots, k-1, \\ 0 & \text{otherwise.} \end{cases}$$

Moreover, if χ_ψ is the character of the representation $\psi : FS_k \to End(V^{\otimes k})$ defined above, then
$$\langle \chi_\psi, \chi_{(k-i,1^i)} \rangle = \binom{2k-i-1}{k}\binom{k-1}{i}.$$

PROOF. The first statement follows from the Murnaghan-Nakayama formula (see [**JK**, 2.4.7]). The second follows from the formula
$$\langle \chi_\psi, \chi_{(k-i,1^i)} \rangle = \binom{2k-i-1}{k} \chi_{(k-i,1^i)}(1)$$

given in [**JK**, 5.2.20] and from the hook formula. □

LEMMA 5.7.3. *Let $\varphi_0(1) = \sum_{\sigma \in S_k} \alpha_\sigma \sigma \in FS_k$. Then, if σ is an k-cycle, $\alpha_\sigma = (-1)^{k+1} \frac{C_k}{(k-1)!k!(2k-1)}$.*

PROOF. For every $i = 0, \ldots, k-1$, let $\varepsilon_i = (k-i, 1^i)$. Then, if σ is a k-cycle, from the previous lemmas we get

$$\alpha_\sigma = \frac{C_k}{(k!)^2} \sum_{i=0}^{k-1} \frac{\chi_{\varepsilon_i}(1)^2 \chi_{\varepsilon_i}(\sigma)}{\langle \chi_\psi, \chi_{\varepsilon_i} \rangle} = \frac{C_k}{(k!)^2} \sum_{i=0}^{k-1} (-1)^i \binom{k-1}{i} \binom{2k-i-1}{k}^{-1}$$

$$= \frac{C_k}{(k!)^2} \binom{2k-1}{k}^{-1} \sum_{i=0}^{k-1} (-1)^i \binom{2k-1}{i} = (-1)^{k+1} \frac{C_k}{(k!)^2} \frac{k}{2k-1}.$$

$\square$

We are now in a position to prove that the polynomials $L_k(x; y)$ are central in $M_k(F)$, char $F = 0$.

THEOREM 5.7.4 (Formanek [**F3**]). *The polynomial*

$$L_k(x; y) = \sum_{\sigma, \tau \in S_{k^2}} (\operatorname{sgn} \sigma\tau) x_{\sigma(1)} y_{\tau(1)} x_{\sigma(2)} x_{\sigma(3)} x_{\sigma(4)} y_{\tau(2)} y_{\tau(3)} y_{\tau(4)}$$
$$\cdots x_{\sigma(k^2-2k+2)} \cdots x_{\sigma(k^2)} y_{\tau(k^2-2k+2)} \cdots y_{\tau(k^2)}$$

is central in $M_k(F)$.

PROOF. We have already mentioned that

$$L_k(\xi_1, \ldots, \xi_{k^2}; \eta_1, \ldots, \eta_{k^2}) = \lambda I \Delta(\xi) \Delta(\eta)$$

where I is the $k \times k$ identity matrix and $\lambda \in F$. Hence the proof of the theorem amounts to show that λ is a non-zero constant. To this end, let $\varphi_0(1) = \sum_{\sigma \in S_k} \alpha_\sigma \sigma$ so that $\varphi(T_1) = \sum_{\sigma \in S_k} \alpha_\sigma T_\sigma(\xi_1, \ldots, \xi_k)$.

Now let $\mu^{-1} = (12 \ldots k)$. Since φ is a left S_k-isomorphism, we have that $\varphi(T_\mu) = \sum_{\sigma \in S_k} \alpha_\sigma T_{\mu\sigma}(\xi_1, \ldots, \xi_k)$. Recalling the definition of the map Γ, we obtain

$$\sum (\operatorname{sgn} \tau) T_\mu(\xi_1 \eta_{\tau(1)}, \xi_2 \eta_{\tau(2)} \eta_{\tau(3)} \eta_{\tau(4)}, \ldots, \xi_k \eta_{\tau(k^2-2k+2)} \cdots \eta_{\tau(k^2)})$$

(5.4)
$$= \Gamma(T_\mu(\xi_1, \ldots, \xi_k))$$
$$= \varphi(T_\mu)(\xi_1, \ldots, \xi_k) \Delta(\eta) = \sum_{\sigma \in S_k} \alpha_\sigma T_{\mu\sigma}(\xi_1, \ldots, \xi_k) \Delta(\eta).$$

We now make the substitution

$$\xi_1 \to \xi_1, \ \xi_2 \to \xi_2 \xi_3 \xi_4, \ \ldots, \ \xi_k \to \xi_{k^2-2k+2} \cdots \xi_{k^2}$$

and we alternate over all permutations of $\xi_1, \ldots, \xi_{k^2}$. Then the left-hand side of (5.4) becomes

$$\operatorname{tr}(L_k(\xi_1, \ldots, \xi_{k^2})).$$

We now compute the right-hand side of (5.4). By the definition of Γ, $T_{\mu\sigma}(\xi_1, \ldots, \xi_k)$ becomes

$$\sum_{\pi \in S_{k^2}} (\operatorname{sgn} \pi) T_{\mu\sigma}(\xi_{\pi(1)}, \xi_{\pi(2)} \xi_{\pi(3)} \xi_{\pi(4)}, \ldots, \xi_{\pi(k^2-2k+2)} \cdots \xi_{\pi(k^2)})$$

$$= \Gamma(T_{\mu\sigma})(I, \ldots, I) = \hat{T}_{\mu\sigma}(I, \ldots, I) \Delta(\xi_1, \ldots, \xi_{k^2})$$
$$\varphi(T_{\mu\sigma})(I, \ldots, I) = \Delta(\xi_1, \ldots, \xi_{k^2}).$$

Hence, by Lemma 5.5.3, we have that

$$\sum_{\pi \in S_{k^2}} \sum_{\sigma \in S_k} (\operatorname{sgn} \pi) \alpha_\sigma T_{\mu\sigma}(\xi_{\pi(1)}, \xi_{\pi(2)} \xi_{\pi(3)} \xi_{\pi(4)}, \ldots, \xi_{\pi(k^2-2k+2)} \cdots \xi_{\pi(k^2)})$$

$$= \sum_{\sigma \in S_k} \alpha_\sigma \varphi(T_{\mu\sigma})(I,\ldots,I)\Delta(\xi_1,\ldots,\xi_{k^2}) = \alpha_{\mu^{-1}} C_k \Delta(\xi).$$

Summing up we obtain from the above

$$\operatorname{tr}(L_k(\xi_1,\ldots,\xi_{k^2},\eta_1,\ldots,\eta_{k^2})) = \alpha_{\mu^{-1}} C_k \Delta(\xi)\Delta(\eta).$$

Since L_k takes only central values on $M_k(F)$, the proof is now completed by recalling the formula of $\alpha_{\mu^{-1}}$ given in Lemma 5.7.3. □

5.8. Cocharacters and trace cocharacters

Recall that in Section 5.2 we defined MT_n as the space of multilinear mixed trace polynomials in the indeterminates $x_1,\ldots,x_n$. Let $MT_n(k)$ denote the subspace of mixed trace identities of $M_k(F)$, i.e., mixed trace polynomials vanishing when evaluated on $M_k(F)$.

Recall also that $C(k,n)^{mult}$ is the space of multilinear pure trace polynomials in the $k \times k$ generic matrices $\xi_1,\ldots,\xi_n$. We now define $\bar{C}(k,n)^{mult}$ as the space of multilinear mixed trace polynomials in the $k \times k$ generic matrices $\xi_1,\ldots,\xi_n$. It is readily seen (see Section 5.3) that the evaluation map $\nu : MT_n \to \bar{C}(k,n)^{mult}$ such that $x_i \to \xi_i$, $1 \leq i \leq n$, induces the following isomorphisms of S_n-modules:

$$\frac{MT_n}{MT_n(k)} \cong \bar{C}(k,n)^{mult}, \quad \frac{PT_n}{PT_n(k)} \cong C(k,n)^{mult}.$$

The connection between pure trace identities and mixed trace identities is given as follows. Let

$$\Psi : \bar{C}(k,n)^{mult} \to C(k,n+1)^{mult}$$

be the map defined by $g(\xi_1,\ldots,\xi_n) \to \operatorname{tr}(g(\xi_1,\ldots,\xi_n)\xi_{n+1})$.

It is readily seen that Ψ is a linear isomorphism. In fact, given $f(\xi_1,\ldots,\xi_{n+1}) \in C(k,n+1)^{mult}$, clearly there exists

$$g(\xi_1,\ldots,\xi_n) \in \bar{C}(k,n)^{mult}$$

such that $f(\xi_1,\ldots,\xi_{n+1}) = \operatorname{tr}(g(\xi_1,\ldots,\xi_n)\xi_{n+1})$ and φ is onto. On the other hand, since tr is a non-degenerate bilinear form, $\operatorname{tr}(g(x_1,\ldots,x_n)x_{n+1})$ is a trace identity of $M_k(F)$ if and only $g(x_1,\ldots,x_n)$ is.

We now let S_n act on $\bar{C}(k,n)^{mult}$ by the permutation action and this corresponds under Ψ to the permutation action of S_n on $C(k,n+1)^{mult}$ where S_n embeds into S_{n+1} as the subgroup of permutations fixing $n+1$. Hence Ψ is an S_n-module isomorphism.

We now define the mixed trace cocharacter of $M_k(F)$ as the S_n-character of $\bar{C}(k,n)^{mult} \cong MT_n/MT_n(k)$ and we denote it by $\chi_n^{mtr}(M_k(F))$. From the above discussion we have that

$$(5.5) \qquad \chi_n^{mtr}(M_k(F)) = \chi_{n+1}^{tr}(M_k(F)) \downarrow S_n$$

where $\chi_{n+1}^{tr}(M_k(F)) \downarrow S_n$ is the restriction of the $(n+1)$th pure trace cocharacter $\chi_{n+1}^{tr}(M_k(F))$ of $k \times k$ matrices to S_n. By standard arguments, since $\dim_F M_k(F) = k^2$ (cf. Theorem 4.6.2), it is easy to prove that

$$\chi_n^{mtr}(M_k(F)) = \sum_{\substack{\lambda \vdash n \\ h(\lambda) \leq k^2}} \bar{m}_\lambda \chi_\lambda$$

for some multiplicities $\bar{m}_\lambda \geq 0$.

The connection between mixed trace cocharacters and ordinary cocharacters of $M_k(F)$ is given in the following theorem that we shall only state here without proof.

THEOREM 5.8.1 (Formanek [**F2**]). *Let*
$$\chi_n(M_k(F)) = \sum_{\lambda \vdash n} m_\lambda \chi_\lambda$$
be the nth cocharacter of $M_k(F)$ and let
$$\chi_n^{mtr}(M_k(F)) = \sum_{\substack{\lambda \vdash n \\ h(\lambda) \leq k^2}} \bar{m}_\lambda \chi_\lambda$$
be the nth mixed trace cocharacter of $M_k(F)$. Then, for all $\lambda = (\lambda_1, \ldots, \lambda_{k^2}) \vdash n$ such that $\lambda_{k^2} \geq 2$, we have $m_\lambda = \bar{m}_\lambda$.

Taking into account the equality (5.5), we obtain

COROLLARY 5.8.2. *Let $\chi_n(M_k(F)) = \sum_{\substack{\lambda \vdash n \\ h(\lambda) \leq k^2}} m_\lambda \chi_\lambda$ be the nth cocharacter of $M_k(F)$ and let $\chi_{n+1}^{tr}(M_k(F))$ be the pure trace cocharacter of $M_k(F)$. If*
$$\chi_{n+1}^{tr}(M_k(F)) \downarrow S_n = \sum_{\substack{\lambda \vdash n \\ h(\lambda) \leq k^2}} \bar{m}_\lambda \chi_\lambda,$$
then $m_\lambda = \bar{m}_\lambda$ whenever $\lambda_{k^2} \geq 2$.

The next step in computing the asymptotics of $c_n(M_k(F))$ is to prove that the multiplicities $\bar{m}_\lambda$ are polynomially bounded.

LEMMA 5.8.3. *The multiplicities $\bar{m}_\lambda$ are polynomially bounded, i.e., there exists a polynomial $f(x)$ such that $\bar{m}_\lambda \leq f(n)$, for all n and $\lambda \vdash n$.*

PROOF. Let $\lambda = (\lambda_1, \ldots, \lambda_s) \vdash n$ be such that $s \leq k^2$ and let
$$f = f(x_1, \ldots, x_n) \in MT_n$$
be a multilinear mixed trace polynomial generating an irreducible S_n-submodule of MT_n. Then the set of variables $\{x_1, \ldots, x_n\}$ can be partitioned into the disjoint union of subsets $\{x_1, \ldots, x_n\} = X_1 \cup \ldots \cup X_s$ such that for every $i = 1, \ldots, s$, f is symmetric in the variables of X_i. Now, by replacing f with σf, for suitable $\sigma \in S_n$, we may assume that
$$X_1 = \{x_1, \ldots, x_{\lambda_1}\},$$
(5.6) $$X_2 = \{x_{\lambda_1+1}, \ldots, x_{\lambda_1+\lambda_2}\},$$
$$\ldots$$
$$X_s = \{x_{\lambda_1+\cdots+\lambda_{s-1}+1}, \ldots, x_n\}.$$

Let us denote by ϕ the endomorphism of $F\langle X, \text{Tr} \rangle$ defined by setting $\phi(x) = x_i$ for all $x \in X_i$, $i = 1, \ldots, s$. Then $g = g(x_1, \ldots, x_s) = \phi(f)$ is a homogeneous trace polynomial such that its complete linearization coincides with f. In particular, f is a trace polynomial identity of $M_k(F)$ if and only if g is too.

Let $R = F\{\xi^{(1)}, \ldots, \xi^{(s)}\}$ be the algebra of s generic $k \times k$ matrices $\xi^{(t)} = \sum_{i,j=1}^{k} \xi_{ij}^{(t)} e_{ij}$. By Corollary 1.4.5, R is the relatively free algebra of rank s of var($M_k(F)$) and by Theorem 1.13.5, it has finite Gelfand-Kirillov dimension. This says that the dimension of the space

$$W_n = \text{span } \{\xi^{(i_1)} \cdots \xi^{(i_r)} \mid r \leq n, 1 \leq i_1, \ldots, i_r \leq s\}$$

is polynomially bounded as a function of n, i.e.,

$$\dim W_n \leq C_1 n^{d_1}$$

for some constants C_1, d_1. Since $s \leq k^2$ and k^2 is finite, we may take the same C_1 and d_1 for all $s = 1, \ldots, k^2$.

On the other hand, any evaluation of a pure trace monomial of degree at most n into generic matrices

$$\text{tr}(\xi^{(i_1)} \cdots \xi^{(i_l)}) \cdots \text{tr}(\xi^{(j_1)} \cdots \xi^{(j_m)})$$

is a polynomial in sk^2 indeterminates of total degree at most n. Hence the dimension of the space generated by all these evaluations is bounded by $C_2 n^{d_2}$ for some constants C_2, d_2. It follows that if $\tilde{W}_n$ is the space of expressions of the type $g(\xi^{(1)}, \ldots, \xi^{(s)})$ where $g(x_1, \ldots, x_s)$ is a mixed trace polynomial of total degree n, then

$$q = \dim \tilde{W}_n \leq Cn^d$$

where $C = C_1 C_2$ and $d = d_1 + d_2$ are constants.

We shall prove that $m_\lambda \leq q$ and this will complete the proof of the lemma. Suppose by contradiction that there exist isomorphic S_n-modules $M_1, \ldots, M_{q+1}$ in MT_n with character χ_λ and

$$(M_1 \oplus \cdots \oplus M_{q+1}) \cap MT_n(k) = 0.$$

Let $M_1 = FS_n f_1, \ldots, M_{q+1} = FS_n f_{q+1}$. By the first part of the proof we may assume that for each $i = 1, \ldots, s$, f_i is symmetric in each of the sets $X_1, \ldots, X_s$ defined in (5.6). Let also ϕ be the endomorphism of $F\langle X, \text{Tr}\rangle$ defined above and set $\phi(f_1) = g_1, \ldots, \phi(f_{q+1}) = g_{q+1}$.

Let $\bar{g}_1, \ldots, \bar{g}_{q+1}$ be obtained from $g_1, \ldots, g_{q+1}$ respectively, by evaluating the variables into generic matrices: $x_1 \to \xi^{(1)}, \ldots, x_s \to \xi^{(s)}$. Clearly $\bar{g}_1, \ldots, \bar{g}_{q+1} \in \tilde{W}_n$ and, since $\dim \tilde{W}_n = q$, there exist not all zero scalars $\alpha_1, \ldots, \alpha_{q+1} \in F$ such that $\alpha_1 \bar{g}_1 + \cdots + \alpha_{q+1} \bar{g}_{q+1} = 0$. Since $\xi^{(l)}, \ldots, \xi^{(s)}$ are generic matrices, it follows that

$$g = g(x_1, \ldots, x_s) = \alpha_1 g_1 + \cdots + \alpha_{q+1} g_{q+1} \in MT_n(k).$$

We now completely linearize g to get

$$\alpha_1 f_1 + \cdots + \alpha_{q+1} f_{q+1} \in MT_n(k),$$

a contradiction. This completes the proof of the lemma. $\square$

We refer the reader to the recent monograph of Drensky and Formanek ([**DF**]) for more information on the trace identities of matrix algebras.

5.9. Multialternating polynomials

In the last lemma of the previous section we proved that, as in the case of ordinary identities, the multiplicities $\bar{m}_\lambda$ in the trace cocharacter of $M_k(F)$ are polynomially bounded. This result allows us to compute, through Corollary 5.8.2, the asymptotics of the ordinary codimensions using trace codimensions. We only need to show that the irreducible characters χ_λ with $\lambda_{k^2} \geq 2$ dominate the other characters in the decomposition of $\chi_n(M_k(F))$. The main tool for proving this will be the theory of central polynomials developed in this chapter and the representation theory of the symmetric group.

We first prove some technical results needed for the computation of explicit asymptotics for the matrix algebras. We start with some definitions.

DEFINITION 5.9.1. Let $f(x_1, \ldots, x_n, y_1, \ldots, y_q)$ be a polynomial multilinear in $x_1, \ldots, x_n$ and let $s_1, \ldots, s_k \geq 2$ be integers. We say that f is $(k; s_1, \ldots, s_k)$-multialternating in $x_1, \ldots, x_n$ if $n = s_1 + \cdots + s_k$ and the set $\{x_1, \ldots, x_n\}$ decomposes into the disjoint union

$$\{x_1, \ldots, x_n\} = X^{(1)} \cup \cdots \cup X^{(k)}$$

of k subsets $X^{(i)} = \{x_1^{(i)}, \ldots, x_{s_i}^{(i)}\}$ of order s_i and f is alternating in $x_1^{(i)}, \ldots, x_{s_i}^{(i)}$ for all $i = 1, \ldots, k$.

In what follows we shall say that f if k-multialternating if it is $(k; s_1, \ldots, s_k)$-multialternating for some $s_1, \ldots, s_k \geq 2$. Also, if $s_1 = \cdots = s_k = s$ (and so, $n = ks$), we say that f is $(k; s)$-multialternating.

We start with the following results about the S_n-action on multialternating polynomials.

LEMMA 5.9.2. Let $f = f(x_1, \ldots, x_n, y_1, \ldots, y_q)$ be a k-multialternating polynomial in $x_1, \ldots, x_n$. If H is a subset of $\{1, \ldots, n\}$ such that $\mid H \mid \geq k+1$, then

$$(\sum_{\sigma \in S(H)} \sigma) f = 0$$

where $S(H)$ is the subgroup of S_n of all permutations fixing $\{1, \ldots, n\} \setminus H$.

PROOF. Since $\mid H \mid \geq k+1$, there exist $i, j \in H$ such that f is alternating in x_i, x_j. Then for all $\tau \in S_n$, $\tau f = -\tau(ij) f$. The conclusion of the lemma follows from the decomposition $S(H) = T \cup T(ij)$ where $T \cap T(ij) = \emptyset$ for some subset $T \subseteq S(H)$. □

LEMMA 5.9.3. Let $f = f(x_1, \ldots, x_n)$ be a $(k; s)$-multialternating polynomial in $x_1, \ldots, x_n$, $n = ks$. If $M = FS_n f$ is the left S_n-submodule of P_n generated by f, then M has character

$$\chi(M) = \sum_{\substack{\lambda \vdash n \\ \lambda_1 \leq k}} m_\lambda \chi_\lambda.$$

PROOF. Recall that, given a partition $\lambda = (\lambda_1, \ldots, \lambda_p) \vdash n$, then T_λ denotes a Young tableau of shape λ and e_{T_λ} the corresponding essential idempotent. In order to prove the lemma it is enough to show that for any $\lambda \vdash n$ and for any tableau T_λ we have $e_{T_\lambda} f = 0$ in P_n as soon as $\lambda_1 \geq k+1$.

Let T_λ be a tableau with $\lambda_1 \geq k+1$ and denote by H the set consisting of all entries of the first row of T_λ. Then $|H| = \lambda_1 \geq k+1$ and we can write

$$e_{T_\lambda} = (\sum_{\sigma \in S(H)} \sigma) e$$

for a suitable $e \in FS_n$. In fact, if $\bar{\lambda} = (\lambda_2, \ldots, \lambda_p)$ and $T_{\bar{\lambda}}$ is the tableau obtained from T_λ by erasing the first row, then

$$e = \sum_{\substack{\sigma \in R_{T_{\bar{\lambda}}} \\ \tau \in C_{T_\lambda}}} (\operatorname{sgn} \tau) \sigma \tau.$$

Since f is $(k;s)$-multialternating, also τf is $(k;s)$-multialternating for any $\tau \in S_n$. It follows that ef is a linear combination of $(k;s)$-multialternating polynomials. But then by Lemma 5.9.2 we obtain $e_{T_\lambda} f = (\sum_{\sigma \in S(H)}) e f = 0$ and the proof is complete. $\square$

Recall that $\lambda = (k^d)$ is the partition $\lambda = (k, \ldots, k)$ of $n = kd$. In this case D_λ is a rectangular diagram.

LEMMA 5.9.4. *Let A be a finite dimensional algebra, $\dim_F A = d$ and let $f = f(x_1, \ldots, x_n)$ be a $(k;d)$-multialternating polynomial in $x_1, \ldots, x_n$, $n = kd$. Then the S_n-module $FS_n f/(\mathrm{Id}(A) \cap FS_n f)$ has character*

$$(5.7) \qquad \chi\left(\frac{FS_n f}{\mathrm{Id}(A) \cap FS_n f}\right) = m_\lambda \chi_\lambda$$

where $\lambda = (k^d)$. Moreover, $m_\lambda \neq 0$ if and only if f is not an identity of A.

PROOF. By the previous lemma $\chi(FS_n f) = \sum_{\substack{\lambda \vdash n \\ \lambda_1 \leq k}} m_\lambda \chi_\lambda$. On the other hand, by Theorem 4.6.2, the nth cocharacter of A lies in a strip of height d, i.e., $\mathrm{Id}(A) \cap P_n$ contains all irreducible S_n-submodules of P_n corresponding to partitions $\lambda \vdash n$ with $h(\lambda) \geq d+1$. Hence $FS_n f$, mod $\mathrm{Id}(A) \cap FS_n f$, decomposes into the sum of irreducible S_n-modules corresponding to partitions $\lambda \vdash n$ such that

$$k \geq \lambda_1 \geq \cdots \geq \lambda_d \geq 0, \quad \lambda_1 + \cdots + \lambda_d = n = kd.$$

Then $\lambda_1 = \cdots = \lambda_d = k$ follows and (5.7) is proved. Clearly the multiplicity $m_{(k^d)}$ in (5.7) is zero if and only if f is an identity of A. $\square$

5.10. Asymptotics for the codimensions of $k \times k$ matrices

We start with the computation of a bound on the degree of an irreducible S_n-character corresponding to a rectangular diagram.

LEMMA 5.10.1. *Let d be a fixed positive integer and $\lambda \vdash n$, $\lambda = (k^d)$, $kd = n$. Then, for k large enough, we have*

$$d_\lambda = \deg \chi_\lambda > \frac{d^n}{n^{\frac{1}{2}d(d-1)}}.$$

PROOF. Recall that by the hook formula (Proposition 2.2.8)

$$\deg \chi_\lambda = \frac{n!}{\prod h_{ij}}.$$

Now, the product of the hook numbers of the last row equals

$$\prod_{j=1}^{k} h_{d,j} = k(k-1) \cdots 2 \cdot 1 = k!.$$

Similarly, for the $(d+1-i)$th row, $2 \leq i \leq d$, we have

$$\prod_{j=1}^{k} h_{(d+1-i),j} = \frac{(k+i-1)!}{(i-1)!}.$$

Hence

$$\prod h_{ij} = k! \frac{(k+1)!}{1!} \cdots \frac{(k+d-1)!}{(d-1)!} = \frac{(k!)^d (k+1)^{d-1}(k+2)^{d-2} \cdots (k+d-1)}{C}$$

where C is a constant. Using Stirling formula we obtain

$$\deg \chi_\lambda = \frac{n!}{\prod h_{ij}} = C\frac{(kd)!}{(k!)^d(k+1)^{d-1}\cdots(k+d-1)}$$

$$\simeq C\frac{\sqrt{2\pi kd}}{(\sqrt{2\pi k})^d}\frac{(kd)^{kd}}{e^{kd}}\left(\frac{e^k}{k^k}\right)^d\frac{1}{(k+1)^{d-1}\cdots(k+d-1)}$$

$$= C'\frac{d^{kd}}{k^{\frac{d-1}{2}}(k+1)^{d-1}\cdots(k+d-1)} > C'\frac{d^n}{k^d(k+1)^{d-1}\cdots(k+d-1)}$$

$$\gtrsim \frac{d^n}{n^{d+(d-1)+\cdots+1}} = \frac{d^n}{n^{\frac{1}{2}d(d+1)}}$$

since C' is a positive constant. $\square$

We now prove the first approximation for the codimension growth of $k \times k$ matrices. In the next chapter we shall determine a similar approximation for any PI-algebra.

THEOREM 5.10.2. *There exist constants $\alpha_1 \leq \alpha_2, \gamma_1 \neq 0$ and γ_2 such that*

$$\gamma_1 n^{\alpha_1} k^{2n} \leq c_n(M_k(F)) \leq \gamma_2 n^{\alpha_2} k^{2n}$$

for all $n \geq 1$.

PROOF. The upper bound, with $\alpha_2 = 0$, follows from Theorem 6.1.5, which will be proved later, since $\dim M_k(F) = k^2$.

Let $L_k(x; y)$ be the central polynomial for $M_k(F)$ defined in Section 5.5 (see Theorem 5.7.4). For every $i = 1, 2, \ldots$, let

$$L_k^{(i)} = L_k(x_1^{(i)}, \ldots, x_{k^2}^{(i)}; y_1^{(i)}, \ldots, y_{k^2}^{(i)})$$

where $x_j^{(i)}, y_j^{(i)}$, $i, j \geq 1$ are distinct indeterminates and, for every $m \geq 1$ set

$$f_m = L_k^{(1)} \cdots L_k^{(m)},$$

a multilinear polynomial of degree $2k^2 m$.

Since $L_k(x; y)$ is alternating in $x_1, \ldots, x_{k^2}$ and in $y_1, \ldots, y_{k^2}$, the polynomial f_m is $(2m; k^2)$-multialternating. Now, since $L_k(x; y)$ is a central polynomial, $f_m \notin \mathrm{Id}(M_k(F))$ and, so, by Lemma 5.9.4 for $\lambda = ((2m)^{k^2})$ the character χ_λ appears in the nth cocharacter of $M_k(F)$ with non-zero multiplicity. Thus, by Lemma 5.10.1

$$c_n(M_k(F)) \geq d_\lambda > \frac{k^{2n}}{n^{\frac{k^2(k^2-1)}{2}}} > \frac{k^{2n}}{n^{k^4}}$$

for $n = 2k^2 m$ and $m \geq 1$ large enough.

Now, for an arbitrary n, pick $m \geq 1$ such that $2k^2 m \leq n < 2k^2(m+1)$. If we set $n_0 = 2k^2 m \leq n$, then $n_0 \leq n - 2k^2$ and we have

$$c_n(M_k(F)) \geq c_{n_0}(M_k(F)) > \frac{k^{2(n-2k^2)}}{n^{k^4}}.$$

The proof of the theorem is now complete if we take $\gamma_1 = k^{-4k^2}, \alpha_1 = -k^4$. $\square$

We are now in a position to prove that codimensions and trace codimensions are asymptotically equal ([**Re4**]).

5.10. ASYMPTOTICS FOR $k \times k$ MATRICES

THEOREM 5.10.3. *For every $n \geq 1$ we have*
$$c_n(M_k(F)) \simeq t_{n+1}(M_k(F)),$$
i.e., the pure trace codimensions and the ordinary codimensions of $k \times k$ matrices are asymptotically equal.

PROOF. Recall that
$$c_n(M_k(F)) = \sum_{\substack{\lambda \vdash n \\ h(\lambda) \leq k^2}} m_\lambda d_\lambda \quad \text{and} \quad t_{n+1}(M_k(F)) = \sum_{\substack{\lambda \vdash n \\ h(\lambda) \leq k^2}} \bar{m}_\lambda d_\lambda.$$

Taking into account Corollary 5.8.2, we write
$$c_n(M_k(F)) = a_n + b_n, \quad t_{n+1}(M_k(F)) = a'_n + b_n$$
where
$$a_n = \sum_{\substack{\lambda \vdash n \\ h(\lambda) \leq k^2 \\ \lambda_{k^2} \leq 1}} m_\lambda d_\lambda, \quad a'_n = \sum_{\substack{\lambda \vdash n \\ h(\lambda) \leq k^2 \\ \lambda_{k^2} \leq 1}} \bar{m}_\lambda d_\lambda,$$
and
$$b_n = \sum_{\substack{\lambda \vdash n \\ h(\lambda) \leq k^2 \\ \lambda_{k^2} \geq 2}} m_\lambda d_\lambda = \sum_{\substack{\lambda \vdash n \\ h(\lambda) \leq k^2 \\ \lambda_{k^2} \geq 2}} \bar{m}_\lambda d_\lambda.$$

Our first goal is to show that a_n and a'_n asymptotically behave like $(k^2 - 1)^n$.

First note that the number of partitions $\lambda \vdash n$ such that $h(\lambda) \leq k^2$ does not exceed n^{k^2}. Hence, since both the multiplicities m_λ and $\bar{m}_\lambda$ are polynomially bounded (see Lemma 12.5.2 and Lemma 5.8.3), there exists a polynomial function $\varphi(n)$ such that
$$(5.8) \qquad a_n, a'_n \leq \varphi(n) \cdot \max\{d_\lambda \mid \lambda \vdash n, h(\lambda) \leq k^2, \lambda_{k^2} \leq 1\}.$$

Let $\lambda \vdash n$. Now, if $\lambda_{k^2} = 0$, then, by Corollary 4.4.7, $d_\lambda \leq (k^2 - 1)^n$.

Suppose now that $\lambda_{k^2} = 1$. Recall that on one hand d_λ is the dimension of the irreducible S_n-character corresponding to λ; on the other hand, let $\mu \vdash (n - k^2)$ be a partition whose corresponding Young diagram is obtained from that of λ by erasing the first column. By Corollary 4.4.7, $d_\mu = \deg \chi_\mu \leq (k^2 - 1)^{(n-k^2)} \leq (k^2 - 1)^n$. But from the hook formula it follows that
$$(5.9) \qquad d_\lambda = \deg \chi_\lambda \leq n^{k^2} d_\mu \leq n^{k^2}(k^2 - 1)^n$$

(see also Lemma 6.2.4).

From (5.8) and (5.9) we obtain that $a_n, a'_n \leq \bar{\varphi}(n)(k^2-1)^n$ for some polynomial function $\bar{\varphi}(n)$.

By the previous theorem, $c_n(M_k(F)) \geq \gamma_1 n^{\alpha_1} k^{2n}$, for some constants $\gamma_1 > 0$ and α_1. Hence $b_n \gtrsim \gamma_1 n^{\alpha_1} k^{2n}$ asymptotically and $c_n(M_k(F)) \simeq b_n \simeq t_{n+1}(M_k(F))$. □

Recall that by Corollary 5.2.5
$$t_{n+1}(M_k(F)) = \sum_{\substack{\lambda \vdash n+1 \\ h(\lambda) \leq k}} d_\lambda^2.$$

The asymptotics for the above sum was computed by Regev in [**Re3**]. By applying this result we find the precise asymptotics for the codimension sequence of $M_k(F)$.

THEOREM 5.10.4 (Regev [**Re4**]). *For any field F of characteristic zero we have*
$$c_n(M_k(F)) \simeq \alpha n^g k^{2n}$$
where $g = -\frac{k^2-1}{2}$ and
$$\alpha = \left(\frac{1}{\sqrt{2\pi}}\right)^{k-1} \left(\frac{1}{2}\right)^{\frac{1}{2}(k^2-1)} \cdot 1!2!\cdots(k-1)! k^{\frac{1}{2}(k^2+4)}.$$

CHAPTER 6

The PI-Exponent of an Algebra

The starting point for understanding the asymptotic behaviour of the sequence of the codimensions is Theorem 4.2.4, stating that such a sequence is exponentially bounded. The actual investigation of these asymptotics has been carried out over a field of characteristic zero, due to the extensive use of the representation theory of the symmetric group and throughout this chapter we shall assume that the ground field F is of characteristic zero.

A result of Kemer states that the codimension sequence of a PI-algebra either is polynomially bounded or grows exponentially. So, no intermediate growth is allowed. From the examples worked out explicitly by Berele and Regev, it appears that the exponential growth of this sequence should be an integer and Amitsur in the early 1980's conjectured that this was a general phenomenon. The purpose of this chapter is to present the proof of this conjecture. We actually prove that for any PI-algebra A the nth codimension is sandwiched between two functions of the type $Cn^t d^n$ with the same integer d, where C and t are constants (Theorem 6.5.2). As a consequence we shall deduce Kemer's result on the non-existence of intermediate growth. The integer d is called the PI-exponent of the algebra A or of the variety generated by A.

One of the main advantages of the exponent is that now we have an integral scale allowing us to measure the growth of any non-trivial variety. Besides proving the existence of the exponent, we also present an actual way of computing it. Recall that any non-trivial variety can be generated by the Grassmann envelope of a finite dimensional superalgebra B. Then it turns out that the exponent of the variety is the dimension of a suitable semisimple subalgebra of B. As a consequence, for a finite dimensional algebra A we shall show that the PI-exponent is bounded by the dimension of the algebra and equals its dimension if and only if A is central simple (Theorem 6.6.1). This method is also applied in order to compute explicitly the PI-exponent of the verbally prime algebras and of the upper block triangular matrix algebras (Corollary 6.6.2 and Corollary 6.6.3).

6.1. The exponential growth of the codimensions

Let A be a PI-algebra over the field F of characteristic zero and let $\{c_n(A)\}_{n \geq 1}$ be its sequence of codimensions.

If A is a nilpotent algebra i.e., $x_1 \cdots x_N \equiv 0$ is a polynomial identity of A for some $N \geq 1$, then $c_n(A) = 0$ for any $n \geq N$. But if A is a non-nilpotent PI-algebra, then $c_n(A) \neq 0$ for $n \geq 1$ and by Theorem 4.2.4,

$$1 \leq c_n(A) \leq a^n$$

for some constant a. Hence the sequence of nth roots $\sqrt[n]{c_n(A)}$, $n = 1, 2, \ldots$, is bounded and we can consider its upper and lower limit.

DEFINITION 6.1.1. Let A be any PI-algebra. Then
$$\underline{\exp}(A) = \liminf_{n\to\infty} \sqrt[n]{c_n(A)}$$
is called the lower exponent of A and
$$\overline{\exp}(A) = \limsup_{n\to\infty} \sqrt[n]{c_n(A)}$$
is called the upper exponent of A.

For an arbitrary bounded sequence the lower limit and the upper limit may not coincide. Nevertheless, in case they do, we can define the exponent of A.

DEFINITION 6.1.2. Let A be any PI-algebra. Then the exponent (or PI-exponent) of A is
$$\exp(A) = \lim_{n\to\infty} \sqrt[n]{c_n(A)},$$
provided $\underline{\exp}(A) = \overline{\exp}(A)$. In case $\mathcal{V} = \mathrm{var}(A)$ is a variety of algebras, we write $\exp(\mathcal{V}) = \exp(A)$ and we call $\exp(A)$ the exponent of the variety $\mathcal{V}$.

In the 1980's two main conjectures about the asymptotic behaviour of $c_n(A)$ were made.

CONJECTURE 6.1.3 (Amitsur). *For any PI-algebra A, $\exp(A)$ exists and is a non-negative integer.*

If $f(x)$ and $g(x)$ are two functions of a real argument, then $f(x)$ and $g(x)$ are asymptotically equal and we write $f(n) \simeq g(n)$ if $\displaystyle\lim_{n\to\infty} \frac{f(n)}{g(n)} = 1$.

CONJECTURE 6.1.4 (Regev). *For any PI-algebra A, there exist a constant C, a semi-integer q and an integer $d \geq 0$ such that*
$$c_n(A) \simeq C n^q d^n.$$

In this chapter we shall prove Amitsur's conjecture over a field of characteristic zero. In fact we shall prove that for any PI-algebra A, there exist constants C_1, C_2, q_1, q_2 and an integer $d \geq 0$ such that
$$C_1 n^{q_1} d^n \leq c_n(A) \leq C_2 n^{q_2} d^n$$
for all $n = 1, 2, \ldots$. The integer d will be defined by the structure of some finite dimensional superalgebra B related to A and we shall give a precise procedure for computing d in case B is known, in particular, when $\mathrm{var}(A) = \mathrm{var}(G(B))$.

We start here by giving a first immediate application of the theory developed in the previous chapters. We show that an upper bound of the codimension growth of A can be given in terms of the dimension of the algebra. In fact we prove that $c_n(A)$ does not exceed d^n where $d = \dim A$. In order to see this, let $\chi_n(A)$ be the nth cocharacter of A. Since A is finite dimensional, it satisfies a Capelli identity hence, by Theorem 4.6.1, $\chi_n(A)$ lies in a strip of height d i.e.,
$$\chi_n(A) = \sum_{\substack{\lambda \vdash n \\ h(\lambda) \leq d}} m_\lambda \chi_\lambda.$$

As it was mentioned in Corollary 4.4.7, all degrees $d_\lambda = \chi_\lambda(1) = \deg \chi_\lambda$ are bounded from above by d^n. Moreover, by Theorem 4.9.3, all multiplicities m_λ are polynomially bounded. This implies the required upper bound. Next we give another proof based on some very general argument.

THEOREM 6.1.5. *Let A be a finite dimensional algebra, $\dim A = d$. Then $c_n(A) \leq d^n$.*

PROOF. Let $a_1, \ldots, a_d$ be a basis of A over F. Then a multilinear polynomial $f = f(x_1, \ldots, x_n)$ is an identity of A if and only if it vanishes under any evaluation φ, $\varphi(x_1) = a_{i_1}, \ldots, \varphi(x_n) = a_{i_n}$ i.e., the following linear relation is satisfied

$$(6.1) \qquad f(a_{i_1}, \ldots, a_{i_n}) = \sum_{\sigma \in S_n} \alpha_\sigma a_{\sigma(i_1)} \cdots a_{\sigma(i_n)} = 0.$$

The total number of such evaluations is d^n since any x_i can takes d distinct values. Hence, (6.1) can be regarded as a system of d^n linear equations in $n!$ indeterminates α_σ, $\sigma \in S_n$. The space of solutions of (6.1) has dimension $n! - r$ where r is the rank of (6.1), $r \leq d^n$. Any solution of (6.1) i.e., any $n!$-tuple of coefficients $\alpha_\sigma, \sigma \in S_n$, gives a multilinear identity of A. Moreover, linearly independent solutions give linearly independent identities. Therefore $\dim(P_n \cap \mathrm{Id}(A)) = n! - r$ and $c_n(A) = \dim P_n/(P_n \cap \mathrm{Id}(A)) = r \leq d^n$. □

REMARK 6.1.6. The previous theorem can be generalized to any non-associative finite dimensional algebra A, $\dim A = d$ ([**BD**]). If we denote by $[\rho]$ some arrangement of the brackets $[\cdots]$ on a monomial $x_{\sigma(1)}, \ldots, x_{\sigma(n)}$ then the coefficients $\alpha_{\sigma, [\rho]}$ in (6.1) will depend also on this arrangements. The total number of indeterminate coefficients will be $N = \varphi(n)n!$ where $\varphi(n) = \dfrac{1}{n}\dbinom{2n-2}{n-1}$ is the nth Catalan number ([**Bi**]) and counts the number of distinct arrangements of the brackets on a monomial of degree n. Nevertheless, the rank r of (6.1) is less or equal to d^n, $\dim(P_n \cap \mathrm{Id}(A)) = N - r$ and $c_n(A) = \dim P_n/(P_n \cap \mathrm{Id}(A)) = r \leq d^n$.

As a consequence of Theorem 6.1.5 for an associative algebra we have.

COROLLARY 6.1.7. *If $\dim A = d < \infty$, then $\overline{\exp}(A) \leq d$.*

Later we shall show that in general this upper bound cannot be improved by exhibiting finite dimensional algebras with $\underline{\exp}(A) = \overline{\exp}(A) = \dim A$. We shall also establish that this condition characterizes central simple algebras.

6.2. A candidate for the PI-exponent

Given any finite dimensional superalgebra A over an algebraically closed field, we next define an integer related to the structure of A. In the following sections our effort will be devoted to proving that such integer coincides with the PI-exponent of the Grassmann envelope of the superalgebra A.

Throughout this section F will be an algebraically closed field of characteristic zero and $A = A^{(0)} \oplus A^{(1)}$ a finite dimensional $\mathbb{Z}_2$-graded algebra over F. Let $A = A_{ss} + J$ be the Wedderburn decomposition of A where A_{ss} is a maximal semisimple subalgebra of A and $J = J(A)$ its Jacobson radical. As we have seen in Theorem 3.4.4, J is a graded ideal and we may assume that $A_{ss} = A_1 \oplus \cdots \oplus A_k$, where $A_1, \ldots, A_k$ are simple superalgebras. We fix the notation and we let $m = \dim A$ and $p > 0$ such that $J^p = 0$.

Consider all possible products of the type

$$(6.2) \qquad B_1 J B_2 J \cdots J B_r \neq 0$$

where $B_1, \ldots, B_r$ are distinct subalgebras taken from the set $\{A_1, \ldots, A_k\}$ and $r = 1, 2, \ldots$. If $r = 1$, we mean that $B_1 = A_i$, for some $1 \leq i \leq k$. Then we define

(6.3) $$q = \max \dim_F(B_1 \oplus \cdots \oplus B_r),$$

the maximal dimension of a subalgebra $B_1 \oplus \cdots \oplus B_r$ satisfying (6.2).

LEMMA 6.2.1. *Let $B_1, \ldots, B_t$ be not necessarily distinct superalgebras from the set $\{A_1, \ldots, A_k\}$. If*

(6.4) $$B_1 J B_2 J \cdots J B_t \neq 0,$$

then $\dim(B_1 + \cdots + B_t) \leq q$.

PROOF. If in the product (6.4) some subalgebra B_i appears more than once, then, since $JB_iJ \subseteq J$, we can reduce this product and get a non-zero product of the type (6.4) with the A_i's all distinct. □

We recall some useful definitions from the previous chapters. Given a multilinear polynomial $f(x_1, \ldots, x_n)$ and a tableau T_λ of shape $\lambda \vdash n$, recall that f corresponds to T_λ if $f = e_{T_\lambda} f_0$ for some multilinear polynomial $f_0 \in P_n$ (see Definition 2.5.3). For any integers $d, l, t \geq 0$, $h(d, l, t)$ denotes the partition $((l+t)^d, l^t)$ (see picture).

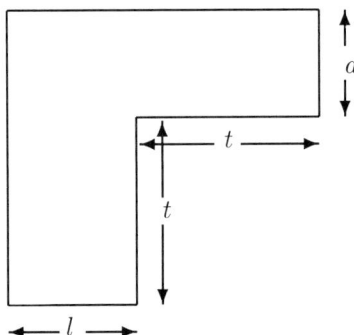

Also, if $\lambda = (\lambda_1, \ldots, \lambda_p) \vdash n$ and $\mu = (\mu_1, \ldots, \mu_q) \vdash m$, then $\lambda \geq \mu$ if and only if $p \geq q$ and $\lambda_i \geq \mu_i$ for all $i = 1, \ldots, p$ (see Section 2.3). Moreover, recall that $G(A) = (G^{(0)} \otimes A^{(0)}) \oplus (G^{(1)} \otimes A^{(1)})$ is the Grassmann envelope of A.

LEMMA 6.2.2. *Let $\lambda \geq h(d, l, t)$ where $d + l > q$, $t > (d+l)m + p$ and m, p, q are defined before Lemma 6.2.1. If f is a multilinear polynomial corresponding to a tableau T_λ, then $f \in \mathrm{Id}(G(A))$.*

PROOF. We first observe that in order to prove the lemma, it is enough to show that there exists an element $r \in FS_n$ such that $0 \neq rf \in \mathrm{Id}(G(A))$. In fact, since f generates an irreducible left S_n-submodule of P_n, we would have $FS_n f = FS_n r f \subseteq \mathrm{Id}(G(A))$ and $f \in \mathrm{Id}(G(A))$ will follow.

An element $r \in FS_n$ with the property that $rf \neq 0$ will be chosen during the proof.

Fix first a homogeneous basis $C = C^{(0)} \cup C^{(1)}$ of A which is the disjoint union of homogeneous bases of $A_1, \ldots, A_k$ and J, respectively. Now, in order to prove that a polynomial of the type rf is an identity of $G(A)$, it is enough to show that rf evaluates to zero on all elements of the type $c \otimes g$ and $c' \otimes g'$, where $c \in C^{(0)}, c' \in C^{(1)}, g \in G^{(0)}, g' \in G^{(1)}$.

Let $c_1, \ldots, c_s$ be distinct homogeneous elements in $C \cap A_{ss}$ and suppose $s > q$. Then any product of elements of A which contains the elements $c_1, \ldots, c_s$ must be equal to zero, as it follows from Lemma 6.2.1. Hence, if we substitute $c_1 \otimes g_1, \ldots, c_s \otimes g_s$ instead of some variables in rf, the resulting evaluation in $G(A)$ will be zero. Therefore, given any $\mathbb{Z}_2$-graded semisimple subalgebra $B = B^{(0)} \oplus B^{(1)}$ of A with $\dim B \leq q$, it is enough to prove that rf takes zero value on elements of the type
$$c_1 \otimes g_1, \ldots, c_s \otimes g_s, e_1 \otimes h_1, e_2 \otimes h_2, \ldots$$
where $c_1, \ldots, c_s \in C \cap B$, $e_1, e_2 \ldots \in J$, $g_i, h_j \in G$ are all homogeneous elements.

Notice that, since by hypothesis $q < d+l$, either $\dim B^{(0)} \leq d-1$ or $\dim B^{(1)} \leq l-1$. Let us examine the two cases.

Suppose first that $\dim B^{(0)} \leq d-1$. By Lemma 2.5.4, there exist $r \in FS_n$ and a subset of variables $Y = Y_1 \cup \ldots \cup Y_d$ such that $rf \neq 0$ is symmetric on each set of variables Y_i for any i; there is also a decomposition
$$rf = f_1 + f_2 + \cdots$$
where $f_1, f_2, \ldots$ are polynomials alternating on suitable disjoint subsets of Y. We claim that for such r, $rf \in \mathrm{Id}(G(A))$. This will complete the proof in this case.

Suppose that there exists some substitution in rf giving a non-zero value in $G(A)$. Then at least one of the summands f_i should have a non-zero evaluation. Let it be f_1. By Lemma 2.5.4, Y can be partitioned as
$$Y = Y_1' \cup \ldots \cup Y_{t+l}'$$
with $|Y_j'| = d$ and f_1 is alternating on the variables of each subset Y_j', $j = 1, \ldots, t+l$. It follows that, if for two variables y_1, y_2 in Y_i' we substitute y_1 with $c \otimes g_1$ and y_2 with $c \otimes g_2$ where $c \in B^{(0)}, g_1, g_2 \in G^{(0)}$, then f_1 will be zero (alternating on y_1 and y_2).

On the other hand, since $\dim B^{(0)} \leq d-1$, in order to get a non-zero value for rf we need to substitute at most $d-1$ elements of the type $c \otimes g$, $c \in B^{(0)}, g \in G^{(0)}$, in each set of variables Y_i'. This means that we need to substitute in rf at least $t+l$ elements of the type $c \otimes g, c \in B^{(1)}, g \in G^{(1)}$ or $c \in J, g \in G$ instead of Y. But $J^p = 0$, hence, we should substitute at least $l+t-p+1 > dm$ elements of the type $c \otimes g$ with $c \in B^{(1)}$. It follows that for some $1 \leq i \leq d$ we will replace more than m variables in Y_i with elements $c \otimes g$ where c is a basis element of $B^{(1)}$. Since $m \geq \dim A_{ss}^{(1)} \geq B^{(1)}$, it follows that there exist two variables $y_1, y_2 \in Y_i$ taking value $c \otimes g_1$ and $c \otimes g_2$ respectively, where $c \in B^{(1)}, g_1, g_2 \in G^{(1)}$. But rf is symmetric on y_1, y_2. Hence, the corresponding value will be zero. This completes the proof of the lemma in this case.

Now assume that $\dim B^{(1)} \leq l-1$. As before by Lemma 2.5.5, there exist $r \in FS_n$ and a set of variables Y for rf such that
$$Y = Y_1 \cup \ldots \cup Y_l,$$
rf is alternating on each $Y_i, i = 1, \ldots, l$, and $rf = f_1 + f_2 + \cdots$ where for any f_i there is a partition $Y = Y_1' \cup \ldots \cup Y_{t+d}'$ on symmetric subsets of order l. If, for example, $f_1 \neq 0$ for some substitution, then we should replace at least $t+d-p+1 > lm$ variables of Y with some $c \otimes g, c \in B^{(0)}, g \in G^{(0)}$ because f_1 is symmetric on $y_1, y_2 \in Y_i'$ and f_1 will be zero if $y_1 = c \otimes g_1, y_2 = c \otimes g_2$ with $c \in B^{(1)}, g_1, g_2 \in G^{(1)}$.

Since $\dim B^{(0)} \leq m$, we should replace some y_1, y_2 in the same alternating set Y_i with tensors $c \otimes g_1, c \otimes g_2$ respectively, where $g_1, g_2 \in G^{(0)}$ and c is one of the basis elements of $B^{(0)}$. Hence rf will take a zero value and the proof is complete also in this case. □

Recall the decomposition $P_n \equiv FS_n = \oplus_{\lambda \vdash n} I_\lambda$ of the group algebra FS_n into its minimal two-sided ideals I_λ. From the above lemma we immediately get the following.

COROLLARY 6.2.3. *Let $d + l = q + 1$ and $t = (m+1)^2$ where $m = \dim A$. Then $\oplus_{\lambda \geq h(d,l,t)} I_\lambda \subseteq \mathrm{Id}(G(A))$.*

PROOF. Let $\lambda \geq h(d,l,t)$. Since $q \leq m$ and $p \leq m$ where $J^p = 0$, then $t = (m+1)m + m + 1 > (q+1)m + p = (d+l)m + p$. Hence, by Lemma 6.2.2 for any tableau T_λ, $e_{T_\lambda} FS_n \subseteq \mathrm{Id}(G(A))$. It follows that $I_\lambda = FS_n e_{T_\lambda} FS_n \subseteq \mathrm{Id}(G(A))$. □

Next we state two results needed in the next proposition.

LEMMA 6.2.4. *Let $\lambda \vdash n$, $\mu \vdash n'$ be such that $\mu \leq \lambda$. If $n - n' \leq c$, then $d_\mu \leq d_\lambda \leq n^c d_\mu$.*

PROOF. Recall that by the hook formula (Proposition 2.2.8),

$$d_\lambda = \frac{n!}{\prod_{i,j} h_{ij}}$$

where the h_{ij}'s are the hook numbers of D_λ. Clearly it is enough to prove the lemma when $c = 1$. Hence we may assume that $n' = n - 1$ and

$$d_\mu = \frac{(n-1)!}{\prod_{i,j} h'_{ij}},$$

where the h'_{ij}'s are the hook numbers of D_μ. Since $\mu \leq \lambda$, it is easily seen that $\prod h'_{ij} < \prod h_{ij}$. Hence

$$d_\lambda = \frac{n!}{\prod h_{ij}} < \frac{n!}{\prod h'_{ij}} = n d_\mu$$

and the first inequality is proved.

The inequality $d_\mu \leq d_\lambda$ clearly follows from the branching theorem (Theorem 2.3.1). □

Recall that $H(d,l)$ denoted the infinite hook of hand d and foot l.

LEMMA 6.2.5. *For some constants $C, r > 0$ the following inequality holds*

$$\sum_{\substack{\lambda \vdash n \\ \lambda \in H(d,l)}} d_\lambda \leq C n^r (d+l)^n.$$

Moreover, for some constants a, b we have the following asymptotic equality

$$d_{h(d,l,k)} \underset{n \to \infty}{\simeq} a n^b (d+l)^n,$$

where $h(d,l,k) \vdash n$.

PROOF. In the proof of Lemma 4.9.1, we showed that the number of partitions $\lambda \vdash n$ lying in the hook $H(d,l)$ is bounded by $(n+1)^{d+l}$ i.e., the number of summands in

$$\sum_{\substack{\lambda \vdash n \\ \lambda \in H(d,l)}} d_\lambda$$

is polynomially bounded. Hence in order to get an upper bound we need only to prove that $d_\lambda \leq (d+l)^n$. As in the proof of the previous lemma we shall use the hook formula. Let $\lambda = (\lambda_1, \ldots, \lambda_t)$ and let $\lambda' = (\lambda'_1, \ldots, \lambda'_l)$ be the conjugate partition of λ. Then, for all $i = 1, \ldots, d$, we have $\prod_j h_{ij} \geq \lambda_i!$. Now we set

$$\lambda''_j = \begin{cases} \lambda'_j - d, & \text{if } \lambda'_j \geq d, \\ 0, & \text{otherwise,} \end{cases}$$

for all $j = 1, \ldots, l$. Then $\lambda_1 + \cdots + \lambda_d + \lambda''_1 + \cdots + \lambda''_l = n$ and

$$\prod_i h_{ij} \geq \lambda''_1! \cdots \lambda''_l!.$$

Since $\lambda \in H(d,l)$, we obtain

$$\prod_{i,j} h_{ij} \geq \lambda_1! \cdots \lambda_d! \lambda''_1! \cdots \lambda''_l!$$

and

$$d_\lambda = \frac{n!}{\prod h_{ij}} \leq \frac{n!}{\lambda_1! \cdots \lambda_d! \lambda''_1! \cdots \lambda''_l!} \leq (d+l)^n$$

since by the multinomial theorem (see [**Bi**]), any generalized binomial coefficient

$$\binom{n}{k_1, \ldots, k_m} = \frac{n!}{k_1! \cdots k_m!}$$

is bounded by n^m. Hence

$$\sum_{\substack{\lambda \vdash n \\ \lambda \in H(d,l)}} d_\lambda \leq Cn^r(d+l)^n,$$

for some constants C, r.

In order to compute the precise asymptotics of $d_{(d,l,t)}$, we split the product $\Pi = \prod h_{ij}$ of the hook numbers into three parts: $\Pi = \Pi_1 \Pi_2 \Pi_3$ where

$$\Pi_1 = \prod_{i=1}^{d} \prod_{j=l+1}^{l+k} h_{ij}, \quad \Pi_2 = \prod_{i=d+1}^{d+k} \prod_{j=1}^{l} h_{ij}, \quad \Pi_3 = \prod_{\substack{1 \leq i \leq d \\ 1 \leq j \leq l}} h_{ij}.$$

Then

$$\Pi_1 = k! \frac{(k+1)!}{1!} \frac{(k+2)!}{2!} \cdots \frac{(k+d-1)!}{(d-1)!},$$

$$\Pi_2 = k! \frac{(k+1)!}{1!} \frac{(k+2)!}{2!} \cdots \frac{(k+l-1)!}{(l-1)!}$$

and

$$\Pi_3 = (2k+1)(2k+2) \cdots (2k+d) \cdot (2k+2)(2k+3) \cdots (2k+d+1) \cdot$$
$$\cdots (2k+l)(2k+l+1) \cdots (2k+l+d-1).$$

Next we apply Stirling formula ([**Bi**])
$$n! \simeq \sqrt{2\pi n}\left(\frac{n}{e}\right)^n.$$
Then, since $n = k(d+l) + dl$, we obtain as $k \to \infty$,
$$n! \simeq k^{ld}(kl+kd)! \simeq C_1\sqrt{k} \cdot k^{ld}\frac{(kl+kd)^{kl+kd}}{e^{kl+kd}},$$
$$\prod h_{ij} = \Pi_1\Pi_2\Pi_3 \simeq C_2 k^{1+2+\cdots+d-1}k^{1+2+\cdots+l-1}(k!)^{d+l}(2k)^{dl}$$
$$\simeq C_3\sqrt{k^{d^2+l^2}}k^{dl}\frac{k^{kl+kd}}{e^{kl+kd}}.$$
Hence
$$d_{h(d,l,k)} = \frac{n!}{\prod h_{ij}} \simeq C_4 k^a (d+l)^k (d+l).$$
Finally, replacing $k \to \infty$ with $n = k(d+l) + dl \to \infty$, we obtain
$$d_{h(d,l,k)} \simeq an^b(d+l)^n$$
for some constants a, b, as required. $\square$

PROPOSITION 6.2.6. *Let A be a finite dimensional superalgebra and let q be defined as in (6.3) before Lemma 6.2.1. Then there exist constants $C_1, r_1 > 0$ depending only on $\dim A$ such that $c_n(G(A)) \leq C_1 n^{r_1} q^n$.*

PROOF. Consider the decomposition of the nth cocharacter
(6.5) $$\chi_n(G(A)) = \sum_{\lambda \vdash n} m_\lambda \chi_\lambda.$$

Suppose that $\lambda \vdash n$ is such that $\lambda \geq h(d,l,t)$ with $d+l = q+1$ and $t = (m+1)^2$ where $m = \dim A$. Then by Corollary 6.2.3, $m_\lambda = 0$ for this λ.

We claim that given any partition $\lambda \vdash n$ such that $m_\lambda \neq 0$, the corresponding diagram lies in $H(d, q-d) \cup (s^u)$, the union of some infinite hook $H(d, q-d)$, $d \geq 0$, and a rectangular diagram (s^u) where $s = (m+1)^2 + m - q + d$ and $u = (m+1)^2 + m - d$ constructed as in the picture below.

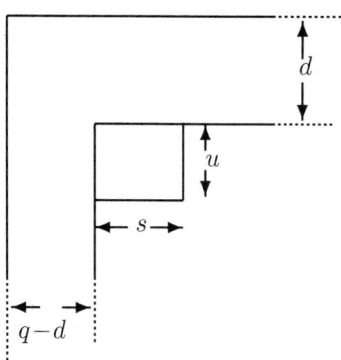

In fact, let $\lambda = (\lambda_1, \lambda_2, \ldots)$ and suppose first that $\lambda_i > (m+1)^2 + m$ for some i. Let k be the largest integer such that $\lambda_k > (m+1)^2 + m$. If $k > q$, then $\lambda \geq h(q+1, 0, (m+1)^2)$ and we get a contradiction by Corollary 6.2.3. Thus $k \leq q$.

If $\lambda_{(m+1)^2+m+1} \geq q - k + 1$, then the Young diagram D_λ contains a subdiagram D_μ where

$$\mu = (\underbrace{(m+1)^2 + m + 1, \ldots, (m+1)^2 + m + 1}_{k}, \underbrace{q-k+1, \ldots, q-k+1}_{(m+1)^2+m+1-k}).$$

Since $(m+1)^2 + m + 1 - (q - k + 1) \geq (m+1)^2$ and $(m+1)^2 + m + 1 - k \geq (m+1)^2$, it is not difficult to see that $\mu \geq h(k, q - k + 1, (m+1)^2)$. Hence $\lambda \geq h(k, q - k + 1, (m+1)^2)$ and again we get a contradiction by the previous corollary. Hence $\lambda_{(m+1)^2+m+1} \leq q - k$ and λ is contained in $H(k, q-k) \cup (s^u)$ as wished.

Therefore we may assume that $\lambda_1 \leq (m+1)^2 + m$. Clearly $\lambda_{(m+1)^2+m+1} \leq q$ since otherwise $\lambda \geq h(0, q+1, (m+1)^2)$, contrary to Corollary 6.2.3. This says that λ is contained in $H(0,q) \cup (s^u)$ and the claim is proved.

The property just proved says that if $m_\lambda \neq 0$, then D_λ, the diagram of λ, contains a subdiagram D_μ such that $D_\mu \subset H(d, q - d)$ and, if $\lambda \vdash n, \mu \vdash n'$ we have that $n - n' \leq T = su$. By Lemma 6.2.4, $d_\lambda \leq n^T d_\mu$. Denote by S the set of all partitions $\lambda \vdash n$ such that $m_\lambda \neq 0$ in (6.5). By Theorem 4.9.3, there exists a constant $k > 0$ such that $m_\lambda \leq n^k$, for all $\lambda \in S$. Therefore, using Lemmas 6.2.4 and 6.2.5 we obtain

$$c_n(G(A)) = \sum_{\lambda \vdash n} m_\lambda d_\lambda \leq n^k \sum_{\substack{\lambda \vdash n \\ \lambda \in S}} d_\lambda \leq$$

$$n^k n^T \sum_{i=0}^{q} \sum_{n'=0}^{n} \sum_{\substack{\mu \vdash n' \\ \mu \in H(i, q-i)}} d_\mu \leq n^{k+T} \sum_{i=0}^{q} \sum_{n'=1}^{n} C(n')^r q^{n'} \leq n^{k+T} C n^r q^n n(q+1).$$

So, $c_n(G(A)) \leq C_1 n^{r_1} q^n$ where $C_1 = (q+1)C, r_1 = k + T + r + 1$. $\square$

6.3. Graded identities and Grassmann envelopes

An important step towards our goal is to relate the graded identities to the ordinary identities of a Grassmann envelope. In this section we will make use of the linear isomorphism $\sim$ constructed in Section 3.7, and the connection between representations of symmetric groups given by the Littlewood-Richardson rule (Theorem 2.3.9).

Let $F\langle Y, Z \rangle$ be the free superalgebra on the set $X = Y \cup Z$ where $Y = \{y_1, y_2, \ldots\}$ and $Z = \{z_1, z_2, \ldots\}$ are of homogeneous degree 0 and 1, respectively. Recall that for a superalgebra A, $\mathrm{Id}^{gr}(A)$ denotes the T_2-ideal of graded identities of A.

Fix $n_1, n_2 \geq 0$, and let P_{n_1, n_2} be the space of multilinear polynomials in $y_1, \ldots, y_{n_1}, z_1, \ldots, z_{n_2}$. Given any polynomial $f \in P_{n_1, n_2}$ write

$$f = \sum_{\substack{\sigma \in S_{n_2} \\ W=(w_0, w_1, \ldots, w_{n_2})}} \alpha_{\sigma, W} w_0 z_{\sigma(1)} w_1 \cdots w_{n_2-1} z_{\sigma(n_2)} w_{n_2}$$

where $w_0, w_1, \ldots, w_{n_2}$ are monomials in $y_1, \ldots, y_{n_1}$. Recall from Section 3.7, that the map $\sim : P_{n_1, n_2} \to P_{n_1, n_2}$ defined by

$$\widetilde{f} = \sum_{\substack{\sigma \in S_{n_2} \\ W=(w_0, w_1, \ldots, w_{n_2})}} (\mathrm{sgn}\,\sigma) \alpha_{\sigma, W} w_0 z_{\sigma(1)} w_1 \cdots w_{n_2-1} z_{\sigma(n_2)} w_{n_2},$$

is a linear isomorphism with the following basic properties:

(1) $f \equiv 0$ is a graded identity of A if and only if $\tilde{f} \equiv 0$ is a graded identity of $G(A)$;

(2) $\tilde{\tilde{f}} = f$.

We now consider the left permutation action of the group $S_{n_1} \times S_{n_2}$ on the space P_{n_1,n_2} by letting S_{n_1} act on $y_1, \ldots, y_{n_1}$ and S_{n_2} on $z_1, \ldots, z_{n_2}$. Let $R_1 = FS_{n_1}$ and $R_2 = FS_{n_2}$ be the two corresponding group algebras. Next we compare the structure of P_{n_1,n_2} as a left module over R_1, R_2 and over $R = R_1 \otimes R_2$. By a natural extension of the above action, for any $b = \sum_{\sigma \in S_{n_2}} \beta_\sigma \sigma$ we write $\tilde{b} = \sum_{\sigma \in S_{n_2}} (\operatorname{sgn} \sigma) \beta_\sigma \sigma$.

LEMMA 6.3.1. *Let* $a \in R_1, b \in R_2, f \in P_{n_1,n_2}$. *Then*

(1) $\widetilde{bf} = \tilde{b}\tilde{f}$, $\widetilde{af} = a\tilde{f}$, $\tilde{\tilde{b}} = b$,

(2) f *is alternating on some variables* $z_1, \ldots, z_m$ *if and only if* $\tilde{f}$ *is symmetric on* $z_1, \ldots, z_m$.

PROOF. Let

$$f = \sum_{\substack{\sigma \in S_{n_2} \\ W=(w_0,w_1,\ldots,w_{n_2})}} \alpha_{\sigma,W} w_0 z_{\sigma(1)} w_1 \cdots w_{n_2-1} z_{\sigma(n_2)} w_{n_2}$$

and $b = \sum_{\tau \in S_{n_2}} \beta_\tau \tau$. Then

$$bf = \sum_{\substack{\sigma,\tau \in S_{n_2} \\ W=(w_0,w_1,\ldots,w_{n_2})}} \alpha_{\sigma,W} \beta_\tau w_0 z_{\tau\sigma(1)} w_1 \cdots w_{n_2-1} z_{\tau\sigma(n_2)} w_{n_2}$$

and

$$\widetilde{bf} = \sum_{\substack{\sigma,\tau \in S_{n_2} \\ W=(w_0,w_1,\ldots,w_{n_2})}} (\operatorname{sgn} \tau\sigma) \alpha_{\sigma,W} \beta_\tau w_0 z_{\tau\sigma(1)} w_1 \cdots w_{n_2-1} z_{\tau\sigma(n_2)} w_{n_2} =$$

$$(\sum_{\tau \in S_{n_2}} (\operatorname{sgn} \tau)\beta_\tau \tau)(\sum_{\substack{\sigma \in S_{n_2} \\ W=(w_0,w_1,\ldots,w_{n_2})}} (\operatorname{sgn} \sigma)\alpha_{\sigma,W} w_0 z_{\sigma(1)} w_1 \cdots w_{n_2-1} z_{\sigma(n_2)} w_{n_2}) = \tilde{b}\tilde{f}.$$

All other statements in 1) and 2) are trivial. □

LEMMA 6.3.2. *Let* $A = A^{(0)} \oplus A^{(1)}$ *be a superalgebra over* F, $d = \dim A^{(0)}$, $l = \dim A^{(1)}$. *Let*

$$f(y_1, \ldots, y_{dr}, z_1, \ldots, z_{ls}) \in P_{dr,ls}$$

be a polynomial which is alternating in the r *disjoint subsets of variables*

$$\{y_1^i, \ldots, y_d^i\} \subseteq \{y_1, \ldots, y_{dr}\}, \ 1 \leq i \leq r,$$

and symmetric in the s *disjoint subsets of variables*

$$\{z_1^i, \ldots, z_l^i\} \subseteq \{z_1, \ldots, z_{ls}\}, \ 1 \leq i \leq s.$$

If $f \notin \operatorname{Id}^{gr}(G(A))$, *then there exist partitions* $\lambda = (r^d)$, $\mu = (l^s)$ *and essential idempotents* $e_{T_\lambda} \in FS_{dr}$, $e_{T_\mu} \in FS_{ls}$ *such that* $e_{T_\lambda} e_{T_\mu} f \notin \operatorname{Id}^{gr}(G(A))$.

PROOF. Set $n_1 = dr$ and $n_2 = ls$ so that $f \in P_{n_1,n_2}$. Consider the left S_{n_1}-submodule of P_{n_1,n_2} generated by f and its decomposition into irreducible S_{n_1}-submodules. Since $f \notin \mathrm{Id}^{gr}(G(A))$, there exists a Young tableau T_λ, $\lambda = (\lambda_1, \lambda_2, \ldots) \vdash n_1$, such that $e_{T_\lambda} f \not\equiv 0$ on $G(A)$.

If $\lambda_1 \geq r+1$, then $e_{T_\lambda} f$ is symmetric on at least $r+1$ variables among $y_1, \ldots, y_{dr}$. But the y_i's are divided into r disjoint alternating subsets. Therefore, $e_{T_\lambda} f = 0$ in $F\langle Y, Z\rangle$ being symmetric and alternating in two variables at the same time.

Assume now that $h(T_\lambda) \geq d+1$, where $h(T_\lambda)$ is the height of the diagram of λ. In this case write $e_{T_\lambda} = e_1 e_2$ where $e_1 = \sum_{\sigma \in R_{T_\lambda}} \sigma$ and $e_2 = \sum_{\tau \in C_{T_\lambda}} (\mathrm{sgn}\,\tau)\tau$. Since the polynomial $e_2 f$ is alternating on some $d+1$ variables y_i's, then also the polynomial $e_2 \widetilde{f}$ is alternating on the same variables. Since $\dim A^{(0)} = d$ it follows that $e_2 \widetilde{f} \in \mathrm{Id}^{gr}(A)$. Hence $e_{T_\lambda} \widetilde{f} \equiv 0$ is also a graded identity of A. By Lemma 6.3.1, $\widetilde{e_{T_\lambda} f} = e_{T_\lambda} f \equiv 0$ is a graded identity of $G(A)$.

Since $n_1 = dr$, we conclude that $e_{T_\lambda} f \notin \mathrm{Id}^{gr}(G(A))$ only if D_λ, the diagram of λ, is a rectangle with d rows and r columns, i.e., $\lambda = (r^d)$.

We now consider the left S_{n_2}-submodule of P_{n_1,n_2} generated by f. As above there exists a Young tableau T_μ with $\mu = (\mu_1, \mu_2, \ldots) \vdash n_2$ such that $e_{T_\mu} f \not\equiv 0$ on $G(A)$.

Suppose first that $h(T_\mu) \geq s+1$ and write, as before, $e_{T_\mu} = e_1 e_2$ where $e_1 = \sum_{\sigma \in R_{T_\mu}} \sigma$ and $e_2 = \sum_{\tau \in C_{T_\mu}} (\mathrm{sgn}\,\tau)\tau$. In this case the action of e_2 on P_{n_1,n_2} is alternating on at least $s+1$ variables z_i's. But all the variables $\{z_1, \ldots, z_{ls}\}$ in f are divided into s symmetric disjoint subsets. Hence, $e_2 f = 0$ in $F\langle Y, Z\rangle$ and $e_{T_\mu} f$ is also zero.

If, on the other hand, $\mu_1 \geq l+1$, then $g = e_{T_\mu} f$ is symmetric on $l+1$ variables z_i's. By Lemma 6.3.1, $\widetilde{g}$ is alternating on the same $l+1$ odd variables; since $\dim A^{(1)} = l$, it follows that $\widetilde{g} \in \mathrm{Id}^{gr}(A)$. By using Lemma 6.3.1 again we then obtain that $e_{T_\mu} f = g = \widetilde{\widetilde{g}} \in \mathrm{Id}^{gr}(G(A))$.

It follows that $e_{T_\mu} f \notin \mathrm{Id}^{gr}(G(A))$ only if the corresponding diagram D_μ is a rectangle with s rows and l columns i.e., $\mu = (l^s)$.

We have proved that $f \notin \mathrm{Id}^{gr}(G(A))$ implies $e_{T_\lambda} e_{T_\mu} f \notin \mathrm{Id}^{gr}(G(A))$ where D_λ and D_μ are two rectangles of size $r \times d$ and $l \times s$ respectively. This completes the proof of the lemma. $\square$

LEMMA 6.3.3. *Let $A = A^{(0)} \oplus A^{(1)}$ be a simple superalgebra over an algebraically closed field F, $d = \dim A^{(0)}, l = \dim A^{(1)}$. Then, for any positive integer t, there exists a partition λ with*

$$h(d, l, 2t - d - l) \leq \lambda \leq h(d, l, 2t),$$

and a tableau T_λ such that $G(A)$ does not satisfy any identity $f \equiv 0$ corresponding to T_λ (i.e., $I_\lambda \not\subseteq \mathrm{Id}(G(A))$).

PROOF. By Theorem 3.5.3, A is isomorphic to either $M_{a,b}(F)$ with $a \geq b \geq 0$ or to $M_N(F \oplus cF)$ where $c^2 = 1$.

Suppose first that $A = M_{a,b}(F)$. Recall that $(a+b)^2 = d+l$ and, for every $1 \leq i \leq t$, let

$$f_i = f_i(x_1^{2i-1}, \ldots, x_{d+l}^{2i-1}, \ldots, x_1^{2i}, \ldots, x_{d+l}^{2i})$$

be the central polynomial of A constructed in Theorem 5.7.4. Recall that f_i is alternating on each set of variables $\{x_1^{2i-1}, \ldots, x_{d+l}^{2i-1}\}$ and $\{x_1^{2i}, \ldots, x_{d+l}^{2i}\}$. Fix

$t > 0$ and set
$$f(x_1^1, \ldots, x_{d+l}^1, \ldots, x_1^{2t}, \ldots, x_{d+l}^{2t}) = f_1 \cdots f_t.$$

Thus f is alternating on each set of variables $\{x_1^i, \ldots, x_{d+l}^i\}$, $i = 1, \ldots, 2t$, and $f \notin \mathrm{Id}(A)$.

Let E be a basis of A which is homogeneous in the $\mathbb{Z}_2$-grading. Notice that, since $|E| = d+l$, for every i, we need to substitute all elements of E for $x_1^i, \ldots, x_{d+l}^i$ in order to get a non-zero evaluation of f. This means that after renaming all the variables in f,
$$\widetilde{f} = f(y_1^1, \ldots, y_d^1, \ldots, y_1^{2t}, \ldots, y_d^{2t}, z_1^1, \ldots, z_l^1, \ldots, z_1^{2t}, \ldots, z_l^{2t})$$

is not a graded identity of A where the y_i^j's are even variables and the z_i^j's are odd variables. But then, by Lemma 3.7.4, $\widetilde{f} \equiv 0$ is not a graded identity of $G(A)$. Since for every $i = 1, \ldots, 2t$, the polynomial f is alternating in $y_1^i, \ldots, y_d^i$ and in $z_1^i, \ldots, z_l^i$, it follows that the polynomial $\widetilde{f}$ is alternating in the variables $y_1^i, \ldots, y_d^i$ and symmetric in the variables $z_1^i, \ldots, z_l^i$.

Let $n_1 = 2td, n_2 = 2tl$. Then $\widetilde{f} \in P_{n_1,n_2}$ and, by the previous lemma, it follows that there exist $e_{T_\lambda} \in R_1 = FS_{n_1}$, $e_{T_\mu} \in R_2 = FS_{n_2}$, $\lambda = ((2t)^d)$, $\mu = (l^{2t})$ such that $g = e_{T_\lambda} e_{T_\mu} \widetilde{f} \not\equiv 0$ on $G(A)$.

If $l = 0$, then $g = e_{T_\lambda} \widetilde{f} = e_{T_\lambda} f$ is the required non-identity since $\lambda = ((2t)^d) = h(d, 0, 2t)$. Therefore, we may assume that $l > 0$.

Let M be the $R_1 \otimes R_2$-submodule of P_{n_1,n_2} generated by g; then M is isomorphic to the tensor product $M_1 \otimes M_2$ where $M_1 = R_1 e_{T_\lambda}, M_2 = R_2 e_{T_\mu}$.

If we write $n = n_1 + n_2$, then $P_{n_1,n_2} \subseteq P_n$ and we let $\bar{M}$ be the S_n-submodule of P_n generated by M. Let
$$\bar{M} = \bar{M}_1 \oplus \cdots \oplus \bar{M}_k$$
be its decomposition into S_n-irreducible submodules. By the Littlewood-Richardson rule, given in Theorem 2.3.9, every $\bar{M}_i$ is associated to a Young diagram D_λ such that
$$h(d, l, 2t - s) \leq \lambda \leq h(d, l, 2t)$$
where $s = \max\{d, l\}$. Therefore $\lambda \geq h(d, l, 2t - d - l)$. Since $\bar{M}$ is not contained in the T-ideal of ordinary (non-graded) identities of $G(A)$, it follows that for some multilinear $u \in \bar{M}$ we must have $e_{T_\lambda} u \not\equiv 0$ on $G(A)$ and the proof of the lemma is completed in case $A = M_{a,b}(F)$.

Now assume $A = A^{(0)} \oplus A^{(1)} = M_N(F) \oplus M_N(cF)$. Then $d = l = N^2 = \dim A^{(0)} = \dim A^{(1)}$. As in the previous case we let
$$f_0 = f_0(x_1^1, \ldots, x_d^1, \ldots, x_1^{2t}, \ldots, x_d^{2t})$$
be a multilinear polynomial which is alternating on the variables $x_1^i, \ldots, x_d^i$, $i = 1, \ldots, 2t$ and $f_0 \notin \mathrm{Id}(A_0)$. If we set
$$f = f_0(y_1^1, \ldots, y_d^1, \ldots, y_1^{2t}, \ldots, y_d^{2t}) f_0(z_1^1, \ldots, z_l^1, \ldots, z_1^{2t}, \ldots, z_l^{2t}),$$
then it is clear that $f \equiv 0$ is not a graded identity of A. The same argument as in the previous case completes the proof of the lemma. $\square$

6.4. Gluing Young tableaux

In this section we shall construct non-identities corresponding to Young tableaux for any PI-algebra. Our technique will involve the polynomials constructed in the previous section for the Grassmann envelope of a finite dimensional simple superalgebra together with some gluing techniques of tableaux that we shall develop in this section.

We recall the notation of Section 6.2. A will be a finite dimensional superalgebra over F with decomposition $A = A_{ss} + J$ where A_{ss} is a maximal stable semisimple subalgebra and J is the Jacobson radical of A. Also, $A_{ss} = A_1 \oplus \cdots \oplus A_k$ where $A_1, \ldots, A_k$ are simple superalgebras. As before, we consider a non-zero product of the type $B_1 J B_2 J \cdots J B_r \neq 0$ where $B_1, \ldots, B_r$ are distinct subalgebras from the set $\{A_1, \ldots, A_k\}$.

We start with the following easy observation.

LEMMA 6.4.1. *Let A be a finite dimensional simple superalgebra over an algebraically closed field F. If $A = M_{k,l}(F)$, then for any non-zero homogeneous element $b \in A$ and for any matrix unit e_{ij} there exist homogeneous elements $a, c_1 \in A$ such that $abc_1 = e_{ij}$. If $A = M_k(F \oplus cF)$, $c^2 = 1$, we can find homogeneous elements $a_1, a_2, c_1, c_2 \in A$ such that $a_1 b c_1 = e_{ij} \in A^{(0)}$ and $a_2 b c_2 = c e_{ij} \in A^{(1)}$.*

PROOF. Clearly one can take, up to scalars, $a = e_{i\alpha}, c_1 = e_{\beta j}$ for suitable α, β when $A = M_{k,l}(F)$. If $A = M_k(F \oplus cF)$, then we can take $a_1 = a_2 = e_{i\alpha}$, $c_1 = e_{\beta j}$ and $c_2 = c e_{\beta j}$. □

LEMMA 6.4.2. *Let A be a finite dimensional superalgebra over an algebraically closed field F and suppose that*
$$B_1 J B_2 J \cdots J B_r \neq 0$$
for some distinct simple $\mathbb{Z}_2$-graded subalgebras $B_1, \ldots, B_r$. Let $f_1, \ldots, f_r$ be multilinear polynomials on distinct sets of variables such that for every $i = 1, \ldots, r$, $f_i \notin \mathrm{Id}(G(B_i))$. Then, if $u_1, v_1, w_1, \ldots, w_{r-1}, u_r, v_r$ are new variables, the multilinear polynomial

(6.6) $$u_1 f_1 v_1 w_1 u_2 f_2 v_2 w_2 \cdots w_{r-1} u_r f_r v_r$$

is not an identity of $G(A)$.

PROOF. Since by hypothesis $B_1 J B_2 J \cdots J B_r \neq 0$, there exist homogeneous elements $b_1 \in B_1, \ldots, b_r \in B_r, e_1, \ldots, e_{r-1} \in J$ such that

(6.7) $$b_1 e_1 b_2 e_2 \cdots e_{r-1} b_r \neq 0.$$

Clearly $b_1, \ldots, b_r$ can be chosen as e_{ij} or $c e_{ij}$ when $B_k = M_l(F \oplus cF)$. For every $i = 1, \ldots, r$, write $f_i = f_i(x_1^i, \ldots, x_{n_i}^i)$. Since f_i is not an identity of $G(B_i)$, there exist homogeneous elements $\bar{x}_1^i, \ldots, \bar{x}_{n_i}^i \in B_i, g_j^i \in G$ such that
$$f_i(\bar{x}_1^i \otimes g_1^i, \ldots, \bar{x}_{n_i}^i \otimes g_{n_i}^i) \neq 0.$$

We regard each polynomial f_i as a graded polynomial by requiring that x_j^i is an even variable in case $\bar{x}_j^i \in B_i^{(0)}$ and x_j^i is an odd variable when $\bar{x}_j^i \in B_i^{(1)}$. Then, recalling the definition of $\widetilde{f_i}$, we have
$$f_i(\bar{x}_1^i \otimes g_1^i, \ldots, \bar{x}_{n_i}^i \otimes g_{n_i}^i) = \widetilde{f_i}(\bar{x}_1^i, \ldots, \bar{x}_{n_i}^i) \otimes g_1^i \cdots g_{n_i}^i.$$

Since $f \notin \mathrm{Id}(G(B_i))$, it follows that $\widetilde{f}_i(\bar{x}_1^i, \ldots, \bar{x}_{n_i}^i) = \bar{b}_i \neq 0$, for some $\bar{b}_i \in B_i$. By Lemma 6.4.1 we can choose homogeneous elements $a_i, c_i \in B_i$ such that $a_i \bar{b}_i c_i = b_i \neq 0$. Therefore the polynomial $u_i \widetilde{f}_i v_i$ takes the value b_i by evaluating the variables $u_i, x_1^i, \ldots, x_{n_i}^i, v_i$ to the elements $a_i, \bar{x}_1^i, \ldots, \bar{x}_{n_i}^i, c_i$, respectively. Let $h_i, h_i', t_i \in G$ be elements of the same homogeneous degree as a_i, e_i, c_i, respectively. Then for $i = 1, \ldots, r-1$, we get

$$(a_i \otimes h_i) f_i(\bar{x}_1^i \otimes g_1^i, \ldots, \bar{x}_{n_i}^i \otimes g_{n_i}^i)(c_i \otimes h_i')(e_i \otimes t_i)$$

$$= a_i \widetilde{f}_i(\bar{x}_1^i, \ldots, \bar{x}_{n_i}^i) c_i e_i \otimes h_i g_1^i \cdots g_{n_i}^i h_i' t_i = b_i e_i \otimes h_i g_1^i \cdots g_{n_i}^i h_i' t_i.$$

Also, if $i = r$ we get

$$(a_r \otimes h_r) f_i(\bar{x}_1^r \otimes g_1^r, \ldots, \bar{x}_{n_r}^r \otimes g_{n_r}^r)(c_r \otimes h_r')$$

$$= b_r \otimes h_r g_1^r \cdots g_{n_r}^r h_r'.$$

Since G is the infinite dimensional Grassmann algebra, we can choose homogeneous elements h_i, h_i', t_i, g_i^j in G such that

(6.8) $\qquad h_1 g_1^1 \cdots g_{n_1}^1 h_1' t_1 h_2 g_1^2 \cdots g_{n_2}^2 h_2' t_2 \cdots t_{r-1} h_r g_1^r \cdots g_{n_r}^r h_r' \neq 0.$

Now if we evaluate the polynomial $u_1 f_1 v_1 w_1 u_2 f_2 v_2 w_2 \cdots w_{r-1} u_r f_r v_r$ by setting $u_i = a_i \otimes h_i, v_i = c_i \otimes h_i', w_i = e_i \otimes t_i, x_i^j = \bar{x}_i^j \otimes g_i^j$, then we obtain the value

$$b_1 e_1 b_2 e_2 \cdots e_{r-1} b_r \otimes h_1 g_1^1 \cdots g_{n_1}^1 h_1' t_1 h_2 g_1^2 \cdots g_{n_2}^2 h_2' t_2 \cdots t_{r-1} h_r g_1^r \cdots g_{n_r}^r h_r'$$

which is non-zero by (6.7) and (6.8). This completes the proof of the lemma. $\square$

Once we have obtained a new polynomial out of the given polynomials f_i we need to act on it with the direct product of symmetric groups. The corresponding irreducible representations can be visualized by a process of gluing Young tableaux as follows.

Let $\lambda_1 \vdash n_1, \lambda_2 \vdash n_2, \ldots, \lambda_r \vdash n_r$ be given partitions and suppose that they satisfy the following conditions:

(6.9) $\qquad h(d_i, l_i, t_i - s_i) \leq \lambda_i \leq h(d_i, l_i, t_i), \ i = 1, \ldots, r,$

and

(6.10) $\qquad t_i - s_i \geq t_{i+1} + l_{i+1}, t_{i+1} + d_{i+1}, \ i = 1, \ldots, r-1.$

Let $D_1 = D_{\lambda_1}, D_2 = D_{\lambda_2}, \ldots, D_r = D_{\lambda_r}$ be the corresponding Young diagrams. Condition (6.9) says that the length of the first d_i rows of D_i is greater or equal to $l_i + t_i - s_i$ and less or equal to $l_i + t_i$. Similarly, the length of the first l_i columns (if $l_i > 0$) is greater or equal to $d_i + t_i - s_i$ and less or equal to $d_i + t_i$. The inequalities (6.10) mean that if we glue the first row of D_{i+1} to the $(d_i + 1)$th row of D_i, the second row of D_{i+1} to the $(d_i + 2)$th row of D_i and so on, as a result, we obtain a new Young diagram denoted $D_i \star D_{i+1}$ with $n_i + n_{i+1}$ boxes.

Consider the diagram $D_\lambda = D_1 \star D_2 \star \cdots \star D_r$ obtained by "gluing" together the diagrams $D_1, D_2, \ldots, D_r$ as described above (see picture below).

6.4. GLUING YOUNG TABLEAUX

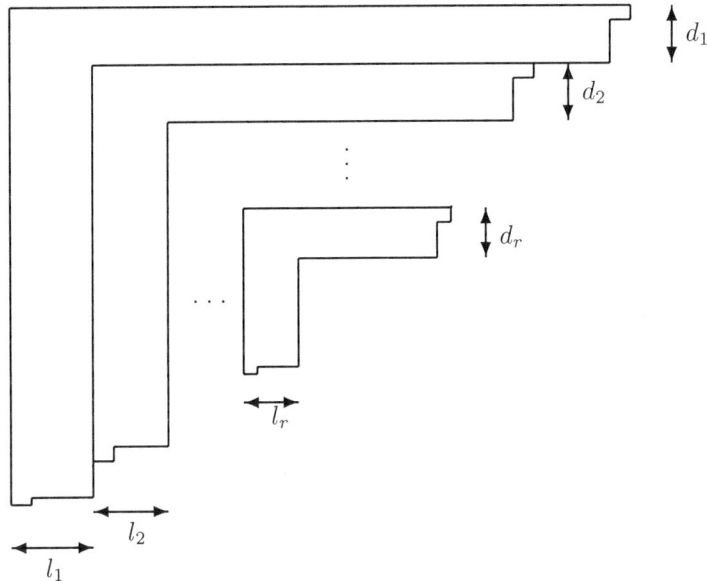

Clearly, $\lambda \leq h(d, l, t)$ where $l = l_1 + \cdots + l_r$, $d = d_1 + \cdots + d_r$ provided that $t \geq t_1 + l_1 - l, t_1 + d_1 - d$. On the other hand, $\lambda \geq h(d, l, t_r - s_r)$.

Now let $T_1, \ldots, T_r$ be Young tableaux corresponding to $\lambda_1, \ldots, \lambda_r$, respectively. If $\lambda_1, \ldots, \lambda_r$ satisfy (6.9) and (6.10) above, we can glue the tableaux in a similar way; if α_{uv} is the entry appearing in the (u, v) position of T_i, we write $T_i = D_i(\alpha_{uv})$. For every $i = 2, \ldots, r$, we now add $n_1 + \cdots + n_{i-1}$ to all the entries of T_i, obtaining in this way a new tableau $D_i(\alpha_{uv} + n_1 + \cdots + n_{i-1})$. If $T_1 = D_1(\alpha_{uv}), T_2 = D_2(\beta_{uv}), \ldots, T_r = D_r(\gamma_{uv})$, we then define

$$T_\lambda = T_1 \star T_2 \star \cdots \star T_r = D_1(\alpha_{uv}) \star D_2(\beta_{uv} + n_1) \star \ldots \star D_r(\gamma_{uv} + n_1 + \cdots + n_{r-1}).$$

It is clear that the tableau obtained in this way is filled up with the distinct integers $1, 2, \ldots, n$, where $n = n_1 + \cdots + n_r$.

Define $N_1 = \{1, \ldots, n_1\}$ and, for $2 \leq i \leq r$, $N_i = \{n_1 + \cdots + n_{i-1} + 1, \ldots, n_1 + \ldots + n_i\}$. Thus $N = \{1, \ldots, n\}$ is the disjoint union $N = N_1 \cup \ldots \cup N_r$. For $i = 1, \ldots, r$, we regard S_{n_i} as the permutation group acting on the set N_i, so that we can consider the group algebras $FS_{n_1}, \ldots, FS_{n_r}$ as embedded in FS_n with one-dimensional intersection.

Next we need to relate the essential idempotent e_{T_λ} to $e_{T_1}, \ldots, e_{T_r}$. We do this in the next lemma.

LEMMA 6.4.3. *Suppose that $\lambda_1, \ldots, \lambda_r$ satisfy conditions (6.9) and (6.10) and let $T_1, \ldots, T_r$ be corresponding tableaux. If $T_\lambda = T_1 \star \cdots \star T_r$, then*

$$e_{T_\lambda} = e_{T_1} \cdots e_{T_r} + b$$

where b is a linear combination of permutations $\sigma \in S_n$ such that $\sigma(N_i) \not\subseteq N_i$ for some $1 \leq i \leq r$.

PROOF. Let $E = \{\sigma \in S_n \mid \sigma(N_i) \subseteq N_i \text{ for all } i = 1, \ldots, r\}$. In our notation, clearly $E = S_{n_1} \times \cdots \times S_{n_r}$. We need to check that

$$e_{T_\lambda} - e_{T_1} \cdots e_{T_r} = \sum_{\sigma \in S_n \setminus E} \alpha_\sigma \sigma$$

for suitable $\alpha_\sigma \in F$. Recall that

(6.11)
$$e_{T_\lambda} = \sum_{\substack{\sigma \in R_{T_\lambda} \\ \tau \in C_{T_\lambda}}} (\operatorname{sgn} \tau) \sigma \tau$$

where R_{T_λ} is the subgroup of S_n of row permutations of T_λ and C_{T_λ} is the subgroup of column permutations T_λ. Denote $R = R_{T_\lambda} \cap E$, $C = C_{T_\lambda} \cap E$ and

$$R_i = \{\sigma \in R|\ \sigma(x) = x \text{ for all } x \in N \setminus N_i\},$$

$$C_i = \{\sigma \in C|\ \sigma(x) = x \text{ for all } x \in N \setminus N_i\}.$$

One can split the sum (6.11) into two parts, $e_{T_\lambda} = u + w$, where

$$u = \sum_{\substack{\sigma \in R \\ \tau \in C}} (\operatorname{sgn} \tau) \sigma \tau = \left(\sum_{\sigma \in R} \sigma \right) \left(\sum_{\tau \in C} (\operatorname{sgn} \tau) \tau \right)$$

and w contains all the remaining terms in the right-hand side of (6.11). We shall prove that $u = e_{T_1} \cdots e_{T_r}$ and w is a linear combination of permutations $\sigma \notin E$.

First note that any $\sigma \in R$ has a unique decomposition of the form $\sigma = \sigma_1 \cdots \sigma_r$ where $\sigma_i \in R_i$, $i = 1, \ldots, r$. Hence,

(6.12)
$$\sum_{\sigma \in R} \sigma = \sum_{\sigma_1 \in R_1, \ldots, \sigma_r \in R_r} \sigma_1 \cdots \sigma_r = \left(\sum_{\sigma_1 \in R_1} \sigma_1 \right) \cdots \left(\sum_{\sigma_r \in R_r} \sigma_r \right).$$

Similarly, any $\tau \in C$ has a unique decomposition of the form $\tau = \tau_1 \cdots \tau_r$ where $\tau_i \in C_i$, $i = 1, \ldots, r$. Thus

$$\sum_{\tau \in C} (\operatorname{sgn} \tau) \tau = \sum_{\tau_1 \in C_1, \ldots, \tau_r \in C_r} (\operatorname{sgn} \tau_1 \cdots \tau_r) \tau_1 \cdots \tau_r =$$

(6.13)
$$\left(\sum_{\tau_1 \in C_1} (\operatorname{sgn} \tau_1) \tau_1 \right) \cdots \left(\sum_{\tau_r \in C_r} (\operatorname{sgn} \tau_r) \tau_r \right).$$

Since for $i \neq j$, FS_{n_i} and FS_{n_j} commute elementwise in FS_n and

$$\left(\sum_{\sigma \in R_i} \sigma \right) \left(\sum_{\tau \in C_i} (\operatorname{sgn} \tau) \tau \right) = e_{T_i}$$

we obtain from (6.12) and (6.13) that $u = e_{T_1} \cdots e_{T_r}$.

Now let $\tau \in C_{T_\lambda} \setminus E$. Then there exists i such that $\tau(x) \in N_j$ for some $x \in N_i$ and $j < i$. Suppose that x belongs to the kth row of T_λ. Then $\tau(x)$ lies in a higher row (say in the mth row) of T_λ since $\tau(x) \in N_j$ and $j < i$. By the construction of T_λ all entries of the mth row belong to $N_1 \cup \ldots \cup N_j$. Hence, for all $\sigma \in R_{T_\lambda}$, we have that $\sigma\tau(x) \notin N_i$ and $\sigma\tau \notin E$ follows.

On the other hand, if $\tau \in C_{T_\lambda} \cap E$ and $\sigma \in R_{T_\lambda} \setminus E$, then $\sigma\tau \notin E$ since E is a subgroup of S_n and $\tau \in E$. It follows that $w = \sum_{\sigma \notin E} \alpha_\sigma \sigma$ and the proof of the lemma is complete. $\square$

In the next lemma we combine the construction of a new polynomial given in Lemma 6.3.3 with the gluing technique developed before.

6.4. GLUING YOUNG TABLEAUX

LEMMA 6.4.4. *Let A be a finite dimensional superalgebra over an algebraically closed field F. Let $B_1, \ldots, B_r$ be distinct simple $\mathbb{Z}_2$-graded subalgebras of A such that $B_1 J B_2 J \cdots J B_r \neq 0$ and let*

$$d = \dim(B_1^{(0)} \oplus \cdots \oplus B_r^{(0)}), \ l = \dim(B_1^{(1)} \oplus \cdots \oplus B_r^{(1)}).$$

Then for any positive integer $t \geq 2 \dim A$ there exist a partition $\lambda \vdash n$ with

$$h(d, l, 2t - s) \leq \lambda \leq h(d, l, 2t),$$

$s = 4 \dim A$, *and a tableau T_λ such that $e_{T_\lambda} f \notin \mathrm{Id}(G(A))$ for some multilinear polynomial $f = f(x_1, \ldots, x_n, \ldots)$ with $\deg f \leq n + 3 \dim A$.*

PROOF. For $i = 1, \ldots, r$, define $d_i = \dim B_i^{(0)}$, $l_i = \dim B_i^{(1)}$ so that $d = d_1 + \cdots + d_r$ and $l = l_1 + \cdots + l_r$. By Lemma 6.3.3, for any integer t_i there exists a partition λ_i such that

$$h(d_i, l_i, 2t_i - s_i) \leq \lambda_i \leq h(d_i, l_i, 2t_i),$$

where $l_i, d_i \leq s_i = \dim B_i$, and a tableau T_i on λ_i such that $g_i \notin \mathrm{Id}(G(B_i))$ for some multilinear polynomial g_i corresponding to T_i.

We choose $t_1, \ldots, t_r$ by the following rule. Let $t_1 = t \geq 2 \dim A$ be arbitrary. Denote $q_i = s_{i-1} + \max\{l_i, d_i\}, i = 2, \ldots, r$, and set $q_i' = q_i$ if q_i is even and $q_i' = q_i + 1$ if q_i is odd. Then define $2t_{i+1} = 2t_i - q_{i+1}', i = 1, \ldots, r-1$. It follows that

$$2t_i - s_i = 2t_{i+1} + q_{i+1}' - s_i \geq 2t_{i+1} + \max\{l_{i+1}, d_{i+1}\}.$$

By the choice made above, we see that $\lambda_1, \ldots, \lambda_r$ satisfy (6.9) and (6.10), with $t_1, \ldots, t_r$ replaced by $2t_1, \ldots, 2t_r$.

We now glue the tableaux $T_1, \ldots, T_r$ (corresponding to the partitions $\lambda_1, \ldots, \lambda_r$, respectively) by the procedure shown in the previous section. We obtain this way a tableau $T_\lambda = T_1 \star \cdots \star T_r$ of shape λ satisfying

(6.14) $$h(d, l, 2t_r - s_r) \leq \lambda \leq h(d, l, u),$$

for every $u \geq 2t_1 + l_1 - l, 2t_1 + d_1 - d$. We now compute

$$2t_1 - 2t_r = \sum_{i=1}^{r-1}(2t_i - 2t_{i+1}) = \sum_{i=1}^{r-1} q_{i+1}' \leq r + \sum_{i=1}^{r-1} q_{i+1}$$

$$\leq r + \sum_{i=1}^{r-1}(s_i + s_{i+1}) \leq r + 2\dim(B_1 \oplus \cdots \oplus B_r) \leq 3 \dim A.$$

Hence $2t_r - s_r \geq 2t - 3\dim A - s_r \geq 2t - 4\dim A$ and the inclusions given in (6.14) become

$$h(d, l, 2t - 4\dim A) \leq \lambda \leq h(d, l, 2t).$$

For every $i = 1, \ldots, r$, set $n_i = \deg g_i$ where $g_i \notin \mathrm{Id}(G(B_i))$ is a multilinear polynomial corresponding to T_i. Write $n = n_1 + \cdots + n_r$ and let $\{1, \ldots, n\} = N_1 \cup \ldots \cup N_r$ where $N_1 = \{1, \ldots, n_1\}$ and, for $2 \leq i \leq r$, $N_i = \{n_1 + \cdots + n_{i-1} + 1, \ldots, n_1 + \ldots + n_i\}$. Finally, for every $i = 1, \ldots, r$, denote by f_i the multilinear polynomial g_i written in the new set of variables $\{x_j \mid j \in N_i\}$.

We now construct the multilinear polynomial

$$f = u_1 f_1 v_1 w_1 u_2 f_2 v_2 w_2 \ldots w_{r-1} u_r f_r v_r$$

where the u_i, v_i, w_i are new variables. Then, by Lemma 6.4.2, $f \notin \mathrm{Id}(G(A))$. Moreover, $\deg f = n + 3r - 1 \leq n + 3\dim A$. Therefore in order to complete the proof of the lemma it is enough to show that $e_{T_\lambda} f \notin \mathrm{Id}(G(A))$.

Now, by the same proof of Lemma 6.4.2, we know that there exists a valuation $\theta : F\langle X \rangle \to A$ such that $\theta(f) \neq 0$ and
$$\theta(w_i) = \bar{w}_i, \ \theta(u_i) = \bar{u}_i, \ \theta(v_i) = \bar{v}_i, \ \theta(x_i) = \bar{x}_i$$
where $\bar{w}_1, \ldots, \bar{w}_{r-1} \in J \otimes G, \bar{u}_i, \bar{v}_i \in B_i \otimes G$ and $\bar{x}_j \in B_i \otimes G$ if $j \in N_i$. Also, by Lemma 6.4.3,
$$\theta(e_{T_\lambda} f) = \theta(e_{T_1} \cdots e_{T_r} f) + \theta(bf)$$
where b is a linear combination of elements $\sigma \in S_n$ such that σ "shuffles" the sets $N_1, \ldots, N_r$. Since e_{T_i} is an essential idempotent, i.e., $e_{T_i}^2 = \mu_i e_{T_i}$ for some $\mu_i \in \mathbb{Z}, \mu_i \neq 0$, and the multilinear polynomial f_i corresponds to T_i, then we have that $e_{T_i} f_i = \mu_i f_i$. Hence
$$\theta(e_{T_1} \cdots e_{T_r} f) = \mu_1 \cdots \mu_r \theta(f) \neq 0.$$

On the other hand, we next show that every summand in $\theta(bf)$ is equal to zero. In fact, let $\sigma \in S_n$ be such that $\sigma(N_i) \not\subset N_i$ for some $1 \leq i \leq r$. Then $\sigma(j) \in N_q$ for some $j \in N_i, q \neq i$. This says that $\theta(\sigma f_i)$ belongs to $G(B_q)$. But then, since $\bar{u}_i \theta(\sigma f_i) \in G(B_i) G(B_q) \subseteq B_i B_q \otimes G = 0$, we have
$$\theta(\sigma f) = \bar{u}_1 \theta(\sigma f_1) \bar{v}_1 \bar{w}_1 \cdots \theta(\sigma f_r) \bar{v}_r = 0,$$
Therefore $\theta(bf) = 0$ and the proof is complete. □

6.5. Existence of the exponent

In this section we prove the main result of this chapter. We shall show that the integer q defined in Section 6.2 for a finite dimensional superalgebra A, coincides with the PI-exponent of its Grassmann envelope. Since any non-trivial variety can be generated by such Grassmann envelope, this will show that the exponent always exists and is a non-negative integer.

PROPOSITION 6.5.1. *Let A be a finite dimensional superalgebra over an algebraically closed field of characteristic zero. Let $q \geq 0$ be the integer defined in (6.3). Then there exist constants C_1, C_2, r_1, r_2 depending only on $\dim A$ such that $C_1 \neq 0$ and*
$$C_1 n^{r_1} q^n \leq c_n(G(A)) \leq C_2 n^{r_2} q^n.$$

PROOF. Recall that $A = A_1 \oplus \cdots \oplus A_k + J$ with $A_1, \ldots, A_k$ simple superalgebras, and $q = \dim_F(B_1 \oplus \cdots \oplus B_r)$ is maximal where $B_1, \ldots, B_r$ are distinct subalgebras from the set $\{A_1, \ldots, A_k\}$ and $B_1 J B_2 J \cdots J B_r \neq 0$. Let
$$d = \dim(B_1^{(0)} \oplus \cdots \oplus B_r^{(0)}), \quad l = \dim(B_1^{(1)} \oplus \cdots \oplus B_r^{(1)})$$
and $m = \dim A$.

For any integer $N > 5m^2 + 3m$ divide $N - dl - 3m$ by $2q$ and write $N = 2tq + dl + 3m + r$, for some $t > 2m$ and $0 \leq r < 2q$.

By Lemma 6.4.4 there exists n with $2tq - 4mq + dl \leq n \leq 2tq + dl$, and a partition $\lambda \vdash n$ with
$$h(d, l, 2t - 4m) \leq \lambda \leq h(d, l, 2t),$$
such that $e_{T_\lambda} f \notin \mathrm{Id}(G(A))$, for a suitable tableau T_λ and a multilinear polynomial f with $n \leq \deg f = c \leq n + 3m$.

We construct the polynomial $f' = f \cdot x_{c+1} \cdots x_N$, where $x_{c+1}, \ldots, x_N$ are new variables distinct from the ones appearing in f. By Lemma 6.4.2 and Lemma 6.4.4 and their proofs, it is easy to see that still $e_{T_\lambda} f' \notin \mathrm{Id}(G(A))$. By the branching theorem (Theorem 2.3.1), it follows that

$$FS_N e_{T_\lambda} f' \subseteq \bigoplus_{\substack{\mu \vdash N \\ \mu \geq \lambda}} I_\mu f'$$

where I_μ is the two-sided ideal of FS_N corresponding to the partition μ. Hence there exists $\mu \geq \lambda$ and a tableau T_μ such that $FS_N e_{T_\mu} f' \not\subseteq \mathrm{Id}(G(A))$.

Recalling that d_μ denotes the degree of the character χ_μ, we obtain that

$$c_N(G(A)) \geq d_\mu \geq d_{h(d,l,2t-4m)}.$$

Since
$$N - |h(d,l,2t-4m)| = N - (2tq - 4mq + dl) < 4m^2 + 5m$$
is bounded, we can apply Lemma 6.2.4. Since by Lemma 6.2.5,

$$d_{h(d,l,2t-4m)} \underset{n\to\infty}{\simeq} a n^b (d+l)^n,$$

it follows that $c_N(G(A)) \geq C_1 N^{r_1} q^N$, for some constants C_1, r_1 depending on m. Since by Proposition 6.2.6, $c_N(G(A)) \leq C_2 N^{r_2} q^N$, for some constants C_2, r_2, the conclusion of the proposition follows. $\square$

Recall that by Theorem 4.1.9, if A is an F-algebra, then the codimensions of A do not change upon extension of the base field F. More precisely, if K is an extension field of F, we regard the algebra $\bar{A} = A \otimes_F K$ as a K-algebra and we denote by $c_n^K(A)$ its nth codimension. Then, for all $n \geq 1$, we have that $c_n^K(A) = c_n(A)$.

We can now state the following result which proves Conjecture 6.1.3 and is also related to Conjecture 6.1.4.

THEOREM 6.5.2 ([**GZ1**], [**GZ2**]). *Let A be a PI-algebra over a field F of characteristic zero. Then there exists an integer $q \geq 0$ and constants C_1, C_2, r_1, r_2 such that $C_1 \neq 0$ and*

$$C_1 n^{r_1} q^n \leq c_n(A) \leq C_2 n^{r_2} q^n.$$

As a consequence $\exp(A)$ exists and is a non-negative integer.

PROOF. As we remarked above, the codimensions do not change by extending the ground field. Therefore, without loss of generality, we may assume that F is algebraically closed. By Corollary 4.8.4, there exists a finite dimensional superalgebra B over F such that $\mathrm{Id}(A) = \mathrm{Id}(G(B))$. But then the conclusion of the theorem follows from Proposition 6.5.1. $\square$

6.6. Computing the exponent of some algebras

In this section we shall exploit the construction of the previous sections and we shall compute the PI-exponent of some significant algebras. We start by examining finite dimensional algebras.

Let A be a finite dimensional algebra over F. Then A can be regarded as a superalgebra with trivial grading $A^{(0)} = A, A^{(1)} = 0$. If we now form its Grassmann envelope, we obtain $G(A) = (G^{(0)} \otimes A) \oplus (G^{(1)} \otimes 0) = G^{(0)} \otimes A$ and, by Lemma 1.4.2, $\mathrm{Id}(G^{(0)} \otimes A) = \mathrm{Id}(A)$ since $G^{(0)}$ is a commutative algebra and F is infinite. This proves that $\mathrm{var}(A) = \mathrm{var}(G(A))$ and we now have a constructive way of computing the PI-exponent of A as follows.

When F is algebraically closed, write $A = A_{ss} + J$ where A_{ss} is a maximal semisimple subalgebra and let $A_{ss} = A_1 \oplus \cdots \oplus A_k$ where $A_1, \ldots, A_k$ are simple subalgebras (recall that the grading on A is trivial!). Then $\exp(A)$ is the maximal dimension of a semisimple subalgebra $A_{i_1} \oplus \cdots \oplus A_{i_r}$ such that $A_{i_1} J \cdots J A_{i_r} \neq 0$ where $A_{i_1}, \ldots, A_{i_r}$ are distinct among $A_1, \ldots, A_k$.

We make use of this reduction in the proof of the next theorem. For an algebra A let us denote by $Z = Z(A)$ the center of A.

THEOREM 6.6.1. *Let A be a finite dimensional algebra over a field F of characteristic zero. Then*
   1) $\exp(A) \leq \dim_F A$.
   2) *If A is semisimple, $\exp(A) = \dim_{Z(B)} B$ where B is a simple subalgebra of A of greatest dimension over its center $Z(B)$. In particular, if A is simple $\exp(A) = \dim_{Z(A)} A$.*
   3) *A is central simple over F if and only if $\exp(A) = \dim_F A$.*

PROOF. Let $\bar{F}$ be the algebraic closure of F and $\bar{A} = A \otimes_F \bar{F}$. By Theorem 4.1.9, the codimensions of A and $\bar{A}$ are the same. Moreover, by the previous discussion $\exp(\bar{A})$ is the dimension of a suitable semisimple subalgebra of $\bar{A}$. Thus $\exp(A) = \exp(\bar{A}) \leq \dim_{\bar{F}} \bar{A} = \dim_F A$ and 1) is proved.

Suppose that A is semisimple and let $A = A_1 \oplus \cdots \oplus A_k$ be the decomposition of A into simple subalgebras. Then
$$\bar{A} = A \otimes_F \bar{F} \cong \oplus_i (A_i \otimes_F \bar{F}) \cong \oplus_i (B_{i1} \oplus \cdots \oplus B_{is_i})$$
where $B_{i1} \cong \ldots \cong B_{is_i}$ are central simple algebras over $\bar{F}$ and $s_i = [Z(A_i) : F]$. Then, by the above discussion, $\exp(A) = \exp(\bar{A}) = \max_i \{\dim_{\bar{F}} B_{i1}\}$. Since for all i,
$$(\dim_{Z(A_i)} A_i) \dim_F Z(A_i) = \dim_F A_i = \dim_{\bar{F}} A_i \otimes_F \bar{F} = (\dim_{\bar{F}} B_{i1}) \dim_F Z(A_i),$$
we deduce that $\dim_{Z(A_i)} A_i = \dim_{\bar{F}} B_{i1}$ and 2) is proved.

Suppose now that $\exp(A) = \dim_F A = \dim_{\bar{F}} \bar{A}$. Since $\exp(A) = \exp(\bar{A}) > 0$, $\bar{A}$ must contain a semisimple subalgebra B. By the description of the exponent, $\dim_{\bar{F}} \bar{A} = \exp(A) \leq \dim_{\bar{F}} B$. Hence $\bar{A} = B$ is semisimple and by the above we obtain that A must be central simple over F. $\square$

An important class of finite dimensional algebras is given by the upper block triangular matrix algebras $UT(d_1, \ldots, d_m)$ (Section 1.9). Recall that

$$UT(d_1, \ldots, d_m) = \begin{pmatrix} M_{d_1}(F) & B_{12} & \cdots & B_{1m} \\ 0 & M_{d_2}(F) & \cdots & B_{2m} \\ \vdots & \vdots & & \vdots \\ 0 & 0 & \cdots & M_{d_m}(F) \end{pmatrix}$$

where all the B_{ij}'s are rectangular matrices over F of corresponding size. Then $UT(d_1, \ldots, d_m) \cong M_{d_1}(F) \oplus \cdots \oplus M_{d_m}(F) + J$ where $\bigoplus_{i,j} B_{ij} \cong J$ is the Jacobson radical. In order to compute the exponent we may assume that F is algebraically closed. Then, since in this case,
$$M_{d_1}(F) B_{12} M_{d_2}(F) B_{23} \cdots B_{m-1,m} M_{d_m}(F) \neq 0,$$

we obtain that $\exp(UT(d_1,\ldots,d_m)) = \dim M_{d_1}(F) + \cdots + \dim M_{d_m}(F)$. Thus we have the following.

COROLLARY 6.6.2. $\exp(UT(d_1,\ldots,d_m)) = d_1^2 + \cdots + d_m^2$.

We next compute the PI-exponent of the prime varieties. Recall (see Section 3.7) that a proper variety is prime if it is generated by either $M_k(F)$ or $M_k(G)$ or $M_{k,l}(G)$ ($0 < l \leq [k/2]$). We already know by the previous theorem (or by Theorem 5.10.4) that $\exp(M_k(F)) = k^2$. The exponent of the other verbally prime algebras was computed in [**Re6**].

COROLLARY 6.6.3.
1) $\exp(M_k(G)) = 2k^2$.
2) $\exp(M_{k,l}(G)) = (k+l)^2$.

PROOF. Assume that F is algebraically closed. In order to prove 1), recall that
$$M_k(G) \cong (M_k(F) \otimes G^{(0)}) \oplus (M_k(F) \otimes G^{(1)}) \cong (M_k(F) \otimes G^{(0)}) \oplus (cM_k(F) \otimes G^{(1)})$$
is the Grassmann envelope of the simple superalgebra $M_k(F \oplus cF)$. Hence
$$\exp(M_k(G)) = \dim M_k(F \oplus cF) = 2k^2.$$

Finally, recall that $M_{k,l}(G) \cong (M_{k,l}^{(0)}(F) \otimes G^{(0)}) \oplus (M_{k,l}^{(1)}(F) \otimes G^{(1)})$ is the Grassmann envelope of the simple superalgebra $M_{k,l}(F)$. Hence $\exp(M_{k,l}(G)) = \dim M_{k,l}(F) = (k+l)^2$. $\square$

CHAPTER 7

Polynomial Growth and Low PI-exponent

This chapter is mainly devoted to the characterization of varieties of polynomial growth (or PI-exponent ≤ 1) over a field of characteristic zero. Here we present three different results: a variety $\mathcal{V}$ has polynomial growth if and only if the Grassmann algebra G and the algebra UT_2 of 2×2 upper triangular matrices do not lie in $\mathcal{V}$. The second gives an explicit description of a generating algebra of $\mathcal{V}$ and the third characterization gives a restriction on the shape of the diagrams of the S_n-characters appearing with non-zero multiplicity in the nth cocharacter of $\mathcal{V}$.

In this setting we address the following question: what kind of polynomial growth can appear for the codimension sequence of a PI-algebra? To this end we construct, for any integer $t \geq 0$, an algebra A_t such that $c_n(A_t)$ grows asymptotically as n^t.

In this chapter we further analyze the basic properties of the algebras G and UT_2. It turns out that the Grassmann algebra actually separates the T-ideals of finitely generated PI-algebras from the other ones. In fact, we prove that for a variety $\mathcal{V}$, $G \notin \mathcal{V}$ if and only if $\mathcal{V}$ is generated by a finitely generated (actually finite dimensional) algebra and this occurs if and only if the standard identity or the Capelli identity of some degree is satisfied by $\mathcal{V}$.

We also give a characterization of varieties $\mathcal{V}$ such that $UT_2 \notin \mathcal{V}$. We conclude this chapter by giving a characterization of varieties of exponent 2.

Throughout this chapter the ground field F is assumed to be of characteristic zero.

7.1. The Grassmann algebra and standard polynomials

A distinguished algebra with small exponential codimension growth is the Grassmann algebra G (see Theorem 4.1.8). This algebra has many interesting properties and in this section we shall prove that a variety $\mathcal{V}$ is generated by a finite dimensional algebra if and only if $G \notin \mathcal{V}$. We start by proving that a variety $\mathcal{V}$ satisfies some standard polynomial if and only if $G \notin \mathcal{V}$.

Recall that, given any integer $k \geq 1$, the essential idempotent $e_{(1^k)}$ corresponding to any tableau of shape (1^k) is given by the formula

$$e_{(1^k)} = \sum_{\tau \in S_k} (\mathrm{sgn}\,\tau)\tau.$$

LEMMA 7.1.1. *Let $r + l + 1 = k$ and $\{i_1, \ldots, i_r, j_1, \ldots, j_l, t\} = \{1, 2, \ldots, k\}$. Then*

(7.1) $$e_{(1^k)}[[x_{i_1} \cdots x_{i_r}, x_{j_1} \cdots x_{j_l}], x_t] = 0$$

in $F\langle X \rangle$.

PROOF. Clearly it is enough to prove the lemma when $i_1 = 1, \ldots, i_r = r, j_1 = r+1, \ldots, j_l = k-1$ and $t = k$. Write
$$[[x_1 \cdots x_r, x_{r+1} \cdots x_{k-1}], x_k] = A + B + C + D$$
where
$$A = x_1 \cdots x_k, \quad B = -x_{r+1} \cdots x_{k-1} x_1 \cdots x_r x_k,$$
$$C = -x_k x_1 \cdots x_{k-1} \quad \text{and} \quad D = x_k x_{r+1} \cdots x_{k-1} x_1 \cdots x_r.$$
By Proposition 1.5.7, the polynomial on the left-hand side of (7.1) is a scalar multiple of the standard polynomial St_k. Hence, in order to prove the lemma, it is enough to compute the coefficient λ of the monomial $x_1 \cdots x_k$ in
$$e_{(1^k)}[[x_1 \cdots x_r, x_{r+1} \cdots x_{k-1}], x_k] = e_{(1^k)}A + e_{(1^k)}B + e_{(1^k)}C + e_{(1^k)}D.$$
First note that if $\lambda_A, \lambda_B, \lambda_C, \lambda_D$ are the coefficients of $x_1 \cdots x_k$ in $e_{(1^k)}A$, $e_{(1^k)}B$, $e_{(1^k)}C$, $e_{(1^k)}D$, respectively, then $\lambda = \lambda_A + \lambda_B + \lambda_C + \lambda_D$ and
$$\lambda_A = 1, \quad \lambda_B = -(-1)^{rp}, \quad \lambda_C = -(-1)^{k-1}, \quad \lambda_D = (-1)^{rp}(-1)^{k-1}$$
where $p = k - 1 - r$. Hence
$$\lambda = 1 - (-1)^{rp} - (-1)^{r+p} + (-1)^{rp}(-1)^{r+p} = (1 - (-1)^{rp})(1 - (-1)^{r+p}).$$
Now, if r or p is even, then $(-1)^{rp} = 1$. If both r and p are odd, then $(-1)^{r+p} = 1$. In any case $\lambda = 0$ and the proof of the lemma is completed. □

THEOREM 7.1.2. *A variety $\mathcal{V}$ satisfies a standard identity if and only if $G \notin \mathcal{V}$.*

PROOF. Recall that G is generated by the elements $1, e_1, e_2, \ldots$, subject to the condition $e_i e_j = -e_j e_i$. Then, since $St_m(e_1, \ldots, e_m) = m! e_1 \cdots e_m$, G does not satisfy any standard identity. It follows that $G \notin \mathcal{V}$ as soon as St_m is an identity of $\mathcal{V}$, for some $m \geq 1$.

Suppose $G \notin \mathcal{V}$ and let $f = f(x_1, \ldots, x_n)$ be a multilinear polynomial such that $f \in \text{Id}(\mathcal{V})$ but $f \notin \text{Id}(G)$.

Now, by Theorem 4.1.8, $\text{Id}(G)$ is generated, as a T-ideal, by the polynomial $[[x_1, x_2], x_3]$. Hence any element of $\text{Id}(G)$ is a linear combination of polynomials of the type

(7.2) $$u[[p,q],r]v$$

where u, v, p, q, r are (eventually empty) words in the alphabet X. For any monomials p, q, a and b, by the Jacobi identity and anticommutativity we have that
$$[[p,q], ab] = [[p,q], a]b + a[[p,q], b].$$
This says that in every polynomial of type (7.2) we may take r to be a monomial of degree one. In other words, every element of $\text{Id}(G)$ can be written as a linear combination of products of the type

(7.3) $$a[[x_{i_1} \cdots x_{i_k}, x_{j_1} \cdots x_{j_l}], x_r]b$$

where a and b are monomials of degree ≥ 0 on X.

Recall that by the proof of Theorem 4.1.8, P_n, the space of multilinear polynomials in $x_1, \ldots, x_n$, is spanned, modulo $P_n \cap \text{Id}(G)$, by elements of the type

(7.4) $$x_{i_1} \cdots x_{i_k}[x_{j_1}, x_{j_2}] \cdots [x_{j_{2m-1}}, x_{j_{2m}}]$$

where

(7.5) $$i_1 < \ldots < i_k, \quad j_1 < j_2 < \ldots < j_{2m}, \quad 2m + k = n,$$

(see (4.7), (4.8)).

Recalling that f is an identity for $\mathcal{V}$ such that $f \in P_n$ and $f \notin \mathrm{Id}(G)$, we write f as a linear combination of polynomials of type (7.3) and (7.4). Moreover, one of the polynomials of type (7.4) must have a non-zero coefficient. Clearly, without loss of generality, we may assume that this polynomial is of the form

$$x_1 \cdots x_k [x_{k+1}, x_{k+2}] \cdots [x_{n-1}, x_n]$$

and appears in f with coefficient 1. Replacing x_i with $[x_i, y_i], i = 1, \ldots, k$, we obtain a new identity $g = g(x_1, \ldots, x_n, y_1, \ldots, y_k)$ of $\mathcal{V}$. Note that, under any evaluation φ, all elements of type (7.4) satisfying (7.5) are mapped to $\mathrm{Id}(G)$ unless $i_1 = 1, \ldots, i_k = k, j_1 = k+1, \ldots, j_{2m} = n$ and $\varphi(\mathrm{Id}(G)) \subseteq \mathrm{Id}(G)$. Hence

$$g = [x_1, y_1] \cdots [x_k, y_k][x_{k+1}, x_{k+2}] \cdots [x_{n-1}, x_n] + h$$

where h is a multilinear polynomial in $x_1, \ldots x_n, y_1, \ldots y_k$ lying in $P_n \cap \mathrm{Id}(G)$. After renaming the indeterminates we write

(7.6) $$g = [x_1, x_2] \cdots [x_{2q-1}, x_{2q}] + h$$

where $h \in P_{2q}$ is a linear combination of polynomials of type (7.3).

Notice that if $\{i_1, \ldots, i_k, j_1, \ldots, j_l, r\} \subseteq \{1, \ldots, 2q\}$ and

$$e = \sum_{\sigma \in S_{\{i_1, \ldots, i_k, j_1, \ldots, j_l, r\}}} (\mathrm{sgn}\,\sigma)\sigma,$$

then $e_{(1^{2q})} = ce$ for some $c \in FS_{2q}$. Hence, if $w = a[[x_{i_1} \cdots x_{i_k}, x_{j_1} \cdots x_{j_l}], x_r]b$ is a multilinear polynomial of type (7.3), appearing in h, by Lemma 7.1.1, we have that $e_{(1^{2q})}w = cew = 0$.

Since the action of $e_{(1^{2q})}$ is that of alternating the first $2q$ variables, as we noticed in Remark 1.7.8, we have that

$$e_{(1^{2q})}[x_1, x_2] \cdots [x_{2q-1}, x_{2q}] = 2^q St_{2q}.$$

We may therefore conclude that

$$\frac{1}{2^q} e_{(1^{2q})} g = \frac{1}{2^q} e_{(1^{2q})} [x_1, x_2] \cdots [x_{2q-1}, x_{2q}] = St_{2q}$$

is an identity of $\mathcal{V}$. $\square$

Our next goal is to find other characterizations of varieties not containing the Grassmann algebra. We start by examining finite dimensional superalgebras whose Grassmann envelope satisfies a standard identity.

LEMMA 7.1.3. *Let $A = A^{(0)} \oplus A^{(1)}$ be a finite dimensional superalgebra such that $G(A)$ satisfies a standard identity. Then the ideal generated by $A^{(1)}$ is nilpotent.*

PROOF. Let $\bar{F}$ be the algebraic closure of F and set $\bar{A} = A \otimes_F \bar{F}$ and $\bar{G} = G \otimes_F \bar{F}$. Then clearly, $G(A)$ satisfies a standard identity if and only if $\bar{G}(\bar{A})$ satisfies a standard identity. Therefore without loss of generality, we may assume that F is algebraically closed.

Let J be the Jacobson radical of A. We shall prove that $A^{(1)} \subseteq J$. Since J is nilpotent, this will finish the proof of the lemma. Suppose to the contrary that $A^{(1)}$ does not lie in J. Since J is a graded ideal, A/J has induced $\mathbb{Z}_2$-grading and $(A/J)^{(1)} \neq 0$. Also $G(A/J)$ still satisfies a standard identity. Hence without loss of generality, we may assume A to be semisimple and $A^{(1)} \neq 0$. By Theorem 3.5.4, A is the direct sum of simple superalgebras of the type $M_k(F), M_{k,l}(F)$ and

$M_k(F \oplus cF)$ with $c^2 = 1$. Since $A^{(1)} \neq 0$, at least one of the simple summands is of type $M_{k,l}(F)$ or $M_k(F \oplus cF)$. In the first case A contains a subalgebra isomorphic to $M_{1,1}(F)$, hence $G(A)$ contains the subalgebra

$$F(e_{11} + e_{22}) \otimes G^{(0)} \oplus F(e_{12} + e_{21}) \otimes G^{(1)}$$

which is isomorphic to G. In the second case $G(A)$ contains the subalgebra $G(F \oplus cF) \simeq G$. Since G does not satisfy any standard identity it follows that also $G(A)$ does not satisfy any standard identity. This contradiction completes the proof of the lemma. □

In the following theorem $H(d, 0)$ denotes the infinite hook with trivial foot, i.e., a strip of width d.

THEOREM 7.1.4. *For a variety of algebras $\mathcal{V}$ the following conditions are equivalent:*

1) $\mathcal{V} = \text{var}(A)$, *for some finite dimensional algebra A.*
2) $\mathcal{V} = \text{var}(A)$, *for some finitely generated algebra A.*
3) $\mathcal{V}$ *satisfies a Capelli identity.*
4) $\chi_n(\mathcal{V}) \subseteq H(d, 0)$, *for some integer $d \geq 1$.*
5) $\mathcal{V}$ *satisfies a standard identity.*
6) $G \notin \mathcal{V}$.

PROOF. The equivalence of 1) and 2) follows from Corollary 4.8.5. The equivalence of 3) and 4) is proved in Theorem 4.6.1. The equivalence of 5) and 6) is the content of Theorem 7.1.2. The implication 3) → 5) is obvious. The implication 1) → 3) follows from Theorem 1.5.8. Hence it is sufficient to check 5) → 1).

Suppose that $\mathcal{V}$ satisfies 5). By Corollary 4.8.4, $\mathcal{V}$ is generated by the Grassmann envelope $G(A)$ of a finite dimensional superalgebra $A = A^{(0)} \oplus A^{(1)}$ and by the previous Lemma $A^{(1)}$ generates a nilpotent ideal in A. This clearly implies that $A^{(1)} \otimes G^{(1)}$ also generates a nilpotent ideal in $G(A)$. In other words, there exists a positive integer m such that any product of elements of $G(A)$ containing at least m factors of the type $a \otimes g, a \in A^{(1)}, g \in G^{(1)}$, must be zero.

Fix a basis $\{a_1, \ldots, a_k\}$ of $A^{(1)}$ and let $t = \max\{k, m\}$. Denote by C the subalgebra of $G(A)$ generated by $A^{(0)} \otimes 1$ and $\{a_i \otimes e_j \mid 1 \leq i \leq k, 1 \leq j \leq t\}$ where $e_1, e_2, \ldots,$ are the standard generators of G. Then C is a homogeneous subalgebra, $C = C^{(0)} \oplus C^{(1)}$, and $\dim C < \infty$.

We claim that $\mathcal{V} = \text{var}(C)$. Since C is a subalgebra of $G(A)$, it satisfies all the identities of $\mathcal{V}$. Conversely, let $f = f(x_1, \ldots, x_n)$ be a multilinear polynomial such that $f \notin \text{Id}(G(A))$. Then we can find $b_1, \ldots, b_s \in A^{(0)}$, $g_1, \ldots, g_s \in G^{(0)}$, $a_{i_1}, \ldots, a_{i_{n-s}} \in \{a_1, \ldots, a_k\}$, $h_1, \ldots, h_{n-s} \in G^{(1)}$ such that

$$f(b_1 \otimes g_1, \ldots, b_s \otimes g_s, a_{i_1} \otimes h_1, \ldots, a_{i_{n-s}} \otimes h_{n-s}) \neq 0$$

in $G(A)$. On the other hand,

$$f(b_1 \otimes g_1, \ldots, a_{i_{n-s}} \otimes h_{n-s}) = f'(b_1, \ldots, b_s, a_{i_1}, \ldots, a_{i_{n-s}}) \otimes g_1 \cdots g_s h_1 \cdots h_{n-s}$$

for some multilinear polynomial f' and the elements $f'(b_1, \ldots, a_{i_{n-s}})$, $g_1 \cdots h_{n-s}$ are non-zero elements of A and G, respectively. By the choice of m and t we conclude that $n - s < m \leq t$. Then one can evaluate $f(x_1, \ldots, x_n)$ on the elements $b_1 \otimes 1, \ldots, b_s \otimes 1, a_{i_1} \otimes e_1, \ldots, a_{i_{n-s}} \otimes e_{n-s}$ of C and

$$f(b_1 \otimes 1, \ldots, b_s \otimes 1, a_{i_1} \otimes e_1, \ldots, a_{i_{n-s}} \otimes e_{n-s}) \neq 0.$$

7.2. Varieties of polynomial growth

Hence f is not an identity of C and the claim is established. We have proved that $\mathcal{V}$ is generated by C, a finite dimensional algebra. □

7.2. Varieties of polynomial growth

In this section we shall give several different characterizations of varieties whose sequence of codimensions is polynomially bounded. We start with two technical results.

LEMMA 7.2.1.
1) If $\mathcal{U} \subseteq \mathcal{V}$ are varieties of algebras, then, for all n, $c_n(\mathcal{U}) \leq c_n(\mathcal{V})$. It follows that $\exp(\mathcal{U}) \leq \exp(\mathcal{V})$.
2) If A and B are PI-algebras, then, for all n, $c_n(A \oplus B) \leq c_n(A) + c_n(B)$. It follows that $\exp(A \oplus B) = \max\{\exp(A), \exp(B)\}$.

PROOF. Recall that for any variety $\mathcal{V}$,
$$c_n(\mathcal{V}) = \dim_F \frac{P_n}{P_n \cap \mathrm{Id}(\mathcal{V})}.$$

Hence, since $\mathrm{Id}(\mathcal{U}) \supseteq \mathrm{Id}(\mathcal{V})$, we have that $c_n(\mathcal{U}) \leq c_n(\mathcal{V})$, for all $n \geq 1$. This proves 1).

Consider the map
$$\varphi \colon P_n \to \frac{P_n}{P_n \cap \mathrm{Id}(A)} \oplus \frac{P_n}{P_n \cap \mathrm{Id}(B)}$$
defined by setting $\varphi(a) = (a + P_n \cap \mathrm{Id}(A), a + P_n \cap \mathrm{Id}(B))$. Clearly φ has kernel $P_n \cap \mathrm{Id}(A) \cap \mathrm{Id}(B)$. Thus, since $\mathrm{Id}(A \oplus B) = \mathrm{Id}(A) \cap \mathrm{Id}(B)$, we have that $\frac{P_n}{P_n \cap \mathrm{Id}(A \oplus B)}$ embeds into $\frac{P_n}{P_n \cap \mathrm{Id}(A)} \oplus \frac{P_n}{P_n \cap \mathrm{Id}(B)}$ and $c_n(A \oplus B) \leq c_n(A) + c_n(B)$ follows. Since $c_n(A), c_n(B) \leq c_n(A \oplus B)$ by computing the limit of the nth roots, we obtain the desired conclusion for the exponents. □

The next theorem gives a characterization of varieties with polynomially bounded codimension sequence in terms of their cocharacter sequence. These results are essentially due to Kemer ([**Ke1**]).

THEOREM 7.2.2. *For a PI-algebra A the following conditions are equivalent:*
1) $c_n(A)$ *is polynomially bounded.*
2) $\exp(A) \leq 1$.
3) *There exists a constant q such that*
$$\chi_n(A) = \sum_{\substack{\lambda \vdash n \\ |\lambda| - \lambda_1 \leq q}} m_\lambda \chi_\lambda$$

for all $n \geq 1$.

PROOF. The implication 1) → 2) is obvious. Let us prove 2) → 3). Suppose first that for any $q > 0$ there exists n and $\lambda \vdash n$ such that $\lambda_2 = m \geq q$ and $m_\lambda \neq 0$ in the decomposition of the nth cocharacter

(7.7) $$\chi_n(A) = \sum_{\lambda \vdash n} m_\lambda \chi_\lambda.$$

Consider any Young tableau T_λ where the integers $1, 3, 5, \ldots, 2m-1$ are placed in the first row from left to right starting with the box $(1,1)$ and the integers $2, 4, 6, \ldots, 2m$ are placed in the second row from left to right starting with the box $(2,1)$. Then by Theorem 2.4.5 and Lemma 2.4.1, there exists a multilinear polynomial $f = f(x_1, \ldots, x_n)$ such that $e_{T_\lambda} f$ is not an identity of A. Moreover, we may assume that f is a monomial on $x_1, \ldots, x_n$.

Consider the canonical embedding of S_{2m} in S_n as the subgroup of permutations fixing the elements $2m+1, \ldots, n$ and denote $\mu = (m,m) \vdash 2m$. Then $\mu \leq \lambda$ and let T_μ be the corresponding subtableau of T_λ. By Lemma 4.5.3, the essential idempotent e_{T_λ} can be written as a linear combination of elements of the type $\sigma e_{T_\mu} \tau$ where $\sigma, \tau \in S_n$. Since $e_{T_\lambda} f$ is not an identity of A and f is a multilinear monomial in $x_1, \ldots, x_n$, there exists a multilinear monomial g in $x_1, \ldots, x_n$ such that $e_{T_\mu} g$ is not an identity of A. Write

$$g = a_0 x_{i_1} a_1 \cdots a_{2m-1} x_{i_{2m}} a_{2m}$$

where $\{i_1, \ldots, i_{2m}\} = \{1, \ldots, 2m\}$ and $a_0, \ldots, a_{2m}$ are monomials on $x_{2m+1}, \ldots, x_n$ of degree ≥ 0. Since $e_{T_\mu} g$ is not an identity, it follows that also

$$e_{T_\mu} h = e_{T_\mu} x_{i_1} c_1 \cdots c_{2m-1} x_{i_{2m}}$$

is not an identity of A where c_j, $1 \leq j \leq 2m-1$, is a new indeterminate in case $\deg a_j \geq 1$ and $c_j = 1$ if $\deg a_j = 0$. Hence h is a multilinear monomial in $x_1, \ldots, x_k$ with $2m \leq k \leq 4m$.

Consider the subspace P_k of multilinear polynomials in $x_1, \ldots, x_k$ and let S_{2m} act on $x_1, \ldots, x_{2m}$. Then, clearly, $e_{T_\mu} h$ generates in P_k an irreducible S_{2m}-submodule with character χ_μ and by Lemma 5.10.1,

$$d_\mu = \deg \chi_\mu > \frac{2^{2m}}{2m}.$$

Hence,

$$c_k(A) = \dim \frac{P_k}{P_k \cap \mathrm{Id}(A)} > \frac{2^{2m}}{2m} > \frac{2^{\frac{k}{2}}}{k}$$

since $\frac{k}{2} < 2m \leq k$.

We have proved that, for any $q > 0$ there exists $k \geq 2m \geq 2q$ such that $c_k(A) > k^{-1} \sqrt{2}^k$. This contradicts the hypothesis $\exp(A) \leq 1$. Hence we conclude that all the irreducible characters appearing in $\chi_\lambda(A)$ with non-zero multiplicity have corresponding diagram with bounded second row, i.e., $\lambda_2 \leq C$ for some constant C.

Since $\exp(A) \leq 1$ and by Theorem 4.1.8, $\exp(G) = 2$, we get $G \notin \mathrm{var}(A)$. Hence by Theorem 7.1.2, A satisfies the standard identity of some degree.

Now by Theorem 7.1.4 there exists d such that $\lambda \in H(d, 0)$ whenever $m_\lambda \neq 0$ in (7.7). Clearly the number of boxes below the first row, i.e., $|\lambda| - \lambda_1$, is bounded by $q = Cd$ and this proves the implication 2) $\to$ 3).

Suppose now that A satisfies 3). Since $|\lambda| - \lambda_1 \leq q$, then $\lambda_1 \geq n - q$ and by the hook formula we immediately get

$$\deg \chi_\lambda \leq \frac{n!}{(n-q)!} \leq n^q.$$

Recall that $P_n \equiv FS_n$ and the multiplicity of χ_λ in the character of the left regular representation of S_n is equal to $d_\lambda = \deg \chi_\lambda$. Hence $m_\lambda \leq \deg \chi_\lambda \leq n^q$ in $\chi_\lambda(A)$

and
$$(7.8) \qquad c_n(A) = \sum_{\lambda \vdash n} m_\lambda \deg \chi_\lambda \leq q^2 n^{2q}$$

since q^2 is an obvious upper bound for the number of partitions $\lambda \vdash n$ with $n - \lambda_1 \leq q$. The inequality (7.8) completes the proof of the theorem. $\square$

As an immediate corollary of the previous theorem we obtain that the codimension growth of a PI-algebra can be either exponential or polynomially bounded.

COROLLARY 7.2.3. *Let A be a PI-algebra such that $c_n(A) < C(2-\varepsilon)^n$ for some constants C and $0 < \varepsilon < 1$. Then $c_n(A)$ is polynomially bounded.*

Combining Theorem 7.2.2 and Theorem 7.1.4 we obtain the following result.

THEOREM 7.2.4. *If $\exp(\mathcal{V}) \leq 1$, then $\mathcal{V} = \text{var}(A)$, for some finite dimensional algebra A.*

Another characterization can be given in terms of the Grassmann algebra G and the algebra UT_2 of 2×2 upper triangular matrices.

THEOREM 7.2.5 ([**Ke2**]). *For a variety $\mathcal{V}$, $\exp(\mathcal{V}) \leq 1$ if and only if $G, UT_2 \notin \mathcal{V}$.*

PROOF. If $\exp(\mathcal{V}) \leq 1$, then clearly $G, UT_2 \notin \mathcal{V}$ since $\exp(G) = \exp(UT_2) = 2$ (see Section 4.1). Suppose now that $G, UT_2 \notin \mathcal{V}$. Since $G \notin \mathcal{V}$ then, by Theorem 7.1.4, there exists a finite dimensional algebra A such that $\mathcal{V} = \text{var}(A)$ and assume, as we may by Theorem 4.1.9, that F is algebraically closed. Write $A = A_1 \oplus \cdots \oplus A_m + J$ where $A_1, \ldots, A_m$ are simple subalgebras and let $A_i \cong M_{n_i}(F)$ for $i = 1, \ldots, m$. If for some i, $n_i > 1$, then $M_{n_i}(F)$, and so A, would contain a copy of UT_2 contrary to the assumption that $UT_2 \notin \mathcal{V}$. Hence $A_1 \cong \cdots \cong A_m \cong F$. But then, recalling the characterization of the exponent, in order to complete the proof of the theorem, it is enough to prove that $A_i J A_k = 0$, for all $i, k \in \{1, \ldots, m\}$, $i \neq k$.

Suppose to the contrary that there exist A_i, A_k, $i \neq k$, such that $A_i J A_k \neq 0$. If $1_1, 1_2$ are the unit elements of A_i and A_k, respectively, it follows that $1_1 j 1_2 \neq 0$ for some $j \in J$. Let B be the subalgebra of A generated by $1_1, 1_2$ and $1_1 j 1_2$. Clearly B is isomorphic to the algebra UT_2 through an isomorphism φ such that $\varphi(1_1) = e_{11}, \varphi(1_2) = e_{22}$ and $\varphi(1_1 j 1_2) = e_{12}$. We have reached a contradiction since by hypothesis $UT_2 \notin \mathcal{V}$ and the proof of the theorem is complete. $\square$

The previous theorem says that the two algebras G and UT_2 share an important property. Let us give the following.

DEFINITION 7.2.6. *A variety $\mathcal{V}$ has polynomial growth if its sequence of codimensions $c_n(\mathcal{V})$, $n = 1, 2, \ldots$, is polynomially bounded. We say that $\mathcal{V}$ has almost polynomial growth if $\mathcal{V}$ has exponential growth but any proper subvariety of $\mathcal{V}$ has polynomial growth.*

By using the notion of exponent, the above definition says that $\mathcal{V}$ has almost polynomial growth if $\exp(\mathcal{V}) \geq 2$ and for any proper subvariety $\mathcal{U} \subset \mathcal{V}$, $\exp(\mathcal{U}) \leq 1$.

We can now describe all varieties with almost polynomial growth.

THEOREM 7.2.7. *$\text{var}(G)$ and $\text{var}(UT_2)$ are the only varieties of almost polynomial growth.*

PROOF. This is an easy consequence of Example 4.1.5 and Theorem 4.1.8, and Theorem 7.2.5. □

Our next objective is to characterize finite dimensional algebras generating varieties of polynomial growth.

PROPOSITION 7.2.8 ([**GZ6**]). *Let $\mathcal{V}$ be a variety of algebras over the algebraically closed field F. Then $\exp(\mathcal{V}) \leq 1$ if and only if $\mathcal{V} = \mathrm{var}(A)$ for some finite dimensional algebra A such that*

1) $A = A_0 \oplus A_1 \oplus \cdots \oplus A_m$ *a vector space direct sum of F-algebras where for $i = 1, \ldots, m$, $A_i = B_i + J_i$, $B_i \cong F$, J_i a nilpotent ideal of A_i and $A_0, J_1, \ldots, J_m$ are nilpotent right ideals of A;*
2) *for all $i, k \in \{1, \ldots, m\}, i \neq k$, $A_i A_k = 0$ and $B_i A_0 = 0$.*

Moreover,
$$\mathrm{Id}(A) = \mathrm{Id}(A_1 \oplus A_0) \cap \ldots \cap \mathrm{Id}(A_m \oplus A_0) \cap \mathrm{Id}(J)$$
where J is the Jacobson radical of A.

PROOF. Suppose that $\exp(\mathcal{V}) \leq 1$. By Theorem 7.2.4, $\mathcal{V} = \mathrm{var}(A)$, for some finite dimensional F-algebra A. Write $A = B_1 \oplus \cdots \oplus B_m + J$ where $B_1, \ldots, B_m$ are simple subalgebras of A and $J = J(A)$ is its Jacobson radical. Since $\exp(A) \leq 1$, by the characterization of the exponent, we must have $B_i J B_k = 0$ for all $i \neq k$ and $\dim_F B_i = 1$, $i, k = 1, \ldots, m$.

Let $e = e_1 + \cdots + e_m$ be the decomposition of the unit element of B into orthogonal (central in B) idempotents. Then $e_i B = B_i \cong F$ and, for all $i = 1, \ldots, m$, define $J_i = e_i J$ and $J_0 = \{x \in J \mid Bx = 0\}$. It is easily seen that
$$A = (B_1 + J_1) \oplus \cdots \oplus (B_m + J_m) \oplus J_0.$$

Set $A_i = B_i + J_i$ and $A_0 = J_0$. Then, for $i, k \in \{1, \ldots, m\}$ distinct, we have $A_i A_k = (B_i + J_i)(B_k + J_k) = 0$ since $e_i e_k = 0$ and $B_i J B_k = 0$. Also, for $i \neq 0$, $B_i A_0 = 0$.

Conversely, suppose $\mathcal{V} = \mathrm{var}(A)$ for some finite dimensional algebra A over the algebraically closed field F satisfying 1) and 2). From the relations $A_i A_k = 0$ and $B_i A_0 = 0$ it follows that $J = A_0 + J_1 + \cdots + J_m$ is a nilpotent two-sided ideal of A and $A = B_1 \oplus \cdots \oplus B_m + J$ where $B_i \cong F$ for all i. Since from the defining relations $A_i A_k = 0$ and $B_i A_0 = 0$ it follows that $B_i J B_k = 0$ for all $i \neq k$, then $\exp(\mathcal{V}) = \exp(A) \leq 1$.

In order to prove the last part of the proposition, take
$$f \in \mathrm{Id}(A_1 \oplus A_0) \cap \ldots \cap \mathrm{Id}(A_m \oplus A_0) \cap \mathrm{Id}(J)$$
and suppose that $f \notin \mathrm{Id}(A)$. Since $\mathrm{char}\, F = 0$, we may assume that f is multilinear and let $r_1, \ldots, r_s \in A$ be such that $f(r_1, \ldots, r_s) \neq 0$. Since $f \in \mathrm{Id}(J)$, among $r_1, \ldots, r_s$ there exists $r_i \notin J$; by linearity we may assume that $r_i \in B_k$ for some k. Recall that for all l, $B_l A_0 = 0$, J_l is a right ideal of A and, when $l \neq k$, $A_k A_l = A_l A_k = 0$. In particular, all $A_1, \ldots, A_k$ are right ideals of A. Then it follows that $r_1, \ldots, r_{i-1}, r_{i+1}, \ldots r_s \in A_k \cup A_0$. Thus $f \notin \mathrm{Id}(A_k \oplus A_0)$, a contradiction. □

COROLLARY 7.2.9. *Let $\mathcal{V}$ be a variety of algebras over the algebraically closed field F. Then $\exp(\mathcal{V}) \leq 1$ if and only if $\mathcal{V} = \mathrm{var}(A_1 \oplus \cdots \oplus A_m)$ where $A_1, \ldots, A_m$ are finite dimensional F-algebras and $\dim A_i/J(A_i) \leq 1$, for all $i = 1, \ldots, m$.*

PROOF. If $\exp(\mathcal{V}) \leq 1$, then by Proposition 7.2.8,
$$\mathrm{Id}(\mathcal{V}) = \mathrm{Id}(A_1 + A_0) \cap \ldots \cap \mathrm{Id}(A_m + A_0) \cap \mathrm{Id}(J)$$
where $B_1 = A_1 + A_0, \ldots, B_m = A_m + A_0$ are of codimension one over their radical and $B_0 = J$ is nilpotent. Since for any two algebras R and S, $\mathrm{Id}(R \oplus S) = \mathrm{Id}(R) \cap \mathrm{Id}(S)$, we get
$$\mathrm{Id}(\mathcal{V}) = \mathrm{Id}(B_0 \oplus \cdots \oplus B_m).$$
Conversely, suppose that $\dim A_i/J(A_i) \leq 1$ and $\dim A_i < \infty$. Then $c_n(A_i)$ is polynomially bounded for all $i = 1, \ldots, m$, by Proposition 7.2.8. Then, by Lemma 7.2.1, $c_n(A_1 \oplus \cdots \oplus A_m)$ is also polynomially bounded. $\square$

In order to prove our final characterization of varieties of polynomial growth over an arbitrary field we first prove two lemmas.

LEMMA 7.2.10. *Let A, A', B, B' be F-algebras. If $\mathrm{var}(A) = \mathrm{var}(A')$ and $\mathrm{var}(B) = \mathrm{var}(B')$, then $\mathrm{var}(A \oplus B) = \mathrm{var}(A' \oplus B')$.*

PROOF. Let $I = \mathrm{Id}(A) = \mathrm{Id}(A')$, $P = \mathrm{Id}(B) = \mathrm{Id}(B')$. Then $\mathrm{Id}(A \oplus B) = I \cap P$, $\mathrm{Id}(A' \oplus B') = I \cap P$ and we are done. $\square$

LEMMA 7.2.11. *Let A be an F-algebra finite dimensional over the algebraic closure $\bar F$ of F. Then there exists a finite dimensional algebra C over F such that $\mathrm{var}(A) = \mathrm{var}(C)$ and $\dim_F C/J(C) = \dim_F A/J(A)$.*

PROOF. By the Wedderburn-Malcev theorem, A as an $\bar F$-algebra can be decomposed into the sum $A = B + J$ where B is a finite dimensional semisimple subalgebra over $\bar F$ and J is the Jacobson radical of A. Since $\bar F$ is algebraically closed, there exists an $\bar F$-basis $\{u_1, \ldots, u_m\}$ of B with rational structure constants. Hence the F-linear span of $u_1, \ldots, u_m$ is a finite dimensional F-algebra.

We now take an arbitrary $\bar F$-basis $\{v_1, \ldots, v_n\}$ of J and denote by C the F-subalgebra of A generated by $\{u_1, \ldots, u_m, v_1, \ldots, v_n\}$. Since J is a nilpotent ideal, $\dim_F C < \infty$. Moreover, $\dim_F C/J(C) = m = \dim_{\bar F} B = \dim_{\bar F} A/J(A)$.

We claim that $\mathrm{var}(A) = \mathrm{var}(C)$. Now, the inclusion $\mathrm{Id}(A) \subseteq \mathrm{Id}(C)$ is obvious. On the other hand, let $f = f(x_1, \ldots, x_k)$ be a multilinear polynomial identity of C. Then $f = f(z_1, \ldots, z_k) = 0$ for any $z_1, \ldots, z_k \in \{u_1, \ldots, u_m, v_1, \ldots, v_n\}$ and by linearity $f \equiv 0$ is an identity of A. This means that C and A satisfy the same multilinear identities as F-algebras and the proof is complete. $\square$

We are now in a position to prove the following result ([**GZ6**]).

THEOREM 7.2.12. *For a variety $\mathcal{V}$, $\exp(\mathcal{V}) \leq 1$ if and only if $\mathcal{V} = \mathrm{var}(A_1 \oplus \cdots \oplus A_m)$ where $A_1, \ldots, A_m$ are finite dimensional algebras over F and $\dim A_i/J(A_i) \leq 1$, for all $i = 1, \ldots, m$.*

PROOF. Suppose that $\exp(\mathcal{V}) \leq 1$. By Theorem 7.2.4, $\mathcal{V} = \mathrm{var}(A)$ where $\dim A < \infty$. Moreover, if $\bar F$ is the algebraic closure of F and $\bar A = A \otimes_F \bar F$, then $\mathrm{var}(A) = \mathrm{var}(\bar A)$. On the other hand, by Theorem 4.1.9 the $\bar F$-codimensions $c_n(\bar A)$ are the same as F-codimensions $c_n(A)$. Hence $\bar A$ generates a variety of algebras over $\bar F$ with polynomially bounded codimension growth. But then by Corollary 7.2.9, there exist finite dimensional $\bar F$-algebras $B_1, \ldots, B_m$ such that $\dim_{\bar F} B_i/J(B_i) \leq 1$ for all $i = 1, \ldots, m$, and the algebra $B_1 \oplus \cdots \oplus B_m$ has the same identities over $\bar F$ as $\bar A$. Thus A and $B_1 \oplus \cdots \oplus B_m$ have the same F-identities, i.e., $\mathrm{var}(B_1 \oplus \cdots \oplus B_m) = \mathrm{var}(A) = \mathcal{V}$. By applying Lemmas 7.2.10 and 7.2.11

we can find finite dimensional F-algebras $B'_1, \ldots, B'_m$ with $\dim B'_i/J(B'_i) \leq 1$ and $\mathrm{var}(B'_1 \oplus \cdots \oplus B'_m) = \mathcal{V}$ as required. □

If in Theorem 7.2.2 we take A to be a finite dimensional algebra (as we may by Theorem 7.2.4), we can find an upper bound for the number of boxes below the first row of any diagram in the nth cocharacter of A (see [**Ke1**], [**GZ6**]).

THEOREM 7.2.13. *Let A be a finite dimensional algebra. Then $\exp(A) \leq 1$ if and only if*
$$\chi_n(A) = \sum_{\substack{\lambda \vdash n \\ |\lambda| - \lambda_1 < q}} m_\lambda \chi_\lambda$$
where q is such that $J(A)^q = 0$.

PROOF. As we remarked in Theorem 4.1.9, the decomposition of $\chi_n(A)$ into irreducible S_n-characters does not change when extending the base field. Therefore if $J(A)^q = 0$, since $J(A \otimes_F \bar{F})^q = 0$, we may assume, without loss of generality, that F is algebraically closed.

By Theorem 7.2.2 we need only to show that $\exp(A) \leq 1$ implies the given decomposition of the nth cocharacter. Suppose $\exp(A) \leq 1$ and let λ be a partition of n such that $|\lambda| - \lambda_1 \geq q$ and $m_\lambda \neq 0$. Then there exists a tableau T_λ and a multilinear polynomial $f(x_1, \ldots, x_n)$ such that $e_{T_\lambda} f(x_1, \ldots, x_n) \notin \mathrm{Id}(A)$. Let $\lambda' = (\lambda'_1, \ldots, \lambda'_t)$ be the conjugate partition of λ. Then $e_{T_\lambda} f(x_1, \ldots, x_n)$ is a linear combination of polynomials each alternating on t disjoint sets of $\lambda'_1, \ldots, \lambda'_t$ variables, respectively. We shall reach a contradiction by proving that each such polynomial g vanishes in A.

let $A = B_1 \oplus \cdots \oplus B_m + J$ where $B_1, \ldots, B_m$ are simple subalgebra and fix a basis of A which is the union of bases of $B_1, \ldots, B_m$ and J, respectively. Then $\dim B_i = 1, i = 1, \ldots, m$ and $B_i B_k = B_i J B_k = 0$ for all $i \neq k$. In order to get a non-zero value of g we must replace all the variables with elements of J and of one simple component, say, B_i. Also, since $\dim B_i = 1$, we can substitute at most one element of B_i in each alternating set. Hence we can substitute in all at most $t = \lambda_1$ elements from B_i. It follows that in order to get a non-zero value, we must substitute at least $|\lambda| - \lambda_1 \geq q$ elements from J. Since $J^q = 0$, we get that $g \equiv 0$ and with this contradiction the proof of the theorem is complete. □

We now collect the results obtained in this section in the following.

THEOREM 7.2.14. *For a variety of algebras $\mathcal{V}$, the following conditions are equivalent:*

1) $c_n(\mathcal{V}) \leq Cn^t$, *for some constants C, t.*
2) $\exp(\mathcal{V}) \leq 1$.
3) $G, UT_2 \notin \mathcal{V}$.
4) $\mathcal{V} = \mathrm{var}(A)$ *where $A = A_1 \oplus \cdots \oplus A_m$ with $A_1, \ldots, A_m$ finite dimensional algebras over F such that $\dim A_i/J(A_i) \leq 1$, for all $i = 1, \ldots, m$.*
5) *There exists a constant q such that*
$$\chi_n(\mathcal{V}) = \sum_{\substack{\lambda \vdash n \\ |\lambda| - \lambda_1 \leq q}} m_\lambda \chi_\lambda$$
for all $n \geq 1$. When $\mathcal{V} = \mathrm{var}(A)$ with A a finite dimensional algebra, q is such that $J(A)^q = 0$.

A finer characterization of varieties of polynomial growth has been recently obtained in [**GL**] in the case of linear growth. It turns out that any such variety is generated by one of the algebras: $N, C \oplus N, M_1 \oplus N, M_2 \oplus N$ and $M_1 \oplus M_2 \oplus N$ where N is a nilpotent algebra, C is a commutative algebra, $M_1 = \begin{pmatrix} F & F \\ 0 & 0 \end{pmatrix}$ and $M_2 = \begin{pmatrix} 0 & F \\ 0 & F \end{pmatrix}$.

We refer the reader interested in a further study of varieties of polynomial growth to the work of Latyshev [**L2**].

7.3. Locally noetherian varieties

In Section 7.2 we proved that a variety $\mathcal{V}$ has polynomial growth if and only if $G, UT_2 \notin \mathcal{V}$. More generally, for the algebra G we proved in Theorem 7.1.2 that a variety $\mathcal{V}$ does not contain G if and only if $\mathcal{V}$ satisfies a standard identity. In this section we want to find an analogous characterization for the varieties not containing the algebra UT_2. This is done in the following theorem.

THEOREM 7.3.1 ([**MRZ1**]). *A variety $\mathcal{V}$ satisfies a non-trivial identity of the type*

(7.9) $$f(x,y) = \sum_{i=0}^{n} \alpha_i y^i x y^{n-i} \equiv 0$$

if and only if $UT_2 \notin \mathcal{V}$.

PROOF. We first check that any non-trivial polynomial of type (7.9) cannot be an identity of UT_2. Set $\bar{x} = e_{12}$ and $\bar{y} = \beta e_{11} + e_{22}$ with $\beta \in F$. Then

$$(\beta e_{11} + e_{22})^i e_{12} (\beta e_{11} + e_{22})^{n-i} = \beta^i e_{12}$$

and, if (7.9) is an identity for UT_2, then $f(\bar{x}, \bar{y}) = 0$ implies

(7.10) $$\sum_{i=0}^{n} \beta^i \alpha_i = 0$$

for all $\beta \in F$. Since F is infinite, by choosing distinct $\beta_1, \ldots, \beta_{n+1} \in F$ we obtain from (7.10) the system of linear equations

$$\Delta \begin{pmatrix} \alpha_0 \\ \vdots \\ \alpha_n \end{pmatrix} = 0$$

where $\Delta = (\beta_i^j)$ is a Vandermonde matrix. It follows that $\alpha_0 = \cdots = \alpha_n = 0$. This says that any variety satisfying an identity of type (7.9), does not contain UT_2.

Now let $UT_2 \notin \mathcal{V}$. In Theorem 4.1.5, we showed that the space of multilinear polynomials $P_n = P_n(x_1, \ldots, x_n)$ is spanned (mod $\mathrm{Id}(UT_2) \cap P_n$), by the polynomials

$$a_{i_1,\ldots,i_k,m} = x_{i_1} \cdots x_{i_k} [x_m, x_{j_1}, \ldots, x_{j_r}]$$

with $i_1 < \cdots < i_k$, $m > j_1 < \cdots < j_r$, $k + r + 1 = n$. Since $\mathcal{V}$ does not contain UT_2, it satisfies an identity of the form

(7.11) $$f = \sum \alpha_{i_1,\ldots,i_k,m} a_{i_1,\ldots,i_k,m} + h \equiv 0$$

where $h \in \mathrm{Id}(UT_2)$. Take k minimal such that $\alpha_{i_1,\ldots,i_k,m} \neq 0$ on the left-hand side of (7.11), and fix the indices $i_1,\ldots,i_k,m$. Consider an endomorphism φ of $F\langle X \rangle$ such that
$$\varphi(x_{i_1}) = \cdots = \varphi(x_{i_k}) = y,$$
$$\varphi(x_m) = \varphi(x_{j_2}) = \cdots = \varphi(x_{j_r}) = y^2 \quad \text{and} \quad \varphi(x_{j_1}) = [x,y].$$
When we apply φ to f and we open all brackets, we obtain
$$\varphi(a_{i_1,\ldots,i_k,m}) = \beta y^k x y^{n-k} + \sum_{i>0} \beta_i y^{k+i} x y^{n-k-i}$$
with $\beta \neq 0$ and in no other polynomial $\varphi(a_{i'_1,\ldots,i'_k,q})$ the monomial $y^k x y^{n-k}$ appears with non-zero coefficient. Note also that, since $h \in \mathrm{Id}(UT_2)$ and $[x_1,x_2][x_3,x_4]$ generates $\mathrm{Id}(UT_2)$ as a T-ideal, then h is a linear combination of polynomials of the type $a[b,c]d[e,g]$. Since the degree of $\varphi(a[b,c]d[e,g])$ on x and y must be 1 and n, respectively, then either $\varphi([b,c]) = 0$ or $\varphi([e,g]) = 0$. Hence $\varphi(h) = 0$ and we obtain that
$$\varphi(f) = \beta y^k x y^{n-k} + \sum_{i>0} \alpha_i y^{k+i} x y^{n-k-i} \equiv 0$$
is an identity of $\mathcal{V}$. With this the proof of the theorem is complete. $\square$

A special case of the identity in (7.9) is the following:

(7.12) $$xy^n \equiv \sum_{i=1}^n \alpha_i y^i x y^{n-i}.$$

Next we show that the above identity characterizes some special type of varieties. We first make the following.

DEFINITION 7.3.2. A variety $\mathcal{V}$ is called locally left noetherian if any finitely generated algebra of $\mathcal{V}$ is left noetherian.

THEOREM 7.3.3 ([**Lv**]). *A variety $\mathcal{V}$ is locally left noetherian if and only if $\mathcal{V}$ satisfies an identity of type (7.12).*

PROOF. It is well known that any homomorphic image of a noetherian algebra is still noetherian. Therefore $\mathcal{V}$ is locally left noetherian if and only if any relatively free algebra of $\mathcal{V}$ of finite rank is left noetherian.

First let $\mathcal{V}$ be locally left noetherian and let $R \cong F\langle x,y\rangle/(\mathrm{Id}(\mathcal{V}) \cap F\langle x,y\rangle)$ be the relatively free algebra of $\mathcal{V}$ with free generators $\bar{x}$ and $\bar{y}$. Consider the ascending chain of left ideals
$$I_1 \subseteq I_2 \subseteq \cdots \subseteq I_m \subseteq \cdots$$
where I_m is the left ideal of R generated by the monomials $\bar{x}, \bar{x}\bar{y}, \bar{x}\bar{y}^2,\ldots,\bar{x}\bar{y}^m$. Since R is noetherian, there exists $n \geq 2$ such that $I_n = I_{n-1}$, hence $\bar{x}\bar{y}^n \in I_{n-1}$. Since R is relatively free, $\bar{x}\bar{y}^n$ is equal to a polynomial $f = f(\bar{x},\bar{y}) \in I_{n-1}$ of degree 1 in $\bar{x}$ and n in $\bar{y}$, i.e.,
$$\bar{x}\bar{y}^n = \sum_{i=1}^n \alpha_i \bar{y}^i \bar{x} \bar{y}^{n-i}.$$
Since R is the relatively free algebra of $\mathcal{V}$ with free generators $\bar{x}$ and $\bar{y}$, any relation on $\bar{x}$ and $\bar{y}$ is an identity of $\mathcal{V}$. Thus $\mathcal{V}$ satisfies an identity of the type (7.12).

Conversely, suppose that $\mathcal{V}$ is a variety satisfying an identity of type (7.12) and let R be the relatively free algebra of $\mathcal{V}$ of rank m with free generators $\bar{x}_1,\ldots,\bar{x}_m$. By Theorem 7.1.4 the algebra R being finitely generated, satisfies a standard identity. Hence $\mathcal{V} = \mathrm{var}(R)$ does not contain the Grassmann algebra. Also, by the previous theorem $UT_2 \not\in \mathcal{V}$. But then by Theorem 7.2.5 and Theorem 7.2.12, $\mathcal{V}$ has polynomially bounded codimension growth, and $\mathcal{V} = \mathrm{var}(A)$ where $A = A_1 \oplus \cdots \oplus A_l$ with $A_1, \ldots, A_l$ finite dimensional algebras such that $\dim A_i/J(A_i) \leq 1$, $i = 1, \ldots, l$. It follows that the algebra A and, so, R satisfies the identity

$$[x_1, y_1] \cdots [x_k, y_k] \equiv 0, \tag{7.13}$$

for some $k \geq 1$.

We proceed by induction on k. If $k = 1$, then R is a finitely generated commutative algebra and R is noetherian.

Let $k > 1$ and denote by T the commutator ideal of R, i.e., the two-sided ideal generated by $[R, R]$. Then $T^k = 0$ and by the inductive hypothesis, R/T^{k-1} is left noetherian. Since an extension of a noetherian module by a noetherian module is also noetherian, we need only to prove that T^{k-1} is a noetherian left module. Actually T^{k-1} is a left module over the noetherian ring R/T, hence it is enough to prove that T^{k-1} is finitely generated as a left ideal of R.

First notice that every element of T^{k-1} is a linear combination of elements of the type

$$a_1[b_1, c_1]a_2 \cdots a_{k-1}[b_{k-1}, c_{k-1}]a_k$$

where a_i, b_i, c_i are monomials in $\bar{x}_1, \ldots, \bar{x}_m$, $\deg a_i \geq 0$, $\deg b_i, \deg c_i \geq 1$. Further, because of the relation $[ab, c] = [a, c]b + a[b, c]$, we may assume that $\deg b_i = \deg c_i = 1$, for all $1 \leq i \leq k-1$. In particular, this implies that the number of distinct commutators $[b_i, c_i]$ is finite. Next, a repeated application of (7.13) implies that each a_i, $i = 1, \ldots, k$, can be written as an ordered monomial of the type $x_1^{\alpha_1} x_2^{\alpha_2} \cdots x_m^{\alpha_m}$, for some $\alpha_1, \ldots, \alpha_m \geq 0$. It follows that R^{k-1} is generated as a left R-module by the elements

$$[x_{i_1}, x_{j_1}] x_1^{\alpha_1} \cdots x_m^{\alpha_m} [x_{i_2}, x_{j_2}] x_1^{\beta_1} \cdots x_m^{\beta_m} [x_{i_{k-1}}, x_{j_{k-1}}] x_1^{\gamma_1} \cdots x_m^{\gamma_m}. \tag{7.14}$$

Finally, a repeated application of (7.12) implies that all degrees $\alpha_1, \ldots, \gamma_m$ can be bounded by $n - 1$. Thus the number of elements of type (7.14) is finite and T^{k-1} is finitely generated as a left R-module. $\square$

In the next section we shall prove that UT_2 characterizes the varieties whose sequence of cocharacters has bounded multiplicities. Here we want to give some motivation by showing that conditions of this type may be important in the study of the structure of the subvarieties of a variety. We first recall some notions and definitions.

From Birkhoff's theorem (Theorem 1.2.3) it easily follows that the set-theoretical intersection $\mathcal{U} \cap \mathcal{V}$ of two varieties $\mathcal{U}$ and $\mathcal{V}$ is still a variety. Moreover, $\mathrm{Id}(\mathcal{U} \cap \mathcal{V}) = \mathrm{Id}(\mathcal{U}) + \mathrm{Id}(\mathcal{V})$ in $F\langle X \rangle$. On the other hand, the set-theoretical union of $\mathcal{U}$ and $\mathcal{V}$ is not in general a variety. Therefore in the theory of varieties, the union $\mathcal{U} \cup \mathcal{V}$ of $\mathcal{U}$ and $\mathcal{V}$ is defined as the smallest variety containing both $\mathcal{U}$ and $\mathcal{V}$. If $\mathcal{U} = \mathrm{var}(A)$ and $\mathcal{V} = \mathrm{var}(B)$, then $\mathcal{U} \cup \mathcal{V} = \mathrm{var}(A \oplus B)$ and $\mathrm{Id}(\mathcal{U} \cup \mathcal{V}) = \mathrm{Id}(\mathcal{U}) \cap \mathrm{Id}(\mathcal{V})$.

DEFINITION 7.3.4. A variety $\mathcal{V}$ is distributive if the lattice of subvarieties of $\mathcal{V}$ is distributive, i.e., for any subvarieties $\mathcal{A}, \mathcal{B}, \mathcal{C}$ of $\mathcal{V}$, $(\mathcal{A} \cap \mathcal{B}) \cup \mathcal{C} = (\mathcal{A} \cup \mathcal{C}) \cap (\mathcal{B} \cup \mathcal{C})$.

It is easily seen that the lattice of subvarieties of $\mathcal{V}$ is anti-isomorphic to the lattice of T-ideals of $F\langle X \rangle$ containing $\mathrm{Id}(\mathcal{V})$. Therefore $\mathcal{V}$ is distributive if and only if for any T-ideals $Q, S, T \supseteq \mathrm{Id}(\mathcal{V})$, $(Q + S) \cap T = Q \cap T + S \cap T$.

Distributive varieties can be characterized as follows ([**An**]).

THEOREM 7.3.5. *Let $\mathcal{V}$ be a variety and let*

$$\chi_n(\mathcal{V}) = \sum_{\lambda \vdash n} m_\lambda \chi_\lambda$$

be its nth cocharacter. Then $\mathcal{V}$ is distributive if and only if $m_\lambda \leq 1$ for all $\lambda \vdash n$ and for all $n \geq 1$.

PROOF. Since char $F = 0$, any T-ideal Q of $F\langle X \rangle$ is completely determined by the spaces $Q \cap P_n$, $n \geq 1$. Therefore $\mathcal{V}$ is distributive if and only if

$$(Q \cap P_n + S \cap P_n) \cap (T \cap P_n) = Q \cap T \cap P_n + S \cap T \cap P_n, \quad n = 1, 2, \ldots$$

for any T-deals Q, S, T containing $I = \mathrm{Id}(\mathcal{V})$. Since $I \cap P_n$ is an S_n-submodule of P_n, by complete reducibility write

(7.15) $$P_n = (I \cap P_n) \oplus W_1 \oplus \cdots \oplus W_k,$$

where $W_1, \ldots, W_k$ are irreducible S_n-modules. If, say, $W_1 \cong W_2$ as S_n-modules, then one can find $W \subseteq W_1 \oplus W_2$ such that $W \cong W_1, W \cap W_1 = W \cap W_2 = 0$. Let Q, S, T be the T-ideals generated by $I + W_1, I + W_2$ and $I + W$, respectively. Then

$$(Q \cap P_n + S \cap P_n) \cap (T \cap P_n) = T \cap P_n = I \cap P_n \oplus W,$$

$$Q \cap T \cap P_n + S \cap T \cap P_n = W_1 \cap W + W_2 \cap W + I \cap P_n = I \cap P_n,$$

and $\mathcal{V}$ is not distributive. This proves that if a variety is distributive, then we must have $m_\lambda \leq 1$.

On the other hand, if $m_\lambda \leq 1$ for all $\lambda \vdash n$, then $W_1, \ldots, W_k$ in (7.15) are pairwise non-isomorphic and any S_n-submodule of P_n containing $I \cap P_n$ must be of the type

$$(I \cap P_n) \oplus W_{i_1} \oplus \cdots \oplus W_{i_p}$$

for some subset $\{i_1, \ldots, i_p\} \subseteq \{1, \ldots, k\}$. Now let $Q, S, T \supseteq I$ be T-ideals and let

$$Q \cap P_n = (I \cap P_n) \oplus W_{i_1} \oplus \cdots \oplus W_{i_a},$$

$$S \cap P_n = (I \cap P_n) \oplus W_{j_1} \oplus \cdots \oplus W_{j_b},$$

$$T \cap P_n = (I \cap P_n) \oplus W_{l_1} \oplus \cdots \oplus W_{l_c}.$$

If we denote $A = \{i_1, \ldots, i_a\}$, $B = \{j_1, \ldots, j_b\}$ and $C = \{l_1, \ldots, l_c\}$, then

$$Q \cap T \cap P_n = (I \cap P_n) \oplus_{i \in A \cap C} W_i,$$

$$S \cap T \cap P_n = (I \cap P_n) \oplus_{i \in B \cap C} W_i$$

and

$$(Q + S) \cap T \cap P_n = (I \cap P_n) \oplus_{i \in (A \cup B) \cap C} W_i.$$

Clearly $Q \cap T \cap P_n + S \cap T \cap P_n = (Q+S) \cap T \cap P_n$ since $(A \cup B) \cap C = (A \cap C) \cup (B \cap C)$ and the proof of the theorem is complete. □

7.4. Polynomial growth and bounded multiplicities

In this section we characterize the varieties whose cocharacter has bounded multiplicities. Among other properties we shall prove that such varieties are precisely those not containing UT_2. We start with the following technical result about the algebra UT_2.

LEMMA 7.4.1. *If $\chi_n(UT_2) = \sum_{\lambda \vdash n} m_\lambda \chi_\lambda$ is the nth cocharacter of UT_2, then $m_{(n-1,1)} = n - 1$.*

PROOF. Define the polynomial
$$f_k = f_k(x_1, \ldots, x_n) = \sum_{\sigma \in S_{n-1}} x_{\sigma(1)} \cdots x_{\sigma(k)}[x_n, x_{\sigma(k+1)}, \ldots, x_{\sigma(n-1)}],$$
for all $k = 0, 1, \ldots, n-2$. Since f_k is symmetric in $n-1$ variables, $e_{T_\lambda} f_k$ is the trivial polynomial of $F\langle X \rangle$, for any tableau T_λ of shape $\lambda = (\lambda_1, \ldots, \lambda_r) \vdash n$, as soon as $\lambda_2 + \cdots + \lambda_r \geq 2$. On the other hand, $e_{T_\lambda} f_k = 0$ if $\lambda = (n)$. It follows that if M is the S_n-submodule of P_n generated by $f_0, \ldots, f_{n-2}$, then M is the direct sum of, say, t isomorphic irreducible S_n-modules
$$M = M_1 \oplus \cdots \oplus M_t$$
and $\chi(M_i) = \chi_{(n-1,1)}$, for all $i = 1, \ldots, t$. Consider the endomorphism $\varphi : F\langle X \rangle \to F\langle X \rangle$ such that $\varphi(x_1) = \cdots = \varphi(x_{n-1}) = y$, $\varphi(x_n) = x$ and write
$$g_k = g_k(x, y) = \frac{1}{(n-1)!} \varphi(f_k) = y^k [x, \underbrace{y, \ldots, y}_{n-k-1}].$$

By Theorem 7.3.1 it easily follows that UT_2 does not satisfy any non-trivial identity of the type
$$\sum_{i=0}^{n-2} \alpha_i g_i \equiv 0.$$

This says that the polynomials $g_0, g_1, \ldots, g_{n-2}$ are linearly independent modulo $\text{Id}(UT_2)$. Then by Theorem 2.4.4, we have that $m_{(n-1,1)} \geq n - 1$. On the other hand, the multiplicity of $\chi_{(n-1,1)}$ in the character of the left regular representation $FS_n \equiv P_n$ is equal to $\deg \chi_{(n-1,1)} = n - 1$. Hence $m_{(n-1,1)} \leq n - 1$ and the proof of the lemma is complete. $\square$

In the next lemma we study a generating algebra for a variety not containing UT_2.

LEMMA 7.4.2. *Let $\mathcal{V}$ be a variety such that $UT_2 \notin \mathcal{V}$. Then $\mathcal{V} = \text{var}(G(A_1) \oplus \cdots \oplus G(A_n))$, where for each i, A_i is a superalgebra, $\dim A_i < \infty$ and either $\dim A_i/J(A_i) \leq 1$ or $A_i/J(A_i) \cong F \oplus cF$, $c^2 = 1$.*

PROOF. Let $\mathcal{V}$ be generated by the Grassmann envelope of some finite dimensional superalgebra B with Jacobson radical J. If C is a maximal semisimple subalgebra homogeneous in the $\mathbb{Z}_2$-grading of B, then
$$C = C_1 \oplus \cdots \oplus C_m$$
and all C_i's are $\mathbb{Z}_2$-graded simple. Hence each C_i is of one of the types $M_k(F)$, $M_{k,l}(F)$ or $M_k(F \oplus cF)$ with $c^2 = 1$.

Note that both the algebras

$$\begin{pmatrix} G^{(0)} & G^{(0)} \\ 0 & G^{(0)} \end{pmatrix} \quad \text{and} \quad \begin{pmatrix} G^{(0)} & G^{(1)} \\ 0 & G^{(0)} \end{pmatrix}$$

contain a subalgebra isomorphic to UT_2. Indeed, if

$$a = \begin{pmatrix} 1 & 0 \\ 0 & 0 \end{pmatrix} \quad b = \begin{pmatrix} 0 & 0 \\ 0 & 1 \end{pmatrix} \quad \text{and} \quad e = \begin{pmatrix} 0 & x \\ 0 & 0 \end{pmatrix}$$

where $x \in G^{(0)}$ or $x \in G^{(1)}$, then span$\{a,b,e\}$ is isomorphic to UT_2, provided $x \neq 0$.

From this remark it follows that $UT_2 \in \text{var}(G(C_i))$ unless $C_i = F$ or $C_i = F \oplus cF, c^2 = 1$. Therefore, $\dim C_i = 1$ or 2, for all $i = 1, \ldots, m$.

Suppose now that $C_i J C_j \neq 0$ for some $i \neq j$. Then one can choose units a, b in C_i and C_j, respectively and an element $0 \neq e \in J$ such that $ae = eb = e$. Since $C_i C_j = 0$, it follows that $e^2 = 0$ and a, b, e generate an algebra isomorphic to UT_2. Hence,

(7.16) $\quad C_i J C_j = 0 \quad \text{and} \quad C_i C_j = 0 \quad \text{for any} \quad i \neq j.$

Set $A_i = C_i + J$, $i = 1, \ldots, m$. Then $B = C_1 \oplus \cdots \oplus C_m + J = (C_1 + J) + \cdots + (C_m + J) = A_1 + \cdots + A_m$. Moreover, $J \subseteq A_i$ is the Jacobson radical of A_i, and $A_i / J \cong C_i$.

The relations in (7.16) imply the same relations for their Grassman envelopes

(7.17) $\quad G(C_i) G(J) G(C_j) = 0 \quad \text{and} \quad G(C_i) G(C_j) = 0 \quad \text{for any} \quad i \neq j.$

We claim that $\text{Id}(G(A_1) + \cdots + G(A_m)) = \text{Id}(G(A_1)) \cap \cdots \cap \text{Id}(G(A_m))$. The inclusion $\subseteq$ is obvious. Conversely, pick

$$f = f(x_1, \ldots, x_n) \in \text{Id}(G(A_1)) \cap \cdots \cap \text{Id}(G(A_m))$$

multilinear, and show that $f \equiv 0$ on $G(A_1) + \cdots + G(A_m)$. In order to do so, it suffices to check for substitutions $x_i \to \bar{x}_i \in G(C_1) \cup \cdots \cup G(C_m) \cup G(J)$. If $\bar{x}_1, \ldots, \bar{x}_n \in G(C_j)$ for a single j, then $f(\bar{x}) = 0$. If, say, $\bar{x}_1 \in G(C_i)$ and $\bar{x}_2 \in G(C_j)$ with $i \neq j$, then any monomial $\bar{x}_{\sigma(1)} \cdots \bar{x}_{\sigma(n)} = 0$ by (7.17). Thus, $f \in \text{Id}(G(A_1) + \cdots + G(A_m))$, as desired.

It follows that $\text{var}(G(B)) = \text{var}(G(A_1) \oplus \cdots \oplus G(A_m))$ and the lemma is proved. $\square$

LEMMA 7.4.3. Let $C = F$ or $C = F \oplus cF$, $c^2 = 1$, and let $A = C + J = (C + J)^{(0)} + (C + J)^{(1)}$ be a finite dimensional superalgebra with $J = J(A)$ its Jacobson radical. If $\chi_n(G(A)) = \sum_{\lambda \vdash n} m_\lambda \chi_\lambda$, then there exists a constant K such that

$$m_\lambda \leq K.$$

PROOF. Clearly, since $G(A) \subseteq A \otimes G$, $G(A)$ is a PI-algebra. Then by Theorem 4.5.1, the cocharacters of $G(A)$ lie in the hook $H(d, d)$ for some positive integer d.

By the assumption, there exist elements $a_0, \ldots, a_{m-1}, b_0, \ldots, b_{t-1}$, homogeneous in the $\mathbb{Z}_2$-grading, such that

$$A^{(0)} = \text{span}\{a_0, a_1, \ldots, a_{m-1}\}, \quad A^{(1)} = \text{span}\{b_0, b_1, \ldots, b_{t-1}\},$$

$a_0 \in C^{(0)}$, $b_0 \in C^{(1)}$ and $a_1, \ldots, a_{m-1}, b_1, \ldots, b_{t-1} \in J$. If $C^{(1)} = 0$, then $b_0 \in J$. Let $k \geq 0$ be an integer such that $J^k = 0$. We denote $N_0 = (2k)^{2d(m+t)}$ and we next prove that every multiplicity m_λ is less or equal to $(m + t) N_0$.

7.4. POLYNOMIAL GROWTH AND BOUNDED MULTIPLICITIES

Consider a partition $\lambda \subseteq H(d,d)$ and take a Young tableau T_λ corresponding to λ. Let e_{T_λ} be the essential idempotent corresponding to T_λ. Clearly, there exists a multilinear polynomial $g = g(x_1, \ldots, x_n)$, such that $f = e_{T_\lambda} g \neq 0$ in $F\langle X \rangle$.

By Lemma 2.5.6, there exists $f' = r e_{T_\lambda} g \neq 0$ such that $FS_n f = FS_n f'$ and the variables of the polynomial f' are partitioned into $2d$ disjoint subsets $X_1, \ldots, X_d$, $Y_1 \ldots, Y_d$, such that g is symmetric on all variables from X_i for any $i = 1, \ldots, d$ and alternating on Y_j, $j = 1, \ldots, d$. We assume for short that $d' = d$ and we observe that X_i may be empty if the length of the ith row of λ is less than or equal to d. If $\lambda = (\lambda_1, \lambda_2, \ldots) \vdash n$, $\lambda_i > d$, then $|X_i| = \lambda_i - d$. For $Y_1, \ldots, Y_d$ we have $|Y_j| = \lambda'_j$, where λ' is the conjugate partiton of λ. For any $\rho \in S_n$ we also have $\rho f' \neq 0$. It follows that $FS_n \rho f' = FS_n f$, and for a suitable ρ it is symmetric on the first $\lambda_1 - d$ variables, the next $\lambda_2 - d$ variables, and so on. A similar condition holds for the alternating sets $Y_1, \ldots, Y_d$.

Let $f_1, \ldots, f_M$ be multilinear polynomials generating in $P_n(A)$ different isomorphic S_n-modules corresponding to λ. By the above, one can find $\rho_1, \ldots, \rho_M \in S_n$ and a decomposition

$$X = X_1 \cup \ldots \cup X_d \cup Y_1 \cup \ldots \cup Y_d$$

such that all $\rho_1 f_1, \ldots, \rho_M f_M$ are simultaneously symmetric on X_i, $i = 1, \ldots, d$, and alternating on Y_j, $j = 1, \ldots, d$. Clearly, we may assume that $f_1, \ldots, f_M$ satisfy this condition.

Now assume that $m_\lambda = M > (m+t)N_0$, and prove that $G(A)$ satisfies an identity of the type

(7.18) $\quad f = \gamma_1 f_1 + \cdots + \gamma_M f_M \equiv 0, \quad \gamma_1, \ldots, \gamma_M \in F, \quad$ not all γ_i equal to 0.

It is sufficient to verify that f has only zero values on elements of the form $a_i \otimes g$, $b_j \otimes h$, $g \in G^{(0)}$, $h \in G^{(1)}$. First define substitutions of a special kind.

Let $0 \leq \alpha_0^j, \alpha_1^j, \ldots, \alpha_{m-1}^j, \beta_0^j, \beta_1^j, \ldots, \beta_{t-1}^j \in \mathbb{Z}$ satisfy

$$\sum_{i=0}^{m-1} \alpha_i^j + \sum_{i=0}^{t-1} \beta_i^j = |X_j|, \quad 1 \leq j \leq d,$$

$$\sum_{i=0}^{m-1} \alpha_i^j + \sum_{i=0}^{t-1} \beta_i^j = |Y_{j-d}| = \lambda'_{j-d}, \quad d+1 \leq j \leq 2d.$$

We say that the substitution φ has type

(7.19) $\quad \left(\alpha_0^j, \alpha_1^j, \ldots, \alpha_{m-1}^j, \beta_0^j, \beta_1^j, \ldots, \beta_{t-1}^j \right), \quad j = 1, \ldots 2d,$

if we replace variables from X in the following way.

First fix j, $1 \leq j \leq d$. We replace the first α_0^j variables from X_j by elements $a_0 \otimes g$ (possibly with different g's for distinct $x \in X_j$), the next α_1^j variables by some $a_1 \otimes g$'s, and so on up to the first $\alpha_0^j + \cdots + \alpha_{m-1}^j$ variables from X_j with all these g's in $G^{(0)}$. Now substitute the next β_0^j variables in X_j by $b_0 \otimes g$'s, the next β_1^j by $b_1 \otimes g$'s, and so on, with all these g's in $G^{(1)}$.

For $j = d+1, \ldots, 2d$ we apply the same procedure in order to replace variables in Y_{j-d} by the elements $a_i \otimes g$'s, $b_i \otimes g$'s. To give a non-zero value in (7.18), any substitution should satisfy the following restrictions:

1) If $1 \leq j \leq d$, then $\beta_i^j \leq 1$, $0 \leq i \leq t-1$, $\alpha_1^j + \cdots + \alpha_{m-1}^j \leq k-1$, and
$$\alpha_0^j = |X_j| - \left(\alpha_1^j + \cdots + \alpha_{m-1}^j + \beta_0^j + \cdots + \beta_{t-1}^j\right).$$

2) If $d+1 \leq j \leq 2d$, then $\alpha_i^j \leq 1$, $0 \leq i \leq m-1$, $\beta_1^j + \cdots + \beta_{m-1}^j \leq k-1$, and $\beta_0^j = |Y_{d-j}| - \left(\beta_1^j + \cdots + \beta_{t-1}^j + \alpha_0^j + \cdots + \alpha_{m-1}^j\right).$

These restrictions follow from the nilpotency of degree k of the Jacobson radical and from the alternating and the symmetry properties of the polynomials $f_1, \ldots, f_M$. For example, f is symmetric on X_i. Hence, it has zero value on $x_1, x_2 \in X_i$ if $x_1 = b_j \otimes h_1$, $x_2 = b_j \otimes h_2$, and $h_1, h_2 \in G^{(1)}$.

Let $1 \leq j \leq d$. Then, obviously, $\alpha_i^j < k$ for $i \geq 1$ and the number of different t-tuples $\left(\beta_0^j, \ldots, \beta_{t-1}^j\right)$ is less than 2^t. So, for $1 \leq j \leq d$ the number of different $m+t$ tuples (7.19) is less than $k^m \cdot 2^t < (2k)^{m+t}$. Similarly for $d+1 \leq j \leq 2d$: $\beta_i^j < k$ (in the second case) and the number of different $(\alpha_0^j, \ldots, \alpha_{m-1}^j)$'s is bounded by 2^m. Hence $k^t 2^m$ is an upper bound for these substitutions. In particular, $(2k)^{m+t}$ is an upper bound, for each $1 \leq j \leq 2d$.

It follows that the number N of distinct types of substitutions (7.19) is less than $((2k)^{m+t})^{2d} = N_0$.

Note that if φ, φ' are two substitutions of the same type, and $\varphi(z) = u \otimes p$ for some $z \in X = X_1 \cup \ldots \cup X_d \cup Y_1 \cup \ldots \cup Y_d$, $u \in \{a_0, \ldots, a_{m-1}, b_0, \ldots, b_{t-1}\}$, $p \in G$, then $\varphi'(z) = u \otimes p'$ with the same $\mathbb{Z}_2$-grading of the elements p, p'. Hence, if $X = \{z_1, \ldots, z_n\}$, $\varphi(z_i) = u_i \otimes p_i$, $\varphi'(z_i) = u_i \otimes p_i'$, then

$$(7.20) \qquad \varphi(f) = f(u_1 \otimes p_1, \ldots, u_n \otimes p_n) = w \otimes p_1 \ldots p_n,$$

$$(7.21) \qquad \varphi'(f) = f(u_1 \otimes p_1', \ldots, u_n \otimes p_n') = w \otimes p_1' \ldots p_n',$$

with the same w. In that case we say that φ and φ' are similar. We choose one such substitution from each similarity class: let $\varphi_1, \ldots, \varphi_N$ be all these substitutions of the distinct types (7.19). Let $\varphi : \{x_1, \ldots, x_n\} \to G(A)$ be one of these special substitutions, and let h_1, h_2 be two multilinear polynomials in $x_1, \ldots, x_n$. Then, by multilinearity and supercommutativity, $\varphi(h_1) = r_1 \otimes p_1 \cdots p_n$ and $\varphi(h_2) = r_2 \otimes p_1 \cdots p_n$ with the same $p_1 \ldots p_n$ ($r_1, r_2 \in A$).

Now consider all these N substitutions of distinct types $\varphi_1, \ldots, \varphi_N$. Then

$$(7.22) \qquad \varphi_j(f_i) = a_{ij} \otimes p_{j1} \cdots p_{jn},$$

where $a_{ij} \in A$ and $p_{j1} \ldots p_{jn}$ depends on φ_j only (and not on f_i).

The matrix (a_{ij}), $1 \leq i \leq M$, $1 \leq j \leq N$, has M rows and N columns of elements from A, and $\dim A \leq m+t$. Therefore the rows of (a_{ij}) are linearly dependent: there exist $\gamma_1, \ldots, \gamma_M \in F$, not all equal to zero, such that

$$(7.23) \qquad \sum_{i=1}^{M} \gamma_i a_{ij} = 0, \quad 1 \leq j \leq N.$$

This, together with (7.22), implies that $\varphi_j\left(\sum_{i=1}^{M} \gamma_i f_i\right) = 0$, $1 \leq j \leq N$. We claim that this implies that $f = \sum_{i=1}^{M} \gamma_i f_i$ is an identity on $G(A)$: for any substitution $\varphi : \{x_1, \ldots, x_n\} \to G(A)$, $\varphi(f) = 0$. Indeed, by multilinearity of f, it suffices to check only substitutions φ^* such that $\varphi^*(x_\ell) = u \otimes p$, where $u \in \{a_0, \ldots, a_{m-1}, b_0, \ldots, b_{t-1}\}$ and $p \in G^{(0)} \cup G^{(1)}$. Given such φ^*, there exists $\sigma \in S_n$ "rearranging" the variables, such that $\varphi^* \sigma = \varphi'$ is similar to some φ_j.

Thus $\varphi'(f_i) = a_{ij} \otimes p'_{j1} \cdots p'_{jn}$ (compare with (7.20) and (7.21)) and by (7.23), $\varphi'(f) = 0$. Note that the above $\sigma : \{x_1, \ldots, x_n\} \to \{x_1, \ldots, x_n\}$ satisfies $\sigma(X_i) = X_i$, $\sigma(Y_i) = Y_i$, $i = 1, \ldots, d$. Since f is symmetric on X_i and alternating on Y_i, $i = 1, \ldots, d$, therefore $\varphi'(f) = \varphi^*\sigma(f) = \varphi^*(\sigma f) = \pm\varphi^*(f)$, hence $\varphi^*(f) = 0$.

This shows that modulo the identities of $G(A)$, the polynomials $f_1, \ldots, f_M$ are dependent over FS_n, which is equivalent to saying that $m_\lambda \leq M$ for any λ. The proof of the lemma is complete. $\square$

Recall that given a partition $\lambda \vdash n$ we denote by λ' the conjugate partition of n.

LEMMA 7.4.4. *Let A be as in Lemma 7.4.3 and let $\chi_n(G(A)) = \sum_{\lambda \vdash n} m_\lambda \chi_\lambda$. Then there exists a constant M such that*

$$n - \lambda_1 - \lambda'_1 \leq M$$

whenever $\lambda \vdash n$ is such that $m_\lambda \neq 0$.

PROOF. Let $J^k = 0$, for some k. First consider a partition $\lambda \vdash n$ with $\lambda_2 \geq k+2$ and verify that $m_\lambda = 0$. It is sufficient to check that $e_{T_\lambda} f = 0$ on $G(A)$ for any multilinear polynomial $f = f(x_1, \ldots, x_n)$ and for any Young tableau T_λ of shape λ.

Assume that $e_{T_\lambda} f$ has non-zero value on $G(A)$, i.e., there exists a substitution $\phi : x_i \to \bar{x}_i \in G(A)$ with $\phi(e_{T_\lambda} f) \neq 0$. This inequality implies that for at least one row permutation $\tau \in R_{T_\lambda}$ the same condition holds for

$$q = \tau \sum_{\sigma \in C_{T_\lambda}} (\text{sgn}\,\sigma)\sigma f,$$

$\phi(q) = (\tau \sum_{\sigma \in C_{T_\lambda}} (\text{sgn}\,\sigma)\sigma f)(\bar{x}_1, \ldots, \bar{x}_n) \neq 0$. Let the tableau T_λ contain the integers $i_1, \ldots, i_{k+2}$ in the first $k+2$ boxes of the first row and the integers $j_1, \ldots, j_{k+2}$ in the first $k+2$ boxes of the second row. Then q is alternating on pairs of variables

$$\{x_{\tau(i_1)}, x_{\tau(j_1)}\}, \ldots, \{x_{\tau(i_{k+2})}, x_{\tau(j_{k+2})}\}.$$

Recall that the semisimple part C of A is one- or two-dimensional, $C = C^{(0)} = F$ or $C = C^{(0)} \oplus C^{(1)} = F \oplus cF$, $c^2 = 1$. Therefore it is sufficient to evaluate $e_{T_\lambda} f$ only on elements of the form $1 \otimes g, c \otimes h, a \otimes g, b \otimes h$ where $g \in G^{(0)}, h \in G^{(1)}, a, b \in J$. If at least k pairs $\{\bar{x}_{\tau(i_s)}, \bar{x}_{\tau(j_s)}\}$ contain elements of the type $a \otimes g, a \in J$, then $\phi(q) = 0$ since $J^k = 0$. Hence we have at least three pairs $\{\bar{x}_{\tau(i_s)}, \bar{x}_{\tau(j_s)}\}$ without radical elements. Due to the skew symmetry of q one has $\phi(q) = 0$ if $\bar{x}_{\tau(i_s)} = 1 \otimes g, \bar{x}_{\tau(j_s)} = 1 \otimes g'$. Otherwise we have two i's, for example, i_1 and i_2, or two j's, for example, j_1 and j_2, such that $\bar{x}_{\tau(i_1)} = c \otimes h, \bar{x}_{\tau(i_2)} = c \otimes h'$ or $\bar{x}_{\tau(j_1)} = c \otimes h, \bar{x}_{\tau(j_2)} = c \otimes h'$ where $h, h' \in G^{(1)}$. But the element $e_{T_\lambda} f$ itself is symmetric on $x_{i_1}, \ldots, x_{i_{k+2}}$ and on $x_{j_1}, \ldots, x_{j_{k+2}}$. Since $\tau \in R_{T_\lambda}$, $e_{T_\lambda} f$ is also symmetric on $x_{\tau(i_1)}, \ldots, x_{\tau(i_{k+2})}$ and on $x_{\tau(j_1)}, \ldots, x_{\tau(j_{k+2})}$. Therefore the value $e_{T_\lambda} f$ on

$$\bar{x}_{\tau(i_1)} = c \otimes h, \ \bar{x}_{\tau(i_2)} = c \otimes h' \quad (\text{or } \bar{x}_{\tau(j_1)} = c \otimes h, \ \bar{x}_{\tau(j_2)} = c \otimes h')$$

with $h, h' \in G^{(1)}$ is equal to zero, a contradiction. Hence, $(e_{T_\lambda} f)(\bar{x}_1, \ldots, \bar{x}_n) = 0$ in all cases and $e_{T_\lambda} f = 0$ on $G(A)$.

Now consider a partition $\lambda \vdash n$ with two sufficiently large columns. More precisely, suppose that the conjugate partition λ' has a long second row, $\lambda'_2 \geq 2k+1$.

The same argument as before shows that $e_{T_\lambda} f = 0$ on $G(A)$ since $\sum_{\sigma \in C_{T_\lambda}} (\operatorname{sgn} \sigma) \sigma f$ is alternating on two sets with $2k + 1$ elements. In any skew set we can replace at most k elements with values different from $c \otimes h, h \in G^{(1)}$. We must then replace two symmetric variables in $e_{T_\lambda} f$ with $c \otimes h_1, c \otimes h_2, h_1, h_2 \in G^{(1)}$. Hence, $e_{T_\lambda} f$ has only zero value on $G(A)$.

We have seen that $m_\lambda = 0$ if $\lambda_2 \geq k + 2$ or $\lambda_2' \geq 2k + 1$. Hence, m_λ may be non-zero only if $n - \lambda_1 - \lambda_1' \leq M = 2k^2$, and the proof is complete. $\square$

LEMMA 7.4.5. *Let A, B be two PI-algebras and let*
$$\chi_n(A) = \sum_{\lambda \vdash n} m_\lambda \chi_\lambda, \quad \chi_n(B) = \sum_{\lambda \vdash n} m_\lambda' \chi_\lambda, \quad \chi_n(A \oplus B) = \sum_{\lambda \vdash n} \overline{m}_\lambda \chi_\lambda.$$
If $m_\lambda \leq C$, $m_\lambda' \leq C'$ for some C, C', then $\overline{m}_\lambda \leq C + C'$.

PROOF. This follows from the equality $\operatorname{Id}(A \oplus B) = \operatorname{Id}(A) \cap \operatorname{Id}(B)$. $\square$

Now we can prove the main result of this section due to Mishchenko, Regev and Zaicev.

THEOREM 7.4.6 ([**MRZ1**]). *Let $\mathcal{V}$ be a variety of algebras and let*
$$\chi_n(\mathcal{V}) = \sum_{\lambda \vdash n} m_\lambda \chi_\lambda$$
be its nth cocharacter. Then the following conditions are equivalent.

1) *There exists a constant K such that for all n and for any $\lambda \vdash n$, the inequality*
$$m_\lambda \leq K$$
holds.
2) $UT_2 \notin \mathcal{V}$.
3) *There exists a constant M such that for all n and for any $\lambda \vdash n$, the inequality*
$$n - \lambda_1 - \lambda_1' \leq M$$
holds whenever $m_\lambda \neq 0$, where λ' is the conjugate partition of λ.
4) $\mathcal{V}$ *satisfies an identity of the type*
$$\sum_{i=0}^{n} \alpha_i y^i x y^{n-i} \equiv 0.$$
5) $\mathcal{V} = \mathcal{U} \cup \mathcal{W}$ *where $\mathcal{U}$ is a variety of polynomial growth and either $\mathcal{W} = 0$ or $\mathcal{W} = \operatorname{var}(G(A))$, where A is a finite dimensional superalgebra with $A/J(A) = F \oplus cF, c^2 = 1$.*

PROOF. The equivalence of 1) and 2) follows from Lemmas 7.4.1, 7.4.2, 7.4.3 and 7.4.5. By Lemma 7.4.4, 2) implies 3). On the other hand, consider $\mathcal{V} = \operatorname{var}(UT_2)$ and assume that $\mathcal{V}$ satisfies 3), i.e., $n - \lambda_1 - \lambda_1' \leq M$ for some constant M if $m_\lambda \neq 0$. Since $\dim UT_2 = 3$, by Theorem 4.6.2, $\chi_n(UT_2) \subset H(3, 0)$ for all n. Then $n - \lambda_1$ is bounded and $c_n(UT_2)$ is polynomially bounded which contradicts Theorem 7.2.14. This means that $\operatorname{var}(UT_2)$ does not satisfy 3), and 3) implies 2). The equivalence of 2) and 4) was proved in Theorem 7.3.1. By Lemma 7.4.2, 2) implies 5). Finally, by Lemmas 7.4.3 and 7.4.5, 3) follows from 5). $\square$

Now we can give an additional characterization of varieties of polynomial growth. We say that an algebra A (or a variety) is of finite colength if there exists a constant K such that $l_n(A) \leq K$ for all $n \geq 1$.

COROLLARY 7.4.7. *An algebra A has finite colength if and only if its codimension growth is polynomially bounded.*

PROOF. Assume first that $c_n(A)$ is polynomially bounded. By Theorem 7.2.5, $UT_2 \notin \mathrm{var}(A)$. Hence by the previous theorem all non-zero multiplicities m_λ in

$$\chi_n(A) = \sum_{\lambda \vdash n} m_\lambda \chi_\lambda$$

are bounded by a constant K. On the other hand, by Theorem 7.2.2, $n - \lambda_1 \leq K'$ as soon as $m_\lambda \neq 0$. Since the number of partitions with $n - \lambda_1 \leq K'$ is less than K'^2, we get

$$l_n(A) = \sum_{\lambda \vdash n} m_\lambda < K \cdot K'^2 = const.$$

Conversely, let $l_n(A)$ be bounded by some constant. In this case the Grassmann algebra G and the upper-triangular matrix algebra UT_2 do not lie in $\mathrm{var}(A)$ by Theorem 4.1.8, and Lemma 7.4.1. Hence by Theorem 7.2.5, $c_n(A)$ is polynomially bounded. $\square$

7.5. Types of polynomial growth

In this section we consider the question of which polynomials can be realized as codimension growth functions of an algebra. In order to prove the next result, we need to introduce a new type of multilinear polynomials as follows.

Let $\Gamma_n = \Gamma_n(x_1, \ldots, x_n)$ be the subspace of the space P_n of multilinear polynomials in $x_1, \ldots, x_n$ consisting of all linear combinations of products of length n of the type

(7.24) $$[x_{\sigma(1)}, \ldots, x_{\sigma(i)}][x_{\sigma(i+1)}, \ldots x_{\sigma(j)}] \cdots [x_{\sigma(k)}, \ldots, x_{\sigma(n)}]$$

where $\sigma \in S_n$ and $[y_1, \ldots, y_n] = [[y_1, y_2], \ldots, y_n]$ is the left-normed Lie commutator of $y_1, \ldots, y_n$. Γ_n is called the space of multilinear proper polynomials of degree n.

For a PI-algebra A we also define

$$\gamma_n(A) = \dim \frac{\Gamma_n}{\Gamma_n \cap \mathrm{Id}(A)},$$

the nth proper codimension of A.

Proper polynomials and proper identities are very useful when dealing with algebras with 1. For instance, it can be shown that in characteristic zero every ideal of identities of an algebra with 1 can be generated by its proper polynomials. We refer the reader to the book of Drensky ([**D10**]) for an account of results on proper polynomial identities and their connection to the ordinary ones. A basic tool in this setting is the Witt theorem and the Poincaré-Birkhoff-Witt theorem. As a consequence one gets that every polynomial in $F\langle X \rangle$ can be written as a linear combination of polynomials of the type

$$x_1^{a_1} \cdots x_m^{a_m}[x_{i_1}, \ldots, x_{i_r}] \cdots [x_{j_1}, \ldots, x_{j_p}]$$

where $a_1, \ldots, a_m, b, \ldots, c \geq 0$ (see [**D10**]).

LEMMA 7.5.1. *Let A be a unitary algebra such that $\exp(A) = 1$. Then*

1) *there exists an integer $k \geq 1$ such that $\gamma_{k+j}(A) = 0$ for all $j \geq 1$;*

2) $c_n(A) = 1 + \sum_{j=2}^{k} \binom{n}{j} \gamma_j(A)$.

PROOF. Let us prove first that

(7.25) $$c_n(A) = 1 + \sum_{j=2}^{n} \binom{n}{j} \gamma_j(A).$$

By the above discussion, any multilinear polynomial in $x_1, \ldots, x_n$ can be written as a linear combination of elements of the type

$$x_{i_1} \cdots x_{i_k} w$$

where $i_1 < \ldots < i_k$ and w is a polynomial of Γ_{n-k} in the indeterminates $x_{j_1}, \ldots, x_{j_{n-k}}$ where $\{j_1, \ldots, j_{n-k}\} = \{1, \ldots, n\} \setminus \{i_1, \ldots, i_k\}$. Let $m_k = \gamma_k(A) = \dim \Gamma_k/(\Gamma_k \cap \mathrm{Id}(A))$ and let $f_k^{(1)}, \ldots, f_k^{(m_k)}$ form a basis of Γ_k (mod $\Gamma_k \cap \mathrm{Id}(A)$). Given $n \geq k \geq 0$, $k \neq n-1$, and a subset $\{i_1, \ldots, i_k\} \subseteq \{1, \ldots, n\}$ with $i_1 < \ldots < i_k$, define

$$f_{i_1,\ldots,i_k}^{(j)} = f_{n-k}^{(j)}(x_{j_1}, \ldots, x_{j_{n-k}}), \quad j = 1, \ldots, m_{n-k}$$

where $j_1 < \ldots < j_{n-k}$, $\{j_1, \ldots, j_{n-k}\} = \{1, \ldots, n\} \setminus \{i_1, \ldots, i_k\}$ and let

(7.26) $$a_{i_1,\ldots,i_k}^{(j)} = x_{i_1} \cdots x_{i_k} f_{n-k}^{(j)}.$$

If $k = n$, we assume that the only element in (7.26) is $a_{1,\ldots,n} = x_1 \cdots x_n$. We claim that the polynomials $a_{i_1,\ldots,i_k}^{(j)}$ are linearly independent modulo $\mathrm{Id}(A)$. Suppose that

$$\sum_{i_1,\ldots,i_k,j} \lambda_{i_1,\ldots,i_k}^{(j)} a_{i_1,\ldots,i_k}^{(j)} \equiv 0$$

in A and $\lambda_{i_1,\ldots,i_k}^{(j)} \neq 0$ for some $i_1, \ldots, i_k, j$. By replacing all the variables $x_{i_1}, \ldots, x_{i_k}$ with $1 \in A$ and all other x_t's with arbitrary elements of A, we obtain that

$$\sum_{j=1}^{m_{n-k}} \lambda_{i_1,\ldots,i_k}^{(j)} f_{i_1,\ldots,i_k}^{(j)} \equiv 0$$

is an identity of A, a contradiction.

Since the polynomials in (7.26) are linearly independent (mod $\mathrm{Id}(A)$), and their number is

$$1 + \sum_{k=0}^{n-2} \binom{n}{k} \gamma_{n-k}(A) = 1 + \sum_{j=2}^{n} \binom{n}{j} \gamma_j(A),$$

the equality (7.25) holds.

Suppose first that $\gamma_n(A) \neq 0$ for all $n \geq 1$. Then by (7.25)

$$c_n(A) \geq \sum_{j=0}^{n} \binom{n}{j} - n = 2^n - n$$

and this contradicts the hypothesis $\exp(A) = 1$. Hence there exists $p > 0$ such that $\gamma_p(A) = 0$. In particular, this says that A satisfies the identity

(7.27) $$[x_1, \ldots, x_p] \equiv 0.$$

Since $\exp(A) = 1$, by Theorem 7.2.12, $\mathrm{Id}(A) = \mathrm{Id}(A_1 \oplus \cdots \oplus A_t)$, where $A_1, \ldots, A_t$ are finite dimensional algebras and $\dim A_i/J(A_i) \leq 1$, $i = 1, \ldots, t$. Since any algebra A_i satisfies the identity

(7.28) $$[x_1, y_1] \cdots [x_r, y_r] \equiv 0,$$

it follows that also A satisfies the same identity. From (7.27) and (7.28) it easily follows that $\gamma_k(A) = 0$ as soon as $k > rp$, and the first part of the lemma is proved.

The second part easily follows from part 1) and (7.25). □

In light of the previous lemma we can describe the asymptotic behaviour of the codimensions of a unitary algebra in case they are polynomially bounded.

THEOREM 7.5.2 (Drensky [**D6**]). *Let A be a unitary PI-algebra whose sequence of codimensions is polynomially bounded. Then*

$$c_n(A) = rn^k + O(n^{k-1}),$$

for some integer $k > 0$ and for some rational number r where $O(n^{k-1})$ is a polynomial of degree $\leq k - 1$.

PROOF. By Lemma 7.5.1 there exists a largest k with $\gamma_k(A) \neq 0$. Then

$$c_n(A) = 1 + \sum_{j=2}^{k} \binom{n}{j} \gamma_j(A) = \frac{\gamma_k(A)}{k!} n^k + O(n^{k-1}).$$

□

We close this section with an example giving a positive solution to the so-called inverse problem: given a positive integer t, does there exist a PI-algebra A such that $c_n(A)$ behaves asymptotically like a polynomial of degree t?

Given two positive integers k and r, let $A_{k,r}$ be the subalgebra of the algebra of $k \times k$ matrices over F with basis $\{e_{ij} \mid 1 \leq i < j \leq n\} \cup \{e_{rr}\}$, where the e_{ij}'s are the usual matrix units. Given an F-algebra R, denote by $A_{k,r}(R)$ the tensor product $A_{k,r} \otimes R$. Clearly, $A_{k,r}(R)$ is isomorphic to the algebra

$$\begin{pmatrix} 0 & R & \cdots & & & R \\ & \ddots & & & & \\ & & 0 & & & \\ & & & R & & \vdots \\ & & & 0 & & \\ & & & & \ddots & R \\ & & & & & 0 \end{pmatrix}.$$

The following result is due to Guterman and Regev.

THEOREM 7.5.3 ([**GuR**]). *Let R be an F-algebra with 1. Then*

1) *the T-ideal $\mathrm{Id}(A_{k,r}(R))$ is generated as a T-ideal by the polynomials*

(7.29) $$x_1 \cdots x_{r-1} f x_r \cdots x_{k-1}$$

where $f = f(y_1, \ldots, y_m) \in \mathrm{Id}(R)$ is an identity of R;

2) $c_n(A_{k,r}) = n(n-1) \cdots (n-k+2) c_{n-k+1}(R).$

PROOF. First we prove that any polynomial in (7.29) is an identity of $A_{k,r}(R)$. For any $\bar{x}_1, \ldots, \bar{x}_{r-1} \in A_{k,r}(R)$ the product $\bar{x}_1 \cdots \bar{x}_{r-1}$ is a linear combination of $e_{ij} \otimes a$ with $a \in R$ and $i \geq r$. Similarly, $\bar{x}_r \cdots \bar{x}_{k-1}$ is a linear combination of $e_{\alpha\beta} \otimes b$ with $\alpha \leq r$, $b \in R$. Since f is an identity of R, any evaluation $\bar{f}$ of f in $A_{k,r}(R)$ is a linear combination of elements $e_{pq} \otimes c$, $c \in R$, with $p < q$. Hence

$$\bar{x}_1 \cdots \bar{x}_{r-1} \bar{f} \bar{x}_r \cdots \bar{x}_{k-1} = 0$$

as required, i.e., any polynomial in (7.29) is an identity of $A_{k,r}(R)$.

Denote by I the T-ideal generated by all polynomials of the type (7.29). We first prove that any identity $g \equiv 0$ of $A_{k,r}(R)$ belongs to I. Since char $F = 0$, we may assume g to be multilinear.

Let $\deg g = n$ and let $m = n - k + 1$. Fix a basis $\{f_i = f_i(x_1, \ldots, x_m) \mid 1 \leq i \leq N\}$ of P_n (mod $P_n \cap I$). Then modulo I, the polynomial g is a linear combination of products of the type

$$(7.30) \qquad x_{i_1} \cdots x_{i_{r-1}} f_s(x_{l_1}, \ldots, x_{l_m}) x_{j_1} \cdots x_{j_{k-r}}$$

where $1 \leq s \leq N$ and $\{l_1, \ldots, l_m\} = \{1, \ldots, n\} \setminus \{i_1, \ldots, i_{r-1}, j_1, \ldots, j_{k-r}\}$ with $l_1 < \ldots < l_m$. If we prove that the elements (7.30) are linearly independent modulo the T-ideal $\mathrm{Id}(A_{k,r}(R))$, then this will imply the equality $I = \mathrm{Id}(A_{k,r}(R))$. Suppose

$$(7.31) \qquad \sum \lambda_{i_1,\ldots,i_{r-1},j_1,\ldots,j_{k-r},s} x_{i_1} \cdots x_{i_{r-1}} f_s(x_{l_1}, \ldots, x_{l_m}) x_{j_1} \cdots x_{j_{k-r}} \equiv 0$$

is an identity of $A_{k,r}(R)$ with non-trivial coefficients. Fix $i_1, \ldots, i_{r-1}, j_1, \ldots, j_{k-r}$ such that $\lambda_{i_1,\ldots,i_{r-1},j_1,\ldots,j_{k-r},s} \neq 0$ for some $1 \leq s \leq N$ and replace

$$x_{i_1}, \ldots, x_{i_{r-1}}, x_{j_1}, \ldots, x_{j_{k-r}}$$

with

$$e_{21} \otimes 1, \ldots, e_{r-1,r} \otimes 1, e_{r,r+1} \otimes 1, \ldots, e_{k-1,k} \otimes 1,$$

respectively. Also replace $x_{l_1}, \ldots, x_{l_m}$ with $e_{rr} \otimes a_1, \ldots, e_{rr} \otimes a_m$ where $a_1, \ldots, a_m$ are arbitrary elements of R. Then from (7.31) it follows that

$$\sum_{s=1}^{N} \mu_s f_s(a_1, \ldots, a_m) = 0$$

for any $a_1, \ldots, a_m \in R$, where $\mu_s = \lambda_{i_1,\ldots,i_{r-1},j_1,\ldots,j_{k-r},s}$, $1 \leq s \leq N$. Hence $\sum \mu_s f_s \equiv 0$ is an identity of R, a contradiction, and the proof of the first part of the theorem is complete.

In order to prove the second part, it is enough to count the elements in (7.30). Since $N = c_{n-k+1}(R)$, by definition of codimension we have

$$c_n(A_{k,r}(R)) = (r-1)!(k-r)!\binom{n}{r-1}\binom{n-r+1}{k-r} c_{n-k+1}(R)$$

$$= \frac{n!}{(n-r+1)!} \frac{(n-r+1)!}{(n-k+1)!} c_{n-k+1}(R) = n(n-1)\cdots(n-k+2) c_{n-k+1}(R)$$

and the proof is complete. $\square$

COROLLARY 7.5.4. *For any integer $t \geq 1$, there exists an algebra A with $c_n(A) \simeq n^t$.*

PROOF. It is enough to take $k = t + 1$ and $R = F$. Then by the previous theorem $c_n(A_{k,r}(F)) = c_n(A_{k,r}) = n(n-1)\cdots(n-k+2) \simeq n^t$. $\square$

7.6. Varieties of exponent two

In this section we shall characterize varieties $\mathcal{V}$ of exponent two. As in the case of polynomial growth we shall describe them by exhibiting a finite number of algebras not contained in any such variety.

Starting with the Grassmann algebra $G = G^{(0)} \oplus G^{(1)}$ we first define the following five algebras over F:

1) $A_1 = \begin{pmatrix} G & G \\ 0 & G^{(0)} \end{pmatrix}$;
2) $A_2 = \begin{pmatrix} G^{(0)} & G \\ 0 & G \end{pmatrix}$;
3) $A_3 = UT_3(F)$, the algebra of 3×3 upper triangular matrices over F;
4) $A_4 = M_2(F)$, the algebra of 2×2 matrices over F;
5) $A_5 = M_{1,1}(G) = \begin{pmatrix} G^{(0)} & G^{(1)} \\ G^{(1)} & G^{(0)} \end{pmatrix}$.

The main result of this section is the following.

THEOREM 7.6.1 ([**GZ5**]). *Let F be a field of characteristic zero and $\mathcal{V}$ a variety of F-algebras. Then $\exp(\mathcal{V}) > 2$ if and only if $A_i \in \mathcal{V}$ for some $i \in \{1, \ldots, 5\}$.*

For every $i = 1, \ldots, 5$ let $\mathcal{V}_i = \text{var}(A_i)$ be the variety generated by the algebra A_i. The above list of algebras cannot be reduced; in fact, we shall give the proof of the following.

PROPOSITION 7.6.2. *For all $i \neq j$, $\mathcal{V}_i \nsubseteq \mathcal{V}_j$.*

Hence $\mathcal{V}_1, \ldots, \mathcal{V}_5$ are the only minimal varieties of exponent > 2 in the sense that, for every i, $\exp(\mathcal{V}_i) > 2$ and for every subvariety $\mathcal{W}$ of $\mathcal{V}_i, \exp(\mathcal{W}) \leq 2$ (see Chapter 8). Notice that A_1 is the Grassmann envelope of the superalgebra

$$B_1 = \begin{pmatrix} F \oplus cF & F \oplus cF \\ 0 & F \end{pmatrix}$$

with grading

$$B_1^{(0)} = \begin{pmatrix} F & F \\ 0 & F \end{pmatrix}, B_1^{(1)} = \begin{pmatrix} cF & cF \\ 0 & 0 \end{pmatrix}$$

whereas $A_2 = G(B_2)$ where

$$B_2 = \begin{pmatrix} F & F \oplus cF \\ 0 & F \oplus cF \end{pmatrix}$$

with grading

$$B_2^{(0)} = \begin{pmatrix} F & F \\ 0 & F \end{pmatrix}, B_1^{(1)} = \begin{pmatrix} 0 & cF \\ 0 & cF \end{pmatrix}.$$

From the basic properties of the exponent given in Chapter 6, it follows that $\exp(\mathcal{V}_1) = \exp(\mathcal{V}_2) = \exp(\mathcal{V}_3) = 3$ and $\exp(\mathcal{V}_4) = \exp(\mathcal{V}_5) = 4$.

Invoking Theorem 7.2.5 we get

COROLLARY 7.6.3. *Let $\mathcal{V}$ be a variety of algebras over a field of characteristic zero. Then $\exp(\mathcal{V}) = 2$ if and only if $A_1, \ldots, A_5 \notin \mathcal{V}$ and either $G \in \mathcal{V}$ or $UT_2(F) \in \mathcal{V}$.*

PROOF OF THEOREM 7.6.1. Suppose that $\exp(\mathcal{V}) = p > 2$. By Corollary 4.8.4, there exists a finite dimensional superalgebra $B = B^{(0)} + B^{(1)}$ such that $\mathcal{V} = \mathrm{var}(G(B))$ where $G(B)$ is the Grassmann envelope of B. Let $B = B_1 \oplus \cdots \oplus B_k + J$ be the Wedderburn-Malcev decomposition of B with $B_1, \ldots, B_k$ simple subalgebras homogeneous in the $\mathbb{Z}_2$-grading. For each $i = 1, \ldots k$, let $B_i = B_i^{(0)} \oplus B_i^{(1)}$ and $J = J^{(0)} \oplus J^{(1)}$ be the induced $\mathbb{Z}_2$-grading.

Now let $\bar{F}$ be the algebraic closure of the field F and $\bar{B} = B \otimes_F \bar{F}$. Then $G(B) \otimes_F \bar{F} \cong G(B \otimes_F \bar{F}) = G(\bar{B})$ and by Theorem 4.1.9, the nth codimension of $G(\bar{B})$ over $\bar{F}$ equals the nth codimension of $G(B)$ over F, for all n. It follows that the exponent of $G(B)$ over F coincides with the exponent of $G(\bar{B})$ over $\bar{F}$. Since $G(\bar{B}) \in \mathrm{var}(G(B)) = \mathcal{V}$, in order to prove that $A_i \in \mathcal{V}$ for some i, it is enough to show that $G(\bar{B})$ contains a copy of A_i for some i. In particular, we may assume that F is algebraically closed.

Recall that each algebra B_i is isomorphic to $M_k(F)$ or to $M_{k,l}(F)$ or to $M_k(F \oplus cF)$, $c^2 = 1$.

Now, if B contains one simple component of type $M_{k,l}(F)$ with $k + l \geq 2$ or of type $M_k(F \oplus cF)$ with $k \geq 2$, then we will get that $G(B)$ contains an algebra isomorphic to either A_4 or A_5 and in this case we are done.

Recall that by the results of Section 6.5, $\exp(\mathcal{V})$ is the greatest dimension of a semisimple subalgebra $B_{i_1} \oplus \cdots \oplus B_{i_r}$ such that $B_{i_1} J \cdots J B_{i_r} \neq 0$. Hence, since $\exp(\mathcal{V}) > 2$, we may assume that one of the following possibilities occurs:

1) for some $i \neq l$, $B_i J B_l \neq 0$ where $B_i \cong F + cF, c^2 = 1$ and $B_l \cong F$;
2) for some $i \neq l$, $B_i J B_l \neq 0$ where $B_i \cong F$ and $B_l \cong F + cF, c^2 = 1$;
3) there exist distinct B_i, B_l, B_m such that $B_i J B_l J B_m \neq 0$ and $B_i \cong B_l \cong B_m \cong F$.

Suppose 1) holds. Then there exists $a + cb \in B_i$ such that $(a+cb)j1_3 \neq 0$ where 1_3 is the unit element of B_l and $j \in J$ is homogeneous. By eventually multiplying by c on the left, we may assume that $(a + cb)j_0 1_3 \neq 0$ for some $j_0 \in J^{(0)}$. Write $a + cb = u_{11}(a+b) + u_{22}(a-b)$ where $u_{11} = (1+c)/2, u_{22} = (1-c)/2$ and $1 = 1_{B_i}$ is the unit element of B_i. Set $u_{33} = 1_3$.

The elements $j_0 u_{33}$ and $cj_0 u_{33}$ being of different homogeneous degree, are linearly independent over F. Hence $u_{11}, u_{22}, u_{33}, u_{13} = u_{11} j_0 u_{33}$ and $u_{23} = u_{22} j_0 u_{33}$ are linearly independent and form a subalgebra in B isomorphic to

$$D'' = \begin{pmatrix} F & 0 & F \\ 0 & F & F \\ 0 & 0 & F \end{pmatrix}$$

with $\mathbb{Z}_2$-grading

$$D''^{(0)} = \left\{ \begin{pmatrix} \lambda & 0 & \nu \\ 0 & \lambda & \nu \\ 0 & 0 & \mu \end{pmatrix} \right\}, \quad D''^{(1)} = \left\{ \begin{pmatrix} \lambda & 0 & \nu \\ 0 & -\lambda & -\nu \\ 0 & 0 & 0 \end{pmatrix} \right\}.$$

Hence $G(D'') \in \mathcal{V}$ and

$$G(D'') = \begin{pmatrix} a+b & 0 & z+w \\ 0 & a-b & z-w \\ 0 & 0 & t \end{pmatrix}$$

where $a, z, t \in G^{(0)}, b, w \in G^{(1)}$.

We construct an algebra isomorphism $G(D'') \cong A_1$ by setting

$$\begin{pmatrix} a+b & 0 & z+w \\ 0 & a-b & z-w \\ 0 & 0 & t \end{pmatrix} \mapsto \begin{pmatrix} a-b & z-w \\ 0 & t \end{pmatrix} \quad \text{where } a,z,t \in G^{(0)}, b,w \in G^{(1)}.$$

If 2) holds, then the same procedure as above shows that $A_2 \in \mathcal{V}$.

Finally, suppose that 3) holds. Then there exist $j_0, j'_0 \in J^{(0)}, j_1, j'_1 \in J^{(1)}$ such that $1_1(j_0+j_1)1_2(j'_0+j'_1)1_3 \neq 0$ where $1_1, 1_2, 1_3$ are the unit elements of B_i, B_l, B_m respectively. In this case at least one of the products $1_1 j_r 1_2 j'_s 1_3$, $r,s \in \{0,1\}$ is non-zero. Then, for fixed r and s set

$$u_{11} = 1_1, \ u_{22} = 1_2, \ u_{33} = 1_3, \ u_{12} = 1_1 j_r 1_2, \ u_{23} = 1_2 j'_s 1_3, \ u_{13} = 1_1 j_r 1_2 j'_s 1_3$$

and let D_{rs} be the $\mathbb{Z}_2$-graded subalgebra of B generated by $u_{11}, u_{22}, u_{33}, u_{12}, u_{23}, u_{13}$. By taking the Grassmann envelope of the algebra D_{rs}, we get that $\mathcal{V}$ must contain at least one of the following four algebras denoted E_1, E_2, E_3, E_4 respectively,

$$UT_3(F), \begin{pmatrix} G^{(0)} & G^{(0)} & G^{(1)} \\ 0 & G^{(0)} & G^{(1)} \\ 0 & 0 & G^{(0)} \end{pmatrix}, \begin{pmatrix} G^{(0)} & G^{(1)} & G^{(1)} \\ 0 & G^{(0)} & G^{(0)} \\ 0 & 0 & G^{(0)} \end{pmatrix}, \begin{pmatrix} G^{(0)} & G^{(0)} & G^{(0)} \\ 0 & G^{(0)} & G^{(1)} \\ 0 & 0 & G^{(0)} \end{pmatrix}.$$

Note that by Theorem 1.9.1, the T-ideal of identities of the upper triangular matrix algebra $UT_n = UT(1,\ldots,1)$ is generated by the polynomial

$$[x_1,x_2]\cdots[x_{2n-1},x_{2n}].$$

It is easy to check that each E_i satisfies the identity $[x_1,x_2][x_3,x_4][x_5,x_6] \equiv 0$ and, according to the previous remark, all the identities of $UT_3(F)$. On the other hand, each one of the algebras E_2, E_3, E_4 has a subalgebra isomorphic to $UT_3(F)$. In the case of E_4 this subalgebra is generated by the elements

$$e_{11}, \ e_{22}, \ e_{33}, \ xe_{12}, \ ye_{23}, \ xye_{13}$$

where x and y are two distinct generators of G. For E_3 it is the subalgebra generated by the elements

$$e_{11}, \ e_{22}, \ e_{33}, \ e_{23}, \ xe_{12}\ xe_{13}.$$

For E_2 we take $e_{11}, e_{22}, e_{33}, e_{12}, xe_{13}$ and xe_{23}. Hence $UT_3(F) \in \mathcal{V}$ and we are done. From the results of Section 6.5, it follows that $\exp(\mathcal{V}_1) = \exp(\mathcal{V}_2) = \exp(\mathcal{V}_3) = 3$ and $\exp(\mathcal{V}_4) = \exp(\mathcal{V}_5) = 4$. Hence if $\mathcal{W} \ni A_i$ for some $i \in \{1,\ldots,5\}$, then $\exp(\mathcal{W}) > 2$. $\square$

PROOF OF PROPOSITION 7.6.2. It is clear that if $\mathcal{W} \subseteq \mathcal{V}$ are varieties, then $\exp(\mathcal{W}) \leq \exp(\mathcal{V})$; hence $\mathcal{V}_4 \not\subseteq \mathcal{V}_i$ and $\mathcal{V}_5 \not\subseteq \mathcal{V}_i$ for all $i = 1,2,3$.

Since $UT_3(F)$ and $M_2(F)$ are the only two algebras among the A_i's satisfying a standard identity, we get that $\mathcal{V}_1, \mathcal{V}_2, \mathcal{V}_5 \not\subseteq \mathcal{V}_i$, $i = 3,4$. Also, the algebra $M_2(F)$ satisfies the standard identity $S_4 \equiv 0$ but $S_4 \not\equiv 0$ on $UT_3(F)$, hence $\mathcal{V}_3 \not\subseteq \mathcal{V}_4$.

The algebra $M_{1,1}(G)$ is the only algebra among the A_i's satisfying the identity $[[x_1,x_2],[x_3,x_4],x_5] \equiv 0$; hence $\mathcal{V}_i \not\subseteq \mathcal{V}_5$ for $i = 1,2,3,4$.

The algebra A_1 satisfies the identity $f_1 = [x_1,x_2,x_3][x_4,x_5] \equiv 0$ and the algebra A_2 satisfies the identity $f_2 = [x_1,x_2][x_3,x_4,x_5] \equiv 0$. Since $f_1 \not\equiv 0$ on A_2 and $f_2 \not\equiv 0$ on A_1, we get that $\mathcal{V}_1 \not\subseteq \mathcal{V}_2$ and $\mathcal{V}_2 \not\subseteq \mathcal{V}_1$. Moreover, since f_1 and f_2 do not vanish on $UT_3(F)$ we get that $\mathcal{V}_3 \not\subseteq \mathcal{V}_1, \mathcal{V}_2$. Finally, $\mathcal{V}_3$ satisfies $[x_1,x_2][x_3,x_4][x_5,x_6] \equiv 0$, hence $\mathcal{V}_4 \not\subseteq \mathcal{V}_3$. $\square$

CHAPTER 8

Classifying Minimal Varieties

Having at hand a scale provided by the exponent it becomes important to study T-ideals with the same exponent and to determine those with the most distinguished properties. In this chapter, for any given integer $d > 1$, we determine all maximal T-ideals of exponent d. We prove that for any given exponent, there are only a finite number of them and each is the product of verbally prime T-ideals. To this end we introduce the notion of minimal superalgebra and we relate it to that of minimal variety.

We prove that a proper T-ideal I is a product of verbally prime T-ideals if and only if $I = \mathrm{Id}(G(A))$ for some minimal superalgebra A. The exponent of the Grassmann envelope of a minimal superalgebra is readily computed: if $A = A_{ss} + J$ is the Wedderburn-Malcev decomposition of A where A_{ss} is a maximal semisimple $\mathbb{Z}_2$-stable subalgebra of A and J is the Jacobson radical, then $\exp(G(A)) = \dim A_{ss}$.

The minimal superalgebras have a nice representation as upper block triangular matrices over the algebraically closed field F with possibly various different $\mathbb{Z}_2$-gradings. In case A is a superalgebra with trivial grading, such representation is uniquely determined.

Another important concept related to the minimal superalgebras is that of minimal variety of given exponent. A variety $\mathcal{V}$ (or the corresponding T-ideal $I = \mathrm{Id}(\mathcal{V})$) is minimal of exponent $d \geq 2$ if $\exp(\mathcal{V}) = d$ and $\exp(\mathcal{U}) < d$ for all proper subvarieties $\mathcal{U}$ of $\mathcal{V}$.

Here we classify such varieties by showing that a variety $\mathcal{V}$ is minimal of exponent d if and only if $\mathcal{V}$ is generated by the Grassmann envelope of a minimal superalgebra A. As a consequence it follows that surprisingly there are only a finite number of minimal varieties of given exponent.

In the last section of this chapter we give an application of the theory developed on minimal superalgebras to the asymptotic computation of the codimension sequence of any PI-algebra.

Throughout this chapter the ground field F is assumed to be of characteristic zero.

8.1. Minimal superalgebras

Throughout this section we shall assume that F is an algebraically closed field. Recall that according to Theorem 3.5.3, any finite dimensional simple superalgebra over F is isomorphic to either $M_k(F)$ with trivial grading or to $M_{k,l}(F)$, $k \geq l > 0$, or to $M_k(F \oplus cF)$, where $c^2 = 1$.

We start with the following.

LEMMA 8.1.1. *If $A = A^{(0)} \oplus A^{(1)}$ is a simple superalgebra, there exist orthogonal idempotents $e_1, \ldots, e_n \in A^{(0)}$ such that $e_1 + \ldots + e_n = 1$ and for every $i = 1, \ldots, n$, Ae_i (e_iA) is a minimal left (resp. right) graded ideal of A.*

PROOF. For the algebra $M_k(F)$ or $M_{k,l}(F)$ we can take $e_1, \ldots, e_k$ as the diagonal matrix units. For the algebra $M_k(F \oplus cF)$ we take $e_1, \ldots, e_k$ as the diagonal matrix units of $M_k(F \oplus cF)^{(0)} = M_k(F)$. □

The above lemma suggests that we make the following definition.

DEFINITION 8.1.2. The idempotents of Lemma 8.1.1 are called the minimal graded idempotents of the simple superalgebra A.

We now fix some notation. If $A = A^{(0)} \oplus A^{(1)}$ is a finite dimensional superalgebra, we write $A = A_{ss} + J$ where A_{ss} is a maximal semisimple subalgebra homogeneous in the $\mathbb{Z}_2$-grading and $J = J(A) = J^{(0)} \oplus J^{(1)}$ is the Jacobson radical of A. Also we write $A_{ss} = A_1 \oplus \cdots \oplus A_m$ where $A_1, \ldots, A_m$ are simple superalgebra.

DEFINITION 8.1.3. Let A be a finite dimensional superalgebra, and let $A = A_1 \oplus \cdots \oplus A_m + J$ with $A_1, \ldots, A_m$ simple superalgebras. Then A is a minimal superalgebra if

1) there exist homogeneous elements $w_{12}, \ldots, w_{m-1,m} \in J^{(0)} \cup J^{(1)}$ and minimal graded idempotents $e_1 \in A_1, \ldots, e_m \in A_m$ such that
$$e_i w_{i,i+1} = w_{i,i+1} e_{i+1} = w_{i,i+1}, \quad i = 1, \ldots, m-1$$
and
$$w_{12} w_{23} \cdots w_{m-1,m} \neq 0;$$

2) $w_{12}, \ldots, w_{m-1,m}$ generate J as a two-sided ideal of A.

We remark that in our notation since $A_1 J A_2 J \cdots J A_m \neq 0$, the order of the simple components $A_1, \ldots, A_m$ is important. Hence throughout this chapter and the next chapter we shall tacitly agree that if A is a minimal superalgebra with radical J and semisimple component $A_1 \oplus \cdots \oplus A_m$, where $A_1, \ldots, A_m$ are simple superalgebras, then $A_1 J A_2 J \cdots J A_m \neq 0$.

The first important feature of minimal superalgebras is given in the next two lemmas.

LEMMA 8.1.4. *Let B be a finite dimensional superalgebra over the algebraically closed field F and let $B = B_{ss} + J$ where $B_{ss} = A_1 \oplus \cdots \oplus A_n$ with $A_1, \ldots, A_n$ simple superalgebras. If $m \leq n$ is such that $A_1 J A_2 J \cdots J A_m \neq 0$, then there exists a minimal superalgebra A contained in B such that $A_{ss} = A_1 \oplus \cdots \oplus A_m$.*

PROOF. By hypothesis there exist $x_1, \ldots, x_{m-1} \in J$ and $a_1 \in A_1, \ldots, a_m \in A_m$ such that

(8.1) $$a_1 x_1 a_2 \cdots a_{m-1} x_{m-1} a_m \neq 0.$$

For each i write $a_i = a_i^{(0)} + a_i^{(1)}$ and $x_i = x_i^{(0)} + x_i^{(1)}$ with $a_i^{(0)}, x_i^{(0)} \in A_i^{(0)}, a_i^{(1)}, x_i^{(1)} \in A_i^{(1)}$. Then from (8.1) it follows that there exist $\varepsilon_1, \eta_1, \eta_{m-1}, \varepsilon_m \in \{0,1\}$ such that
$$a_1^{(\varepsilon_1)} x_1^{(\eta_1)} a_2^{(\varepsilon_2)} \cdots a_{m-1}^{(\varepsilon_{m-1})} x_{m-1}^{(\eta_{m-1})} a_m^{(\varepsilon_m)} \neq 0.$$

In other words, we may assume that the elements $x_1, \ldots, x_{m-1}, a_1, \ldots, a_m$ are homogeneous in the $\mathbb{Z}_2$-grading. Let $1_1, \ldots, 1_m$ be the unit elements of the algebras $A_1, \ldots, A_m$, respectively. Then from the above inequality we can write
$$1_1(a_1 x_1 a_2) 1_2 (x_2 a_3) 1_3 \cdots 1_{m-1}(x_{m-1} a_m) 1_m \neq 0.$$
Decompose now each unit element 1_i, $i = 1, \ldots, m$, into the sum of minimal graded idempotents (cfr. Lemma 8.1.1). Then we obtain that for some minimal graded idempotents $e_1 \in A_1, \ldots, e_m \in A_m$,
$$e_1(a_1 x_1 a_2) e_2 (x_2 a_3) \cdots e_{m-1}(x_{m-1} a_m) e_m \neq 0.$$
Now define $w_{12} = e_1(a_1 x_1 a_2) e_2$, $w_{23} = e_2(x_2 a_3) e_3, \ldots, w_{m-1,m} = e_{m-1} x_{m-1} a_m e_m$.

Then the elements $w_{12}, \ldots, w_{m-1,m}$ belong to J and are homogeneous in the $\mathbb{Z}_2$-grading. Also $e_i w_{i,i+1} = w_{i,i+1} e_{i+1} = w_{i,i+1}$, for all $i = 1, \ldots, m-1$. Let A be the algebra generated by $A_1, \ldots, A_m, w_{12}, \ldots, w_{m-1,m}$ over F. Clearly $w_{12}, \ldots, w_{m-1,m}$ generate the maximal nilpotent ideal of A and $A_{ss} = A_1 \oplus \cdots \oplus A_m$. This says that A is a minimal superalgebra. $\square$

LEMMA 8.1.5. *Let $\mathcal{V}$ be a variety of algebras over the algebraically closed field F. If $\exp(\mathcal{V}) \geq 2$, then there exists a minimal superalgebra A with maximal semisimple subalgebra A_{ss} such that $G(A) \in \mathcal{V}$ and $\exp(\mathcal{V}) = \dim_F A_{ss}$.*

PROOF. Let B be a finite dimensional superalgebra such that $\mathcal{V} = \text{var}(G(B))$ and let $B = B_{ss} + J$ where $B_{ss} = A_1 \oplus \cdots \oplus A_n$ with $A_1, \ldots, A_n$ simple superalgebras. Then, by the characterization of the exponent given in Chapter 6, there exists $1 \leq m \leq n$ such that $A_1 J A_2 J \cdots J A_m \neq 0$ and $\exp(\mathcal{V}) = \dim(A_1 \oplus \cdots \oplus A_m)$.

By Lemma 8.1.4 there exists a minimal superalgebra A contained in B such that $A_{ss} = A_1 \oplus \cdots \oplus A_m$. Hence $\dim A_{ss} = \exp(\mathcal{V})$ and $G(A) \subseteq G(B)$ implies that $G(A) \in \mathcal{V}$. $\square$

In the next two lemmas we shall describe the structure of a minimal superalgebra. Until the end of the section A will be a minimal superalgebra, such that $A = A_{ss} + J$ with $A_{ss} = A_1 \oplus \cdots \oplus A_m$, $A_1, \ldots, A_m$ simple superalgebras, and $w_{12}, \ldots, w_{m-1,m} \in J$ will be the homogeneous elements satisfying the conditions 1) and 2) of the definition.

LEMMA 8.1.6. *The minimal superalgebra A has the following vector space decomposition*
$$A = \bigoplus_{1 \leq i \leq j \leq m} A_{ij}$$
where $A_{11} = A_1, \ldots, A_{mm} = A_m$ and, for all $i < j$,
$$A_{ij} = A_i w_{i,i+1} A_{i+1} \cdots A_{j-1} w_{j-1,j} A_j.$$
Moreover, $J = \oplus_{i<j} A_{ij}$ and $A_{ij} A_{kl} = \delta_{jk} A_{il}$ where δ_{jk} is the Kronecker delta.

PROOF. By condition 1) of the definition of minimal superalgebra, the product
$$a_1 w_{i_1, i_1+1} a_2 w_{i_2, i_2+1} \cdots a_k w_{i_k, i_k+1} a_{k+1}$$
with $a_1, \ldots, a_k \in A_{ss}$ is non-zero only if $i_1 + 1 = i_2, \ldots, i_{k-1} + 1 = i_k$ and $a_1 \in A_{i_1}, \ldots, a_{k+1} \in A_{i_k+1}$; in this case it belongs to A_{i_1, i_k+1}. Then from the multiplication rule of the w_{ij}'s, it follows that A is the sum of the subspaces A_{ij} with $1 \leq i \leq j \leq m$. Since the elements $w_{12}, \ldots, w_{m-1,m}$ generate J as an ideal of A, it is also clear that $J = \sum_{i<j} A_{ij}$. Finally, the orthogonality of the graded

idempotents $e_1 \in A_1, \ldots, e_m \in A_m$ guarantees that the sum $A = \oplus_{1 \leq i \leq j \leq m} A_{ij}$ is direct and that $A_{ij} A_{kl} = \delta_{jk} A_{il}$. □

If $m \geq 3$, i.e., there are at least three simple components, one can determine the structure of the subspaces $w_{i-1,i} A_i w_{i,i+1}$, for $i = 2, \ldots, m-1$, as follows.

Suppose first that the simple superalgebra A_i is isomorphic to either $M_k(F)$ or $M_{k,l}(F)$. Then, in this case $e_i A_i e_i = \text{span}\{e_i\}$ and

$$w_{i-1,i} A_i w_{i,i+1} = w_{i-1,i} e_i A_i e_i w_{i,i+1} = \text{span}\{w_{i-1,i} w_{i,i+1}\}.$$

If $A_i = M_{n_i}(F \oplus c_i F)$ with $c_i^2 = 1$, then $e_i A_i e_i$ is spanned by the elements e_i and $c_i e_i$. Hence

$$w_{i-1,i} A_i w_{i,i+1} = \text{span}\{w_{i-1,i} w_{i,i+1}, w_{i-1,i} c_i w_{i,i+1}\}.$$

Note that from the inequality $w_{12} w_{23} \cdots w_{m-1,m} \neq 0$ it follows that $w_{i-1,i} w_{i,i+1} \neq 0$. On the other hand, the product $w_{i-1,i} c_i w_{i,i+1}$ might be zero.

By extending this procedure for any i, j with $j - i \geq 2$, we define

$$w_{ij}(q_{i+1}, \ldots, q_{j-1}) = w_{i,i+1} q_{i+1} w_{i+1,i+2} \cdots w_{j-2,j-1} q_{j-1} w_{j-1,j}$$

where $q_r = 1$ if A_r is a simple superalgebra of type $M_k(F)$ or $M_{k,l}(F)$ and $q_r = 1$ or $q_r = c_r$ in case $A_r = M_{n_r}(F \oplus c_r F)$. For short we shall also write w_{ij} for $w_{ij}(1, \ldots, 1)$.

The next lemma gives us the structure of A_{ij} as an (A_i, A_j)-bimodule.

LEMMA 8.1.7. *Let A be a minimal superalgebra. Then, the elements*

$$w_{ij}(q_{i+1}, \ldots, q_{j-1}),$$

for all possible values of $q_{i+1}, \ldots, q_{j-1}$, generate A_{ij} as an (A_i, A_j)-bimodule. Moreover, $w_{ij} = w_{ij}(1, \ldots, 1)$ generates a non-zero irreducible graded (A_i, A_j)-bimodule.

PROOF. The first statement of the lemma is clear since

$$A_{ij} = A_i w_{i,i+1} A_{i+1} \cdots A_{j-1} w_{j-1,j} A_j$$

and $w_{i,i+1} A_{i+1} \cdots A_{j-1} w_{j-1,j}$ is spanned by the elements $w_{ij}(q_{i+1}, \ldots, q_{j-1})$, for all possible values of $q_{i+1}, \ldots, q_{j-1}$. Now, for any $w = w_{ij}(q_{i+1}, \ldots, q_{j-1}) \neq 0$, we have that $e_i w = w e_j = w$. Hence $A_i w A_j$ is a non-zero irreducible graded (A_i, A_j)-bimodule. □

8.2. Some examples

We remark that any minimal superalgebra A can be realized as a subalgebra of some block triangular matrix algebra with suitable $\mathbb{Z}_2$-grading. We do not prove this embedding here in general but we present the following significant special case.

Suppose that A is a minimal superalgebra with trivial $\mathbb{Z}_2$-grading. Then $A_{ss} = A_1 \oplus \cdots \oplus A_m$ and for all $j = 1, \ldots, m$, $A_j \cong M_{d_j}(F)$ is an ordinary matrix algebra with trivial $\mathbb{Z}_2$-grading. We claim that

$$A \cong UT(d_1, \ldots, d_m) = \begin{pmatrix} M_{d_1}(F) & & & * \\ 0 & \ddots & & \\ \vdots & & & \\ 0 & \cdots & 0 & M_{d_m}(F) \end{pmatrix},$$

an algebra of upper block triangular matrices with trivial $\mathbb{Z}_2$-grading.

In fact, first identify the algebra $A_1 \oplus \cdots \oplus A_m$ with a subalgebra of $M_q(F)$ of block-diagonal matrices where $q = d_1 + \cdots + d_m$. This is done by setting

$$q_0 = 0, q_1 = d_1, \ldots, q_k = d_1 + \cdots + d_k, \ldots, q_m = d_1 + \cdots + d_m = q$$

and A_j is the span of all matrix units $e_{\alpha\beta}$ with $q_{j-1} + 1 \leq \alpha, \beta \leq q_j$, $j = 1, \ldots, m$.

By our assumption there exist minimal idempotents $e_{\alpha_1\alpha_1}, e_{\alpha_2\alpha_2}, \ldots, e_{\alpha_m\alpha_m}$ with $q_{i-1} + 1 \leq \alpha_i \leq q_i, i = 1, \ldots, m$ and $w_{i,i+1} \in J$ with

$$e_{\alpha_i\alpha_i}w_{i,i+1} = w_{i,i+1}e_{\alpha_{i+1}\alpha_{i+1}} = w_{i,i+1}.$$

In order to prove the claim we shall define elements x_{ij} in A and we shall show that they form a basis of A with the same structural constants as $UT(d_1, \ldots, d_m)$.

If $q_{k-1} + 1 \leq i, j \leq q_k$ for some $1 \leq k \leq m$, then we define $x_{ij} = e_{ij} \in A_k$. Now let $1 \leq i < j \leq q$ and

$$q_{k-1} + 1 \leq i \leq q_k; \quad q_{k+s-1} + 1 \leq j \leq q_{k+s}$$

where $1 \leq k \leq m, s \geq 1$ and $k + s \leq m$. Then define

$$x_{ij} = e_{i\alpha_k}e_{\alpha_k\alpha_k}w_{k,k+1}\cdots w_{k+s-1,k+s}e_{\alpha_{k+s}\alpha_{k+s}}e_{\alpha_{k+s},j}.$$

By the defining properties of minimal superalgebra, $x_{ij} \neq 0$ and

(8.2) $$x_{ij}x_{lk} = \delta_{jk}x_{ik}.$$

We now verify that the x_{ij}'s are linearly independent. Indeed, if $\sum \lambda_{ij}x_{ij} = 0$, then from (8.2) it follows that for any fixed i_0, j_0 one has

$$0 = e_{i_0i_0}\left(\sum \lambda_{ij}x_{ij}\right)e_{j_0j_0} = x_{i_0i_0}\left(\sum \lambda_{ij}x_{ij}\right)x_{j_0j_0} = \lambda_{i_0j_0}x_{i_0j_0}$$

and $\lambda_{i_0j_0} = 0$.

From (8.2) it follows that $\{x_{ij}\}_{i,j}$ is a basis of A. Finally, since the x_{ij}'s satisfy (8.2), the natural map $\varphi(\sum a_{ij}x_{ij}) = \sum a_{ij}e_{ij}$ is an isomorphism of A onto $UT(d_1, \ldots, d_m)$.

The previous discussion proves the following.

THEOREM 8.2.1 ([**GZ7**]). *Let A be a finite dimensional algebra over the algebraically closed field F and let $A = A_{ss} + J$ where $A_{ss} = A_1 \oplus \cdots \oplus A_n$ with $A_i \cong M_{d_i}(F)$, $1 \leq i \leq n$. If for some $m \leq n$, $A_1JA_2J\cdots JA_m \neq 0$, then A contains a subalgebra isomorphic to $UT(d_1, \ldots, d_m)$.*

PROOF. By Lemma 8.1.4 there exists a minimal superalgebra B contained in A such that $B_{ss} = A_1 \oplus \cdots \oplus A_m$. Since the algebra B has trivial $\mathbb{Z}_2$-grading, by the above discussion $B \cong UT(d_1, \ldots, d_m)$. □

More generally, if two minimal superalgebras A and B have isomorphic maximal semisimple $\mathbb{Z}_2$-stable subalgebras with simple components of type $M_k(F)$ or $M_{k,l}(F)$, then $A \cong B$ as non-graded algebras; but they might be equipped with distinct $\mathbb{Z}_2$-gradings and might be non-isomorphic as superalgebras.

For instance, the algebra

$$UT(k+l, m) = \begin{pmatrix} M_{k+l}(F) & * \\ 0 & M_m(F) \end{pmatrix}$$

can be equipped with the following two $\mathbb{Z}_2$-gradings:

$$\left(\begin{pmatrix} M_k(F) & 0 & M_{k\times m}(F) \\ 0 & M_l(F) & 0 \\ 0 & 0 & M_m(F) \end{pmatrix}, \begin{pmatrix} 0 & M_{k\times l}(F) & 0 \\ M_{l\times k}(F) & 0 & M_{l\times m}(F) \\ 0 & 0 & 0 \end{pmatrix}\right)$$

and

$$\left(\begin{pmatrix} M_k(F) & 0 & 0 \\ 0 & M_l(F) & M_{l\times m}(F) \\ 0 & 0 & M_m(F) \end{pmatrix}, \begin{pmatrix} 0 & M_{k\times l}(F) & M_{k\times m}(F) \\ M_{l\times k}(F) & 0 & 0 \\ 0 & 0 & 0 \end{pmatrix}\right)$$

where $M_{k\times l}(F)$ is the vector space of $k \times l$ rectangular matrices over F. The Grassmann envelope of these two algebras are

$$R_1 = \begin{pmatrix} M_{k,l}(G) & M_{k\times m}(G^{(0)}) \\ & M_{l\times m}(G^{(1)}) \\ 0 & M_m(G^{(0)}) \end{pmatrix}, \quad R_2 = \begin{pmatrix} M_{k,l}(G) & M_{k\times m}(G^{(1)}) \\ & M_{l\times m}(G^{(0)}) \\ 0 & M_m(G^{(0)}) \end{pmatrix}$$

and we shall see in Theorem 8.3.4 that $\mathrm{Id}(R_1) = \mathrm{Id}(R_2) = \mathrm{Id}(M_{k,l}(G))\mathrm{Id}(M_m(F))$.

If simple superalgebras of type $M_k(F \oplus cF)$ appear, then the corresponding minimal superalgebras may not even be isomorphic in the ordinary sense. We give an example to illustrate this fact.

We identify the algebra $M_k(F \oplus cF)$, $c^2 = 1$, with the algebra of block diagonal matrices

$$\begin{pmatrix} M_k(F) & 0 \\ 0 & M_k(F) \end{pmatrix}$$

with grading

$$\left(\begin{pmatrix} A & 0 \\ 0 & A \end{pmatrix}, \begin{pmatrix} B & 0 \\ 0 & -B \end{pmatrix}\right),$$

where $A, B \in M_k(F)$.

Then set $M_k(F) = M$ and let

$$R_1 = \begin{pmatrix} M & 0 & M & M \\ 0 & M & M & M \\ 0 & 0 & M & 0 \\ 0 & 0 & 0 & M \end{pmatrix}$$

with grading

$$\left(\begin{pmatrix} A & 0 & C & D \\ 0 & A & D & C \\ 0 & 0 & B & 0 \\ 0 & 0 & 0 & B \end{pmatrix}, \begin{pmatrix} A' & 0 & C' & D' \\ 0 & -A' & -D' & -C' \\ 0 & 0 & B' & 0 \\ 0 & 0 & 0 & -B' \end{pmatrix}\right),$$

and

$$R_2 = \begin{pmatrix} M & 0 & 0 & M \\ 0 & M & M & 0 \\ 0 & 0 & M & 0 \\ 0 & 0 & 0 & M \end{pmatrix}$$

with grading

$$\left(\begin{pmatrix} A & 0 & 0 & C \\ 0 & A & C & 0 \\ 0 & 0 & B & 0 \\ 0 & 0 & 0 & B \end{pmatrix}, \begin{pmatrix} A' & 0 & 0 & D \\ 0 & -A' & -D & 0 \\ 0 & 0 & B' & 0 \\ 0 & 0 & 0 & -B' \end{pmatrix}\right),$$

where $A, B, C, D, A', B', C', D' \in M_k(F)$. Clearly $R_1 \not\cong R_2$; nevertheless, as we shall see in Theorem 8.3.4, $\mathrm{Id}(G(R_1)) = \mathrm{Id}(G(R_2))$.

8.3. The superenvelope of a minimal superalgebra

In this section we keep the notation of the previous sections. Hence F will be an algebraically closed field unless otherwise stated. Recall that by Lemma 1.3.10, if $f = f(x_1, \ldots, x_n)$ is a polynomial of $F\langle X \rangle$ and B is an F-algebra, then $f(B)$, the subspace of B generated by all valuations of f in B, is a Lie ideal of B. Also recall that if A is a $\mathbb{Z}_2$-graded algebra, $S(A) = (A^{(0)} \otimes S^{(0)}) \oplus (A^{(1)} \otimes S^{(1)})$ denotes the superenvelope of A where $S = F[U, V]$ is the free supercommutative algebra over F (see Section 3.8). Also, by Proposition 3.8.4, $S(A)$ and $G(A)$ satisfy the same identities.

LEMMA 8.3.1. *Let $A = A_1 \oplus \cdots \oplus A_m + J$ be a minimal superalgebra and suppose that $f = f(x_1, \ldots, x_n)$ is a multilinear polynomial which is not an identity of $S(A)$, the superenvelope of A.*

1) *If $m \geq 2$, then $f(S(A))$ contains the element $w_{1m} \otimes p$, for some non-zero monomial $p \in F[U, V]$.*
2) *If $m = 1$, then either $E \otimes p \in f(S(A))$ for some non-zero monomial $p \in F[U, V]$, where E is the unit element of A or, for any two minimal graded idempotents $e_1, e_2 \in A$, there exists a non-zero monomial $p \in F[U, V]$ such that $(e_1 - e_2) \otimes p \in f(S(A))$ or $(e_1 + e_2) \otimes p \in f(S(A))$.*

PROOF. Suppose first that $m = 1$, i.e., A is a simple superalgebra. Since f is not an identity of $S(A)$, there exist homogeneous elements $a_1, \ldots, a_n \in A$ and monomials $p_1 \ldots p_n \in F[U, V]$ such that $f(a_1 \otimes p_1, \ldots, a_n \otimes p_n) = a \otimes q \neq 0$ for some homogeneous elements $a \in A$, $q = p_1 \cdots p_n \in F[U, V]$.

We first prove that if a is a central element of A, then the conclusion of the lemma follows. In fact, if A is a simple superalgebra of the type $M_k(F)$ with suitable $\mathbb{Z}_2$-grading, then $a = \alpha E$, for some $\alpha \in F$ and we are done. If $A = M_k(F \oplus cF)$, then since a is homogeneous, either $a = \alpha E$ of $a = \alpha cE$. In the latter case one of the a_i's, say a_1 must be of the type $a_1 = ca_1'$ with $a_1' \in A^{(0)}$. But then $f(a_1' \otimes p_1, \ldots, a_n \otimes p_n) = \alpha E \otimes q$ and we are done also in this case.

Therefore we may assume that a in not central in A. Recall that by Lemma 1.3.10, $f(S(A))$ is a Lie ideal of $S(A)$. Suppose first that $A = M_k(F)$ is a matrix algebra with trivial $\mathbb{Z}_2$-grading. Let the e_{ij}'s be the usual matrix units and suppose that there exists a matrix $a = \sum_{i,j} a_{ij} e_{ij} \in M_k(F)$ with at least one off-diagonal non-zero entry a_{ij}. Then

$$[a \otimes p, e_{ii} \otimes 1, e_{jj} \otimes 1] = -a_{ij} e_{ij} \otimes p \in f(S(A)).$$

Now, for any $t \neq i$,

$$[e_{ij} \otimes p, e_{jt} \otimes 1] = e_{it} \otimes p \in f(S(A))$$

and we can easily deduce that $e_{rt} \otimes p \in f(S(A))$, for any $r \neq t$. It follows that

$$(e_{rr} - e_{tt}) \otimes p = [e_{rt} \otimes p, e_{tr} \otimes 1] \in f(S(A))$$

and we are done in this case. Therefore we may assume that there exists $a = \sum_i a_{ii} e_{ii} \in M_k(F)$, a non-central element, and let $i \neq j$ be such that $a_{ii} \neq a_{jj}$. Then

$$[a \otimes p, e_{ij} \otimes 1] = (a_{ii} - a_{jj}) e_{ij} \otimes p \in f(S(A))$$

and again as above we obtain that $(e_{rr} - e_{tt}) \otimes p$ lies in $f(S(A))$, for any $r \neq t$. Since $e_{11}, \ldots, e_{kk}$ are precisely all minimal graded idempotents of $M_k(F)$, the proof is complete when $A = M_k(F)$.

If $A = M_k(F \oplus cF)$, then we may clearly assume that there exists a non-central element a of $A^{(0)} = M_k(F)$. By applying the same argument as in the case $A = M_k(F)$, we complete the proof in case $A = M_k(F \oplus cF)$.

Suppose now that $A = M_{k,l}(F)$. If there exists a non-diagonal matrix $a = \sum_{i,j} a_{ij} e_{ij}$ with, say, $a_{ij} \neq 0$, then as in the case $A = M_k(F)$, we obtain $e_{ij} \otimes p \in f(S(A))$. If $t \neq j$ and $e_{jt} \in A^{(0)}$ is an even element, then

$$e_{it} \otimes p = [e_{ij} \otimes p, e_{jt} \otimes 1] \in f(S(A)).$$

If $e_{jt} \in A^{(1)}$ is odd, then we obtain

$$[e_{ij} \otimes p, e_{jt} \otimes q] = e_{it} \otimes pq \in f(S(A)),$$

for any odd $q \in S$. So, for any $r \neq s$ we can find a homogeneous element $p' \in S$ such that $e_{rs} \otimes p' \in f(S(A))$. If a is a non-central diagonal element, then, as before $a_{ii} \neq a_{jj}$ and $e_{ij} \otimes p' \in f(S(A))$, for some $i \neq j, p' \in S^{(0)} \cup S^{(1)}$. Finally, given $r \neq s$ with $e_{rs} \otimes p' \in f(S(A))$ we can take $q' \in S^{(0)} \cup S^{(1)}$ such that $p'q' \neq 0$ and then

$$[e_{rs} \otimes p', e_{sr} \otimes q'] = (e_{rr} \pm e_{ss}) \otimes p'q' \in f(S(A)),$$

where the sign $\pm$ depends on the parity of p' and q'. This completes the proof in the case $m = 1$.

Suppose now that $m \geq 2$ and we proceed by induction on m. Suppose first that f is not an identity of one of the simple summands, say A_i, $1 \leq i \leq m$. By the definition of minimal superalgebra, there exists a minimal idempotent $e_i \in A_i$ such that $e_i w_{i,i+1} = w_{i,i+1}$, when $i < m$ and $w_{m-1,m} e_m = w_{m-1,m}$ when $i = m$. By the first part of the proof there exists $a \otimes p \in f(S(A_i))$ such that either $a = E$, the unit element of A_i, or $a = e_i - e_i'$ for some idempotent $e_i' \in A_i$ orthogonal to e_i. In any case, for $i < m$ we have

$$a w_{i,i+1} \cdots w_{m-1,m} = w_{i,i+1} \cdots w_{m-1,m},$$

$$w_{i,i+1} \cdots w_{m-1,m} a = 0$$

and for $i = m$ we have

$$w_{1m} a = w_{12} \cdots w_{m-1,m} a = w_{12} \cdots w_{m-1,m} = w_{1m},$$

$$a w_{1m} = a w_{12} \cdots w_{m-1,m} = 0.$$

Let $p', p'' \in S$ be such that $p' p p'' \neq 0$. Then, since $f(S(A_i)) \subseteq f(S(A))$ and $f(S(A))$ is a Lie ideal of $S(A)$, it follows that

$$w_{1m} \otimes p' p p'' = [w_{12} \cdots w_{i-1,i} \otimes p', [a \otimes p, w_{i,i+1} \cdots w_{m-1,m} \otimes p'']] \in f(S(A))$$

when $i < m$ and $w_{1m} \otimes p'p = [a \otimes p, w_{1m} \otimes p'] \in f(S(A))$ when $i = m$, proving the lemma.

Therefore we may assume that f is an identity for $A_{ss} = A_1 \oplus \cdots \oplus A_m$. Consider the following two subalgebras of A,

$$B = \bigoplus_{1 \leq i \leq j \leq m-1} A_{ij}, \quad C = \bigoplus_{2 \leq i \leq j \leq m} A_{ij}.$$

Now, if f is not an identity of $S(B)$, then by induction either $w_{1,m-1} \otimes p \in f(S(B))$ for some $p \neq 0$ or $E_B \otimes p \in f(S(B))$ where E_B is the unit element of B or

$(e_1 - e_2) \otimes p \in f(S(B))$ for any two minimal graded idempotents of B. Thus in the first case, if $p' \in S$ is such that $pp' \neq 0$, we get
$$w_{1m} \otimes pp' = [w_{1,m-1} \otimes p, w_{m-1,m} \otimes p'] \in f(S(A))$$
and we are done. In the second case, $aw_{12} = w_{12}$ and $w_{12}a = 0$ for $a = E_B$ or $a = e_1 - e_2$. This implies that
$$w_{1m} \otimes pp'p'' = [[a \otimes p, w_{12} \otimes p'], w_{23} \cdots w_{m-1,m} \otimes p''] \in f(S(A))$$
and we are also done if we take $p', p'' \in S$ such that $pp'p'' \neq 0$. Similarly, if f is not an identity for $S(C)$, we also get $w_{1m} \otimes \bar{p} \in f(S(A))$ for some $\bar{p} \in S$.

Therefore, we may assume that f is an identity of $S(B)$ and $S(C)$. In this case all non-zero values of f belong to $S(A_{1m})$. Thus there exist homogeneous elements $u_1, \ldots, u_n \in A$, $p_1, \ldots, p_n \in S$ such that
$$(8.3) \qquad f(u_1 \otimes p_1, \ldots, u_n \otimes p_n) = u \otimes p \neq 0$$
and by Lemma 8.1.7 we have that $u = aw_{1m}(q_1, \ldots, q_m)b$ for some $q_1, \ldots, q_m$ and for some $a \in A_1, b \in A_m$. Moreover, some of the elements $u_1, \ldots, u_n$, say $u_1, \ldots, u_k$, belong to $J = \oplus_{i<j} A_{ij}$ and $u_{k+1}, \ldots, u_n \in A_{ss}$. Let
$$u_1 = a_{i_1} w_{i_1 j_1}(q^{(1)}_{i_1+1}, \ldots, q^{(1)}_{j_1-1}) b_{j_1}$$
$$\vdots$$
$$u_k = a_{i_k} w_{i_k j_k}(q^{(k)}_{i_k+1}, \ldots, q^{(k)}_{j_k-1}) b_{j_k}$$
where $a_{i_1} \in A_{i_1}, b_{j_1} \in A_{j_1}, \ldots, a_{i_k} \in A_{i_k}, b_{j_k} \in A_{j_k}$.

Now replace in (8.3) the element u_1 with $\bar{u}_1 = a_{i_1} w_{i_1 j_1}(1, \ldots, 1) b_{j_1}, \ldots$, the element u_k with $\bar{u}_k = a_{i_k} w_{i_k j_k}(1, \ldots, 1) b_{j_k}$. Then, by the multiplication rules of A we obtain
$$f(\bar{u}_1 \otimes p_1, \ldots, \bar{u}_k \otimes p_k, u_{k+1} \otimes p_{k+1}, \ldots, u_n \otimes p_n) = \bar{u} \otimes p$$
where $\bar{u} = aw_{1m}(1, \ldots, 1)b = aw_{1m}b \neq 0$.

Recall that by the choice of w_{ij} there exist minimal graded idempotents $e_1 \in A_1, e_m \in A_m$ such that $e_1 w_{1m} = w_{1m} e_m = w_{1m}$. Hence since $e_1 \in A_1 a e_1$ and $e_m \in e_m b A_m$, by taking $p', p'' \in S$ such that $p'pp'' \neq 0$, we get
$$0 \neq w_{1m} \otimes p'pp'' \in A_1 a e_1 w_{1m} e_m b A_m \otimes p'pp''$$
$$= [[A_1 \otimes p', aw_{1m}b \otimes p], A_m \otimes p''] \subseteq f(S(A)).$$
This completes the proof of the lemma. $\square$

LEMMA 8.3.2. *If $A = A_1 \oplus \cdots \oplus A_m + J$ is a minimal superalgebra, then $\mathrm{Id}(G(A_1)) \cdots \mathrm{Id}(G(A_m)) \subseteq \mathrm{Id}(G(A))$.*

PROOF. Let $f_1 \in \mathrm{Id}(G(A_1)), \ldots, f_m \in \mathrm{Id}(G(A_m))$ and $f = f_1 \cdots f_m$. If $m = 1$, the conclusion of the lemma is clear. Suppose $m > 1$ and let A' be the subalgebra of A generated by $A_1, \ldots, A_{m-1}$ and $w_{12}, \ldots, w_{m-2,m-1}$. Then, by the inductive hypotheses, $f_1 \cdots f_{m-1}(G(A')) = 0$ and this says that $f_1 \cdots f_{m-1}(G(A)) \subseteq G(A_{1m}) \oplus \cdots \oplus G(A_{mm})$.

On the other hand, $f_m(G(A)) \subseteq G(A_1) \oplus \cdots \oplus G(A_{m-1}) + G(J)$. Thus $f(G(A)) \subseteq (G(A_{1m}) \oplus \cdots \oplus G(A_{mm}))(G(A_1) \oplus \cdots \oplus G(A_{m-1}) + G(J)) = 0$ since $A_{im} J = A_{im} A_t = 0$ for all $i = 1, \ldots, m$, $t = 1, \ldots, m-1$. $\square$

LEMMA 8.3.3. *Let $A = A_1 \oplus \cdots \oplus A_m + J$ be a minimal superalgebra. Then $S(A)$ contains a relatively free algebra of the variety determined by the T-ideal $\mathrm{Id}(G(A_1)) \cdots \mathrm{Id}(G(A_m))$.*

PROOF. We proceed by induction on m. If $m = 1$, the result follows from Proposition 3.8.5. Let $m > 1$ and partition each of the sets U and V into three disjoint countable subsets

$$U = \widetilde{U} \cup \overline{U} \cup \widehat{U}, \quad V = \widetilde{V} \cup \overline{V} \cup \widehat{V}.$$

Hence the free supercommutative algebra $S = F[U, V]$ contains three isomorphic copies $\widetilde{S} = F[\widetilde{U}, \widetilde{V}]$, $\overline{S} = F[\overline{U}, \overline{V}]$ and $\widehat{S} = F[\widehat{U}, \widehat{V}]$.

Let C be the subalgebra of A generated by $A_1, \ldots, A_{m-1}$ and by the elements $w_{12}, \ldots, w_{m-2,m-1}$. Write $C = C^{(0)} \oplus C^{(1)}$ and fix a homogeneous basis $\{c_1^{(0)}, \ldots, c_{r_1}^{(0)}\}$ of $C^{(0)}$ and $\{c_1^{(1)}, \ldots, c_{s_1}^{(1)}\}$ of $C^{(1)}$.

By the inductive hypothesis the algebra $\widetilde{S}(C)$ contains a relatively free algebra $\widetilde{C}$ of the variety determined by the T-ideal $\mathrm{Id}(G(A_1)) \cdots \mathrm{Id}(G(A_{m-1})) = \mathrm{Id}(S(A_1)) \cdots \mathrm{Id}(S(A_{m-1}))$. But, since C is a minimal superalgebra, by Lemma 8.3.2

$$\mathrm{Id}(\widetilde{C}) = \mathrm{Id}(S(A_1)) \cdots \mathrm{Id}(S(A_{m-1})) \subseteq \mathrm{Id}(G(C)),$$

and, by Proposition 3.8.4, $\mathrm{Id}(G(C)) = \mathrm{Id}(\widetilde{S}(C))$. It follows that $\mathrm{Id}(\widetilde{C}) = \mathrm{Id}(\widetilde{S}(C))$, i.e., $\widetilde{C}$ is the relatively free algebra of the variety $\mathrm{var}(G(C))$.

Now, for $k = 1, 2, \ldots$, set

$$X_k = \sum_j c_j^{(0)} \otimes \widetilde{u}_{kj} + \sum_j c_j^{(1)} \otimes \widetilde{v}_{kj}$$

where $\widetilde{u}_{kj} \in \widetilde{U}$, $\widetilde{v}_{kj} \in \widetilde{V}$ are distinct elements. Then by Proposition 3.8.5, we can suppose that $\widetilde{C} = F\{X_1, X_2, \ldots\}$ and $X_1, X_2, \ldots$ are free generators of the algebra $\widetilde{C}$.

Write $A_m = B = B^{(0)} \oplus B^{(1)}$ and take $\{b_1^{(0)}, \ldots, b_{r_2}^{(0)}\}$ as a basis of $B^{(0)}$ and $\{b_1^{(1)}, \ldots, b_{s_2}^{(1)}\}$ as a basis of $B^{(1)}$. For $k = 1, 2, \ldots$ set

$$Y_k = \sum_j b_j^{(0)} \otimes \widehat{u}_{kj} + \sum_j b_j^{(1)} \otimes \widehat{v}_{kj}$$

with $\widehat{u}_{kj} \in \widehat{U}, \widehat{v}_{kj} \in \widehat{V}$ distinct.

If $\widehat{B}$ denotes the relatively free algebra of the variety $\mathrm{var}(G(A_m))$, then by Proposition 3.8.5, $\widehat{B} = F\{Y_1, Y_2, \ldots\}$ and $Y_1, Y_2, \ldots$ are free generators of $\widehat{B}$ over F. Notice that $\widehat{B} \subseteq \widehat{S}(B)$ and $\widetilde{C} \subseteq \widetilde{S}(C)$.

Now consider the subspace $A_{m-1} w_{m-1,m} A_m = A_{m-1,m}$ and let

$$d_1^{(0)}, \ldots, d_{r_3}^{(0)} \in A_{m-1,m}^{(0)}, \quad d_1^{(1)}, \ldots, d_{s_3}^{(1)} \in A_{m-1,m}^{(1)}$$

be a homogeneous basis. For $k = 1, 2, \ldots$, set

$$Z_k = \sum_j d_j^{(0)} \otimes \bar{u}_{kj} + \sum_j d_j^{(1)} \otimes \bar{v}_{kj}$$

with $\bar{u}_{kj} \in \bar{U}, \bar{v}_{kj} \in \bar{V}$ distinct.

We prove next that the elements $Z_1, Z_2, \ldots$ generate a free $(\widetilde{C}, \widehat{B})$-bimodule in $S(A)$.

Suppose to the contrary that there exists a non-trivial relation

(8.4) $$\sum_{i,j} a_{ij} Z_j b_{ij} = 0$$

where $a_{ij} \in \widetilde{C}$, $b_{ij} \in \widehat{B}$. Since any two distinct Z_i and Z_j depend on distinct sets of variables $\bar{u}_{kj}$ and $\bar{v}_{kj}$, from (8.4) we get a non-trivial relation of the type

(8.5) $$a_1 Z_1 b_1 + \cdots + a_n Z_1 b_n = 0$$

where the elements $a_1, \ldots, a_n \in \widetilde{C}$ are linearly independent and $b_1, \ldots, b_n \in \widehat{B}$ are all non-zero.

Recall that the elements $b_1, \ldots, b_n$ in the relation (8.5) are polynomials in $Y_1, Y_2, \ldots$ and next we claim that we may assume that these polynomials are multilinear.

To see this, suppose that $b_1, \ldots, b_n$ are polynomials in $Y_1, \ldots, Y_k$ and pick $\gamma_1, \ldots, \gamma_k \in F$. Define a map $\varphi : U \cup V \to S$ by setting $\varphi(\widehat{u}_{ij}) = \gamma_i \widehat{u}_{ij}$, $\varphi(\widehat{v}_{ij}) = \gamma_i \widehat{v}_{ij}$ for all $i = 1, \ldots, k$ and for all $j \geq 1$ and φ is the identity map on the remaining elements of $U \cup V$. Then φ extends to a homomorphism $\bar{\varphi} : A \otimes S \to A \otimes S$ such that

$$\bar{\varphi}(Y_i) = \begin{cases} \gamma_i Y_i & \text{for } 1 \leq i \leq k, \\ Y_i & \text{for } i \geq k+1, \end{cases}$$

$$\bar{\varphi}(X_i) = X_i, \quad \bar{\varphi}(Z_i) = Z_i, \quad \text{for } i \geq 1,$$

and $\bar{\varphi}$ is the identity map on A. Since F is infinite, a Vandermonde determinant argument shows that we may take $b_1, \ldots, b_n$ to be homogeneous in each variable Y_i.

In order to make $b_1, \ldots, b_n$ multilinear, we fix Y_i, $1 \leq i \leq k$ and we construct an endomorphism $\psi : A \otimes S \to A \otimes S$, which is fixed on A, such that

$$\psi(Y_i) = Y_i + Y_{k+1}, \quad \psi(Y_j) = Y_j, \text{ for } j \neq i,$$

and

$$\psi(X_j) = X_j, \quad \psi(Z_j) = Z_j, \quad \text{for all } j \geq 1.$$

Now in each b_i appearing in (8.5) we replace Y_i with $Y_i + Y_{k+1}$. A standard multilinearization process leads to the linearity of the polynomials $b_1, \ldots, b_n$ as claimed.

We have proved that in relation (8.5) we may take $b_i = b_i(Y_1, \ldots, Y_k)$, $i = 1, \ldots, n$, to be non-zero multilinear polynomials in $Y_1, \ldots, Y_k$.

Let $e_m \in A_m = B$ be the minimal graded idempotent such that $w_{m-1,m} e_m = w_{m-1,m}$. We know, by Lemma 8.1.1, that $e_m = e_{\alpha\alpha} \in B^{(0)}$ is a diagonal matrix unit of the simple algebra B. By Lemma 8.3.1, there exists a non-zero monomial $p \in \widehat{F}[\widehat{U}, \widehat{V}]$ such that either $E \otimes p \in b_1(S(B))$ or $(e_{\alpha\alpha} - e_{\beta\beta}) \otimes p \in b_1(S(B))$ for any $\beta \neq \alpha$. Hence there exists a valuation $\varphi : Y_i \to h_i \otimes p_i$, $i = 1, \ldots, k$, where $h_i \in B^{(0)} \cup B^{(1)}$ are homogeneous elements and $p_i \in \widehat{S}$ are monomials such that

(8.6) $$\varphi(b_j) = (\sum_{i,l} \gamma_{il}^j e_{il}) \otimes p$$

with $\gamma_{\alpha\alpha}^1 \neq 0$. Now, notice that since $a_1, \ldots, a_n \in \widetilde{C}$ are linearly independent and $\gamma_{\alpha\alpha}^1 \neq 0$, the polynomial

(8.7) $$f = f(X_1, \ldots, X_t) = \sum_j \delta_{\alpha\alpha}^j a_j$$

is not an identity of $\widetilde{S}(C)$.

Since the elements X_i, Y_i and Z_i depend on disjoint sets of indeterminates, we obtain that for any evaluation $\eta : \widetilde{C} \to \widetilde{S}(C)$ there exists a homomorphism $\theta : A \otimes S \to A \otimes S$ which is the identity on A, such that

$$\theta(X_i) = \eta(X_i), \ i = 1, 2, \ldots$$
$$\theta(Z_1) = w_{m-1,m} \otimes p'$$

where $p' = 1$ if $w_{m-1,m}$ is even, $p' \in \overline{V}$ if $w_{m-1,m}$ is odd and θ takes arbitrary values on $Y_1, \ldots, Y_k$ in $\widehat{S}(B)$. Let

$$\theta(b_j) = (\sum_{i,l} \gamma_{il}^j e_{il}) \otimes p, \ j = 1, \ldots, n.$$

Then from (8.5) it follows that

$$0 = \theta(a_1 Z_1 b_1 + \cdots + a_n Z_1 b_n) = \sum_j \theta(a_j)(w_{m-1,m} \otimes p')((\sum_{i,l} \gamma_{il}^j e_{il}) \otimes p).$$

Since $w_{m-1,m} = w_{m-1,m} e_{\alpha\alpha}$, we obtain

$$0 = \sum_j \theta(a_j)(w_{m-1,m} \otimes p')((\sum_l \gamma_{\alpha l}^j e_{\alpha l}) \otimes p).$$

We now multiply the above relation by $e_{\alpha\alpha} \otimes 1$ on the right and we get

$$0 = \sum_j \theta(a_j)(w_{m-1,m} \otimes p')((\gamma_{\alpha\alpha}^j e_{\alpha\alpha}) \otimes p) = \sum_j \gamma_{\alpha\alpha}^j \theta(a_j) w_{m-1,m} \otimes p'p$$

(8.8) $$= \theta(\sum_j \gamma_{\alpha\alpha}^j a_j) w_{m-1,m} \otimes p'p = \eta(\sum_j \gamma_{\alpha\alpha}^j a_j) w_{m-1,m} \otimes p'p.$$

Since $\eta : \widetilde{C} \to \widetilde{S}(C)$ is an arbitrary evaluation and recalling the definition of f given in (8.7), from (8.8) and (8.6) it follows that

$$f(\widetilde{S}(C)) \cdot w_{m-1,m} \otimes p'p = 0.$$

If $m > 2$, then, since f is not an identity of $\widetilde{S}(C)$, by Lemma 8.3.1, $f(\widetilde{S}(C))$ contains the element $w_{1,m-1} \otimes p''$ for some monomial $p'' \in \widetilde{S}$. Thus

$$0 \neq w_{1,m} \otimes p''p'p = (w_{1,m-1} \otimes p'')(w_{m-1,m} \otimes p'p) \in f(\widetilde{S}(C)) \cdot w_{m-1,m} \otimes p'p$$

and this is a contradiction.

If $m = 2$, there exists a minimal idempotent $e_1 \in A$ such that $e_1 w_{1,2} = w_{1,2}$ and, by Lemma 8.3.1, either $E_1 \otimes p'' \in f(\widetilde{S}(C))$ or $(e_1 - e_1') \otimes p'' \in f(\widetilde{S}(C))$ where E_1 is the unit element of A_1 and $e_1' \neq e_1$ is a minimal idempotent of A_1. We get

$$0 \neq w_{1,2} \otimes p''p'p \in f(\widetilde{S}(C)) w_{1,2} \otimes p'p,$$

a contradiction.

We have proved that the elements $Z_1, Z_2, \ldots$ generate a free $(\widetilde{C}, \widehat{B})$-bimodule in $S(A)$. By Corollary 1.8.2, it follows that the subalgebra of $S(A)$ generated by $X_i + Y_i + Z_i$ $i = 1, 2, \ldots$, is the relatively free algebra determined by the T-ideal $\mathrm{Id}(S(C))\mathrm{Id}(S(B)) = \mathrm{Id}(S(A_1)) \cdots \mathrm{Id}(S(A_m))$. By Proposition 3.8.4, the proof of the lemma is complete. $\square$

Let A be a minimal superalgebra. Since by Proposition 3.8.4, $G(A)$ and $S(A)$ satisfy the same identities, by combining Lemma 8.3.2 and Lemma 8.3.3 we obtain that $\mathrm{Id}(G(A)) = \mathrm{Id}(G(A_1)) \cdots \mathrm{Id}(G(A_m))$. We note this in the following.

THEOREM 8.3.4 ([**GZ9**]). *If $A = A_1 \oplus \cdots \oplus A_m + J$ is a minimal superalgebra, then* $\mathrm{Id}(G(A)) = \mathrm{Id}(G(A_1)) \cdots \mathrm{Id}(G(A_m))$.

Consider two minimal superalgebras $A = A_1 \oplus \cdots \oplus A_m + J$ and $B = B_1 \oplus \cdots \oplus B_m + J'$ where $A_1, \ldots, A_m, B_1, \ldots, B_m$ are simple superalgebras and J and J' are the Jacobson radical of A and B, respectively. If $A_i \cong B_i$, $i = 1, \ldots, m$ and the $\mathbb{Z}_2$-grading on A and B is trivial, then $A \cong B$, as it was remarked in Section 2. In general the isomorphisms $A_i \cong B_i$, $i = 1, \ldots, m$, do not imply that A and B are isomorphic as superalgebras. Nevertheless, by the previous lemma $G(A)$ and $G(B)$ generare the same variety.

COROLLARY 8.3.5. *Let $A = A_1 \oplus \cdots \oplus A_m + J$ and $B = B_1 \oplus \cdots \oplus B_m + J'$ be two minimal superalgebras. If $A_i \cong B_i$, for all $i = 1, \ldots, m$, then $\mathrm{Id}(G(A)) = \mathrm{Id}(G(B))$.*

In particular, if $A = A_1 \oplus \cdots \oplus A_m + J$ and $A_1, \ldots, A_m$ are matrix algebras with trivial $\mathbb{Z}_2$-grading, then the grading on J may also be taken to be trivial and we immediately obtain (cfr. Section 8.2).

COROLLARY 8.3.6. *Let $A = A_1 \oplus \cdots \oplus A_m + J$ be a minimal superalgebra such that $A_1 \cong M_{d_1}(F), \ldots, A_m \cong M_{d_m}(F)$ are simple superalgebras with trivial $\mathbb{Z}_2$-grading. Then $\mathrm{Id}(G(A)) = \mathrm{Id}(UT(d_1, \ldots, d_m))$.*

8.4. Products of verbally prime T-ideals

Throughout this section F will be an arbitrary field of characteristic zero, not necessarily algebraically closed.

DEFINITION 8.4.1. Let $\mathcal{V}$ be a variety of algebras. We say that $\mathcal{V}$ is minimal of exponent $d \geq 2$ if $\exp(\mathcal{V}) = d$ and for every proper subvariety $\mathcal{U} \subset \mathcal{V}$ we have that $\exp(\mathcal{U}) < d$.

As a consequence of the results in Section 8.1 and Section 8.3 we first obtain a characterization of minimal varieties in terms of generating algebras and T-ideals. In fact, we have

THEOREM 8.4.2. *If a variety $\mathcal{V}$ is minimal of exponent $d \geq 2$, then $\mathcal{V} = \mathrm{var}(G(A))$ for some minimal superalgebra A. Hence there exist verbally prime T-ideals $I_1, \ldots, I_m$ such that $\mathrm{Id}(\mathcal{V}) = I_1 \cdots I_m$.*

PROOF. Let B be an algebra such that $\mathcal{V} = \mathrm{var}(B)$. If $\overline{F}$ is the algebraic closure of the field F we form $\overline{B} = B \otimes_F \overline{F}$. Since by Theorem 4.1.9, the nth codimension $c_n(B)$ of B over F coincides with the nth codimension $c_n(\overline{B})$ of $\overline{B}$ over $\overline{F}$, it follows that $\exp(\overline{B}) = d$ over $\overline{F}$. By Lemma 8.1.5, $\mathrm{var}(\overline{B})$ contains the Grassmann envelope $G(A)$ of a minimal superalgebra A and $\exp(\overline{B}) = \exp(G(A))$. Since $\overline{B}$ as an F-algebra belongs to $\mathcal{V}$, it follows that $G(A) \in \mathcal{V}$. By the minimality of $\mathcal{V}$ we get that $\mathcal{V} = \mathrm{var}(G(A))$. Hence $\mathrm{Id}(\mathcal{V}) = \mathrm{Id}(G(A))$ and by Theorem 8.3.4, $\mathrm{Id}(\mathcal{V}) = I_1 \cdots I_m$ with $I_k = \mathrm{Id}(G(A_k))$, $k = 1, \ldots, m$, a verbally prime T-ideal since A_k is a simple superalgebra. □

The previous result has a converse as the following theorem shows.

THEOREM 8.4.3. *For any finite dimensional simple superalgebras $A_1, \ldots, A_m$, there exists a minimal superalgebra A such that $A_{ss} = A_1 \oplus \cdots \oplus A_m$. Hence, if $I_1, \ldots, I_m$ are proper verbally prime T-ideals of $F\langle X \rangle$, then there exists a minimal superalgebra A such that $\mathrm{Id}(G(A)) = I_1 \cdots I_m$.*

PROOF. Let $A_1, \ldots, A_m$ be finite dimensional simple superalgebras such that $I_j = \mathrm{Id}(G(A_j))$, $j = 1, \ldots, m$. For $j = 1, \ldots, m$, we define subsets Q_j as follows: if $A_j = M_k(F)$ with suitable $\mathbb{Z}_2$-grading, then we set $Q_j = \{1_j\}$ where 1_j is the unit element of A_j; if $A_j = M_{n_j}(F \oplus c_j F)$ with $c_j^2 = 1$, we set $Q_j = \{1_j, c_j\}$ where 1_j is the unit element of $M_{n_j}(F)$.

Let us now denote by $e_{ij}^{(\alpha)}$ the matrix units of the algebra A_α when $A_\alpha = M_k(F)$. If $A_\alpha = M_k(F \oplus cF)$, then we let $e_{ij}^{(\alpha)}$ be the matrix units of $A_\alpha^{(0)}$.

Now let $A_{i,i+1} \cong A_i e_{11}^{(i)} \otimes e_{11}^{(i+1)} A_{i+1}$ be the $\mathbb{Z}_2$-graded irreducible (A_i, A_{i+1})-bimodule with even generator $w_{i,i+1}$ satisfying the equalities

$$w_{i,i+1} = e_{11}^{(i)} w_{i,i+1} = w_{i,i+1} e_{11}^{(i+1)}.$$

More generally, for any $j \geq i+2$ and for any $q_{i+1} \in Q_{i+1}, \ldots, q_{j-1} \in Q_{j-1}$, let $A_{ij}(q_{i+1}, \ldots, q_{j-1})$ be the $\mathbb{Z}_2$-graded irreducible (A_i, A_j)-bimodule with homogeneous generator $w_{i,j}(q_{i+1}, \ldots, q_{j-1})$ satisfying the equalities

$$(8.9) \quad w_{ij}(q_{i+1}, \ldots, q_{j-1}) = e_{11}^{(i)} w_{ij}(q_{i+1}, \ldots, q_{j-1}) = w_{ij}(q_{i+1}, \ldots, q_{j-1}) e_{11}^{(j)}.$$

The degree of the element $w_{ij}(q_{i+1}, \ldots, q_{j-1})$ in the $\mathbb{Z}_2$-grading is a function of $q_{i+1}, \ldots, q_{j-1}$: $w_{ij}(q_{i+1}, \ldots, q_{j-1})$ is even or odd accordingly as the number of $q_\alpha = t_\alpha \in Q_\alpha$, $i+1 \leq \alpha \leq j-1$, is even or odd, respectively.

Now define $A_{11} = A_1, \ldots, A_{mm} = A_m$ and

$$A_{ij} = \bigoplus_{q_{i+1} \in Q_{i+1}, \ldots, q_{j-1} \in Q_{j-1}} A_{ij}(q_{i+1}, \ldots, q_{j-1}).$$

Then set

$$A = \bigoplus_{1 \leq i \leq j \leq m} A_{ij}.$$

We make A into an algebra by defining the following multiplication. If $a, b \in A_{ii}$ for some i, then ab is the ordinary product in the algebra A_{ii}. For any $1 \leq i \leq j \leq m$, $1 \leq k \leq l \leq m$ with $j \neq k$ we set $A_{ij} A_{kl} = 0$. We now define a multiplication ab where $a \in A_{ij}, b \in A_{jk}$ as follows: from (8.9) it follows that the elements

$$q_i e_{\alpha 1}^{(i)} w_{ij}(q_{i+1}, \ldots, q_{j-1}) q_j e_{1\beta}^{(j)}$$

with $q_i \in Q_i, \ldots, q_j \in Q_j$, $1 \leq \alpha \leq n_i$, $1 \leq \beta \leq n_j$, form a linear basis of A_{ij}. Denote for short $\bar{q}_{ij} = (q_{i+1}, \ldots, q_{j-1})$ and define

$$q_i e_{\alpha 1}^{(i)} w_{ij}(\bar{q}_{ij}) q_j e_{1\beta}^{(j)} \cdot q'_j e_{\gamma 1}^{(j)} w_{jk}(\bar{q}'_{jk}) q'_k e_{1\delta}^{(k)}$$
$$= \begin{cases} q_i e_{\alpha 1}^{(i)} w_{ik}(\bar{q}_{ij}, q_j q'_j, \bar{q}'_{jk}) q'_k e_{1\delta}^{(k)}, & \text{if } \beta = \gamma, \\ 0, & \text{otherwise.} \end{cases}$$

With this definition A becomes an associative finite dimensional superalgebra with maximal semisimple subalgebra $A_1 \oplus \cdots \oplus A_m$ and Jacobson radical

$$J = \oplus_{1 \leq i < j \leq m} A_{ij}.$$

Clearly A is a minimal superalgebra and, by Theorem 8.3.4 we have $\mathrm{Id}(G(A)) = I_1 \cdots I_m$. □

In the next theorem we generalize the defining property of a verbally prime T-ideal to products of verbally prime T-ideals.

THEOREM 8.4.4. *Let $I_1, \ldots, I_m$ be verbally prime T-ideals and set $I = I_1 \cdots I_m$. If P, Q are T-ideals such that $PQ \subseteq I$, then either $P \subseteq I$ or $Q \subseteq I$ or there exists $1 \leq k \leq m-1$ such that $P \subseteq I_1 \cdots I_k$ and $Q \subseteq I_{k+1} \cdots I_m$.*

PROOF. Suppose that $P \not\subseteq I$ and $Q \not\subseteq I$. Since I is the product of verbally prime T-ideals, by Theorem 8.4.3 there exists a minimal superalgebra A such that $I = \mathrm{Id}(G(A))$. Moreover, if $A = A_{ss} + J$ with $A_{ss} = A_1 \oplus \cdots \oplus A_m$ maximal semisimple $\mathbb{Z}_2$-stable subalgebra, then $I_k = \mathrm{Id}(G(A_k))$ for $k = 1, \ldots, m$.

Since $P \not\subseteq I$, there exists a smallest integer $k \in \{1, \ldots, m\}$ such that $P \not\subseteq I_1 \cdots I_k$. If $Q \subseteq I_k \cdots I_m$, then we are done. Therefore, we may assume that $Q \not\subseteq I_k \cdots I_m$.

Let $f \in P \setminus I_1 \cdots I_k$ and $g \in Q \setminus I_k \cdots I_m$ be multilinear polynomials in disjoint sets of variables. Since $fg \in PQ \subseteq I$, we shall reach a contradiction by proving that $fg \notin I_1 \cdots I_m = I = \mathrm{Id}(G(A))$.

Clearly $m \geq 2$. Suppose first that $k \neq 1, m-1$. Let B be the subalgebra of A generated by $A_1, \ldots, A_k, w_{12}, \ldots, w_{k-1,k}$ and let C be the subalgebra generated by $A_k, \ldots, A_m, w_{k,k+1}, \ldots, w_{m-1,m}$. Since B and C are both minimal superalgebras, by Lemma 8.3.1, there exist monomials $p, q \in S = F[U, V]$ such that

$$w_{1k} \otimes p \in f(S(B)) \quad \text{and} \quad w_{km} \otimes q \in g(S(C)).$$

Since we may clearly assume that $pq \neq 0$, then $f(S(B))f(S(C)) \neq 0$, i.e., fg is not an identity of $S(A)$. Since by Proposition 3.8.4, $S(A)$ and $G(A)$ have the same identities, we obtain that $fg \notin \mathrm{Id}(G(A))$ and we are done in this case.

Suppose now that $k = 1$. Then as above, invoking Lemma 8.3.1, there exist monomials $p, q \in S = F[U, V]$ such that

$$w_{1m} \otimes q \in g(S(A))$$

and either $(e_1 - e_1') \otimes p \in f(S(A_1))$ where $e_1 w_{1m} = w_{1m}, e_1' w_{1m} = 0$ or $E \otimes p \in f(S(A_1))$ with E the unit element of A_1. Again, $f(S(A_1))g(S(A)) \neq 0$ and fg is not an identity of $S(A)$. The case $k = m$ is similar. □

The argument of the previous theorem can be applied in order to prove the following.

PROPOSITION 8.4.5. *Let $I_1, \ldots, I_m$ be proper verbally prime T-ideals and $m \geq 2$. If for some T-ideals P, Q we have $[P, Q] \subseteq I_1 \cdots I_m$, then $PQ + QP \subseteq I_1 \cdots I_m$.*

PROOF. If either P or Q is contained in $I = I_1 \cdots I_m$, then the conclusion is clearly true. Hence we may assume that $P \not\subseteq I$ and $Q \not\subseteq I$. But then by the same arguments as in the previous theorem we obtain that $PQ \subseteq I$ and $QP \subseteq I$ as desired. □

8.5. Classifying minimal varieties of exponential growth

In this section we shall classify all minimal varieties of algebras of exponent ≥ 2. Our aim is to prove that a variety $\mathcal{V}$ is minimal if and only if its T-ideal of identities $\mathrm{Id}(\mathcal{V})$ is a product of verbally prime T-ideals. This result was conjectured by Drensky in [**D4**]. Partial results were obtained in [**D4**], [**D5**] and [**SV**].

From Chapter 7 it is clear that the Grassmann algebra G and the algebra of 2×2 upper triangular matrices UT_2 generate the only two minimal varieties of exponent 2.

Recall that we have already proved in Theorem 8.4.2 that if $\mathcal{V}$ is a minimal variety of exponent ≥ 2, then there exist verbally prime T-ideals $I_1, \ldots, I_m$ such that $\mathrm{Id}(\mathcal{V}) = I_1 \cdots I_m$. In order to prove the other direction we shall next compare products of verbally prime T-ideals.

Recall that in case I is a T-ideal and $I = \mathrm{Id}(\mathcal{V})$ for some variety of algebras $\mathcal{V}$, then $\exp(I) = \exp(\mathcal{V})$. We start by comparing the exponent of verbally prime T-ideals.

LEMMA 8.5.1. *Let $I \subseteq Q$ be two verbally prime T-ideals. Then $\exp(I) = \exp(Q)$ if and only if $I = Q$.*

PROOF. Let $I = \mathrm{Id}(G(A))$, $Q = \mathrm{Id}(G(B))$ where $A = A^{(0)} \oplus A^{(1)}$, $B = B^{(0)} \oplus B^{(1)}$ are finite dimensional simple superalgebras. Denote $d = \dim A^{(0)}, l = \dim A^{(1)}, s = \dim B^{(0)}, r = \dim B^{(1)}$. Then, by the characterization of the exponent, $\exp(I) = d + l$ and $\exp(Q) = s + r$. Also by Theorem 4.8.7,

$$\chi_n(G(A)) \subseteq H(l, d) \quad \text{and} \quad \chi_n(G(B)) \subseteq H(r, s).$$

Since $I \subseteq Q$, then $\chi_n(G(B)) \subseteq \chi_n(G(A)) \subseteq H(l, d)$ and, by Lemma 6.3.3, this is possible only if $r \leq l$ and $s \leq d$.

Suppose that $\exp(I) = \exp(Q)$. Then we obtain $r = l$ and $s = d$, i.e., $\dim A^{(0)} = \dim B^{(0)}, \dim A^{(1)} = \dim B^{(1)}$. Now if, say, $A = M_n(F)$ with trivial $\mathbb{Z}_2$-grading, then also $B^{(1)} = 0$ and $B \cong M_n(F)$ with trivial grading. Thus $A \cong B$ in this case. If $A = M_n(F \oplus cF)$, then $\dim A = 2n^2$. Since all simple superalgebras of type $M_k(F)$ have order a square, it follows that B must also be of type $M_k(F \oplus cF)$; hence $A \cong B$ in this case also. Finally, if $A = M_{p,q}(F)$, then B must also be of this type. Say $B = M_{a,b}(F)$. Since $\dim A^{(0)} = p^2 + q^2 = a^2 + b^2 = \dim B^{(0)}$ and $\dim A^{(1)} = 2pq = 2ab = \dim B^{(1)}$, we obtain $a = p, b = q$ and $A \cong B$ follows. This completes the proof of the lemma. □

In the next lemma we analyze the more general situation when a verbally prime T-ideal is contained in a product of such T-ideals.

LEMMA 8.5.2. *Let $m \geq 2$ and let $I_1, \ldots, I_m, Q$ be proper verbally prime T-ideals such that $Q \subseteq I_1 \cdots I_m$. Then $\exp(Q) > \exp(I_1 \cdots I_m)$.*

PROOF. Since $\exp(Q) \geq \exp(I_1 \cdots I_m)$, we only need to show that $\exp(Q) \neq \exp(I_1 \cdots I_m)$. Suppose to the contrary that $\exp(Q) = \exp(I_1 \cdots I_m)$. By Theorem 8.4.3 there exists a minimal superalgebra $A = A^{(0)} \oplus A^{(1)}$ with maximal semisimple $\mathbb{Z}_2$-stable subalgebra $A_{ss} = A_1 \oplus \cdots \oplus A_m$ such that $I = I_1 \cdots I_m = \mathrm{Id}(G(A))$ and $I_j = \mathrm{Id}(G(A_j))$, $j = 1, \ldots, m$. Let B be a finite dimensional simple superalgebra such that $\mathrm{Id}(G(B)) = Q$. By Theorem 4.8.7, $\chi_n(G(B)) \subseteq H(l, d)$ where $d = \dim B^{(0)}$ and $l = \dim B^{(1)}$; moreover, $\exp(Q) = \exp(G(B)) = d + l$. Since by hypothesis $\mathrm{Id}(G(B)) = Q \subseteq I = \mathrm{Id}(G(A))$, then

$$\chi_n(G(A)) \subseteq H(l, d).$$

Moreover, by assumption, $\exp(G(A)) = \exp(G(B)) = d + l$. For $i = 1, \ldots, m$, denote $d_i = \dim A_i^{(0)}$, $l_i = \dim A_i^{(1)}$ and $l' = l_1 + \cdots + l_m$, $d' = d_1 + \cdots + d_m$.

For every $n \geq 1$, let W_{n+m-1} be the space of all multilinear polynomials in the variables $x_1, \ldots, x_n, y_1, \ldots, y_{m-1}$ of the type

$$f(x_1, \ldots, x_n, y_1, \ldots, y_{m-1}) = f_1 y_1 f_2 \cdots f_{m-1} y_{m-1} f_m$$

where $f_1, \ldots, f_m$ are polynomials in the x_i's. The permutation action of S_n on the variables $x_1, \ldots, x_n$ turns W_{n+m-1} into an S_n-module. Since A is a minimal superalgebra, $A_1 J \cdots J A_m \neq 0$ where $J = J(A)$ and we can apply Lemma 6.4.4. Therefore for any positive integer $t \geq 2 \dim A$, there exists n such that

$$l'd' + (2t-s)(l'+d') \leq n \leq l'd' + 2t(l'+d'), \quad s = 4\dim A$$

and a polynomial $f \in W_{n+m-1}$ which is not an identity of $G(A)$. Moreover, $FS_n f = M$, the S_n-module generated by f, is irreducible with character χ_λ where

(8.10) $$h(l', d', 2t-s) \leq \lambda \leq h(l', d', 2t).$$

If we now regard W_{n+m-1} as an S_{n+m-1}-module (via the permutation action) then, by the Littlewood-Richardson rule (Theorem 2.3.9), any irreducible component of $M \uparrow S_{n+m-1}$, the S_{n+m-1}-module generated by f, is associated to a partition $\mu \vdash n+m-1$ with $\mu \geq \lambda$. But then, since $l+d = l'+d'$, from (8.10) we obtain that $l' = l$ and $d' = d$.

We shall reach a contradiction by proving that $\chi_n(G(A)) \not\subseteq H(l,d)$. Now, in Lemma 6.4.4, it was also shown that there exists an evaluation

$$\varphi : x_1, \ldots, x_n \to G(A_1 \oplus \cdots \oplus A_m),$$

$$\varphi : y_1, \ldots, y_{m-1} \to G(J)$$

such that $\varphi(f) \neq 0$ and $\varphi(f) \in A_1 J A_2 J \cdots J A_m$. Since $FS_n f = M$ is irreducible with character χ_λ, there exists a Young tableau T_λ such that

$$e_{T_\lambda} f = \gamma f,$$

for some $0 \neq \gamma \in F$. We rename the variables $y_1 = x_{n+1}, \ldots, y_{m-1} = x_{n+m-1}$ and we construct a partition $\mu \vdash n+m-1$ and a corresponding Young tableau T_μ by the following rules: we let $\mu_i = \lambda_i$ for all $i \neq d+1$ and $\mu_{d+1} = \lambda_{d+1} + m - 1 = l + m - 1$. Since D_λ is a subdiagram of D_μ we let T_μ to be the tableau containing the tableau T_λ and with the entries $n+1, \ldots, n+m-1$ in the boxes $(d+1, l+1), \ldots, (d+1, l+m-1)$, respectively.

We wish to show that $e_{T_\mu} f \notin \mathrm{Id}(G(A))$. To this end, since T_λ is a subtableau of T_μ then, by considering the canonical embedding $S_n \subseteq S_{n+m-1}$, we get that $C_{T_\lambda} \subseteq C_{T_\mu}$ and $R_{T_\lambda} \subseteq R_{T_\mu}$.

Let φ be the above evaluation of f; notice that from the multiplication rules of A, it easily follows that for any permutation $\rho \in S_{n+m-1}$, $\varphi(\rho f)$ is non-zero only if $\rho(n+1) = n+1, \ldots, \rho(n+m-1) = n+m-1$. On the other hand, if $\rho \in C_{T_\mu}$ is a column permutation of T_λ, and $\rho(n+1) = n+1, \ldots, \rho(n+m-1) = n+m-1$, then $\rho \in C_{T_\lambda}$. Hence

$$\varphi\left(\sum_{\tau \in C_{T_\mu}} (\operatorname{sgn}\tau)\tau f\right) = \varphi\left(\sum_{\rho \in C_{T_\lambda}} (\operatorname{sgn}\rho)\rho f\right).$$

If we apply the same argument to R_{T_λ}, R_{T_μ} and $g = \sum_{\rho \in C_{T_\lambda}} (\operatorname{sgn}\rho)\rho f$, we obtain

$$\varphi(e_{T_\mu} f) = \varphi\left(\sum_{\sigma \in R_{T_\lambda}} \sigma g\right) = \varphi(e_{T_\lambda} f) = \gamma\varphi(f) \neq 0.$$

This proves that $e_{T_\mu} f \notin \mathrm{Id}(G(A))$ and, since $\mu \notin H(l,d)$, we get $\chi_n(G(A)) \not\subseteq H(l,d)$. With this contradiction the proof is complete. $\square$

Next we generalize Lemma 8.5.2 to the case of several factors. To this end, we need to recall the following result of Berele and Regev which we state here without proof.

THEOREM 8.5.3 ([**BR4**]). *Let $A, A_1, \ldots, A_m$ be PI-algebras and suppose that $\mathrm{Id}(A) = \mathrm{Id}(A_1) \cdots \mathrm{Id}(A_m)$. If $c_n(A_j) \simeq \alpha_j n^{e_j} d_j^n$, for all $j = 1, \ldots, m$, then*
$$c_n(A) \simeq \alpha n^e d^n$$
where $d = d_1 + \cdots + d_m$, $e = e_1 + \cdots + e_m + m - 1$ and
$$\alpha = \alpha_1 \cdots \alpha_m \frac{d_1^{e_1} \cdots d_m^{e_m}}{(d_1 + \cdots + d_m)^e}.$$

LEMMA 8.5.4. *Let $I_1, \ldots, I_m, Q_1, \ldots, Q_n$ be proper verbally prime T-ideals such that*
$$Q_1 \cdots Q_n \subseteq I_1 \cdots I_m.$$
Then either $\exp(Q_1 \cdots Q_n) > \exp(I_1 \cdots I_m)$ or $m = n$ and $I_1 = Q_1, \ldots, I_m = Q_m$.

PROOF. The case $n = 1$ was settled in Lemma 8.5.2. Suppose $n \geq 2$ and denote $Q' = Q_2 \cdots Q_n$. If either Q_1 or Q' is contained in $I_1 \cdots I_m$, then, since by Theorem 8.5.3, $\exp(Q_1 Q') = \exp(Q_1) + \exp(Q')$, we obtain by induction on n that $\exp(Q_1 Q') > \exp(I_1 \cdots I_m)$ and we are done in this case. Therefore we may assume that $Q_1, Q' \nsubseteq I_1 \cdots I_m$.

By Theorem 8.4.4, there exists $k \geq 1$ such that $Q_1 \subseteq I_1 \cdots I_k$ and $Q' \subseteq I_{k+1} \cdots I_m$. Now, if $k > 1$, by Lemma 8.5.2, $\exp(Q_1) > \exp(I_1 \cdots I_k)$; hence
$$\exp(Q_1 Q') = \exp(Q_1) + \exp(Q')$$
$$> \exp(I_1 \cdots I_k) + \exp(I_{k+1} \cdots I_m) = \exp(I_1 \cdots I_m)$$
and we are done. Therefore we may assume that $k = 1$, i.e., $Q_1 \subseteq I_1$. A repeated application of this process leads to $m = n$ and $Q_1 \subseteq I_1, \ldots, Q_m \subseteq I_m$. The conclusion of the lemma now follows from Lemma 8.5.1. □

As an immediate consequence of the above lemma we get the following.

COROLLARY 8.5.5. *Let $I_1, \ldots, I_m, Q_1, \ldots, Q_n$ be proper verbally prime T-ideals. Then*
$$I_1 \cdots I_m = Q_1 \cdots Q_n$$
if and only if $m = n$ and $I_1 = Q_1, \ldots, I_m = Q_m$. □

Corollary 8.5.5 says that all verbally prime T-ideals of the free algebra $F\langle X \rangle$ generate a free semigroup. But this result can also be deduced from [**BL**, Theorem 7] where it was proved that the semigroup of all non-zero T-ideals of $F\langle X \rangle$ is free. In fact, any verbally prime T-ideal by definition cannot be decomposed into the product of two T-ideals properly contained in it.

We are now in a position to put together the results about minimal varieties. This is the main theorem of [**GZ9**]. The equivalence of 2) and 3) can also be deduced from [**BR5**].

THEOREM 8.5.6 ([**GZ9**]). *Let $\mathcal{V}$ be a variety of algebras over a field F of characteristic zero such that $\exp(\mathcal{V}) \geq 2$. Then the following properties are equivalent:*

1) $\mathcal{V}$ *is a minimal variety of exponent d.*

2) $\text{Id}(\mathcal{V})$ *is a product of verbally prime T-ideals.*

3) $\mathcal{V} = \text{var}(G(A))$, *for some minimal superalgebra A such that* $\dim A_{ss} = d$.

PROOF. Theorem 8.4.2 says that 1) implies 2) and 3). Also, by Theorem 8.4.3 and Theorem 8.3.4, 2) and 3) are equivalent. We now prove that 2) implies 1). Let $\mathcal{V}$ be a variety of exponent d such that $\text{Id}(\mathcal{V}) = I_1 \cdots I_m$, for some verbally prime T-ideals $I_1, \ldots, I_m$ and let $\mathcal{U}$ be a proper subvariety of $\mathcal{V}$. By the proof of Theorem 8.4.2, there exists a minimal superalgebra A such that $\text{var}(G(A)) \subseteq \mathcal{U}$ and $\exp(G(A)) = \exp(\mathcal{U})$. Let $\text{Id}(\mathcal{U}) = \text{Id}(G(A)) = Q_1 \cdots Q_n$, a product of verbally prime T-ideals. Since $\mathcal{U} \subsetneq \mathcal{V}$, then $I_1 \cdots I_m \subsetneq Q_1 \cdots Q_n$ and by Lemma 8.5.4 we get that
$$\exp(\mathcal{U}) = \exp(Q_1 \cdots Q_n) < \exp(I_1 \cdots I_m) = \exp(\mathcal{V}).$$
Thus $\mathcal{V}$ is a minimal variety. □

COROLLARY 8.5.7. *For any integer $d \geq 2$, there exists only a finite number of minimal varieties of exponent d.*

8.6. Some consequences

The classification of minimal varieties is an efficient tool for the study of the growth of a variety. The main application will be given in the next chapter where we shall introduce an effective way for computing the exponent of a variety whose T-ideal is generated by a given polynomial. Here we want to study the general problem of determining which functions can appear as the growth function of a codimension sequence. Namely, we ask for which functions $f(n)$ of a natural argument n there exists a PI-algebra A such that $f(n) \simeq c_n(A)$ for some equivalence $\simeq$ of functions. Problems of this type have been considered in various areas.

For instance, about the Gelfand-Kirillov dimension of an algebra, as it was mentioned in Section 1.13, for any real number $\alpha \geq 2$ there exists an algebra A such that $\text{GK-dim}(A) = \alpha$ whereas the only real numbers < 2 that can be realized as the GK-dimension of an algebra are 0 and 1 [**KL**].

In group theory the so-called distorsion functions were considered allowing the embedding of an arbitrary group in a finitely presented group. The problem of describing such functions was posed by Gromov in [**Gr**]. Ol'shanskii in [**O**] showed that any "reasonable" function can occur as a distortion function. In particular, n^θ where $\theta \geq 1$ is any "computable" real number can be realized as the distortion function of the embedding of the infinite cyclic group into a finitely presented group.

Concerning the theory of PI-algebras, recall that Conjecture 6.1.4 states that for any PI-algebra A, there exist a constant α, a semi-integer e and an integer d such that
$$c_n(A) \simeq \alpha n^e d^n.$$

The precise asymptotics have been computed only for very few algebras and in all cases the conjecture was confirmed. Maybe the most important case is Theorem 5.10.4, showing that the conjecture holds for $A = M_k(F)$. It can also be checked that the above asymptotic equality is valid for block-triangular matrix algebras.

In fact, if $A = UT(d_1, \ldots, d_m)$ is an upper block-triangular matrix algebra, then A is a minimal superalgebra with trivial grading and by Theorem 8.3.6, $\text{Id}(A) = \text{Id}(M_{d_1}(F)) \cdots \text{Id}(M_{d_m}(F))$. Hence from Theorem 8.5.3 and Theorem 5.10.4, we easily deduce.

THEOREM 8.6.1. *If $A = UT(d_1, \ldots, d_m)$, then*
$$c_n(A) \simeq \alpha n^g d^n$$
with $d = d_1^2 + \cdots + d_m^2$, $g = -\frac{1}{2}\left(\sum_{i=1}^m d_i^2 - 3m + 2\right)$ and
$$\alpha = \alpha_1 \cdots \alpha_m \frac{d_1^{g_1} \cdots d_m^{g_m}}{(d_1^2 + \cdots + d_m^2)^g},$$
where $g_i = -(d_i^2 - 1)$,
$$\alpha_i = \left(\frac{1}{\sqrt{2\pi}}\right)^{d_i - 1} \cdot \left(\frac{1}{2}\right)^{\frac{1}{2}(d_i^2 - 1)} \cdot 1! \cdot 2! \cdots (d_i - 1)! d_i^{\frac{1}{2}(d_i^2 + 4)}.$$

More generally, we can prove the following result.

THEOREM 8.6.2. *For any positive integer d there exists a non-zero constant α such that*
$$c_n(A) \geq \alpha n^{\frac{1-d}{2}} d^n$$
for any PI-algebra A with $\exp(A) = d$.

PROOF. Recall that by Theorem 5.10.4,
$$c_n(M_k(F)) \simeq \alpha n^{-\frac{k^2-1}{2}} k^{2n}.$$
Moreover, the best known approximation of the asymptotics for the other verbally prime T-ideals was computed by Berele and Regev in [**BR3**]. Namely,
$$\alpha_1 n^{-\frac{k^2+l^2-1}{2}} (k+l)^{2n} \leq c_n(M_{k,l}(G)) \leq \alpha_2 n^{-\frac{k^2+l^2-1}{2}} (k+l)^{2n}$$
and
$$\beta_1 n^{-\frac{2k^2-1}{2}} (2k^2)^n \leq c_n(M_k(G)) \leq \beta_2 n^{-\frac{k^2-1}{2}} (2k^2)^n$$
for some constants $\alpha, \alpha_1, \alpha_2, \beta_1, \beta_2$ such that $\alpha, \alpha_1, \beta_1 > 0$.

Note that as a consequence,
$$(8.11) \qquad \frac{c_n(\mathcal{V})}{d^n} \geq \alpha n^{\frac{1-d}{2}},$$
for any prime variety $\mathcal{V}$ with $\exp(\mathcal{V}) = d$, for some constant α.

Now let A, A_1, A_2 be PI-algebras such that $\mathrm{Id}(A) = \mathrm{Id}(A_1)\mathrm{Id}(A_2)$. Then in [**Re7**, Lemma 7.6], it was prove that if
$$\alpha_{1,i} n^{v_i} d_i^n \leq c_n(A_i) \leq \alpha_{2,i} n^{u_i} d_i^n$$
for some $\alpha_{1,i}, \alpha_{2,i} > 0$, $v_i \leq u_i$, $i = 1, 2$, then
$$(8.12) \qquad \beta_1 n^{v_1+v_2+1}(d_1+d_2)^n \leq c_n(A) \leq \beta_2 n^{u_1+u_2+1}(d_1+d_2)^n$$
for some constants $\beta_1, \beta_2 > 0$. Now let $\mathcal{V}$ be any minimal variety with $\mathrm{Id}(\mathcal{V}) = I_1 \cdots I_k$ where $I_1, \ldots, I_k$ are verbally prime T-ideals. Then, combining (8.11) and (8.12) we obtain
$$\frac{c_n(\mathcal{V})}{d^n} \geq \alpha n^{(\frac{1-d_1}{2} + \cdots + \frac{1-d_k}{2}) + k - 1} = \alpha n^{\frac{k - (d_1 + \cdots + d_k)}{2} + k - 1} \geq \alpha n^{\frac{1-d}{2}}$$
where for $j = 1, \ldots, k$ $d_j = \exp(I_j)$ and $d = d_1 + \cdots + d_k = \exp(\mathcal{V})$. This proves the theorem. $\square$

Concerning the inverse problem of what integer $d \geq 2$ can be realized as the PI-exponent of an algebra, we can prove more generally the following.

THEOREM 8.6.3 ([**GZ7**]). *For any positive integer k there exists a variety $\mathcal{V}$ such that $\exp(\mathcal{V}) = d$ but $\exp(\mathcal{U}) \leq d - k$ for any proper subvariety $\mathcal{U} \subset \mathcal{V}$.*

PROOF. Let $\mathcal{V} = \text{var}(M_m(F))$ and let $\mathcal{U} \subset \mathcal{V}$ be a proper subvariety with $\exp(\mathcal{U}) = e \leq d = m^2$. Then $\mathcal{U}$ contains some upper block-triangular matrix algebra $UT(q_1, \ldots, q_s)$ with $q_1^2 + \cdots + q_s^2 = e$. Also, one has $q_1 + \cdots + q_s \leq m$. Since $\mathcal{U}$ is proper, then $s > 1$ or $q_1 < m$. Hence $\exp(\mathcal{U}) = e = q_1^2 + \cdots + q_s^2 \leq (m-1)^2 + 1^2 = m^2 - 2m + 2$. Now it is enough to choose m large enough, and the proof of the theorem is complete. □

The problem of what constants α and half integers e can be realized among codimension growth functions was discussed in Section 7.5. Further results can be found in [**DR**].

For the second approximation we need to compute the polynomial factor n^e. In this case the inverse problem can be formulated as follows: describe a subset E of the real numbers such that for any $e \in E$ there exists a PI-algebra A with $c_n(A) \simeq Cn^e d^n$. More precisely,

$$e = \lim_{n \to \infty} \log_n \frac{c_n(A)}{d^n}$$

exists and $e \in E$ where $d = \exp(A)$. According to Regev's conjecture, $E \subseteq \frac{1}{2}\mathbb{Z}$. Next we construct varieties with the same exponent d but with distinct polynomial factors.

In the following theorem we use the well-known Lagrange theorem (see [**D**]) stating that any positive integer can be written as the sum of at most four squares. In order to simplify the notation we write $d = k_1^2 + \cdots + k_j^2$ if d is the sum of j squares but it cannot be written as the sum of $i < j$ squares.

THEOREM 8.6.4. *Given any integer $d \geq 2$, there exists a constant $0 \leq \gamma \leq \frac{9}{2}$ such that, for any integer $r \geq 0$ we can find a variety $\mathcal{V}$ with the property that*

$$c_n(\mathcal{V}) \simeq \beta n^e d^n$$

where β is a constant and $e = \frac{1-d}{2} + \gamma + r$. Moreover, $\gamma = 0$, $\frac{3}{2}$, 3, $\frac{9}{2}$ if $d = k^2$, $k_1^2 + k_2^2$, $k_1^2 + k_2^2 + k_3^2$, $k_1^2 + k_2^2 + k_3^2 + k_4^2$, respectively.

PROOF. Let $R = UT(d_1, \ldots, d_m)$ be the upper block triangular matrix algebra

$$UT(d_1, \ldots, d_m) = \begin{pmatrix} M_{d_1}(F) & & * \\ 0 & \ddots & \\ \vdots & & \\ 0 & \cdots & 0 & M_{d_m}(F) \end{pmatrix}.$$

Then by Section 2, R generates a minimal variety and $c_n(R) \simeq \alpha n^g d^n$ where α is a constant, $d = d_1^2 + \cdots + d_m^2$ and $g = -\frac{1}{2}(d - 3m + 2) = \frac{1-d}{2} + \frac{3m-3}{2}$.

Now, if d can be written as a sum of $m = 1, 2, 3,$ or 4 squares, we obtain that $g = \frac{1-d}{2} + \gamma$ where $\gamma = 0$ or $\frac{3}{2}$ or 3 or $\frac{9}{2}$, respectively. If we now apply to R the second part of Theorem 7.5.3, we obtain an algebra A such that $c_n(A) \simeq \beta n^e d^n$ with $e = \frac{1-d}{2} + \gamma + r$ for any $r = 1, 2, \ldots$. □

CHAPTER 9

Computing the Exponent of a Polynomial

This chapter is devoted to the study of the asymptotic behaviour of the codimensions of the varieties defined by a polynomial or a set of polynomials in some significant cases. More precisely, after defining the exponent of a polynomial as the exponent of the corresponding variety, we compute the exponent of a standard polynomial, a Capelli polynomial, a Lie nilpotent monomial, etc. It turns out that the exponent of the standard polynomial of degree n is the largest possible among all polynomials of degree n. Motivated by Amitsur's theorem (see Theorem 4.5.4) stating that any PI-algebra satisfies some power of a standard identity, we also compute the exponent of any such power as well as the exponent of any Amitsur polynomial (see Section 4.7).

In the second part of the chapter we make a more detailed study of the variety generated by a standard or a Capelli polynomial. We introduce the notion of reduced superalgebra and via a thorough study of the identities of these algebras we determine a generating algebra for some significant varieties. As a consequence we prove that the standard polynomial of degree $2k$, the Capelli polynomial of rank $k^2 + 1$ and the algebra of $k \times k$ matrices have the same asymptotic behavior of the codimensions.

A highlight of this chapter is the introduction of the so-called essential hook for any PI-algebra A. This is the smallest union of a hook and a square containing the cocharacter of A. We prove that any PI-algebra has an essential hook and this is a characterization finer than the one given in Theorem 4.5.1.

Throughout this chapter the ground field F is assumed to be of characteristic zero.

9.1. The exponent of standard and Capelli polynomials

In this section we shall compute the exponent of a variety defined by a single polynomial. To this end, given a polynomial $f \in F\langle X \rangle$, we denote by $\text{var}(f)$ the variety whose T-ideal is $\langle f \rangle_T$, the T-ideal generated by f. We then define

DEFINITION 9.1.1. The exponent of the variety $\text{var}(f)$ is called the exponent of the polynomial f and is denoted $\exp(f) = \exp(\text{var}(f))$.

Similarly, we define the exponent of any set of polynomials as the exponent of the corresponding variety.

Our basic strategy for computing the exponent of a polynomial is the following. By the definition of minimal variety (Definition 8.4.1), the exponent of a polynomial f is equal to the maximum of the exponent of a minimal variety $\mathcal{V}$ lying in $\text{var}(f)$. Hence $f \equiv 0$ must be an identity of $\mathcal{V}$ and we can write

$$\exp(f) = \max\{\exp(\mathcal{V}) \mid \mathcal{V} \text{ is a minimal variety satisfying } f \equiv 0\}.$$

Our first result will be the computation of the exponent of a standard polynomial. To this end we need to estimate the minimal degree of a standard identity satisfied by a block-triangular matrix algebra.

LEMMA 9.1.2. *The block-triangular matrix algebra $UT(d_1, \ldots, d_m)$ satisfies the standard identity $St_k \equiv 0$ if and only if $k \geq 2(d_1 + \cdots + d_m)$.*

PROOF. Set $A = UT(d_1, \ldots, d_m)$ and $n = d_1 + \cdots + d_m$. Since A is a subalgebra of $M_n(F)$, by Theorem 1.7.7, A satisfies the standard identity of degree $2n$. On the other hand, A contains all matrix units of the principal staircase $e_{11}, e_{12}, e_{22}, e_{23}, \ldots, e_{n-1,n}, e_{nn}$ of $M_n(F)$ and

$$St_{2n-1}(e_{11}, e_{12}, e_{22}, e_{23}, \ldots, e_{n-1,n}, e_{nn}) = e_{1n} \neq 0.$$

Hence $St_{2n} \equiv 0$ is a standard identity of minimal degree for A.

Since $St_p \equiv 0$ implies $St_q \equiv 0$ as soon as $p < q$, the conclusion of the lemma is clear. □

We can now prove the following.

THEOREM 9.1.3 ([**BR5**]). *For all $m \geq 1$,*

$$\exp(St_{2m}) = \exp(St_{2m+1}) = m^2.$$

PROOF. As we remarked above, the exponent of St_k is equal to the exponent of some minimal variety $\mathcal{V}$ lying in $\mathrm{var}(St_k)$. Since the Grassmann algebra G does not satisfy any standard identity, then the algebras $M_{k,l}(G)$ and $M_q(G)$ do not satisfy any standard identity either. It follows that any minimal variety $\mathcal{V}$ satisfying $St_k \equiv 0$ is generated by the Grassamnn envelope of a minimal superalgebra with trivial grading (see Corollary 8.3.6). Thus $\mathcal{V}$ is generated by an algebra of the type $UT(d_1, \ldots, d_r)$ and

(9.1) $\exp(St_{2k}) = \max\{\exp(UT(d_1, \ldots, d_r)) \mid UT(d_1, \ldots, d_r) \text{ satisfies } St_k \equiv 0\}.$

By Lemma 9.1.2, $UT(d_1, \ldots, d_r)$ satisfies St_k if and only if $k \geq 2(d_1 + \cdots + d_r)$. On the other hand, by Corollary 6.6.2,

$$\exp(UT(d_1, \ldots, d_r)) = d_1^2 + \cdots + d_r^2.$$

Suppose first that $k = 2m$ is even. Since $M_m(F)$ satisfies $St_k = St_{2m}$, and $\exp(M_m(F)) = m^2$, by (9.1) $\exp(St_{2m}) \geq m^2$. On the other hand, if $r \geq 2$ and $2m \geq 2(d_1 + \cdots + d_r)$ clearly $m^2 > d_1^2 + \cdots + d_r^2 = \exp(UT(d_1, \ldots, d_r))$. It follows that $\exp(St_{2m}) = m^2$.

Now let $k = 2m+1$ be odd and suppose that $UT(d_1, \ldots, d_r)$ satisfies St_k where $r \geq 2$. Then $2m + 1 \geq 2(d_1 + \cdots + d_r)$ implies that

$$m + \frac{1}{2} \geq d_1 + \cdots + d_r.$$

Since m and $d_1, \ldots, d_r$ are all integers, we have that $m \geq d_1 + \cdots + d_r$ and $m^2 > d_1^2 + \cdots + d_r^2$. Hence the maximal value of the right-hand side of (9.1) is m^2 and the proof is complete. □

We now turn our attention to Capelli polynomials. As in the proof of the previous theorem, any minimal variety satisfying the Capelli polynomial Cap_k for some $k \geq 2$, is generated by some block-triangular matrix algebra and we need to find the minimal degree of a Capelli identity satisfied by $UT(d_1, \ldots, d_r)$.

9.1. THE EXPONENT OF STANDARD AND CAPELLI

LEMMA 9.1.4. *If $d = d_1^2 + \cdots + d_m^2$, the algebra $UT(d_1, \ldots, d_m)$ satisfies the Capelli identity $Cap_{d+m} \equiv 0$ but does not satisfy $Cap_{d+m-1} \equiv 0$.*

PROOF. Write $A = UT(d_1, \ldots, d_m) = B + J$ where J is the Jacobson radical of A, $J^m = 0$ and $B = M_{d_1}(F) \oplus \cdots \oplus M_{d_m}(F)$, $\dim B = d$.

Consider an evaluation $\varphi : F\langle X \rangle \to A$ of the Capelli polynomial
$$Cap_{d+m} = \sum_{\sigma \in S_{d+m}} (\operatorname{sgn} \sigma) y_1 x_{\sigma(1)} y_2 \cdots y_{d+m} x_{\sigma(d+m)} y_{d+m+1}.$$

Since Cap_{d+m} is multilinear, it is enough to check evaluations on a basis of A, say on the matrix units e_{ij}. If at least $d+1$ values among $\varphi(x_1), \ldots, \varphi(x_{d+m})$ lie in B, then $\varphi(Cap_{d+m}) = 0$, since Cap_{d+m} is alternating on $x_1, \ldots, x_{d+m}$ and $\dim B = d$. But if at most d elements among $\varphi(x_1), \ldots, \varphi(x_{d+m})$ belong to B, then at least m elements among them lie in J. In this case $\varphi(Cap_{d+m}) = 0$ since $J^m = 0$. Thus $Cap_{d+m} \equiv 0$ is an identity of A.

Next we show that A does not satisfy Cap_{d+m-1}. Suppose first that $m = 1$ and, so, $A = M_{d_1}(F)$. Write $n = d_1^2 = \dim A$, and take a basis $\{v_1, \ldots, v_n\}$ of A consisting of all matrix units e_{ij}, $1 \leq i, j \leq d_1$, ordered in an arbitrary fixed way. We claim that there exist $a_0, a_1, \ldots, a_n \in A$ such that

(9.2) $$a_0 v_1 a_1 \cdots a_{n-1} v_n a_n = e_{11}$$

and

(9.3) $$a_0 v_{\sigma(1)} a_1 \cdots a_{n-1} v_{\sigma(n)} a_n = 0,$$

as soon as $\sigma \in S_n, \sigma \neq 1$. Indeed, if $v_1 = e_{i_1 j_1}, \ldots, v_n = e_{i_n j_n}$, then we take
$$a_0 = e_{1 i_1}, a_1 = e_{j_1 i_2}, \ldots, a_{n-1} = e_{j_{n-1} i_n}, a_n = e_{j_n 1}.$$

Clearly $a_0, \ldots, a_n$ and $v_1, \ldots, v_n$ satisfy (9.2). On the other hand, given $2 \leq k \leq n$, any product $a_{k-1} v_p a_k$ is equal to zero as soon as $p \neq k$. Hence (9.3) also holds.

From (9.2) and (9.3) it follows that
$$Cap_n(v_1, \ldots, v_n; a_0, \ldots, a_n) = e_{11} \neq 0.$$

Now let $m \geq 2$ and let $A = UT(d_1, \ldots, d_m)$. Write $r_1 = 0, r_2 = d_1, r_3 = d_1 + d_2, \ldots, r_m = d_1 + \cdots + d_{m-1}$. Then
$$B_j = \operatorname{span}\{e_{r_j+p, r_j+q} \mid 1 \leq p, q \leq d_j\}$$

is isomorphic to the $d_j \times d_j$ matrix algebra of dimension $n_j = d_j^2$, $j = 1, \ldots, m$. By the previous part of the proof, for any $1 \leq j \leq m$, there exist $a_0^j, \ldots a_{n_j}^j, v_1^j, \ldots, v_{n_j}^j \in B_j$ such that
$$a_0^j v_1^j a_1^j \cdots a_{n_j-1}^j v_{n_j}^j a_{n_j}^j = e_{r_j+1, r_j+1} = c_j \neq 0$$

and

(9.4) $$a_0^j v_{\sigma(1)}^j a_1^j \cdots a_{n_j-1}^j v_{\sigma(n_j)}^j a_{n_j}^j = 0,$$

for any $\sigma \in S_{n_j}$, $\sigma \neq 1$. It is not difficult to see that $B_1 J B_2 \cdots B_{m-1} J B_m \neq 0$ and there exist $w_1, \ldots, w_{m-1} \in J$ such that

(9.5) $$c_1 w_1 c_2 \cdots c_{m-1} w_{m-1} c_m \neq 0.$$

Moreover, $w_i \in J_{i,i+1} = B_i J B_{i+1}$ and $1_i w_i 1_{i+1} = w_i$ where $1_i \in B_i$ is the unit element of B_i. We rewrite (9.5) in the form
$$(c_1 1_1) w_1 (1_2 c_2 1_2) w_2 \cdots w_{m-1} (1_m c_m) \neq 0.$$

Notice that

(9.6) $$1_i r 1_{i+1} = 0,$$

for any $r \in J_{j,j+1}$ with $j \neq i$, and for any $r \in B_1 + \cdots + B_m$. Finally, we get the non-zero product

$$(a_0^1 v_1^1 a_1^1 \cdots a_{n_1-1}^1 v_{n_1}^1 a_{n_1}^1 1_1) w_1 (1_2 a_0^2 v_1^2 a_1^2 \cdots a_{n_2-1}^2 v_{n_2}^2 a_{n_2}^2 1_2) w_2$$
$$\cdots w_{m-1} (1_m a_0^m v_1^m a_1^m \cdots a_{n_m-1}^m v_{n_m}^m a_{n_m}^m 1_m)$$

such that any permutation of

$$v_1^1, \ldots, v_{n_1}^1, w_1, v_1^2, \ldots, v_{n_2}^2, w_2, \ldots, w_{m-1}, v_1^m, \ldots, v_{n_m}^m$$

inside this expression gives a zero value. This is easily seen by (9.4), (9.6) and the relation $a_s^j v_q^p a_{s+1}^j = 0$, where $p \neq j$.

It follows that $UT(d_1, \ldots, d_m)$ does not satisfy the Capelli identity Cap_k with $k = n_1 + \cdots + n_m + m - 1 = d_1^2 + \cdots + d_m^2 + m - 1$ and the proof of the lemma is complete. $\square$

We are now in a position to compute the exponent of a Capelli polynomial.

THEOREM 9.1.5 ([**MRZ3**]). *For any $k \geq 1$ we have*

$$\exp(Cap_k) = \max\{a_1, a_2, a_3, a_4\}$$

where

$$a_j = \max\{d_1^2 + \cdots + d_j^2 \mid d_1, \ldots, d_j \in \mathbb{Z}, d_1, \ldots, d_j > 0, d_1^2 + \cdots + d_j^2 + j \leq k\}.$$

In particular, $k - 4 \leq \exp(Cap_k) \leq k - 1$.

PROOF. As we remarked in the proof of Theorem 9.1.3, we have that

$$\exp(Cap_k) = \max\{\exp(UT(d_1, \ldots, d_m)) \mid UT(d_1, \ldots, d_m) \text{ satisfies } Cap_k \equiv 0\}.$$

Also, the exponent of $UT(d_1, \ldots, d_m)$ is equal to $d_1^2 + \cdots + d_m^2$ (see Corollary 6.6.2). By Lemma 9.1.4, the algebra $UT(d_1, \ldots, d_m)$ satisfies the Capelli polynomial $Cap_k \equiv 0$ if and only if $d_1^2 + \cdots + d_m^2 + m \leq k$. Hence

$$\exp(Cap_k) = \max\{d_1^2 + \cdots + d_m^2 \mid d_1^2 + \cdots + d_m^2 \leq k - m\}.$$

By Lagrange Theorem (see [**D**]) any positive integer is the sum of at most four squares. Denote

$$a_j = \max\{d_1^2 + \cdots + d_j^2 \mid d_1^2 + \cdots + d_j^2 + j \leq k\}.$$

Then $a_j \leq k - j$ and, so, $a_5, a_6, \ldots \leq k - 5$. On the other hand, $k - 4 = d_1^2 + \cdots + d_s^2$, with $s \leq 4$. Then either $a_1, a_2, a_3,$ or a_4 is not less than $k - 4$. Hence

$$\max\{a_1, a_2, a_3, a_4\} \geq k - 4,$$

and the proof of the theorem follows. $\square$

Straightforward computations allow us to give the exponent of Capelli polynomials in the following form.

COROLLARY 9.1.6.
1) *If $k - 1 = n^2$, then $\exp(Cap_k) = k - 1$.*
2) *If $k - 1 \neq n^2$ and $k - 2 = n_1^2 + n_2^2$, then $\exp(Cap_k) = k - 2$.*
3) *If $k - 1 \neq n^2$, $k - 2 \neq n_1^2 + n_2^2$ and $k - 3 = n_1^2 + n_2^2 + n_3^2$, then $\exp(Cap_k) = k - 3$.*

4) If $k-1 \neq n^2$, $k-2 \neq n_1^2 + n_2^2$ and $k-3 \neq n_1^2 + n_2^2 + n_3^2$, then $\exp(Cap_k) = k-4$.

For instance, $\exp(Cap_5) = 4$, $\exp(Cap_4) = 2$, $\exp(Cap_9) = 5$ and $\exp(Cap_{58}) = 54$.

9.2. An upper bound for the exponent of a polynomial

In this section we prove that the standard identity is, in some way, the weakest identity among all identities of the same degree. More precisely, we prove a result of Berele and Regev stating that for any polynomial f of degree n, $\exp(St_n) \geq \exp(f)$.

We start by studying some natural embedding of minimal superalgebras in matrices. Recall that the exponent of a variety $\mathcal{V}$ coincides with the exponent of some minimal variety contained in $\mathcal{V}$. Moreover, any minimal variety is generated by the superenvelope $S(A)$ of a minimal superalgebra A (see Theorem 8.5.6). Recall that for an algebra A, $J(A)$ always denotes the Jacobson radical of A.

LEMMA 9.2.1. *The simple superalgebra $R = M_k(F \oplus cF)$, $c^2 = 1$, contains two minimal superalgebras A and B such that*

$$A/J(A) = M_{k-1}(F \oplus cF) \oplus (F \oplus cF),$$

and

$$B/J(B) = \underbrace{(F \oplus cF) \oplus \cdots \oplus (F \oplus cF)}_{k}.$$

Hence $\mathrm{Id}(M_k(G)) = I_k \subseteq I_{k-1}I_1 \subseteq I_1^k$, *where* $I_j = \mathrm{Id}(M_j(G))$, $1 \leq j \leq k$.

PROOF. Set $A_1 = \mathrm{span}\{e_{ij}, ce_{ij} \mid 1 \leq i, j \leq k-1\}$ and $A_2 = \mathrm{span}\{e_{kk}, ce_{kk}\}$. Then $A_1 \cong M_{k-1}(F \oplus cF)$, $A_2 \cong F \oplus cF$ and define A to be the algebra generated by A_1, A_2 and $w_{12} = e_{1k}$. It follows that A satisfies all conditions of Definition 8.1.3, of minimal superalgebra.

The existence of B easily follows by induction, the basis of the induction being the case $k = 1$, i.e., $B = A$. The inclusions $I_k \subseteq I_{k-1}I_1 \subseteq I_1^k$ are a consequence of Theorem 8.3.4 and the fact that $\mathrm{Id}(G(R)) = \mathrm{Id}(M_k(G))$. $\square$

LEMMA 9.2.2. *The simple superalgebra $R = M_{k,l}(F)$ contains a minimal superalgebra A such that*

$$A/J(A) = \underbrace{M_{1,1}(F) \oplus \cdots \oplus M_{1,1}(F)}_{l}.$$

In particular, $I_{k,l} \subseteq I_{1,1}^l$ *where* $I_{a,b} = \mathrm{Id}(S(M_{a,b}(F)))$.

PROOF. We identify the superalgebra $M_{k,l}(F)$ with the algebra of linear transformations of a graded vector space $V = V^{(0)} \oplus V^{(1)}$ with $\dim V^{(0)} = k$, $\dim V^{(1)} = l$. A linear map $f : V \to V$ is even if $f(V^{(0)}) \subseteq V^{(0)}$, $f(V^{(1)}) \subseteq V^{(1)}$ and f is odd if $f(V^{(0)}) \subseteq V^{(1)}$, $f(V^{(1)}) \subseteq V^{(0)}$. The standard $\mathbb{Z}_2$-grading on $M_{k,l}(F)$, $k \geq l$, can be obtained by choosing a basis $\{v_1, \ldots, v_n\}$ such that $v_1, \ldots, v_k \in V^{(0)}$ and $v_{k+1}, \ldots, v_n \in V^{(1)}$.

If we take the following ordered homogeneous basis

$$\{v_1', \ldots, v_n'\} = \{v_1, v_{k+1}, v_2, v_{k+2}, \ldots, v_l, v_{k+l}, v_{l+1}, v_{l+2}, \ldots, v_k\},$$

then we get another realization of $M_{k,l}(F)$. In this new grading all matrix units e_{ij} are still homogeneous and $e_{11}, \ldots, e_{2l,2l} \in R^{(0)}$, $e_{12}, e_{21}, \ldots, e_{2l-1,2l}, e_{2l,2l-1} \in R^{(1)}$.

Then the subspace of R spanned by e_{ij}, $1 \leq i \leq j \leq 2l$, and by $e_{21}, e_{32}, \ldots, e_{2l,2l-1}$ is a minimal superalgebra according to Definition 8.1.3, with

$$A_1 = \text{span}\{e_{11}, e_{12}, e_{21}, e_{22}\}, \ldots, A_l = \text{span}\{e_{2l-1,2l-1}, e_{2l-1,2l}, e_{2l,2l-1}, e_{2l,2l}\},$$

and

$$w_{12} = e_{13}, w_{23} = e_{35}, \ldots, w_{l-1,l} = e_{2l-3,2l-1}.$$

The inclusion $I_{k,l} = \text{Id}(S(R)) \subseteq I_{1,1}^l$ now follows from Theorem 8.3.4. The proof is therefore complete. □

In order to compute a lower bound for the exponent of an arbitrary polynomial, we need to estimate the minimal degree of an identity for a prime variety. In other words, given a verbally prime T-ideal I, we need to bound from below the degree of a polynomial in I. If $I = \text{Id}(M_k(F))$, by Theorem 1.7.2, $\deg(f) \geq 2k$ as soon as $f \in I$.

Our next goal is to get a lower bound for the degree of an identity of $M_k(G)$ and of $M_{k,l}(G)$, i.e., for an identity of the T-ideal $\text{Id}(G(A))$ where $A = M_k(F \oplus cF)$ or $A = M_{k,l}(F)$, respectively. Since, according to Proposition 3.8.4, the superenvelope $S(A)$ and the Grassmann envelope $G(A)$ of a superalgebra A satisfy the same identities, in the sequel we shall work with $S(A)$ instead of the corresponding Grassmann envelope. We start with the following.

LEMMA 9.2.3. *The algebra $M_2(G)$ does not satisfy any identity of degree* 6.

PROOF. By Lemma 9.2.1, the superalgebra $A = M_2(F \oplus cF)$ contains a minimal superalgebra B with semisimple part $(F \oplus cF) \oplus (F \oplus cF)$. Hence the T-ideal $\text{Id}(M_2(G)) = \text{Id}(G(A)) = \text{Id}(S(A))$ is contained in I^2 where I is the T-ideal of identities of $S(F \oplus cF)$, i.e., $I = \text{Id}(G)$. By Theorem 4.1.8, I is generated by the polynomial $[x_1, x_2, x_3]$. Hence any identity of $S(A)$ of degree 6 is a linear combination of polynomials of the form

(9.7) $$[x_{i_1}, x_{i_2}, x_{i_3}][x_{i_4}, x_{i_5}, x_{i_6}].$$

Note that the superenvelope $S(A) = S(M_2(F \oplus cF))$ is isomorphic to $M_2(S)$. Suppose that f is a multilinear identity of $M_2(S)$ of degree 6. Then f is a linear combination of polynomials of type (9.7) with $\{i_1, \ldots, i_6\} = \{1, \ldots, 6\}$. We can assume that f generates an irreducible S_6-module corresponding to a partition $\lambda \vdash 6$ and a Young tableau T_λ. Since $M_2(G)$ does not satisfy the identity $x^6 \equiv 0$, it also does not satisfy its complete linearization $\sum_{\sigma \in S_6} x_{\sigma(1)} \cdots x_{\sigma(6)} \equiv 0$. Hence we must have $\lambda = (\lambda_1, \ldots, \lambda_k)$ with $k \geq 2$. In particular, the column stabilizer $C_{T_\lambda} \subseteq S_6$ is non-trivial and by Lemma 2.5.1 the element

$$\left(\sum_{\sigma \in C_{T_\lambda}} (\text{sgn}\,\sigma)\sigma\right) e_{T_\lambda} f$$

is non-zero and alternating in at least two indeterminates. Without loss of generality, we may assume that f itself is alternating on x_1 and x_2.

Using Jacobi identity and anticommutativity we can write any element in (9.7), so f, as a linear combination of polynomials of the type

(9.8) $\quad [x_{i_1}, x_{i_2}, x_{i_3}][x_{i_4}, x_{i_5}, x_{i_6}]$ with $i_1 > i_2 < i_3$ and $i_4 > i_5 < i_6$.

Notice that the elements in (9.8) are linearly independent.

Suppose first that f is a linear combination of polynomials (9.8) with either $i_1 = 2, i_2 = 1$ or $i_4 = 2, i_5 = 1$. That is, f is a linear combination of products of the type

(9.9) $$[x_2, x_1, x_i][x_j, x_k, x_l] \quad \text{and} \quad [x_j, x_k, x_l][x_2, x_1, x_i].$$

Write
$$a_1 = [x_2, x_1, x_4], \ a_2 = [x_2, x_1, x_5], \ a_3 = [x_2, x_1, x_6],$$
$$b_1 = [x_5, x_3, x_6], \ b_2 = [x_6, x_3, x_5], \ b_3 = [x_4, x_3, x_6],$$
$$b_4 = [x_6, x_3, x_4], \ b_5 = [x_4, x_3, x_5], \ b_6 = [x_5, x_3, x_4]$$

and set
$$f_1 = a_1 b_1, \ f_2 = a_1 b_2, \ f_3 = a_2 b_3, \ f_4 = a_2 b_4, \ f_5 = a_3 b_5, \ f_6 = a_3 b_6,$$
$$f_1' = b_1 a_1, \ f_2' = b_2 a_1, \ f_3' = b_3 a_2, \ f_4' = b_4 a_2, \ f_5' = b_5 a_3, \ f_6' = b_6 a_3.$$

If $f_1, \ldots, f_6, f_1', \ldots, f_6'$ do not appear in the decomposition of f, then f is a linear combination of the polynomials
$$g_1 = [x_2, x_1, x_3][x_6, x_4, x_5], \ g_2 = [x_2, x_1, x_3][x_5, x_4, x_6],$$
$$g_1' = [x_6, x_4, x_5][x_2, x_1, x_3], \ g_2' = [x_5, x_4, x_6][x_2, x_1, x_3].$$

But it is easily seen that any linear combination $g = \alpha_1 g_1 + \alpha_2 g_2 + \alpha_1' g_1' + \alpha_2' g_2'$ is not an identity of $M_2(S)$. For instance, the evaluation

$$\varphi(x_2) = e_{21}, \ \varphi(x_1) = e_{11}, \ \varphi(x_3) = e_{11}, \ \varphi(x_6) = e_{12}, \ \varphi(x_4) = e_{22}, \ \varphi(x_5) = e_{22}$$

is such that $\varphi(g_2) = \varphi(g_2') = 0$ and $\varphi(g) = \alpha_1 e_{22} + \alpha_1' e_{11}$. Thus $\alpha_1 = \alpha_1' = 0$. Similarly, one gets that $\alpha_2 = \alpha_2' = 0$.

Suppose now that some polynomial among $f_1, \ldots, f_6, f_1', \ldots, f_6'$ appears in the decomposition of f and suppose, for instance, that

$$f = \alpha_1 f_1 + \cdots + \alpha_6 f_6 + f''$$

where at least one among $\alpha_1, \ldots, \alpha_6$ is non-zero and f'' is a linear combination of all other polynomials in (9.8).

Consider the evaluation φ such that
$$\varphi(x_1) = e_{22}, \ \varphi(x_2) = e_{12}, \ \varphi(x_3) = e_{21},$$
$$\varphi(x_4) = s_1 e_{11}, \ \varphi(x_5) = s_2 e_{11}, \ \varphi(x_6) = s_3 e_{11}$$

where $s_1, s_2, s_3 \in S$ are homogeneous elements. Then $\varphi([x_i, x_j, x_k]) = 0$ as soon as $\{i, j, k\} = \{4, 5, 6\}$ and therefore $\varphi(g_1) = \varphi(g_1') = \varphi(g_2) = \varphi(g_2') = 0$. Further, for any h of the type (9.8) the value $\varphi(h)$ is equal to $t_1 e_{11}$ or $t_2 e_{22}$, for some $t_1, t_2 \in S$. Moreover, $\varphi(f_i)$ is a multiple of e_{22}, for all $i = 1, \ldots, 6$. Hence by computing explicitly $\varphi(f)$ and by choosing all possible parities for s_1, s_2, s_3, we get some linear relations among $\alpha_1, \ldots, \alpha_6$.

By direct calculations it follows that
$$\varphi(f_1) = s_1 s_2 s_3 e_{11}, \ \varphi(f_2) = s_1 s_3 s_2 e_{11}, \ \varphi(f_3) = s_2 s_1 s_3 e_{11},$$
$$\varphi(f_4) = s_2 s_3 s_1 e_{11}, \ \varphi(f_5) = s_3 s_1 s_2 e_{11}, \ \varphi(f_6) = s_3 s_2 s_1 e_{11}.$$

If $s_1 s_2 s_3 \neq 0$ and s_1, s_2, s_3 are all even, then we get

$$\alpha_1 + \cdots + \alpha_6 = 0.$$

If s_1, s_2, s_3 are odd, we obtain the relation

(9.10) $$r_1 \alpha_1 + \cdots + r_6 \alpha_6 = 0$$

with $(r_1, \ldots, r_6) = (1, -1, -1, 1, 1, -1)$. Similarly, if s_1, s_2 are odd and s_3 is even, then we get the relation (9.10) with coefficients $(1, 1, -1, -1, 1, -1)$. If s_i, s_3 are odd and s_2 is even, we get $(1, 1, 1, -1, -1, -1)$ and if s_1 is even and s_2, s_3 are odd, we get $(1, -1, 1, 1, -1, -1)$. Hence $\alpha_1, \ldots, \alpha_6$ satisfy the homogeneous system of linear equations with the associated matrix

$$T = \begin{pmatrix} 1 & 1 & 1 & 1 & 1 & 1 \\ 1 & -1 & -1 & 1 & 1 & -1 \\ 1 & 1 & -1 & -1 & 1 & -1 \\ 1 & 1 & 1 & -1 & -1 & -1 \\ 1 & -1 & 1 & 1 & -1 & -1 \end{pmatrix}.$$

If we now add the second and the fourth equation, we get $\alpha_1 = \alpha_6$. By subtracting the third equation from the fourth we get $\alpha_3 = \alpha_5$ and by subtracting the fifth from the fourth we get $\alpha_2 = \alpha_4$.

In order to get one more relation, we consider the evaluation ϕ such that

$$\psi(x_i) = \varphi(x_i), \quad i = 1, 2, 3$$

and

$$\psi(x_4) = t_1 e_{22}, \quad \psi(x_5) = t_2 e_{22}, \quad psi(x_6) = t_3 e_{22}.$$

Then

$$\psi(f_1) = -t_1 t_3 t_2 e_{11}, \quad \psi(f_2) = -t_1 t_2 t_3 e_{11}, \quad \psi(f_3) = -t_2 t_3 t_1 e_{11},$$
$$\psi(f_4) = -t_2 t_1 t_3 e_{11}, \quad \psi(f_5) = -t_3 t_2 t_1 e_{11}, \quad \psi(f_6) = -t_3 t_1 t_2 e_{11}.$$

By choosing the parities of t_1, t_2, t_3 as $(1, 1, 0), (1, 0, 1)$ and $(1, 1, 1)$, we obtain three linear equations among $\alpha_1, \ldots, \alpha_6$ with coefficients

$$(1, 1, -1, -1, -1, 1), \ (1, 1, -1, 1, -1, -1) \text{ and } (-1, 1, 1, -1, -1, 1),$$

respectively. From this we obtain $\alpha_4 = \alpha_6$ and $\alpha_2 + \alpha_6 = \alpha_4 + \alpha_5$. Taking into account the previous relations we obtain $\alpha_1 = \cdots = \alpha_6 = 0$, a contradiction.

We have proved that f cannot be a linear combination of the polynomials in (9.9). Hence f is a linear combination of the polynomials in (9.8) and some polynomial of the type

$$f_{ijkl} = [x_i, x_1, x_j][x_k, x_2, x_l], \quad g_{ijkl} = [x_i, x_2, x_j][x_k, x_1, x_l],$$
$$p_{ijkl} = [x_i, x_1, x_2][x_j, x_k, x_l], \quad q_{ijkl} = [x_i, x_j, x_k][x_l, x_1, x_2].$$

Recall that f is alternating on x_1 and x_2, i.e.,

$$f(x_1, x_2, x_3, \ldots, x_6) = -f(x_2, x_1, x_3, \ldots, x_6).$$

Since $f_{ijkl}, g_{ijkl}, p_{ijkl}, q_{ijkl}$ are linearly independent, it follows that the coefficients of f_{ijkl} and g_{ijkl} are opposite, for any fixed (i, j, k, l). On the other hand, if we exchange x_1 and x_2 in all polynomials, the polynomial p_{ijkl} becomes

$$[x_i, x_2, x_1][x_j, x_k, x_l] = [x_i, x_1, x_2][x_j, x_k, x_l] - [x_2, x_1, x_i][x_j, x_k, x_l]$$

and the only other polynomials giving rise to the same products of Lie commutators are $[x_2, x_1, x_\alpha][x_\beta, x_\gamma, x_\delta]$ which give the same element with coefficient 2. It follows that the polynomials p_{ijkl} (and similarly q_{ijkl}) have zero coefficients in the decomposition of f.

We now fix i, j, k, l and consider the evaluation

$$\varphi(x_i) = s_1 e_{12}, \quad \varphi(x_1) = e_{11}, \quad \varphi(x_j) = t_1 e_{22},$$
$$\varphi(x_k) = e_{21}, \quad \varphi(x_2) = s_2 e_{11}, \quad \varphi(x_l) = t_2 e_{22}$$

with odd $s_1, s_2 \in S$, $t_1 t_2 \in S^{(0)} \cup S^{(1)}$ and t_1, t_2 have the same parity. Under this evaluation $\varphi(f_{i'j'k'l'} - g_{i'j'k'l'})$ can have the summand e_{11} only if $i \in \{i', j'\}$ and $k \in \{k', l'\}$. Since $\varphi(x_1 x_j) = \varphi(x_1 x_k) = \varphi(x_2 x_k) = \varphi(x_2 x_l) = 0$, it follows that $i = i'$, $k = k'$ and $(i', j', k', l') = (i, j, k, l)$ or (i, l, k, j). Note also that $\varphi(g_1) = \varphi(g'_1) = \varphi(g_2) = \varphi(g'_2) = \varphi(f_m) = \varphi(f'_m) = 0$ for all $1 \leq m \leq 6$, since $\varphi([x_1, x_2]) = 0$ and f does not contain summands of the type p_{ijkl}, q_{ijkl}. Hence only $h_1 = f_{ijkl} - g_{ijkl}$ and $h_2 = f_{ilkj} - g_{ilkj}$ may have the summand e_{11}. By computing $\varphi(h_1)$ and $\varphi(h_2)$ we obtain

$$\varphi(h_1) = (s_1 t_1 t_2 s_2 - s_2 s_1 t_1 t_2) e_{11} \quad \text{and} \quad \varphi(h_2) = (s_1 t_2 t_1 s_2 - s_2 s_1 t_2 t_1) e_{11}.$$

Since $t_1 t_2$ is even, we have

$$\varphi(h_1) = 2 s_1 s_2 t_1 t_2 e_{11}, \varphi(h_2) = 2 s_1 s_2 t_2 t_1 e_{11}.$$

Let α_1 and α_2 be the coefficients of h_1 and h_2, respectively in the decomposition of f. By choosing the parities of t_1, t_2 as $(0, 0)$ and $(1, 1)$, we obtain $\alpha_1 + \alpha_2 = 0$ and $\alpha_1 - \alpha_2 = 0$. Hence $\alpha_1 = \alpha_2 = 0$. Since i, j, k, l are arbitrary, all f_{ijkl}, g_{ijkl} have zero coefficient. This contradiction completes the proof of the lemma. $\square$

An easy induction allows us to generalize the previous lemma to the case $k \geq 3$ as follows.

LEMMA 9.2.4. *If $k \geq 2$, the algebra $M_k(G)$ does not satisfy any identity of degree $3k$.*

PROOF. For $k = 2$ this is the content of Lemma 9.2.3. If $k \geq 3$, by Lemma 9.2.1, $I_k = \mathrm{Id}(S(M_k(F \oplus cF))) = \mathrm{Id}(M_k(G))$ is contained in the product $I_{k-1} I_1$. By induction the minimal degree of a polynomial in I_{k-1} is at least $3(k-1) + 1 = 3k - 2$. Since I_1 is generated by a polynomial of degree 3 (Theorem 4.1.8), we have $\deg(f) \geq 3k + 1$ for any $f \in I_{k-1} I_1$, and the proof is complete. $\square$

For simple superalgebras of type $M_{k,l}(F)$ the following lemma gives a lower bound of an identity of its Grassmann envelope.

LEMMA 9.2.5. *If $f \equiv 0$ is an identity of $M_{k,l}(G)$, then $\deg(f) \geq 2(k+l)$.*

PROOF. By Theorem 1.3.8, we may assume f to be multilinear and let

$$f = x_1 \cdots x_n + \sum_{\substack{\sigma \in S_n \\ \sigma \neq 1}} \alpha_\sigma x_{\sigma(1)} \cdots x_{\sigma(n)}.$$

If $n \leq 2m - 1$ with $m = k + l$, then, by eventually multiplying f on the right by a monomial in new variables, we may assume that $n = 2m - 1$. Since $\mathrm{Id}(M_{k,l}(G)) = \mathrm{Id}(S(M_{k,l}(F)))$, we evaluate f in $S(M_{k,l}(F))$. Let

$$\bar{x}_1 = e_{11} \otimes s_1, \quad \bar{x}_2 = e_{12} \otimes s_2, \quad \bar{x}_3 = e_{22} \otimes s_3,$$
$$\ldots, \bar{x}_{n-1} = e_{m-1,m} \otimes s_{n-1}, \quad \bar{x}_n = e_{mm} \otimes s_n,$$

where all s_i's except s_{2k} are equal to 1 and s_{2k} is an odd generator of S. Then $f(\bar{x}_1, \ldots, \bar{x}_n) = e_{1m} \otimes s_{2k} \neq 0$ and the lemma is proved. $\square$

We now have all the ingredients for finding the lower bound of the exponent of an arbitrary polynomial.

THEOREM 9.2.6 ([**BR5**]). *Let f be a homogeneous polynomial of degree $n \geq 4$. Then $\exp(f) \leq [\frac{n}{2}]^2 = \exp(St_n)$.*

PROOF. As we remarked in Section 1, the exponent of f is the largest exponent of $S(A)$ where A runs among all minimal superalgebras with $S(A)$ satisfying $f \equiv 0$. Hence it is enough to prove that $\exp(S(A)) \leq [\frac{n}{2}]^2$ as soon as $S(A)$ satisfies an identity of degree n. Let A be one such minimal superalgebra and let

$$A/J(A) \cong M_{a_1}(F) \oplus \cdots \oplus M_{a_r}(F) \oplus M_{p_1,q_1}(F) \oplus$$
$$\cdots \oplus M_{p_t,q_t}(F) \oplus M_{b_1}(F \oplus cF) \oplus \cdots \oplus M_{b_s}(F \oplus cF).$$

Since A is minimal and $\mathrm{Id}(G(A)) = \mathrm{Id}(S(A))$, we have that $\exp(S(A)) = \exp(G(A)) = \dim A/J(A)$. Hence

(9.11) $\exp(S(A)) = a_1^2 + \cdots + a_r^2 + (p_1 + q_1)^2 + \cdots + (p_t + q_t)^2 + 2(b_1^2 + \cdots + b_s^2).$

Denote by s' the number of b_i's among $b_1, \ldots, b_s$, satisfying the inequality $b_i \geq 2$. Since $\mathrm{Id}(S(A)) = \mathrm{Id}(M_{a_1}(F)) \cdots \mathrm{Id}(M_{b_s}(S))$ and the minimal degree of an identity of $S(F \oplus cF)$ is 3, by invoking Theorem 1.7.2, and Lemmas 9.2.4 and 9.2.5, we obtain that the degree of a polynomial in $\mathrm{Id}(S(A))$ is at least

$$2(a_1 + \cdots + a_r + p_1 + q_1 + \cdots p_t + q_t) + 3(b_1 + \cdots + b_s) + s'.$$

Now, if $\deg(f) = n = 2m$ is even, then

$$\frac{n}{2} = m \geq a_1 + \cdots + a_r + p_1 + q_1 + \cdots p_t + q_t + \frac{3}{2}(b_1 + \cdots + b_s)$$

and

$$[\frac{n}{2}]^2 = \frac{n^2}{4} \geq a_1^2 + \cdots + a_r^2 + (p_1 + q_1)^2 + \cdots + (p_t + q_t)^2 + \frac{9}{4}(b_1^2 + \cdots + b_s^2).$$

Since $\frac{9}{4} > 2$, by comparing with (9.11), we obtain

$$\exp(S(A)) \leq [\frac{n}{2}]^2,$$

as desired.

Suppose now that $n = 2m + 1$ is odd. Then since

$$2m + 1 \geq 2(a_1 + \cdots + a_r + p_1 + q_1 + \cdots p_t + q_t) + 3(b_1 + \cdots + b_s) + s',$$

if $s' \neq 0$, we obtain

$$2m \geq 2(a_1 + \cdots + a_r + p_1 + q_1 + \cdots p_t + q_t) + 3(b_1 + \cdots + b_s)$$

and $\exp(S(A)) \leq m^2 = [\frac{n}{2}]^2$, as before.

Suppose now that $s' = 0$. Then either $s = 0$ or $b_1 = \ldots = b_s = 1$. If $s = 0$,

$$2m + 1 \geq 2(a_1 + \cdots + a_r + p_1 + q_1 + \cdots p_t + q_t)$$

and, since $2m + 1$ is odd and $2(a_1 + \cdots + a_r + p_1 + q_1 + \cdots p_t + q_t)$ is even,

$$2m \geq 2(a_1 + \cdots + a_r + p_1 + q_1 + \cdots p_t + q_t).$$

Hence

$$[\frac{n}{2}]^2 = m^2 \geq (a_1 + \cdots + a_r + p_1 + q_1 + \cdots p_t + q_t)^2$$
$$\geq a_1^2 + \cdots + a_r^2 + (p_1 + q_1)^2 + \cdots (p_t + q_t)^2 = \exp(S(A)),$$

and the proof is complete in this case.

Now let $s \neq 0$ and $b_1 = \cdots = b_s = 1$. Then

$$[\frac{n}{2}] = m \geq a_1 + \cdots + a_r + p_1 + q_1 + \cdots p_t + q_t + \frac{3}{2}s - \frac{1}{2}.$$

Also, by (9.11),
$$\exp(S(A)) = a_1^2 + \cdots + a_r^2 + (p_1 + q_1)^2 + \cdots + (p_t + q_t)^2 + 2s.$$
If $s \geq 2$, then
$$(\frac{3}{2}s - \frac{1}{2})^2 = \frac{9}{4}s^2 - \frac{3}{2}s + \frac{1}{4} > \frac{9}{4}s^2 - \frac{3}{2}s = s(\frac{9}{4}s - \frac{3}{2}) \geq s(\frac{9}{2} - \frac{3}{2}) = 3s > 2s$$
and $m^2 > \exp(S(A))$, in this case.

We are left with the case $s = 1$, $b_1 = 1$. Write
$$h = a_1 + \cdots + a_r + p_1 + q_1 + \cdots + p_t + q_t.$$
If $h \geq 1$, then
$$[\frac{n}{2}]^2 = m^2 \geq (h+1)^2 > h^2 + 2h$$
$$\geq a_1^2 + \cdots + a_r^2 + (p_1 + q_1)^2 + \cdots + (p_t + q_t)^2 + 2 \geq \exp(S(A)),$$
and we are done.

The case $h = 0$ is trivial since then $A \cong F \oplus cF$, $\exp(S(A)) = \exp(G) = 2$ and by hypothesis $n \geq 4 > 2$. $\square$

We remark that the conclusion of the previous theorem is no longer true for $n = 3$. In fact, the Grassmann algebra G satisfies $[x_1, x_2, x_3] \equiv 0$, an identity of degree 3, but $\exp(G) = 2$.

9.3. Powers of standard polynomials

In this section we compute the PI-exponent of the polynomial St_n^m for arbitrary n and m. As in the previous sections we need to study the minimal varieties satisfying St_n^m. Recalling that the T-ideal of identities of a minimal variety is a product of verbally prime T-ideals (Theorem 8.4.2), we need to check the inclusion $St_n^m \in I_1 \cdots I_k$, where $I_1, \ldots, I_k$ are verbally prime T-ideals.

LEMMA 9.3.1. *Let $I_1, \ldots, I_m, I_{m+1}, \ldots, I_k$ be verbally prime T-ideals with $1 \leq m < k$, and set $Q_1 = I_1 \cdots I_m$, $Q_2 = I_{m+1} \cdots I_k$. If for some $r, l \geq 1$, $St_n^r \notin Q_1$ and $St_n^l \notin Q_2$, then $St_n^{r+l} \notin Q_1 Q_2$.*

PROOF. By Theorem 8.4.3, there exists a minimal superalgebra A such that $\mathrm{Id}(G(A)) = I_1 \cdots I_k$. Also, A decomposes as $A = A_1 \oplus \cdots \oplus A_k + J$, where $A_1, \ldots, A_k$ are simple superalgebras and $\mathrm{Id}(G(A_i)) = I_i$ for all $i = 1, \ldots, k$. Since A is a minimal superalgebra, there exist $w_{12}, \ldots, w_{k-1,k} \in J$ such that $w_{12}, \ldots, w_{k-1,k}$ generate J as an ideal of A and
$$w_{i,i+1} e_{i+1} = e_i w_{i,i+1} = w_{i,i+1},$$
for all $i = 1, \ldots, k-1$, and for suitable minimal idempotents $e_i \in A_i$. We next consider two subalgebras of A.

Let B_1 be the subalgebra generated by the simple algebras $A_1, \ldots, A_m$ and by $w_{12}, \ldots, w_{m-1,m}$ and let B_2 be the subalgebra generated by $A_{m+1}, \ldots, A_k$ and $w_{m+1,m+2}, \ldots, w_{k-1,k}$. Then $\mathrm{Id}(G(B_1)) = Q_1$ and $\mathrm{Id}(G(B_2)) = Q_2$. Now, as above we work with $S(A)$, $S(B_1)$, $S(B_2)$ instead of the corresponding Grassmann envelopes.

Write $M = rn$ and let $f(x_1, \ldots, x_M)$ be the complete linearization of St_n^r. Then f has the form
$$f = \mathrm{Symm}_1 \cdots \mathrm{Symm}_r (St_n(x_1, \ldots, x_n) \cdots St_n(x_{n(r-1)+1}, \ldots, x_M))$$

where Symm_1 denotes symmetrization on the variables $x_1, x_{n+1}, \ldots, x_{n(r-1)+1}$, Symm_2 denotes symmetrization on the variables $x_2, x_{n+2}, \ldots, x_{n(r-1)+2}$ and so on. Since by hypothesis $St_n^r \notin Q_1$, then $f \notin Q_1$, i.e., f is not an identity of $S(B_1)$. Similarly, write $N = nl$ and let $g = g(y_1, \ldots, y_N)$ be the complete linearization of St_n^l, i.e.,

$$g = \text{Symm}_1 \cdots \text{Symm}_l (St_n(y_1, \ldots, y_n) \cdots St_n(y_{n(l-1)+1}, \ldots, y_N)).$$

Then also g is not an identity of $S(B_2)$. This means that there exist $\bar{x}_1, \ldots, \bar{x}_M \in S(B_1)$, $\bar{y}_1, \ldots, \bar{y}_N \in S(B_2)$ such that $f(\bar{x}_1, \ldots, \bar{x}_M) \neq 0$ and $g(\bar{y}_1, \ldots, \bar{y}_N) \neq 0$.

We recall the structure of a minimal superalgebra. If $m = 1$, then $B_1 = A_1$ contains a minimal idempotent e_1 such that $e_1 w_{12} = w_{12}$. Since $S(B_1)$ does not satisfy $St_n^r \equiv 0$, $A_1 \ncong F$. Hence either A_1 contains at least one more minimal graded idempotent $e_1' \neq e_1$ or $A_1 = F \oplus cF$. By Lemma 8.3.1, the space $f(S(A_1))$ contains the element $(e_1 - e_1') \otimes p$ or the element $1 \otimes p$, for some $p \in S$. Since $e_1' w_{12} = e_1' e_1 w_{12} = 0$, there exists $\bar{z} = w_{12} \otimes h = w_{1,m+1} \otimes h$, for some $h \in S$ such that

(9.12) $$(\bar{f}_1 + \cdots + \bar{f}_\alpha)\bar{z} = w_{1,m+1} \otimes ph \neq 0$$

in $S(A)$, where $\bar{f}_1, \ldots, \bar{f}_\alpha$ are some evaluations of f in $S(A_1)$. In case $m > 1$, by Lemma 8.3.1, the relation (9.12) still holds for a suitable $\bar{z} = w_{m,m+1} \otimes h$, $h \in S$. Similarly we obtain that there exist evaluations $\bar{g}_1, \ldots, \bar{g}_\beta$ of g in $S(B_2)$ such that

$$(w_{1,m+1} \otimes ph)(\bar{g}_1 + \cdots + \bar{g}_\beta) = w_{1,k} \otimes phq \neq 0.$$

Hence $\bar{f}_i \bar{z} \bar{g}_j \neq 0$ for at least one pair (i, j), and we may assume without loss of generality, that

(9.13) $$f(\bar{x}_1, \ldots, \bar{x}_M)\bar{z}g(\bar{y}_1, \ldots, \bar{y}_N) \neq 0.$$

Notice that since for all $i = 1, \ldots, M$, $\bar{z}\bar{x}_i = 0$, we have that $[\bar{x}_i, \bar{z}] = \bar{x}_i \bar{z}$. Similarly $f(\bar{x}_1, \ldots, \bar{x}_M)\bar{z} = [f(\bar{x}_1, \ldots, \bar{x}_M), \bar{z}]$. On the other hand, from the linearity of f it follows that

$$[f(\bar{x}_1, \ldots, \bar{x}_M), \bar{z}] = \sum_{i=1}^{M} f(\bar{x}_1, \ldots, [\bar{x}_i, \bar{z}], \ldots, \bar{x}_M) = \sum_{i=1}^{M} f(\bar{x}_1, \ldots, \bar{x}_i \bar{z}, \ldots, \bar{x}_M).$$

Hence from (9.13) it follows that at least one product of the type

$$f(\bar{x}_1, \ldots, \bar{x}_\gamma \bar{z}, \ldots, \bar{x}_M) g(\bar{y}_1, \ldots, \bar{y}_N)$$

is non-zero. We obtain that

(9.14) $$a = f(\bar{x}_1, \ldots, \bar{x}_M) g(\bar{y}_1, \ldots, \bar{y}_N) \neq 0,$$

where $\bar{y}_1, \ldots, \bar{y}_N \in S(B_2)$, $\bar{x}_1, \ldots, \bar{x}_{\gamma-1}, \bar{x}_{\gamma+1}, \ldots, \bar{x}_M \in S(B_1)$ and $\bar{x}_\gamma = \bar{x}_\gamma' \bar{z}$ with $\bar{x}_\gamma' \in S(B_1)$, $\bar{z}\bar{x}_j = 0$, $j = 1, \ldots, M$, $\bar{y}_i \bar{z} = 0$, $i = 1, \ldots, N$.

Now recall that since f and g are the complete linearizations of St_n^r and St_n^l, respectively, the set $\{x_1, \ldots, x_M\}$ is partitioned into n subsets $\{x_1, \ldots, x_M\} = X_1 \cup \ldots \cup X_n$, where each

$$X_i = \{x_i, x_{n+i}, x_{2n+i}, \ldots\}, \quad 1 \leq i \leq n,$$

has order r and f is symmetric on the variables of each set X_i. Similarly, g is symmetric on the variables of each set Y_i, $1 \leq i \leq n$, where

$$Y_i = \{y_i, y_{n+i}, y_{2n+i}, \ldots\}.$$

If we symmetrize the product $f(x_1,\ldots,x_M)g(y_1,\ldots,y_N)$ first on the set $X_1 \cup Y_1$, then on $X_2 \cup Y_2$, and so on, then we obtain a multilinear polynomial $h = h(x_1,\ldots,x_M,y_1,\ldots,y_N)$ which is the complete linearization of St_n^{r+l}.

We claim that h is not an identity of $S(A)$. Indeed, any permutation of $\bar{x}_i$ with $\bar{y}_i$ on the left-hand side of (9.14) gives zero value since $\bar{x}_\alpha \bar{y}_\beta = \bar{y}_\beta \bar{x}_\alpha = \bar{z}\bar{x}_\alpha = \bar{y}_\beta \bar{z} = 0$ in $S(A)$, for any α, β. Hence the value of $h(x_1,\ldots,y_N)$ on $\bar{x}_1,\ldots,\bar{x}_M,\bar{y}_1,\ldots,\bar{y}_N$ is equal to $(r!)^n(l!)^n a \neq 0$. Since the linearization of St_n^{r+l} is not an identity of $S(A)$, the polynomial St_n^{r+l} does not lie in $Q_1 Q_2 = \mathrm{Id}(S(A))$, and the proof is complete. □

We remark that the superenvelope of $F \oplus cF$, containing the Grassmann algebra G, does not satisfy any standard identity. Hence by Lemma 9.2.1 and Lemma 9.3.1 we have that the superenvelope (and the Grassmann envelope) of $M_k(F \oplus cF)$ does not satisfy $St_n^k \equiv 0$, for $n \geq 1$. We denote for short $St_1(x) = x$. We record this in the following.

LEMMA 9.3.2. *The algebra $M_k(G)$ does not satisfy the identity $St_n^k \equiv 0$, for all $n \geq 1$.*

Now we prove that the superenvelope of the simple superalgebra $M_{1,1}(F)$ does not satisfy the square of any standard identity.

LEMMA 9.3.3. *The algebra $M_{1,1}(G)$ does not satisfy the identity $St_n^2 \equiv 0$, for all $n \geq 1$, and the identity*
$$St_i(x_1,\ldots,x_i)St_{i+j+1}(x_1,\ldots,x_{i+j+1})St_j(x_{i+1},\ldots,x_{i+j}) \equiv 0,$$
for all $i,j \geq 0$.

PROOF. Let $A = M_{1,1}(F)$. Since $\mathrm{Id}(M_{1,1}(G)) = \mathrm{Id}(G(A)) = \mathrm{Id}(S(A))$, we shall prove the lemma for the superenvelope $S(A)$. Consider the complete linearization $g = g(x_1,\ldots,x_{2n})$ of St_n^2. Thus
$$g = \sum_{\sigma \in H} St_n(x_{\sigma(1)},\ldots,x_{\sigma(n)}) St_n(x_{\sigma(n+1)},\ldots,x_{\sigma(2n)}),$$

where H is the subgroup of S_{2n} generated by the transpositions $(1,n+1)$, $(2,n+2),\ldots,(n,2n)$. In other words,
$$g = e_{T_\lambda}(x_1 \cdots x_{2n}) = \sum_{\sigma \in R_{T_\lambda}} \sum_{\tau \in C_{T_\lambda}} (\mathrm{sgn}\,\tau) x_{\sigma\tau(1)} \cdots x_{\sigma\tau(2n)},$$

where $\lambda = (2^n) \vdash 2n$ and T_λ is the standard tableau whose first column is filled up from top to bottom with the integers $1,2,\ldots,n$ and the second column is filled up from top to bottom with $n+1,\ldots,2n$.

In order to prove the lemma, we shall verify that g does not vanish on odd elements of $S(A)$. To this end, recall that by Lemma 4.8.6, it is enough to show that
$$f(x_1,\ldots,x_{2n}) = \pm e^*_{T_{\lambda'}}(x_1 \cdots x_{2n})$$

is not an identity of the odd component $A^{(1)}$ of A, where $T_{\lambda'}$ is the conjugate Young tableau of T_λ, that is,

$$T_{\lambda'} = \begin{array}{|c|c|c|} \hline 1 & \cdots & n \\ \hline n+1 & \cdots & 2n \\ \hline \end{array}$$

Now,

$$\pm f(x_1,\ldots,x_{2n}) = \sum_{\sigma\in H}(\operatorname{sgn}\sigma)\left(\sum_{\tau\in S_n,\tau'\in S_n'} x_{\tau(1)}\cdots x_{\tau(n)}x_{\tau'(n+1)}\cdots x_{\tau'(2n)}\right),$$

where S_n' is the symmetric group on $\{n+1,\ldots,2n\}$ and $H\subseteq S_n$ is the subgroup generated by $(1,n+1),\ldots,(n,2n)$.

We now evaluate f on the odd matrices

$$\bar{x}_i = \begin{pmatrix}0 & \alpha_i\\ 1 & 0\end{pmatrix},\quad \bar{x}_{n+i} = \begin{pmatrix}0 & 1\\ \beta_i & 0\end{pmatrix},\quad i=1,\ldots,n.$$

Then for any $\tau\in S_n$, $\tau'\in S_n'$, the product $x_{\tau(1)}\cdots x_{\tau(n)}x_{\tau'(n+1)}\cdots x_{\tau'(2n)}$ takes value

$$\begin{pmatrix}0 & \alpha_{i_1}\\ 1 & 0\end{pmatrix}\cdots\begin{pmatrix}0 & \alpha_{i_n}\\ 1 & 0\end{pmatrix}\begin{pmatrix}0 & 1\\ \beta_{j_1} & 0\end{pmatrix}\cdots\begin{pmatrix}0 & 1\\ \beta_{j_n} & 0\end{pmatrix}$$

and in the subgroup H there exists at most one element ε with the following property: in the product $\bar{x}_{\varepsilon\tau(1)}\cdots\bar{x}_{\varepsilon\tau(n)}\bar{x}_{\varepsilon\tau'(n+1)}\cdots\bar{x}_{\varepsilon\tau'(2n)}$ every element of the type $\bar{x}_j$ with $1\le j\le n$, is followed by an element of the type $\bar{x}_{j'}$, $n\le j'\le 2n$.

We call the pair (τ,τ') "good" if such an ε exists. Note that for any n, the number $N(n)$ of good pairs is non-zero. For example, if $n=2k+1$ is odd, then (e,e) is a good pair since we can take $\varepsilon=(2,n+2)(4,n+4)\cdots(2k,n+2k)$. If $n=2k$ is even, then $(1,\tau')$ is a good pair for the n-cycle $\tau'=(2n,2n-1,\cdots,n+1)$. In this case $\varepsilon=(2,n+2)(4,n+4)\cdots(n,2n)$. Note also that $\operatorname{sgn}\varepsilon=(-1)^{[\frac{n}{2}]}$.

Since

$$\begin{pmatrix}0 & \alpha_i\\ 1 & 0\end{pmatrix}\begin{pmatrix}0 & 1\\ \beta_j & 0\end{pmatrix} = \begin{pmatrix}\alpha_i\beta_j & 0\\ 0 & 1\end{pmatrix},\quad \begin{pmatrix}0 & 1\\ \beta_i & 0\end{pmatrix}\begin{pmatrix}0 & \alpha_j\\ 1 & 0\end{pmatrix} = \begin{pmatrix}1 & 0\\ 0 & \alpha_j\beta_i\end{pmatrix},$$

and

$$\begin{pmatrix}0 & \alpha_i\\ 1 & 0\end{pmatrix}\begin{pmatrix}0 & \alpha_j\\ 1 & 0\end{pmatrix} = \begin{pmatrix}\alpha_i & 0\\ 0 & \alpha_j\end{pmatrix},\quad \begin{pmatrix}0 & 1\\ \beta_i & 0\end{pmatrix}\begin{pmatrix}0 & 1\\ \beta_j & 0\end{pmatrix} = \begin{pmatrix}\beta_j & 0\\ 0 & \beta_i\end{pmatrix},$$

then

$$(\operatorname{sgn}\varepsilon)\bar{x}_{\varepsilon\tau(1)}\cdots\bar{x}_{\varepsilon\tau(n)}\bar{x}_{\varepsilon\tau'(n+1)}\cdots\bar{x}_{\varepsilon\tau'(2n)} = (-1)^{[\frac{n}{2}]}\begin{pmatrix}\alpha_{i_1}\cdots\alpha_{i_n}\beta_{j_1}\cdots\beta_{j_n} & 0\\ 0 & 1\end{pmatrix}$$

$$= (-1)^{[\frac{n}{2}]}\begin{pmatrix}\alpha_1\cdots\alpha_n\beta_1\cdots\beta_n & 0\\ 0 & 1\end{pmatrix}$$

for any "good" pair (τ,τ'). All other products $\bar{x}_{\sigma\tau(1)}\cdots\bar{x}_{\sigma\tau(n)}\bar{x}_{\sigma\tau'(n+1)}\cdots\bar{x}_{\sigma\tau'(2n)}$, with $\sigma\in H$, are of the type

$$\begin{pmatrix}\alpha_{i_1}\cdots\alpha_{i_k}\beta_{j_1}\cdots\beta_{j_m} & 0\\ 0 & *\end{pmatrix},$$

with $k+m\le 2n-1$. Hence

$$f(\bar{x}_1,\ldots,\bar{x}_{2n}) = \begin{pmatrix}\varphi & 0\\ 0 & \psi\end{pmatrix},$$

where $\varphi = \varphi(\alpha_1,\ldots,\alpha_n,\beta_1,\ldots,\beta_n)$ is a polynomial in $\alpha_1,\ldots,\alpha_n,\beta_1,\ldots,\beta_n$ of degree $2n$ and the monomial

$$(-1)^{[\frac{n}{2}]}N(n)\alpha_1\cdots\alpha_n\beta_1\cdots\beta_n$$

is the only monomial of φ of highest degree. Thus φ is non-zero as a polynomial and it has a non-zero value on some scalars $\alpha_1,\ldots,\alpha_n,\beta_1,\ldots,\beta_n$. This completes the proof of the first part of the lemma.

The proof of the second part is based on the same idea. We consider the complete linearization f of $St_i St_{i+j+1} St_j$. Hence

$$f = \sum_{\sigma \in P} \sigma St_i(x_{i+j+2}\ldots,x_{2i+j+1}) St_{i+j+1}(x_1,\ldots,x_{i+j+1})$$

$$\cdot St_j(x_{2i+j+2},\ldots,x_{2i+2j+1}) = \sum_{\sigma \in P}\sum_{\tau \in Q}(\operatorname{sgn}\tau)\sigma\tau g,$$

where

$$g = (x_{i+j+2}\cdots x_{2i+j+1})(x_1 \cdots x_{i+j+1})(x_{2i+j+2}\cdots x_{2i+2j+1}),$$

P is the subgroup of $S_{2i+2j+1}$ generated by the $i+j$ transpositions

$$(1,i+j+2),\ldots,(i,2i+j+1),(i+1,2i+j+2),\ldots,(i+j,2i+2j+1),$$

and $Q = Q_1 \times Q_2 \times Q_3$ with $Q_1 = S_i(i+j+2,\ldots,2i+j+1)$, $Q_2 = S_{i+j+1}(1,\ldots,i+j+1)$, $Q_3 = S_j(2i+j+2,\ldots,2i+2j+1)$, where $S_a(i_1,\ldots,i_a)$ denotes the symmetric group acting on the set $\{i_1,\ldots,i_a\}$. By Lemma 3.7.4, $f \equiv 0$ is an identity of $S(M_{1,1}(F))$ if and only if $\tilde{f} \equiv 0$ is an identity of $M_{1,1}(F)$.

If we replace the variables $x_1,\ldots,x_{2i+2j+1}$ with odd elements $\bar{x}_1,\ldots,\bar{x}_{2i+2j+1}$ of $M_{1,1}(F)$, we obtain

$$\tilde{f} = \sum_{\sigma \in P}\sum_{\tau \in Q}(\operatorname{sgn}\sigma)\sigma\tau g,$$

by Lemma 4.8.6. For $k = 1,\ldots,i+j+1$ take

$$\bar{x}_k = \begin{pmatrix} 0 & 1 \\ \beta_k & 0 \end{pmatrix}$$

and, for $k = 1,\ldots,i+j$ take

$$\bar{x}_{i+j+1+k} = \begin{pmatrix} 0 & \alpha_k \\ 1 & 0 \end{pmatrix}.$$

For any $\tau \in Q$ the product $\tau g(\bar{x}_1,\ldots,\bar{x}_{2i+2j+1})$ has the form

$$\begin{pmatrix} 0 & \alpha_{k_1} \\ 1 & 0 \end{pmatrix}\cdots\begin{pmatrix} 0 & \alpha_{k_i} \\ 1 & 0 \end{pmatrix}\begin{pmatrix} 0 & 1 \\ \beta_{r_1} & 0 \end{pmatrix}\cdots\begin{pmatrix} 0 & 1 \\ \beta_{r_{i+j+1}} & 0 \end{pmatrix}\begin{pmatrix} 0 & \alpha_{s_1} \\ 1 & 0 \end{pmatrix}\cdots\begin{pmatrix} 0 & \alpha_{s_j} \\ 1 & 0 \end{pmatrix}.$$

As before, for any $\tau \in Q$, there exists at most one permutation $\varepsilon \in P$ such that

$$\varepsilon\tau g(\bar{x}_1,\ldots,\bar{x}_{2i+2j+1})$$

$$= \begin{pmatrix} 0 & 1 \\ \beta_{l_1} & 0 \end{pmatrix}\begin{pmatrix} 0 & \alpha_{t_1} \\ 1 & 0 \end{pmatrix}\cdots\begin{pmatrix} 0 & 1 \\ \beta_{l_{i+j}} & 0 \end{pmatrix}\begin{pmatrix} 0 & \alpha_{t_{i+j}} \\ 1 & 0 \end{pmatrix}\begin{pmatrix} 0 & 1 \\ \beta_{l_{i+j+1}} & 0 \end{pmatrix}$$

$$= \begin{pmatrix} 0 & 1 \\ \alpha_1\cdots\alpha_{i+j}\beta_1\cdots\beta_{i+j+1} & 0 \end{pmatrix}.$$

If the permutation ε exists, we call τ a "good" permutation. If τ is not good or $\sigma \neq \varepsilon$, then

$$\sigma\tau g(\bar{x}_1,\ldots,\bar{x}_{2i+2j+1}) = \begin{pmatrix} 0 & y \\ * & 0 \end{pmatrix},$$

where y is a product of α's and β's with at most $2i + 2j$ factors. Hence

$$f(\bar{x}_1,\ldots,\bar{x}_{2i+2j+1}) = \begin{pmatrix} 0 & \varphi \\ \psi & 0 \end{pmatrix},$$

where $\varphi = \varphi(\alpha_1, \ldots, \alpha_{i+j}, \beta_1, \ldots, \beta_{i+j+1})$, with monomial of highest degree

$$(\operatorname{sgn}\varepsilon) N(i,j) \alpha_1 \cdots \alpha_{i+j} \beta_1 \cdots \beta_{i+j+1},$$

where $N(i,j)$ is the number of good permutations in Q. Note that $\operatorname{sgn}\varepsilon$ does not depend on τ but only on the parity of i and j and we shall next give its precise value.

In order to complete the proof we need only to show that for any pair (i,j) there exists at least one good permutation. We write $\tau = \tau_1 \tau_2 \tau_3$ with $\tau_1 \in Q_1$, $\tau_2 \in Q_2$, $\tau_3 \in Q_3$ and we obtain good permutations as follows.

If both i and j are odd, then $\tau_1 \tau_2 \tau_3$ is a good permutation for $\tau_1 = 1 \in Q_1$, $\tau_3 = 1 \in Q_3$ and $\tau_2 = (2i+1, 2i+2, \ldots, 2i+j+1)$, a $(j+1)$-cycle. The corresponding ε is equal to the following product of transpositions

$$(1, i+1)(3, i+3) \cdots (i, 2i)(2i+1, 2i+j+2)(2i+3, 2i+j+4) \cdots (2i+j, 2i+2j+1)$$

and $\operatorname{sgn}\varepsilon = (-1)^{[\frac{i}{2}]+[\frac{j}{2}]+2}$.

If i is odd and j is even, then we can take $\tau_1 = 1$, $\tau_2 = 1$, $\tau_3 = 1$ and

$$\varepsilon = (1, i+1)(3, i+3) \cdots (i, 2i)(2i+2, 2i+2j+3)(2i+4, 2i+2j+5) \cdots (2i+j, 2i+2j+1).$$

In this case $\operatorname{sgn}\varepsilon = (-1)^{[\frac{i}{2}]+[\frac{j}{2}]+1}$.

For i even and j odd, we may take $\tau_3 = 1$, $\tau_1 = (12)(34) \cdots (i-1, i)$, $\tau_2 = (2i+j+1, 2i+j, \ldots, 2i+1)$, a $(j+1)$-cycle, and

$$\varepsilon = (2, i+2)(4, i+4) \cdots (i, 2i)(2i+1, 2i+j+2)(2i+3, 2i+j+4) \cdots (2i+j, 2i+2j+1)$$

with $\operatorname{sgn}\varepsilon = (-1)^{[\frac{i}{2}]+[\frac{j}{2}]+1}$.

Finally, if both i and j are even, then $\tau_2 = 1$, $\tau_3 = 1$ and $\tau_1 = (12)(34) \cdots (i-1, i)$ give a good permutation with

$$\varepsilon = (2, i+2)(4, i+4) \cdots (i, 2i)(2i+2, 2i+j+3)(2i+4, 2i+j+5) \cdots (2i+j, 2i+2j+1)$$

and $\operatorname{sgn}\varepsilon = (-1)^{[\frac{i}{2}]+[\frac{j}{2}]}$. $\square$

We need a result similar to the previous one for the Grassmann algebra G. Recall that G has the same identities as the superenvelope $S(F \oplus cF)$.

LEMMA 9.3.4. *For any $i, j \geq 0$, the Grassmann algebra G does not satisfy the identity*

$$St_i(x_1, \ldots, x_i) St_j(x_{i+1}, \ldots, x_{i+j}) \equiv 0.$$

PROOF. Clearly if $g_1, \ldots, g_{i+j}$ are odd elements such that $g_1 \cdots g_{i+j} \neq 0$, then

$$St_i(g_1, \ldots, g_i) St_j(g_{i+1}, \ldots, g_{i+j}) = i! j! g_1 \cdots g_{i+j} \neq 0.$$

$\square$

As an application of Lemma 9.3.3 we obtain the following.

LEMMA 9.3.5. *For all $n \geq 1$, $St_n^{2l} \notin \operatorname{Id}(M_{k,l}(G))$.*

PROOF. We prove the lemma for the superenvelope $S(A)$ of $A = M_{k,l}(F)$. By Lemma 9.2.2, A contains a minimal superalgebra B with maximal semisimple subalgebra of the type $(M_{1,1}(F))^l = M_{1,1}(F) \oplus \cdots \oplus M_{1,1}(F)$. Since by Lemma 9.3.3, $M_{1,1}(F)$ does not satisfy $St_n^2 \equiv 0$, for any n, from Lemma 9.3.1 it follows that $St_n^{2l} \notin \operatorname{Id}(S(A))$. $\square$

The strategy of the proof of the next lemma is similar to that of Lemma 9.3.1. We use the notation $x_1 \cdots \hat{x}_i \cdots x_n = x_1 \cdots x_{i-1} x_{i+1} \cdots x_n$ to denote that the variable x_i is omitted.

LEMMA 9.3.6. *For $k \geq 3$, let $I_1, \ldots, I_k$ be verbally prime T-ideals such that $St_n^m \in I_1 \cdots I_k$. Suppose that for some $1 \leq t \leq m$ and $2 \leq s \leq k-1$, we have*

$$St_i(x_1, \ldots, x_i) St_n^{t-1}(x_1, \ldots, x_n) St_{n-i-1}(x_{i+1}, \ldots, x_{n-1}) \notin I_s, \quad i = 0, \ldots, n-2,$$

and

$$St_n^t \notin I_s.$$

Then $St_n^{m-t} \in I_1 \cdots \hat{I}_s \cdots I_k$, the product of all $I_1, \ldots, I_k$ except I_s.

PROOF. By Theorem 8.4.3, there exists a minimal superalgebra A such that $\mathrm{Id}(S(A)) = I_1 \cdots I_k$. Also $A = A_1 \oplus \cdots \oplus A_k + J$ with $A_1, \ldots, A_k$ simple superalgebras and $I_j = \mathrm{Id}(S(A_j))$, for all $j = 1, \ldots, k$. By Lemma 8.1.6, J has the following vector space decomposition

$$J = \bigoplus_{1 \leq i < j \leq k} J_{ij},$$

where $J_{ij} = A_i J A_j$. Let $C = A_s$ and set

$$B = A_1 \oplus \cdots \oplus A_{s-1} + \bigoplus_{1 \leq i < j \leq s-1} J_{ij},$$

$$D = A_{s+1} \oplus \cdots \oplus A_k + \bigoplus_{s+1 \leq i < j \leq k} J_{ij}.$$

Then B and D are also minimal superalgebras with $\mathrm{Id}(S(B)) = I_1 \cdots I_{s-1}$ and $\mathrm{Id}(S(D)) = I_{s+1} \cdots I_k$.

Fix integers $a, b \geq 0$ such that $a + b = m - t - 1$, and let $0 \leq i \leq n - 1$. Denote by $f = f(x_1, \ldots, x_{an+n-i-1})$ the complete linearization of the polynomial $St_n^a(x_1, \ldots, x_n) St_{n-i-1}(x_1, \ldots, x_{n-i-1})$. Hence

$$f = \sum_{\sigma \in P} \sigma St_n(x_1, \ldots, x_n)$$

$$\cdots St_n(x_{(a-1)n+1}, \ldots, x_{an}) St_{n-i-1}(x_{an+1}, \ldots, x_{an+n-i-1}),$$

where $P = P_1 \times \cdots \times P_n$ with P_j the symmetric group on $\{j, n+j, 2n+j, \ldots\}$, $1 \leq j \leq n$, and $|P_1| = \cdots = |P_{n-i-1}| = (a+1)!$, $|P_{n-i}| = \cdots = |P_n| = a!$. Similarly, let $g = g(y_1, \ldots, y_{bn+i})$ be the complete linearization of the polynomial $St_i(y_1, \ldots, y_i) St_n^b(y_1, \ldots, y_n)$. Then

$$g = \sum_{\sigma \in Q} \sigma St_i(y_{bn+1}, \ldots, y_{bn+i}) St_n(y_1, \ldots, y_n) \cdots St_n(y_{(b-1)n+1}, \ldots, y_{bn}),$$

where $Q = Q_1 \times \cdots \times Q_n$ with Q_j the symmetric group on $\{j, n+j, 2n+j, \ldots\}$, $1 \leq j \leq n$, and $|Q_1| = \cdots = |Q_i| = (b+1)!$, $|Q_{i+1}| = \cdots = |Q_n| = b!$.

We claim that either $f \equiv 0$ on $S(B)$ or $g \equiv 0$ on $S(D)$. Suppose to the contrary that either f is not an identity of $S(B)$ or g is not an identity of $S(D)$. Denote by $h = h(z_1, \ldots, z_{tn-1})$ the complete linearization of $St_i St_n^{t-1} St_{n-i-1}$. Hence

$$h = \sum_{\sigma \in T} \sigma St_i(z_{tn-i}, \ldots, z_{tn-1}) St_n(z_1, \ldots, z_n)$$

$$\cdots St_n(z_{(t-2)n+1}, \ldots, z_{(t-1)n}) St_{n-i-1}(z_{(t-1)n+1}, \ldots, z_{tn-i-1}),$$

where $T = T_1 \times \cdots \times T_n$ is the subgroup of S_{tn-1} with T_j the symmetric group on $\{j, n+j, 2n+j, \ldots\}$, $1 \leq j \leq n$, and $|T_1| = \cdots = |T_{n-1}| = t!$, $|T_n| = (t-1)!$. Thus by hypothesis, $h \equiv 0$ is not an identity of $S(C) = S(A_s)$.

Recall that according to Definition 8.1.3, the minimal superalgebra A is generated by $A_1, \ldots, A_k$ and by the elements $w_{12}, \ldots, w_{k-1,k} \in J$. Hence, by making use of Lemma 8.3.1, we can find evaluations $\bar{f}_1, \ldots, \bar{f}_\alpha, \bar{h}_1, \ldots, \bar{h}_\beta$ and $\bar{g}_1, \ldots, \bar{g}_\gamma$ of f, h and g in $S(B), S(C)$ and $S(D)$, respectively, such that

$$(\bar{f}_1 + \cdots + \bar{f}_\alpha)w_{s-1,s}(\bar{h}_1 + \cdots + \bar{h}_\beta)w_{s,s+1}(\bar{g}_1 + \cdots + \bar{g}_\gamma) \neq 0$$

in $S(A)$. Hence

(9.15) $\qquad f(\bar{x}_1, \ldots, \bar{x}_{an+n-i-1})\bar{u}_1 h(\bar{z}_1, \ldots, \bar{z}_{tn-1})\bar{u}_2 g(\bar{y}_1, \ldots, \bar{y}_{bn+i}) \neq 0,$

where $\bar{u}_1 = w_{s-1,s}$, $\bar{u}_2 = w_{s,s+1}$, $\bar{x}_1, \ldots, \bar{x}_{an+n-i-1} \in S(C)$, $\bar{z}_1, \ldots, \bar{z}_{tn-1} \in S(C)$, $\bar{y}_1, \ldots, \bar{y}_{bn+i} \in S(D)$.

Recall that the polynomials f, h and g are symmetric on the sets of variables corresponding to the subgroups Q, T and P, respectively. Hence $\sigma_1\sigma_2\sigma_3 f'h'g' \neq 0$, for some $\sigma_1 \in P$, $\sigma_2 \in T$, $\sigma_3 \in Q$, where f', g', h' are products of standard polynomials. It follows that, without loss of generality, we may assume that

$$St_n(\bar{x}_1, \ldots, \bar{x}_n) \cdots St_n(\bar{x}_{(a-1)n}, \ldots, \bar{x}_{an})St_{n-i-1}(\bar{x}_{an+1}, \ldots, \bar{x}_{an+n-i-1})\bar{u}_1$$

(9.16) $\qquad \cdot St_i(\bar{z}_{tn-i}, \ldots, \bar{z}_{tn-1})St_n(\bar{z}_1, \ldots, \bar{z}_n) \cdots$

$$St_n(\bar{z}_{(t-2)n+1}, \ldots, \bar{z}_{(t-1)n})St_i(\bar{z}_{(t-1)n+1}, \ldots, \bar{z}_{tn-i-1})\bar{u}_2$$

$$\cdot St_i(\bar{y}_{bn+1}, \ldots, \bar{y}_{bn+i})St_n(\bar{y}_1, \ldots, \bar{y}_n) \cdots St_n(\bar{y}_{(b-1)n+1}, \ldots, \bar{y}_{bn}) \neq 0.$$

Denote by $H = H(x_1, \ldots, z_{tn-1}, u_1, u_2)$ the polynomial whose evaluation appears on the left hand side of (9.16). Next we construct a Young tableau corresponding to the linearization of the polynomial St_n^m as follows.

Take $\lambda = (m, \ldots, m) = (m^n) \vdash mn$ and let D_λ be the corresponding rectangular Young diagram with n rows and m columns. If we now consider the mn variables

$$x_1, \ldots, x_{an+n-i-1}, u_1, z_{tn-i}, \ldots, z_{tn-1}, z_1,$$
$$\ldots, z_{tn-i-1}, u_2, y_{bn+1}, \ldots, y_{bn+i}, y_1, \ldots, y_{bn},$$

we can form, going from left to right, m consecutive disjoint subsets each of order n. We then fill up the columns of the diagram D_λ with these subsets from left to right and from top to bottom (we speak for short of variables instead of the corresponding indices). Now, if e_{T_λ} is the corresponding essential idempotent of FS_{mn}, then $e_{T_\lambda}H$ is the complete linearization of the polynomial St_n^m, up to a non-zero scalar factor.

We next check that $e_{T_\lambda}H(\bar{x}_1, \ldots, \bar{z}_{tn-1}, \bar{u}_1, \bar{u}_2) \neq 0$ in $S(A)$. In fact, the polynomial $e_{T_\lambda}H$ is the sum of polynomials of the form $(\text{sgn}\,\tau)\sigma\tau H$. From the relations

$$BC = CB = BD = DB = CD = DC = CJ_1 = DJ_1 = J_2B = J_2C = J_2J_1 = 0,$$

where $J_1 = BJC$, $J_2 = CJD$, it easily follows that

$$\sigma\tau H(\bar{x}_1, \ldots, \bar{z}_{tn-1}, \bar{u}_1, \bar{u}_2) = 0$$

unless

$$\sigma\tau(\{x_1, \ldots, x_{an+n-i-1}\}) = \{x_1, \ldots, x_{an+n-i-1}\},$$
$$\sigma\tau(\{y_1, \ldots, y_{bn+i}\}) = \{y_1, \ldots, y_{bn+i}\}$$
$$\sigma\tau(\{z_1, \ldots, z_{tn-1}\}) = \{z_1, \ldots, z_{tn-1}\}, \quad \sigma\tau(u_1) = u_1, \quad \sigma\tau(u_2) = u_2.$$

But under the above conditions σ and τ preserve each of the subsets
$$\{x_1,\ldots,x_{an+n-i-1}\},\ \{y_1,\ldots,y_{bn+i}\},\ \{z_1,\ldots,z_{tn-1}\},\ \{u_1,u_2\}.$$
Therefore $(\operatorname{sgn}\tau)\tau H(\bar{x}_1,\ldots,\bar{z}_{tn-1},\bar{u}_1,\bar{u}_2)$ equals the element on the left-hand side of (9.16) and $e_{T_\lambda}H(\bar{x}_1,\ldots,\bar{z}_{tn-1},\bar{u}_1,\bar{u}_2)$ equals the element on the left-hand side of (9.15), up to the scalar factor
$$(n!)^a (n-i-1)!\,i!\,(n!)^{t-1}i!(n-i-1)!(n!)^b \neq 0.$$
Therefore $e_{T_\lambda}H$ and St_n^m are not identities of $S(A)$, a contradiction.

We have proved that either $f \equiv 0$ is an identity of $S(B)$ or $g \equiv 0$ is an identity of $S(D)$, for all a,b,i such that $a+b = m-t-1$ and $0 \leq i \leq n-1$. Write $f = f_{a,i}$ and $g = g_{b,i}$. Next we deduce that $St_n^{m-t} \in I_1 \cdots \hat{I}_s \cdots I_k$.

By Theorem 8.4.3 there exists a minimal superalgebra
$$A' = A_1 \oplus \cdots \oplus A_{s-1} \oplus A_{s+1} \oplus \cdots \oplus A_k + J'$$
such that $\operatorname{Id}(S(A')) = I_1 \cdots \hat{I}_s \cdots I_k$. Let us define
$$B' = A_1 \oplus \cdots \oplus A_{s-1} + \sum_{1 \leq i < j \leq s-1} A_i J' A_j,$$
$$D' = A_{s+1} \oplus \cdots \oplus A_k + \sum_{s+1 \leq i < j \leq k} A_i J' A_j,$$
$$J'' = \sum_{\substack{1 \leq i \leq s-1 \\ s+1 \leq j \leq k}} A_i J' A_j.$$
Then $A' = B' + D' + J''$, $\operatorname{Id}(S(B')) = \operatorname{Id}(S(B)) = I_1 \cdots I_{s-1}$, $\operatorname{Id}(S(D')) = \operatorname{Id}(S(D)) = I_{s+1} \cdots I_k$ and $J''^2 = 0$. Hence $f_{a,i} \equiv 0$ is an identity of $S(B')$ or $g_{b,i} \equiv 0$ is an identity of $S(D')$, for all $a+b = m-t-1$, $0 \leq i \leq n-1$.

We recall that in order to complete the proof of the lemma we need to show that St_n^{m-t} is not an identity of $S(A')$.

We claim that both $S(B')$ and $S(D')$ satisfy the identity $St_n^{m-t} \equiv 0$. In fact, $L_1 = B+C+(B+C)J(B+C)$ and $L_2 = D+C+(D+C)J(D+C)$ are subalgebras of the minimal superalgebra A and, so, by the assumption they both satisfy $St_n^m \equiv 0$. Since $\operatorname{Id}(S(L_1)) = \operatorname{Id}(S(B))\operatorname{Id}(S(C))$, and by hypothesis $St_n^t \notin \operatorname{Id}(S(C))$, we can invoke Lemma 9.3.1 and conclude that $St_n^{m-t} \in \operatorname{Id}(S(B)) = \operatorname{Id}(S(B'))$. Similarly $St_n^{m-t} \in \operatorname{Id}(S(B)) = \operatorname{Id}(S(B'))$.

As before, we now consider the complete linearization h of the polynomial St_n^{m-t} and we write
$$h = h(x_1,\ldots,x_{(m-t)n}) = \sum_{\sigma \in T} \sigma h_1 \cdots h_{m-t},$$
where $h_j = St_n(x_{(j-1)n+1},\ldots,x_{jn})$ and $T = T_1 \times \cdots \times T_{m-t}$ is the subgroup of $S_{(m-t)n}$ with T_j the symmetric group on $\{j, n+j, 2n+j,\ldots\}$, $1 \leq j \leq n$. Since h is multilinear and $A' = B' + D' + J''$, it is enough to evaluate the variables in $S(B') \cup S(D') \cup S(J'')$. By the previous claim $h \equiv 0$ is an identity of $S(B')$ and of $S(D')$, hence we need only to check that h takes a non-zero value under an evaluation $x_j \to \bar{x}_j$, $1 \leq j \leq (m-t)n$, with at least one $\bar{x}_j$ in $S(J'')$. Since $J''^2 = 0$, we must have $\bar{x}_j \in S(J'')$ for exactly one j. Write such j as $j = na+i+1$, $0 \leq$

$a \leq m-t-1$. Then if we denote $\bar{h}_i = h_i(\bar{x}_{(i-1)n+1}, \ldots, \bar{x}_{in})$, $1 \leq i \leq m-t$, we have

$$h(\bar{x}_1, \ldots, \bar{x}_{(m-t)n}) = \left(\sum_{\sigma_1 \in T'} \sigma_1 h_1 \cdots h_a St_i(\bar{x}_{na+1}, \ldots, \bar{x}_{na+i}) \right) \bar{z}$$

$$\cdot \left(\sum_{\sigma_2 \in T''} \sigma_2 St_{n-i-1}(\bar{x}_{na+i+2}, \ldots, \bar{x}_{(a+1)n}) h_{a+2} \cdots h_{(m-t)n} \right),$$

where T', T'' are the subgroups of T preserving the subsets $\{1, \ldots, na+i\}$ and $\{na+i+2, \ldots, (m-t)n\}$, respectively and $\bar{z} = \bar{x}_{na+i+1} \in S(J'')$. Moreover, $\bar{x}_1, \ldots, \bar{x}_{na+i} \in S(B')$ and $\bar{x}_{na+i+2}, \ldots, \bar{x}_{(m-t)n} \in S(D')$. But then

$$\bar{h} = h(\bar{x}_1, \ldots, \bar{x}_{(m-t)n}) = f_{a,i}(\bar{x}_1, \ldots, \bar{x}_{na+i}) \bar{z} g_{b,i}(\bar{x}_{na+i+2}, \ldots, \bar{x}_{(m-t)n}),$$

where $b = m-t-1-a$. Since by the first part of the proof either $f_{a,i} \equiv 0$ is an identity of $S(B')$ or $g_{b,i} \equiv 0$ is an identity of $S(D')$, we get $\bar{h} = 0$, and the proof of the lemma is complete. $\square$

As a consequence of the previous lemmas we obtain the following.

LEMMA 9.3.7. *Let $I_1, \ldots, I_k$ be verbally prime T-ideals, $k \geq 2$. If $St_n^m \in I_1 \cdots I_k$ and $I_s = \mathrm{Id}(M_p(G))$ or $I_s = \mathrm{Id}(M_{p,l}(G))$, then $I_1 \cdots \hat{I}_s \cdots I_k$ contains St_n^{m-p} or St_n^{m-2l}, respectively.*

PROOF. Suppose first that $I_s = \mathrm{Id}(M_p(G)) = \mathrm{Id}(S(M_p(F \oplus cF)))$. If $s = 1$ or $s = k$, then the conclusion follows from Lemma 9.3.1 and Lemma 9.3.2. If $1 < s < k$, then, by Lemma 9.2.1, $I_s \subseteq I_0^p$, where $I_0 = \mathrm{Id}(M_1(S)) = \mathrm{Id}(G)$. Hence

$$St_n^m \in I_1 \cdots I_{s-1} I_0^p I_{s+1} \cdots I_k.$$

Since the Grassmann algebra does not satisfy the standard identity, by applying p times Lemma 9.3.6 we obtain

$$St_n^{m-p} \in I_1 \cdots I_{s-1} I_{s+1} \cdots I_k.$$

If $I_s = \mathrm{Id}(M_{p,l}(G)) = \mathrm{Id}(S(M_{p,l}(F)))$ with $p \geq l \geq 1$, then the same argument as above together with Lemmas 9.3.1, 9.3.5, 9.2.2, 9.3.4, 9.3.6, completes the proof of the lemma. $\square$

LEMMA 9.3.8. *The algebra $M_{2,1}(G)$ does not satisfy the identities $St_4^4 \equiv 0$ and*

$$St_i(x_1, \ldots, x_i) St_4^3(x_1, \ldots, x_4) St_{3-i}(x_{i+1}, \ldots, x_3) \equiv 0,$$

for any $i = 0, \ldots, 3$.

PROOF. We have $\mathrm{Id}(M_{2,1}(G)) = \mathrm{Id}(S(M_{2,1}(F)))$. Now, the algebra $S(M_{2,1}(F))$ is isomorphic to the algebra of 3×3 matrices of the type

$$\begin{pmatrix} A & B \\ C & D \end{pmatrix},$$

where A and D are 2×2 and 1×1 matrices, respectively, with entries in $S^{(0)}$ and B, C are 2×1 and 1×2 matrices, respectively, with entries in $S^{(1)}$.

First we make the following evaluations of $St_4(x_1, \ldots, x_4)$ on matrix units:

$$St_4(e_{11}, e_{12}, e_{22}, e_{13}) = St_3(e_{11}, e_{12}, e_{22}) e_{13} = 0,$$

$$St_4(e_{11}, e_{21}, e_{22}, e_{13}) = St_3(e_{11}, e_{21}, e_{22}) e_{13} = -e_{22} e_{21} e_{11} e_{13} = -e_{23}$$

and
$$St_4(e_{11}, e_{12}, e_{22}, e_{23}) = e_{13}, \quad St_4(e_{11}, e_{21}, e_{22}, e_{23}) = 0.$$
Hence
$$St_4(e_{11}, e_{12} + e_{21}, e_{22}, e_{13}) = -e_{23}, \quad St_4(e_{11}, e_{12} + e_{21}, e_{22}, e_{23}) = e_{13}.$$
If we replace e_{13} and e_{23} with e_{31} and e_{32} respectively, we get
$$St_4(e_{11}, e_{12} + e_{21}, e_{22}, e_{31}) = -e_{32} \text{ and } St_4(e_{11}, e_{12} + e_{21}, e_{22}, e_{32}) = e_{31}.$$
Now let $\alpha, \beta, \gamma, \delta \in S^{(1)}$ be odd elements. Then, if we set
$$\bar{x}_1 = e_{11}, \quad \bar{x}_2 = e_{12} + e_{21}, \quad \bar{x}_3 = e_{22}, \quad \bar{x}_4 = \alpha e_{13} + \beta e_{23} + \gamma e_{31} + \delta e_{32},$$
we obtain
$$St_4(\bar{x}_1, \bar{x}_2, \bar{x}_3, \bar{x}_4) = -\alpha e_{23} + \beta e_{13} - \gamma e_{32} + \delta e_{31} = \begin{pmatrix} 0 & 0 & \beta \\ 0 & 0 & -\alpha \\ \delta & -\gamma & 0 \end{pmatrix} = Q,$$
and
$$Q^2 = \begin{pmatrix} \beta\delta & -\beta\gamma & 0 \\ -\alpha\delta & \alpha\gamma & 0 \\ 0 & 0 & \delta\beta + \gamma\alpha \end{pmatrix}.$$
Hence Q^4 has the non-zero entry $\delta\beta\gamma\alpha + \gamma\alpha\delta\beta = 2\delta\beta\gamma\alpha$ on the $(3,3)$ position. This means that $St_4^4 \equiv 0$ is not an identity of $S(M_{2,1}(F))$.

In order to prove the second part of the lemma, we compute
$$Q^3 = \begin{pmatrix} 0 & 0 & \beta\gamma\alpha \\ 0 & 0 & -\alpha\delta\beta \\ \gamma\alpha\delta & -\delta\beta\gamma & 0 \end{pmatrix}.$$
For the values of $\bar{x}_1, \bar{x}_2, \bar{x}_3, \bar{x}_4$ given above, we have $St_2(\bar{x}_3, \bar{x}_4) = \beta e_{23} - \delta e_{32}$ and $St_1(\bar{x}_1) = e_{11}$. Hence
$$St_1(\bar{x}_1) St_4^3(\bar{x}_1, \bar{x}_3, \bar{x}_4, \bar{x}_2) St_2(\bar{x}_3, \bar{x}_4) = St_1(\bar{x}_1) St_4^3(\bar{x}_1, \bar{x}_2, \bar{x}_3, \bar{x}_4) St_2(\bar{x}_3, \bar{x}_4)$$
$$= e_{11} Q^3 (\beta e_{23} - \delta e_{32}) = -\beta\gamma\alpha\delta e_{12} \neq 0$$
and
$$St_2(\bar{x}_3, \bar{x}_4) St_4^3(\bar{x}_3, \bar{x}_4, \bar{x}_1, \bar{x}_2) St_1(x_1) = St_2(\bar{x}_3, \bar{x}_4) St_4^3(\bar{x}_1, \bar{x}_2, \bar{x}_3, \bar{x}_4) St_1(\bar{x}_1)$$
$$= (\beta e_{23} - \delta e_{32}) Q^3 e_{11} = \beta\gamma\alpha\delta e_{31} \neq 0.$$
It follows that $St_i St_4^3 St_{3-i} \not\equiv 0$ on $S(M_{2,1}(F))$ for $i = 1, 2$.

If $i = 0$ or 1, then
$$St_4^4(\bar{x}_1, \bar{x}_2, \bar{x}_3, \bar{x}_4) = St_4^3(\bar{x}_1, \bar{x}_2, \bar{x}_3, \bar{x}_4)(C_1\bar{x}_1 + \cdots + C_4\bar{x}_4),$$
where C_j denotes, up to sign, the standard polynomial on $\{x_1, \ldots, x_4\} \setminus \{x_j\}$. Therefore $St_4^3 St_3 \equiv 0$ is not an identity of $S(M_{2,1}(F))$ since $St_4^4 \not\equiv 0$ and
$$St_4(x_{\sigma(1)}, x_{\sigma(2)}, x_{\sigma(3)}, x_{\sigma(4)}) = \pm St_4(x_1, x_2, x_3, x_4).$$
Similarly, $St_3 St_4^3 \not\equiv 0$, and the proof is complete. $\square$

LEMMA 9.3.9. *Let A be an algebra with unit element. If A satisfies $St_{2n+1}^m \equiv 0$, for some $n \geq 1$, then A satisfies $St_{2n}^m \equiv 0$.*

PROOF. Note that for any $a_1, \ldots, a_{2n} \in A$ the value of St_{2n+1} on $a_1, \ldots, a_{2n}$ and $1 \in A$ is the sum of $2n+1$ terms of the type $\pm St_{2n}(a_1, \ldots, a_{2n})$. Since $2n+1$ is odd, $St_{2n+1}(a_1, \ldots, a_{2n}, 1) = St_{2n}(a_1, \ldots, a_{2n})$. Hence $St_{2n}(a_1, \ldots, a_{2n})^m = 0$, for any $a_1, \ldots, a_{2n} \in A$. $\square$

By Theorem 4.5.4, any PI-algebra satisfies some power of a standard identity. We are now ready to compute the PI-exponent of any such polynomial.

THEOREM 9.3.10 ([**BR5**]). *If $n \geq 4$, then $\exp(St_n^m) = m[\frac{n}{2}]^2$.*

PROOF. Note that the upper block-triangular matrix algebra $UT(d_1, \ldots, d_m)$ with $d_1 = \cdots = d_m = [\frac{n}{2}]$ satisfies St_n^m, by Theorem 1.7.7 and Theorem 1.9.1. Hence $\exp(St_n^m) \geq \exp(UT(d_1, \ldots, d_m)) = m[\frac{n}{2}]^2$. Therefore it is enough to prove that $\exp(I_1 \cdots I_k) \leq m[\frac{n}{2}]^2$, for any verbally prime T-ideals $I_1, \ldots, I_k$ as soon as $St_n^m \in I_1 \cdots I_k$.

First suppose that $I_j = \text{Id}(M_{d_j}(F))$, for all $j = 1, \ldots, k$. Then any matrix algebra $M_{d_j}(F)$ satisfies $St_n^m \equiv 0$ and, as it was remarked in the proof of Lemma 9.2.1, $M_{d_j}(F)$ satisfies $St_n \equiv 0$. Hence $d_1, \ldots, d_k \leq [\frac{n}{2}]$. By Lemma 9.3.9, we may replace n with $n-1$, as soon as n is odd. So, we may assume $n = 2q$ to be even and $[\frac{n}{2}] = \frac{n}{2} = q$. By Theorem 1.7.2, the minimal degree of an identity of $M_{d_j}(F)$ is $2d_j$, hence the minimal degree of a polynomial in $I_1 \cdots I_k$ is $2(d_1 + \cdots + d_k)$ and

(9.17) $$mq \geq d_1 + \cdots + d_k.$$

Recall also that

(9.18) $$q \geq d_1, \ldots, d_k.$$

Now we check that for any non-negative integers $d_1, \ldots, d_k$ satisfying (9.17) and (9.18), the inequality

$$d_1^2 + \cdots + d_k^2 \leq mq$$

holds. If we fix the sum $d_1 + \cdots + d_k$, then from the relation

$$(a+1)^2 + (b-1)^2 = a^2 + b^2 + 2(a-b+1) > a^2 + b^2, \quad \text{for any } a \geq b,$$

it follows that the maximal value of the sum $d_1^2 + \cdots + d_k^2$ is realized if $d_1 = \cdots = d_j = q$, $d_{j+1} = q' < q$, $d_{j+2} = \cdots = d_k = 0$, for some $j \leq m$. Hence

$$d_1^2 + \cdots + d_k^2 \leq q^2 + \cdots + q^2 = mq^2,$$

for any $d_1, \ldots, d_k$ satisfying (9.17) and (9.18). But $d_1^2 + \cdots + d_k^2$ is precisely the PI-exponent of $I_1 \cdots I_k$. Hence the conclusion of the theorem holds if $I_1, \ldots, I_k$ are all T-ideals of identities of matrix algebras.

Suppose now that at least one among $I_1, \ldots, I_k$ is a non-matrix verbally prime T-ideal. We proceed by induction on k.

If $k = 1$, $I_1 = \text{Id}(S(A))$ where A is a simple superalgebra of type $M_{p,l}(F)$ or $M_p(F \oplus cF)$. If $S(M_{p,l}(F))$ satisfies $St_n^m \equiv 0$, then $m \geq 2l+1$ by Lemma 9.3.5, and $p \leq [\frac{n}{2}]$, since $M_p(F)$ is a subalgebra of $S(M_{p,l}(F))$. Hence the exponent of $S(M_{p,l}(F))$ satisfies

$$(p+l)^2 = p^2 + l^2 + 2pl \leq 4p^2 \leq 4[\frac{n}{2}]^2 \leq m[\frac{n}{2}]^2,$$

as soon as $l \geq 2$, since $p \geq l$. Now let $l = 1$. If $m \geq 4$, then the same argument gives the desired conclusion. Suppose $m = 3$. Then, for $p \geq 2$, we have

$$(p+1)^2 = p^2 + 2p + 1 < 3p^2 \leq m[\frac{n}{2}]^2.$$

If $p = 1$, $A = M_{1,1}(F)$ and
$$\exp(S(M_{1,1}(F)) = 4 \le m[\tfrac{n}{2}]^2 = 3[\tfrac{n}{2}]^2,$$
since $n \ge 4$.

Now let $A = M_p(F \oplus cF)$. Then $[\tfrac{n}{2}] \ge p$ and $m > p$ by Lemma 9.3.2, hence
$$\exp(S(A)) = 2p^2 < m[\tfrac{n}{2}]^2.$$
This completes the proof when $k = 1$.

Suppose now that $k > 1$. Then $I_s = \mathrm{Id}(S(A))$ where A is a simple superalgebra of type $M_p(F \oplus cF)$ or $M_{p,l}(F)$, for some $1 \le s \le k$. In the first case by Lemma 9.3.1 and Lemma 9.3.2 (for $s = 1$ or k) or Lemma 9.3.6 and Lemma 9.3.7 (for $s \ne 1, k$), the T-ideal $I' = I_1 \cdots \hat{I}_s \cdots I_k$ contains St_n^{m-p} and by induction
$$\exp(I') \le (m-p)[\tfrac{n}{2}]^2.$$

As before, $M_p(F)$ is a subalgebra of $S(A)$ and therefore $p \le [\tfrac{n}{2}]$. Hence
$$\exp(I_1 \cdots I_k) = \exp(I_s) + \exp(I')$$
$$\le 2p^2 + (m-p)[\tfrac{n}{2}]^2 = m[\tfrac{n}{2}]^2 + p(2p - [\tfrac{n}{2}]^2) \le m[\tfrac{n}{2}]^2,$$
since $n \ge 4$ and $p \le [\tfrac{n}{2}]$.

Consider the case $A = M_{p,l}(F)$. Again, $p \le [\tfrac{n}{2}]$ and by Lemma 9.3.1 for $s = 1, k$ and by Lemma 9.3.6 and Lemma 9.3.7 for $1 < s < k$, the T-ideal $I' = I_1 \cdots \hat{I}_s \cdots I_k$ contains St_n^{m-2l}. Then by induction, the PI-exponent of I' is bounded by $(m-2l)[\tfrac{n}{2}]^2$ and
$$\exp(I_1 \cdots I_k) \le (p+l)^2 + (m-2l)[\tfrac{n}{2}]^2 = m[\tfrac{n}{2}]^2 + (p+l+\sqrt{2l}[\tfrac{n}{2}])(p+l-\sqrt{2l}[\tfrac{n}{2}]).$$
If $l \ge 2$, then we have $l \le p \le [\tfrac{n}{2}]$ and
$$p + l \le 2p \le \sqrt{2l}[\tfrac{n}{2}].$$
Hence $\exp(I_1 \cdots I_k) \le m[\tfrac{n}{2}]^2$ since $p + l - \sqrt{2l} \le 0$. If $l = 1$ but $p \ge 3$, then $(p+1)^2 \le 2p^2 \le 2[\tfrac{n}{2}]^2$ and $(p+1)^2 + (m-2)[\tfrac{n}{2}]^2 \le m[\tfrac{n}{2}]^2$.

Now let $p = 1$ or 2. If $p = 1$, the exponent of I_s is
$$\exp(S(M_{1,1}(F))) = (1+1)^2 = 4 \le 2[\tfrac{n}{2}]^2$$
since $n \ge 4$, and
$$\exp(I_1 \cdots I_k) \le (m-2)[\tfrac{n}{2}]^2 + 4 \le m[\tfrac{n}{2}]^2.$$

Finally, let $p = 2$, $A = M_{2,1}(F)$ and $\exp(I_s) = (2+1)^2 = 9$. For any $n \ge 6$ we have $9 \le 2[\tfrac{n}{2}]^2$, and
$$\exp(I_1 \cdots I_k) \le (m-2)[\tfrac{n}{2}]^2 + 9 \le m[\tfrac{n}{2}]^2,$$
as required. If $n = 5$, then $St_5^m \in I_1 \cdots I_k$. Since $I_1 \cdots I_k$ is a T-ideal of $S(B)$ for some minimal superalgebra B, by Lemma 9.3.9, $St_4^m \in I_1 \cdots I_k$ since $S(B)$ contains 1. Then by Lemma 9.3.1, Lemma 9.3.6 and Lemma 9.3.8, the T-ideal I' contains St_4^{m-4} and
$$\exp(I') \le (m-4)4,$$

by induction on m. Therefore
$$\exp(I) \leq 4(m-4) + 9 < 4m = m[\frac{n}{2}]^2,$$
and the proof is complete. □

9.4. Essential hooks and reduced algebras

In Chapter 4 we showed that given any PI-algebra A, there exist two integers $k, l \geq 0$, such that the corresponding cocharacters $\chi_n(A)$ lie in an infinite hook $H(k,l)$. In this section we want to sharpen this result by proving that there exists a smallest hook $H(k', l')$ (called the essential hook of A) with the property that $\chi_n(A)$ is contained in the union of $H(k', l')$ and a finite square. We start by proving the following essential result.

LEMMA 9.4.1. *Let A be a finite dimensional superalgebra with Jacobson radical J and maximal semisimple subalgebra A_{ss} with $\dim A_{ss}^{(0)} = p$, $\dim A_{ss}^{(1)} = q$ and $J^m = 0$. Then the cocharacters of the Grassmann envelope $G(A)$ lie in the union of an infinite hook $H(p, q)$ and a finite square of size $p + q + m$. In other words, for all $n \geq 1$,*
$$\chi_n(G(A)) = \sum_{\substack{\lambda \vdash n \\ \lambda \in H(p,q) \cup ((p+q+m)^{p+q+m})}} m_\lambda \chi_\lambda,$$
where

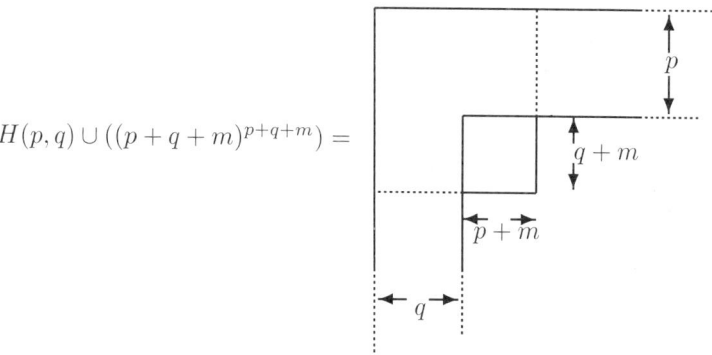

Moreover, any character χ_λ, $\lambda \vdash n$, participating with non-zero multiplicity in $\chi_n(G(A))$ has at most $m - 1$ boxes out of $H(p, q)$.

PROOF. Consider the algebra $G(A)$ as a superalgebra with its natural grading where $G(A)^{(0)} = G^{(0)} \otimes A^{(0)}$ and $G(A)^{(1)} = G^{(1)} \otimes A^{(1)}$ and recall that $\text{Id}^{gr}(G(A))$ denotes the ideal of $\mathbb{Z}_2$-graded identities of $G(A)$. Given any $r \in \{0, \ldots, n\}$, consider the space $P_{r,n-r}$ of multilinear polynomials in r variables of homogeneous degree 0 and $n - r$ variables of homogeneous degree 1. The group $S_r \times S_{n-r}$ acts on $P_{r,n-r}$ by permuting the even and odd variables separately. This action preserves $P_{r,n-r} \cap \text{Id}^{gr}(G(A))$, the graded identities of $G(A)$ lying in $P_{r,n-r}$. Then the $S_r \times S_{n-r}$-character of the quotient space $P_{r,n-r}/(P_{r,n-r} \cap \text{Id}^{gr}(G(A)))$ is called the graded cocharacter of $G(A)$.

Now, the ordinary cocharacter of the algebra $G(A)$ can be induced from the graded cocharacter in the following way. Let $\nu \vdash n$ be a partition whose corresponding S_n-character χ_ν appears with non-zero multiplicity in $\chi_n(G(A))$. Then

9.4. ESSENTIAL HOOKS AND REDUCED ALGEBRAS

there exists a multilinear polynomial $h = h(x_1, \ldots, x_n)$ such that

$$h(a_1, \ldots, a_r, b_1, \ldots, b_{n-r}) \neq 0,$$

in $G(A)$, for some $r \geq 0$ and for some even elements $a_1, \ldots, a_r$ and odd elements $b_1, \ldots, b_{n-r}$. Moreover, $FS_n h$ is an irreducible S_n-module with character χ_ν.

Consider the S_r-permutation action on $x_1, \ldots, x_r$ and the S_{n-r}-permutation action on $x_{r+1}, \ldots, x_n$. Then $FS_n h$ becomes an $S_r \times S_{n-r}$-module which is not irreducible in general. Denote by M one of its irreducible components with character $\chi_\lambda \otimes \chi_\mu$, where $\lambda \vdash r, \mu \vdash n - r$. Then, by Frobenious reciprocity (see Theorem 2.1.12), χ_ν is one of the irreducible S_n-characters induced from $\chi_\lambda \otimes \chi_\mu$. Let $f = f(x_1, \ldots, x_n)$ be an $S_r \times S_{n-r}$-generator of M. In conclusion we have:

1) χ_ν is one of the irreducible S_n-characters induced from the $S_r \times S_{n-r}$-character $\chi_\lambda \otimes \chi_\mu$,
2) f is not a graded identity of $G(A)$ if we consider $x_1, \ldots, x_r$ as even variables and $x_{r+1}, \ldots, x_n$ as odd variables,
3) f generates an irreducible $S_r \times S_{n-r}$-module with character $\chi_\lambda \otimes \chi_\mu$.

For any such $\nu \vdash n$ we next restrict the shape of the corresponding λ and μ. Since f generates an irreducible $S_r \times S_{n-r}$-module with character $\chi_\lambda \otimes \chi_\mu$, there exist two essential idempotents e_{T_λ} and e_{T_μ}, acting non-trivially on f. From Lemma 2.5.1, it follows that

$$(9.19) \qquad f' = \left(\sum_{\tau \in C_{T_\lambda}} (\operatorname{sgn} \tau) \tau \right) f$$

is non-zero, where C_{T_λ} is the group of column permutations of T_λ and f' still generates the same module.

Now let $\tilde{} : P_{r,n-r} \to P_{r,n-r}$ be the map defined in Section 3.7, where $x_1, \ldots, x_r$ are even variables and $y_1, \ldots, y_{r+1}$ are odd variables. Recall that the map $\tilde{}$ relates the graded identities of A and $G(A)$. In fact, if we consider the polynomial

$$\tilde{f}' = \tilde{f}'(x_1, \ldots, x_r, y_1, \ldots, y_{n-r}),$$

then by Lemma 3.7.4, $\tilde{f}'$ is not a graded identity of A.

Let λ' be the conjugate partition of λ. Suppose that $\lambda'_1, \ldots, \lambda'_i > p$ and $\lambda'_{i+1} \leq p$, i.e., the first i columns of λ are longer than p and all others are smaller or equal to p. Similarly, suppose that $\mu_1, \ldots, \mu_j > q, \mu_{j+1} \leq q$, i.e., the first j rows of μ are larger than q and all remaining are shorter than q. Then by (9.19), the polynomial $\tilde{f}'$ is alternating on i sets of indeterminates $X_1, \ldots, X_i$, with $|X_1| = \lambda'_1 > p, \ldots, |X_i| = \lambda'_i > p$. By using Lemma 4.8.6, we obtain that $\tilde{f}'$ is alternating in j sets of odd variables $Y_1, \ldots, Y_j$ with $|Y_1| = \mu_1 > q, \ldots, |Y_j| = \mu_j > q$. Since p and q are the dimensions of the even and the odd components of $A/J(A)$ and f is not an identity, it follows that there exists a non-zero evaluation with at least

$$|X_1| + \cdots + |X_i| - ip = \lambda'_1 + \cdots + \lambda'_i - ip$$

even elements taken from the radical $J(A)$ and at least

$$|Y_1| + \cdots + |Y_j| - jq = \mu_1 + \cdots + \mu_j - jq$$

odd elements taken from $J(A)$. Since $J(A)^m = 0$, this implies that

$$\lambda_1 + \cdots + \lambda_i - ip + \mu_1 + \cdots + \mu_j - jq \leq m - 1,$$

that is the number of boxes of T_λ out of the first p rows plus the number of boxes of T_μ out of the first q columns, is not greater than $m - 1$.

We now split λ into two partitions $\bar\lambda \vdash r'$ and $\tilde\lambda \vdash r - r'$ such that $\bar\lambda = (\lambda_1, \ldots, \lambda_p)$, $\tilde\lambda = (\lambda_{p+1}, \lambda_{p+2}, \ldots)$. Then the S_n-character χ_λ is one of the irreducible characters induced from the irreducible $S_{r'}$-character $\chi_{\bar\lambda}$. Similarly we take $\bar\mu \vdash r''$, $\bar\mu = (\bar\mu_1, \bar\mu_2, \ldots)$ with $\bar\mu_1 = q, \ldots, \bar\mu_j = q, \bar\mu_{j+1} = \mu_{j+1}, \bar\mu_{j+2} = \mu_{j+2}, \ldots$, and the S_{n-r}-character χ_μ may be induced from $\chi_{\bar\mu}$.

Note that the S_n- character χ_ν is one of the irreducible characters induced from the $S_{r'} \times S_{r''}$-character $\chi_{\bar\lambda} \otimes \chi_{\bar\mu}$ with $\chi_{\bar\lambda} \subseteq H(p, 0)$, $\chi_{\bar\mu} \subseteq H(0, q)$ and $n - r' - r'' \leq m - 1$. By the Littlewood-Richardson rule (Theorem 2.3.9), the $S_{r'+r''}$-character induced from $\chi_{\bar\lambda} \otimes \chi_{\bar\mu}$ lies in $H(p, q)$ and therefore the number of boxes of the Young diagram of ν not belonging to $H(p, q)$ does not exceed $n - r' - r'' \leq m - 1$. □

If A is any finite dimensional simple superalgebra, we have that $\dim A^{(0)} \geq \dim A^{(1)}$. In fact, by tensoring with the algebraic closure of F, we obtain a simple superalgebra $\bar A$ over $\bar F$ and by Theorem 3.5.3, $\dim \bar A^{(0)} \geq \dim \bar A^{(1)}$ and the same conclusion holds for A. It follows that if A is finite dimensional semisimple, still $\dim A^{(0)} \geq \dim A^{(1)}$ and, so, by the previous lemma, $\chi_n(G(A)) \subseteq H(k, l)$, with $k \geq l$.

Next we want to show that the same conclusion $k \geq l$ still holds for the smallest infinite hook $H(k, l)$ satisfying the conclusion of the previous lemma. To this end we introduce the notion of reduced algebra.

Let $A = A_1 \oplus \cdots \oplus A_m + J$ be a finite dimensional superalgebra over the algebraically closed field F with radical $J = J(A)$ and maximal semisimple subalgebra $A_1 \oplus \cdots \oplus A_m$, with $A_1, \ldots, A_m$ simple superalgebras. We introduce the following ([**GZ8**]).

DEFINITION 9.4.2. *The superalgebra $A = A_1 \oplus \cdots \oplus A_m + J$ is called reduced if $A_1 J A_2 J \cdots J A_m \neq 0$.*

Next we show how to generate any variety starting from reduced superalgebras.

THEOREM 9.4.3. *Let $\mathcal{V}$ be a proper non-nilpotent variety of algebras over the algebraically closed field F. Then there exists a finite number of reduced superalgebras $B_1, \ldots, B_t$ and a finite dimensional superalgebra D such that*
$$\mathcal{V} = \mathrm{var}(G(B_1) \oplus \cdots \oplus G(B_t) \oplus G(D)),$$
where $\exp(G(B_1)) = \ldots = \exp(G(B_t)) = \exp(\mathcal{V})$ and $\exp(G(D)) < \exp(\mathcal{V})$.

PROOF. Let A be a finite dimensional superalgebra such that $\mathcal{V} = \mathrm{var}(G(A))$. Write $A = A_1 \oplus \cdots \oplus A_s + J$ where $A_1, \ldots, A_s$ are simple superalgebras and $J = J(A)$. Suppose $\exp(\mathcal{V}) = d$ and let $L_1, \ldots, L_t$ be all possible subsets of $\{1, \ldots, s\}$ of the form $\{i_1, \ldots, i_k\}$ with the property that $\dim(A_{i_1} \oplus \cdots \oplus A_{i_k}) = d$ and $A_{\sigma(i_1)} J A_{\sigma(i_2)} J \cdots J A_{\sigma(i_k)} \neq 0$, for some permutation $\sigma \in S_k$. Now, for each $j = 1, \ldots, t$ if $L_j = \{i_1, \ldots, i_k\}$, we define $B_j = A_{i_1} \oplus \cdots \oplus A_{i_k} + J$. Then clearly $\exp(G(B_1)) = \ldots = \exp(G(B_t)) = \exp(G(A))$.

Now let $D_1, \ldots, D_p$ be all subalgebras of A of the type $A_{j_1} \oplus \cdots \oplus A_{j_q} + J$, where $1 \leq j_1 < \ldots < j_q \leq s$ and $\dim(A_{j_1} \oplus \cdots \oplus A_{j_q}) < d$. If we set $D = D_1 \oplus \cdots \oplus D_p$, then $\exp(G(D)) < \exp(G(A))$.

We are left to show that $\mathrm{var}(G(B_1) \oplus \cdots \oplus G(B_t) \oplus G(D)) = \mathrm{var}(G(A))$. Since for $i = 1, \ldots, t$, $G(B_i), G(D) \in \mathrm{var}(G(A))$, then $\mathrm{var}((G(B_1) \oplus \cdots \oplus G(B_t) \oplus G(D)) \subseteq \mathrm{var}(G(A))$.

On the other hand, let $f = f(x_1, \ldots, x_n)$ be a multilinear polynomial and suppose that $f \notin \mathrm{Id}(G(A))$. Let $a_1, \ldots, a_n \in A, g_1, \ldots, g_n \in G$ be such that

$$f(a_1 \otimes g_1, \ldots, a_n \otimes g_n) \neq 0.$$

By the definition of d, it is clear that we must have $a_1, \ldots, a_n \in A_{i_1} \oplus \cdots \oplus A_{i_k} + J$ where $\dim(A_{i_1} \oplus \cdots \oplus A_{i_k}) \leq d$. It follows that f is not an identity of one of the algebras $G(B_1), \ldots, G(B_t), G(D)$. Hence $\mathrm{var}(G(A)) \subseteq \mathrm{var}(G(B_1) \oplus \cdots \oplus G(B_t) \oplus G(D))$, and the proof is completed. □

Applying the same argument to $\mathrm{var}(G(D))$, from the previous theorem we obtain the following.

THEOREM 9.4.4. *Any variety $\mathcal{V}$ over the algebraically closed field F can be generated by the direct sum*

$$G(B_1) \oplus \cdots \oplus G(B_t) \oplus C,$$

where $B_1, \ldots, B_t$ are reduced superalgebras and C is a nilpotent algebra.

We are now in a position to refine Theorem 4.5.1 stating that the cocharacter $\chi_n(\mathcal{V})$ of any proper variety $\mathcal{V}$ lies into an infinite hook $H(k, l)$.

DEFINITION 9.4.5. Let $\mathcal{V}$ be a proper variety. An infinite hook $H(k, l)$ is called an essential hook for $\mathcal{V}$, if there exists a finite square M such that for all n,

$$\chi_n(\mathcal{V}) \subseteq H(k, l) \cup M$$

but for the hooks $H(k-1, l)$ and $H(k, l-1)$ we cannot find finite squares M', M'' such that $\chi_n(\mathcal{V}) \subseteq H(k-1, l) \cup M'$ for all n or $\chi_n(\mathcal{V}) \subseteq H(k, l-1) \cup M''$ for all n.

THEOREM 9.4.6. *For any variety $\mathcal{V}$ there exists an essential hook $H(k, l)$. Moreover, $k \geq l$ and $\exp(\mathcal{V}) \leq k + l$. If $\mathcal{V}$ is generated by $G(A)$ where A is a reduced superalgebra, then $\exp(\mathcal{V}) = k + l$.*

PROOF. Since by extending the base field F, the cocharacter sequence does not change, we may assume that F is algebraically closed.

Suppose first that $\mathcal{V} = \mathrm{var}(G(A))$ with A reduced. Denote $k = \dim(A_1 \oplus \cdots \oplus A_m)^{(0)}$, $l = \dim(A_1 \oplus \cdots \oplus A_m)^{(1)}$ where $A_{ss} = A_1 \oplus \cdots \oplus A_m$ is a maximal semisimple subalgebra of A. Then by Lemma 9.4.1, $\chi_n(G(A)) \subseteq H(k, l) \cup M$ with M finite and $k \geq l$. Moreover, $\exp(G(A)) = k + l$ as it was shown in Theorem 6.5.2. Since the exponent of any variety $\mathcal{W}$ with $\chi_n(\mathcal{W}) \subseteq H(k-1, l) \cup D$ or $\chi_n(\mathcal{W}) \subseteq H(k, l-1) \cup D$ with a finite D is at most $k+l-1$, it follows that $H(k, l)$ is the essential hook for $\mathcal{V} = \mathrm{var}(G(A))$.

In general, by Theorem 9.4.4, a variety $\mathcal{V}$ is generated by $G(B_1) \oplus \cdots \oplus G(B_t) \oplus C$ where $B_1, \ldots, B_t$ are reduced superalgebras and C is nilpotent. If $H(k_i, l_i)$ is an essential hook for $G(B_i)$, $i = 1, \ldots, t$, then clearly $H(k, l)$ is an essential hook for $\mathcal{V}$ where $k = \max\{k_1, \ldots, k_t\}$, $l = \max\{l_1, \ldots, l_t\}$. Since $k_i \geq l_i$ for all $1 \leq i \leq t$, we obtain the inequality $k \geq l$. □

Theorem 9.4.6 says that the cocharacter of any proper variety lies in an essential hook $H(k, l)$, with $k \geq l$. In the next section we shall estimate the PI-exponent of the polynomials related to the most important hooks.

9.5. The exponent of Amitsur polynomials

Let A be a PI-algebra and let $H(k,l)$ be an infinite hook such that $\chi_n(A) \subseteq H(k,l)$, for all n. By Theorem 4.7.2, the inclusion $\chi_n(A) \subseteq H(k,l)$ is equivalent to the fact that $E^*_{k,l} \equiv 0$ is an identity of A, where

$$E^*_{k,l}(x_1,\ldots,x_n;y_1,\ldots,y_{n-1}) = \sum_{\sigma \in S_n} \chi_\lambda(\sigma) x_{\sigma(1)} y_1 x_{\sigma(2)} y_2 \cdots y_{n-1} x_{\sigma(n)}$$

is the Amitsur polynomial corresponding to the rectangle $\lambda = ((l+1)^{k+1}) \vdash n = (k+1)(l+1)$.

As we have already remarked several times in this chapter, in order to compute the exponent of the polynomial $E^*_{k,l}$, we need to describe all minimal superalgebras A such that $G(A)$ satisfies $E^*_{k,l}$.

Consider a minimal superalgebra A generated by the simple superalgebras $A_1,\ldots,A_m$ and by the elements $w_{12},\ldots w_{m-1,m} \in J(A)$ (see Definition 8.1.3) and set

$$p = \dim A_1^{(0)} + \cdots + \dim A_m^{(0)}, \quad q = \dim A_1^{(1)} + \cdots + \dim A_m^{(1)}.$$

LEMMA 9.5.1. *The algebra $G(A)$ satisfies the identity $E^*_{k,l} \equiv 0$ if and only if $a = k - p \geq 0$, $b = l - q \geq 0$ and $(a+1)(b+1) > m - 1$. In particular, if A is a simple superalgebra with $\dim A^{(0)} = k$, $\dim A^{(1)} = l$, then*

$$\chi_n(G(A)) \subseteq H(k,l),$$

and $\chi_n(G(A))$ does not lie in any smaller hook $H(k',l')$ with $k' < k$ or $l' < l$.

PROOF. We first show that if $a,b \geq 0$ and $(a+1)(b+1) \geq m - 1$, then $\chi_n(G(A)) \subseteq H(k,l)$. By Lemma 9.4.1, the cocharacter of $G(A)$ lies in the union of an infinite hook $H(p,q)$ and a finite square of size $p + q + m$. Actually in that proof we showed that if χ_ν participates in $\chi_n(G(A))$ with non-zero multiplicity, the number of boxes of the Young diagram of ν not belonging to $H(p,q)$, does not exceed $m-1$. But then it is not hard to observe (see picture) that any such diagram lies inside the hook $H(k,l)$.

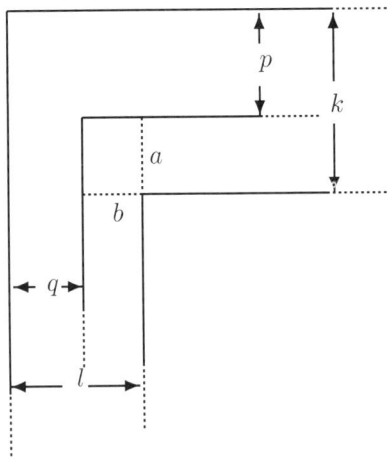

Now suppose that the inequalities $a \geq 0, b \geq 0, (a+1)(b+1) > m - 1$ do not hold. Recall that, since A is a minimal superalgebra,

$$A_1 J A_2 J \cdots J A_m \neq 0.$$

In the proof of Lemma 6.4.4, for any such algebra we constructed a multilinear polynomial of the type

$$f = f(x_1, \ldots, x_n, u_1, \ldots, u_m, v_1, \ldots, v_m, w_1, \ldots, w_{m-1})$$
$$= u_1 f_1 v_1 w_1 v_2 f_2 v_2 w_2 \cdots w_{m-1} u_m f_m v_m,$$

with the following properties:

- under the S_n-action on $x_1, \ldots, x_n$, the polynomial f generates an irreducible module with character χ_λ such that

$$h(p, q, 2t-s) \leq \lambda \leq h(p, q, 2t),$$

where $s = 4 \dim A$ and $t \geq 2 \dim A$ is arbitrary;

- there exists an evaluation

$$\varphi : u_i \to \bar{u}_i, v_i \to \bar{v}_i, f_i \to \bar{f}_i \text{ with } \bar{u}_i, \bar{v}_i, \bar{f}_i \in G(A_i), \ i = 1, \ldots, m,$$
$$w_j \to \bar{w}_j \in G(A_j J A_{j+1}), \ j = 1, \ldots, m-1,$$

such that

$$\varphi(f) = \bar{u}_1 \bar{f}_1 \bar{v}_1 \bar{w}_1 \cdots \bar{w}_{m-1} \bar{f}_m \bar{w}_m \neq 0;$$

- any permutation σ of $\bar{x}_1, \ldots, \bar{x}_n, \bar{w}_1, \ldots, \bar{w}_{m-1} \in G(A)$, such that $\sigma(\bar{w}_i) \neq \bar{w}_i$, for some $1 \leq i \leq m-1$, gives a zero value of f.

Note that $\lambda \subseteq H(p, q)$.

If $a = k - p$ or $b = l - q$ is negative, then, obviously $\lambda \not\subseteq H(k, l)$. If $a, b \geq 0$, then $m - 1 \geq (a+1)(b+1)$. In this case for sufficiently large t, we can construct a partition $\mu \vdash n + (a+1)(b+1)$ with

$$\mu_1 = \lambda_1, \ \ldots, \ \mu_p = \lambda_p, \ \mu_{p+1} = \lambda_{p+1} + b + 1, \ \ldots, \ \mu_{p+a+1} = \lambda_{p+a+1} + b + 1,$$

and $\mu_j = \lambda_j$, $j \geq p + a + 2$. Then $\mu > \lambda$ and we can choose a tableau T_μ containing T_λ and $C_{T_\mu} \supseteq C_{T_\lambda}$, $R_{T_\mu} \supseteq R_{T_\lambda}$.

The groups R_{T_μ} and C_{T_μ} can be partitioned into left and right cosets of the type ρR_{T_λ} and $C_{T_\lambda} \rho'$, respectively. Then

$$(\sigma f)(\bar{x}_1, \ldots, \bar{x}_n, \bar{u}_1, \ldots, \bar{u}_m, \bar{v}_1, \ldots, \bar{v}_m, \bar{w}_1, \ldots, \bar{w}_{m-1}) = 0$$

in $G(A)$ as soon as $\sigma \in \rho R_{T_\lambda} C_{T_\lambda} \rho'$, with $\rho \in R_{T_\lambda}$ or $\rho' \in C_{T_\lambda}$, by the choice of f and φ. It follows that

$$(e_{T_\mu} f)(\bar{x}_1, \ldots, \bar{w}_{m-1}) = (e_{T_\lambda} f)(\bar{x}_1, \ldots, \bar{w}_{m-1}) \neq 0,$$

since $e_{T_\lambda} f$ is a scalar multiple of f.

We have proved that $e_{T_\mu} f$ is not an identity of $G(A)$, where $\mu \vdash n' \leq n + m - 1$, and D_μ contains the rectangle $((q+b+1)^{p+a+1}) = ((l+1)^{k+1})$. Hence $\mu \not\subseteq H(k, l)$, and the proof is complete. $\square$

Notice that $\exp(G(A)) = p + q$. As an immediate consequence of Lemma 9.5.1 we get.

THEOREM 9.5.2 ([**BR5**]). *The PI-exponent of the Amitsur polynomial $E_{k,l}^*$ is the maximal value of*

$$a_1^2 + \cdots + a_r^2 + (p_1 + q_1)^2 + \cdots + (p_s + q_s)^2 + 2b_1^2 + \cdots + 2b_t^2,$$

where $a_1, \ldots, a_r, p_1, q_1, \ldots, p_s, q_s, b_1, \ldots, b_t$ are positive integers satisfying the inequalities

$$u = k - a_1^2 - \cdots - a_r^2 - p_1^2 - \cdots - p_s^2 - b_1^2 - \cdots - b_t^2 \geq 0$$

$$v = l - 2p_1q_1 - \cdots - 2p_sq_s - b_1^2 - \cdots - b_t^2 \geq 0,$$
$$r + s + t \leq (u+1)(v+1).$$

PROOF. The proof is a straightforward consequence of Lemma 9.5.1 and the results on the exponent of a minimal variety. $\square$

When computing the PI-exponent of a Capelli polynomial we used Lagrange theorem about the decomposition of any integer into the sum of at most 4 squares. In order to compute the exponent of the Amitsur polynomials we need to introduce the notion of generalized square.

DEFINITION 9.5.3. Any pair of non-negative integers of the type (a^2, a^2) or $(a^2 + b^2, 2ab)$ is called a generalized square.

The next theorem is due to Cohen and Regev and its proof is given in Appendix A. In the set $\mathbb{Z} \times \mathbb{Z}$ define an addition as follows: $(a, b) + (c, d) = (a+c, b+d)$.

THEOREM 9.5.4 ([**CR**]). For any $k \geq l \geq 0$ any pair (k, l) is the sum of at most 6 generalized squares.

Using generalized squares decompositions, we can give the following bounds.

THEOREM 9.5.5 ([**BR5**]). Let $k \geq l \geq 0$ and let $E_{k,l}^*$ be an Amitsur polynomial. Then
$$k + l - 3 \leq \exp(E_{k,l}^*) \leq k + l.$$

PROOF. By Theorem 4.7.2, $\chi_n(E_{k,l}^*) \subseteq H(k, l)$ and then by Lemma 6.2.5 and Theorem 4.9.3, $\exp(E_{k,l}^*) \leq k + l$. For $l = 0$, the lower bound was obtained in Theorem 9.1.5.

Let $l \geq 2$. Then $k \geq 2$ and by Proposition 9.5.4, we can decompose $(k-1, l-2)$ into the sum of (at most) 6 generalized squares. Let
$$k - 1 = a_1 + \cdots + a_6, \quad l - 2 = b_1 + \cdots + b_6,$$
where either $(a_i, b_i) = (r_i^2, r_i^2)$ or $(a_i, b_i) = (p_i^2 + q_i^2, 2p_iq_i)$, $i = 1, \ldots, 6$. Hence there exist at most six simple superalgebras $A_1, \ldots, A_6$ with $\dim A_i^{(0)} = a_i$, $\dim A_i^{(1)} = b_i$, $i = 1, \ldots, 6$, and a minimal superalgebra $A = A_1 \oplus \cdots \oplus A_6 + J$ with $J^6 = 0$ and $\exp(G(A)) = \dim(A_1 \oplus \cdots \oplus A_6) = k + l - 3$. By Lemma 9.4.1, the cocharacter $\chi_n(G(A))$ lies in $H(k-1, l-2)$ plus at most 5 boxes (see picture).

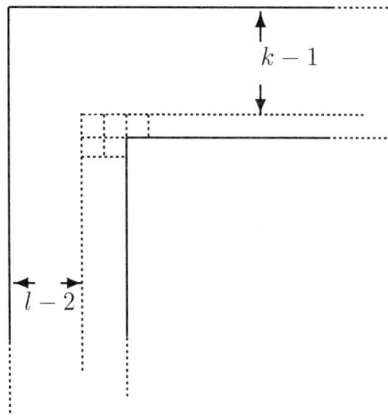

Hence $G(A)$ satisfies $E_{k,l}^* \equiv 0$ and then $\exp(E_{k,l}^*) \geq k + l - 3$.

Let now $l = 1$. If $k \geq 2$, then we decompose $(k - 2, l - 1) = (k - 2, 0)$ into 6 generalized squares and as before we obtain $\exp(E_{k,l}^*) \geq k - 2 = k + l - 3$. Finally, if $k < 2$, then $k + l - 3$ is negative and this completes the proof of the theorem. $\square$

We now consider all remaining hooks $H(k, l)$ with $k < l$.

THEOREM 9.5.6. *Let $E_{k,l}^*$ be an Amitsur polynomial with $k \leq l$. Then:*
1) $2k - 2 \leq \exp(E_{k,l}^*) \leq 2k$.
2) *If $l \geq k + 3$, then $\exp(E_{k,l}^*) = 2k$.*

PROOF. By Theorem 9.4.6, for the polynomial $E_{k,l}^*$ there exists an essential hook $H(k', l')$, i.e., the cocharacter of the variety defined by $E_{k,l}^* \equiv 0$ asymptotically lies in $H(k', l')$, $k' \geq l'$ and $\exp(E_{k,l}^*) \leq k' + l'$. Clearly $k' \leq k$, hence $\exp(E_{k,l}^*) \leq 2k$.

On the other hand, for $k = 1$ there is nothing to prove and for $k \geq 2$, by Lagrange theorem, there is a decomposition $k - 1 = a_1^2 + \cdots + a_4^2$ into the sum of (at most) four squares. Then we may construct a minimal superalgebra $A = A_1 \oplus \cdots \oplus A_4 + J$ with $J^4 = 0$ where $A_i = M_{a_i}(F \oplus cF)$ is a finite dimensional simple superalgebra with $\dim A_i^{(0)} = \dim A_i^{(1)} = a_i^2$, $i = 1, \ldots, 4$. As before, by Lemma 9.4.1, $\chi_n(G(A))$ lies in $H(k-1, k-1)$ plus at most 3 boxes. Hence $\chi_n(G(A)) \subseteq H(k, k) \subseteq H(k, l)$ and $\exp(E_{k,l}^*) \geq 2k - 2 = \dim(A_1 \oplus \cdots \oplus A_4)$.

Similarly, if $l \geq k + 3$, there exists a minimal superalgebra A such that $\dim(A/J(A))^{(0)} = k$, $\dim(A/J(A))^{(1)} = k$ and $J^4 = 0$. Hence $\chi_n(G(A)) \subseteq H(k, k+3)$ and the proof is complete since $\exp(G(A)) = 2k$. $\square$

Similar arguments show that $\exp(E_{k,k+2}^*) \geq 2k - 1$ and by using generalized squares decompositions we can prove that the case $\exp(E_{k,k+2}^*) \neq 2k$ is very exceptional. Since $(2a^2, 2a^2)$ is a generalized square, the PI-exponent of $E_{k,k+2}^*$ is less than $2k$ only if $k \neq a^2, a^2 + b^2, a^2 + b^2 + c^2$ or $k \neq 2a^2 + b^2 + c^2$.

9.6. The exponent of a Lie monomial

In this section we compute the PI-exponent of a polynomial which can be written as a single Lie commutator, i.e., a Lie monomial. In fact, the description of the minimal varieties and minimal superalgebras allows us to compute the exponent of any Lie monomial or more generally of any proper polynomial (see Section 7.5), but we shall consider here only a few types of monomials.

We start with Lie-nilpotent algebras. Recall that the left-normed Lie commutator $[x_1, \ldots, x_n]$ is defined as $[x_1, \ldots, x_n] = [\ldots [[x_1, x_2], x_3], \ldots, x_n]$. Clearly,
$$\exp([x_1, x_2]) = 1.$$
For all other $n \geq 3$ we have.

THEOREM 9.6.1. *For all $n \geq 3$, $\exp([x_1, \ldots, x_n]) = 2$.*

PROOF. Let $\mathcal{V} = \mathrm{var}([x_1, \ldots, x_n])$ and consider the algebras $A_1, \ldots, A_5$ defined before Theorem 7.6.1. It is easy to check that all these algebras do not satisfy $[x_1, \ldots, x_n] \equiv 0$. Hence by Theorem 7.6.1, $\exp(\mathcal{V}) \leq 2$. On the other hand, the Grassmann algebra satisfies $[x_1, x_2, x_3] \equiv 0$ and, by Theorem 4.1.8, $\exp(G) = 2$. Hence $G \in \mathcal{V}$ and $\exp(\mathcal{V}) = 2$. $\square$

We define the Engel polynomial $[x, y^{(n)}]$ as
$$[x, y^{(n)}] = [[x, y^{(n-1)}], y], \text{ for } n \geq 2, \text{ and } [x, y^{(1)}] = [x, y].$$

Recall that a Lie algebra satisfying the identity $[x, y^{(n)}] \equiv 0$, for some $n \geq 1$, is called an Engel Lie algebra. For associative algebras satisfying an Engel identity, Theorem 9.6.1 implies the following.

COROLLARY 9.6.2. *If $n \geq 2$, then $\exp([x, y^{(n)}]) = 2$.*

PROOF. Since $\operatorname{char} F = 0$, it is well known that $[x, y^{(n)}] \equiv 0$ implies the Lie nilpotence $[x_1, \ldots, x_m] \equiv 0$, for some $m \geq n$ (see for instance [**Kt**] or [**Z**]). Hence by Theorem 9.6.1, $\exp([x, y^{(n)}]) \leq 2$. Now, since the Grassmann algebra satisfies $[x, y^{(2)}] \equiv 0$, it follows that $\exp([x, y^{(n)}]) \geq 2$. This completes the proof. □

Our next goal is to compute the exponent of a Lie monomial defining Lie solvability. Define $s_1(x_1, x_2) = [x_1, x_2]$ and inductively

$$s_{m+1}(x_1, \ldots, x_{2^{m+1}}) = [s_m(x_1, \ldots, x_{2^m}), s_m(x_{2^m+1}, \ldots, x_{2^{m+1}})].$$

We have

THEOREM 9.6.3 ([**GZ9**]). *For any $m \geq 1$, $\exp(s_m(x_1, \ldots, x_{2^m})) = 2^{m-1}$.*

PROOF. For any $m \geq 1$, let $L_m = \langle s_m \rangle_T$. Then $L_1 = [F\langle X \rangle, F\langle X \rangle]$ and $L_{m+1} = [L_m, L_m]$, for all $m \geq 1$.

If $m = 1$, then $c_n(L_1) = 1$, for all n, and this implies that $\exp(s_1) = 1 = 2^0$. Let $m \geq 2$ and consider all minimal varieties $\mathcal{V}$ such that $s_m \in \operatorname{Id}(\mathcal{V})$, i.e., in light of Lemma 8.1.5 (see also the proof of Theorem 8.4.2), we should consider all possible products $I_1 \cdots I_k$ of verbally prime T-ideals containing L_m. Note first that if $UT_{2^{m-1}}$ denotes the algebra of $2^{m-1} \times 2^{m-1}$ upper triangular matrices over F, then $UT_{2^{m-1}}$ satisfies $s_m \equiv 0$. Moreover, $UT_{2^{m-1}}$ is a minimal superalgebra with trivial grading and with exponent 2^{m-1}. Hence

(9.20) $$\exp(s_m) \geq \exp(UT_{2^{m-1}}) = 2^{m-1}.$$

Next we shall prove, by induction on m, that $\exp(G(A)) \leq 2^{m-1}$ for any minimal superalgebra A such that $G(A)$ satisfies $s_m \equiv 0$. Let $A = A_{ss} + J$ be a minimal superalgebra where $A_{ss} = A_1 \oplus \cdots \oplus A_k$, A_j simple superalgebras, and $\operatorname{Id}(G(A)) = I_1 \cdots I_k$ with $I_j = \operatorname{Id}(G(A_j))$, $j = 1, \ldots, k$. Suppose that $I_1 \ldots I_k \supseteq L_m$. Then $I_j \supseteq L_m$ and, so, $G(A_j)$ satisfies $s_m \equiv 0$, for all j. Recalling the classification of the simple superalgebras, a direct inspection shows that either $A_j \cong F$ satisfies $s_1 \equiv 0$ or $A_j \cong M_{1,1}(F)$ and $G(A_j)$ satisfies $s_3 \equiv 0$ but $s_2 \not\equiv 0$ or $A_j \cong F \oplus c_j F$, $c_j^2 = 1$, and $G(A_j)$ satisfies $s_2 \equiv 0$ but $s_1 \not\equiv 0$.

First let $k \geq 2$. Since $L_m = [L_{m-1}, L_{m-1}] \subseteq I_1 \cdots I_k$, by Theorem 8.4.4 and Proposition 8.4.5, we obtain that either $L_{m-1} \subseteq I_1 \cdots I_k$ or there exists $1 \leq r \leq k-1$ such that $L_{m-1} \subseteq (I_1 \cdots I_r) \cap (I_{r+1} \cdots I_k)$. In the first case, by induction on m, $\exp(G(A)) \leq 2^{m-2}$. But $\exp(s_m) \geq 2^{m-1}$ by (9.20); hence A can be excluded. Thus, if $k \geq 2$, then

$$\exp(s_m) \leq \exp(I_1 \cdots I_r) + \exp(I_{r+1} \cdots I_k) \leq 2^{m-2} + 2^{m-2} = 2^{m-1}$$

and this, together with (9.20) gives the desired conclusion.

If $k = 1$ and $m \geq 4$, by the discussion above, $\exp(G) = 2 < 2^{m-1}$ and $\exp(M_{1,1}(G)) = 4 < 2^{m-1}$. For $m = 2$ or 3 one has $\exp(s_2) = \exp(G) = 2 = 2^1$, $\exp(s_3) = \exp(M_{1,1}(G)) = 4 = 2^2$. □

9.7. Evaluating polynomials

In this section we study the standard and Capelli polynomials of minimal degree vanishing on a finite dimensional reduced algebra of a special type. These results although technical, will be applied in the next section. Throughout this section we assume that R is a finite dimensional algebra over F such that

$$R = A + J$$

where $A = M_k(F)$ is the algebra of $k \times k$ matrices over F and $J = J(R)$.

We start with the following key lemma about the decomposition of the Jacobson radical of R.

LEMMA 9.7.1. *The Jacobson radical J can be decomposed into the direct sum of four A-bimodules*

$$J = J_{00} \oplus J_{01} \oplus J_{10} \oplus J_{11}$$

where for $p,q \in \{0,1\}$, J_{pq} is a left faithful module or a 0-left module if $p=1$ or $p=0$, respectively. Similarly, J_{pq} is a right faithful module or a 0-right module if $q=1$ or $q=0$, respectively. Moreover, for $p,q,i,l \in \{0,1\}$, $J_{pq}J_{ql} \subseteq J_{pl}$, $J_{pq}J_{il} = 0$ for $q \neq i$ and there exists a finite dimensional nilpotent algebra N such that $J_{11} \cong A \otimes_F N$ (isomorphism of A-bimodules and of algebras).

PROOF. Let E be the unit element of $A = M_k(F)$. Denote by $L_E, R_E \colon J \to J$ the linear transformations of J of left and right multiplication by E, respectively. Since $L_E^2 = L_E$ and $R_E^2 = R_E$, they both are diagonalizable linear transformations with eigenvalues 0 and 1. Moreover, $L_E R_E = R_E L_E$ and J decomposes into the sum of its eigenspaces $J = J_{00} \oplus J_{01} \oplus J_{10} \oplus J_{11}$, as desired. The inclusions $J_{pq}J_{ql} \subseteq J_{pl}$ and the equalities $J_{pq}J_{il} = 0$ for $i \neq q$ are clear.

Now let $J_{11} = V_1 \oplus \cdots \oplus V_m$ be the decomposition of J_{11} into irreducible A-bimodules. Each irreducible V_i is isomorphic to ${}_A A_A$; hence V_i contains a non-zero element d_i (unique up to a scalar) commuting with A and $V_i = Ad_i$. Moreover, for every $i,j \in \{1,\ldots,m\}$, $d_i d_j$ commutes with A, hence $d_i d_j$ is a linear combination of $d_1, \ldots, d_m$. It follows that $N = \mathrm{span}\{d_1, \ldots, d_m\}$ is a subalgebra of R and $J_{11} = AN \cong A \otimes_F N$. □

LEMMA 9.7.2. *Suppose that $J_{01} \oplus J_{10} \neq 0$. Then $St_{2k} \notin \mathrm{Id}(R)$ and $Cap_{k^2+1} \notin \mathrm{Id}(R)$.*

PROOF. Suppose $J_{10} \neq 0$ and let $d \in J_{10}$, $d \neq 0$. Then $dA = 0$ and $e_{ii}d \neq 0$ for some $i \in \{1, \ldots, k\}$.

If $k=1$, the conclusion of the lemma follows. Therefore assume that $k \geq 2$.

Since St_{2k-1} is not an identity of $A = M_k(F)$, there exist $a_1, \ldots, a_{2k-1} \in A$ such that $St_{2k-1}(a_1, \ldots, a_{2k-1}) = e_{ji}$ for some $j \neq i$ (for instance the staircase $e_{i+1,i+1}, e_{i+1,i+2}, \ldots, e_{nn}, e_{n1}, e_{11}, \ldots, e_{ii}$ will do). But then

$$e_{ij}St_{2k}(a_1, \ldots, a_{2k-1}, d)$$
$$= e_{ij}St_{2k-1}(a_1, \ldots, a_{2k-1})d = e_{ij}e_{ji}d = e_{ii}d \neq 0$$

and $St_{2k} \not\equiv 0$ on A.

By Proposition 1.7.1, $M_k(F)$ does not satisfy the Capelli polynomial of rank k^2. Also, there exist $a_1, \ldots, a_{k^2}, b_1, \ldots, b_{k^2-1} \in A$ such that

$$Cap_{k^2}(a_1, \ldots, a_{k^2}; b_1, \ldots, b_{k^2+1}) = e_{ii}.$$

We now compute
$$Cap_{k^2+1}(a_1,\ldots,a_{k^2},d;b_1,\ldots,b_{k^2+1},e)$$
$$= Cap_{k^2}(a_1,\ldots,a_{k^2};b_1,\ldots,b_{k^2+1})ed = e_{kk}ed = e_{kk}d \neq 0.$$
Hence $Cap_{k^2+1} \not\equiv 0$ on A. A similar proof holds in case $J_{01} \neq 0$. □

LEMMA 9.7.3. *Write $J_{11} \cong A \otimes_F N$ as in Lemma 9.7.1. If N is not commutative, then $St_{2k+1} \notin Id(R)$, and $Cap_{k^2+2} \notin Id(R)$.*

PROOF. Let $J_{11} = AN \cong A \otimes_F N$ be as in Lemma 9.7.1 and pick $d_1, d_2 \in N$ such that $d_1 d_2 \neq d_2 d_1$. We claim that for any choice of $x_1,\ldots,x_{2k-1} \in A$,
$$St_{2k+1}(x_1,\ldots,x_{2k-1},d_1,d_2) = \gamma St_{2k-1}(x_1,\ldots,x_{2k-1})[d_1,d_2],$$
for some non-zero constant γ.

In order to simplify the notation, let us write $2k-1 = n$. Denote $\mathcal{A} = Alt_n = \sum_{\sigma \in S_n}(\text{sgn }\sigma)\sigma \in FS_n$ and recall the left action of S_n on the space of multilinear polynomials in $x_1,\ldots,x_n$.

Consider the subgroup $H = S_n \times S_2$ of S_{n+2}, where S_n acts on $\{1,\ldots,n\}$ and S_2 acts on $\{n+1,n+2\}$. Write S_{n+2} as the union of its right cosets $H\tau$ where $\tau = 1$ or $(i,n+1)$ or $(i,n+2)$ or $(i,n+1)(j,n+2)$, with $i < j$. By applying this decomposition we obtain
$$St_{n+2}(x_1,\ldots,x_{2k-1},d_1,d_2) =$$
$$\mathcal{A}(\sum_{1 \leq i < j \leq n} x_1 \cdots x_{i-1} d_1 \cdots x_{j-1} d_2 \cdots x_n x_i x_j - x_1$$
$$\cdots x_{i-1} d_2 \cdots x_{j-1} d_1 \cdots x_n x_i x_j)$$
$$-\mathcal{A}(\sum_{i=1}^{n} x_1 \cdots x_{i-1} d_1 \cdots x_n x_i d_2 + x_1 \cdots x_{i-1} d_2 \cdots x_n d_1 x_i$$
$$- x_1 \cdots x_{i-1} d_2 \cdots x_n x_i d_1 - x_1 \cdots x_{i-1} d_1 \cdots x_n d_2 x_i)$$
$$+\mathcal{A}(x_1 \ldots x_n (d_1 d_2 - d_2 d_1)).$$
Since d_1 and d_2 commute with $x_1,\ldots,x_n$, we obtain that
$$St_{n+2}(x_1,\ldots,x_n,d_1,d_2) =$$
$$\sum_{1 \leq i < j \leq n} St_n(x_1,\ldots,\widehat{x_i},\ldots,\widehat{x_j},\ldots,x_n,x_i,x_j)[d_1,d_2]$$
$$+ St_n(x_1,\ldots,x_n)[d_1,d_2],$$
where $\widehat{x}$ means that the variable x is omitted.

On the other hand,
$$St_n(x_1,\ldots,\widehat{x_i},\ldots,\widehat{x_j},\ldots,x_n,x_i,x_j)$$
$$= (-1)^{n-j+n-i-1} St_n(x_1,\ldots,x_n).$$
Since $(-1)^{n-j+n-i-1} = (-1)^{i+j-1}$ and
$$\sum_{i=1}^{j-1}(-1)^{i+j-1} = \begin{cases} 1 & \text{if } j \text{ is even,} \\ 0 & \text{if } j \text{ is odd,} \end{cases}$$
we get that
$$St_{2k+1}(x_1,\ldots,x_n,d_1,d_2) = \gamma St_n(x_1,\ldots,x_n)[d_1,d_2],$$
where $\gamma = 1 + 1 + 0 + 1 + \cdots \neq 0$ as claimed.

Since $M_k(F)$ does not satisfy St_{2k-1}, it is clear that $St_{2k+1} \notin \mathrm{Id}(R)$.

Next we prove that $Cap_{k^2+2}(x_1, \ldots, x_{x^2+2}; y_1, \ldots, y_{k^2+3})$ is not an identity of R provided N is not commutative.

Let $d_1, d_2 \in N$ with $d_1 d_2 \neq d_2 d_1$. Since $J_{11} = N = AN \cong A \otimes N$ as an A-bimodule, it follows that $a[d_1, d_2] \neq 0$ in R for any non-zero $a \in A$.

As in the proof of Proposition 1.7.1, we can take $2k^2 + 1$ matrix units

$$a_1, \ldots, a_{k^2}, b_1, \ldots, b_{k^2+1},$$

with $a_1, \ldots, a_{k^2}$ distinct, such that

(9.21) $$b_1 a_1 b_2 \cdots b_{k^2} a_{k^2} b_{k^2+1} = e_{11}$$

and

(9.22) $$b_1 a_{\sigma(1)} b_2 \cdots b_{k^2} a_{\sigma(k^2)} b_{k^2+1} = 0,$$

for all $1 \neq \sigma \in S_{k^2}$. Since only one among $a_1, \ldots, a_{k^2}$, say a_i, is equal to e_{11}, we can write the left-hand side of (9.21) in the form

$$c_1 e_{11} c_2,$$

where $c_1 = b_1 a_1 b_2 \cdots b_i$, and $c_2 = b_{i+1} a_{i+1} \cdots b_{k^2} a_{k^2}$.

Now take $a_{k^2+1} = d_1, a_{k^2+2} = d_2$ and compute

(9.23) $$Cap_{k^2+2}(a_1, \ldots, a_{k^2+2}; b_1, \ldots, b_{k^2+1}, e_{11}, e_{11})$$
$$= \sum_{\sigma \in S_{k^2+2}} b_1 a_{\sigma(1)} b_2 \cdots b_{k^2+1} a_{\sigma(k^2+1)} e_{11} a_{\sigma(k^2+2)} e_{11}.$$

From (9.21) it follows that $b_{k^2+1} = e_{\alpha 1}$, hence all summands in (9.23) with $\sigma(k^2+1), \sigma(k^2+2) \leq k^2$ are equal to zero. Moreover, by the construction of the product in (9.21), all summands in (9.23) are equal to zero unless $\sigma \in S_{k^2+2}$ preserves the set $\{i, k^2+1, k^2+2\}$ and $\sigma(r) = r$, for all $r \in \{1, \ldots, k^2\} \setminus \{i\}$. Therefore the left-hand side of (9.23) is equal to

$$c_1 e_{11} c_2 d_1 e_{11} d_2 e_{11} - c_1 e_{11} c_2 d_2 e_{11} d_1 e_{11} - c_1 d_1 c_2 e_{11} e_{11} d_2 e_{11}$$
$$+ c_1 d_2 c_2 e_{11} e_{11} d_1 e_{11} - c_1 d_2 c_2 d_1 e_{11} e_{11} e_{11} + c_1 d_1 c_2 d_2 e_{11} e_{11} e_{11}.$$

Note that $c_1 c_2 = c_1 e_{11} c_2$ and, for instance,

$$c_1 d_1 c_2 d_2 e_{11} = c_1 d_1 c_2 e_{11} d_2 e_{11} = c_1 c_2 d_1 d_2,$$

since d_1 and d_2 commute with A. Hence (9.23) is equal to

$$c_1 c_2 [d_1, d_2] - c_1 c_2 [d_1, d_2] + c_1 c_2 [d_1, d_2] = e_{11}[d_1, d_2] \neq 0$$

and the proof is complete. $\square$

LEMMA 9.7.4. *If $J_{01} J_{10} + J_{10} J_{01} + J_{10} J_{00} + J_{00} J_{01} \neq 0$, then $St_{2k+1} \notin \mathrm{Id}(R)$.*

PROOF. Suppose first that $k = 1$. If $J_{10} J_{01} \neq 0$, let $a \in J_{10}, b \in J_{01}$ be such that $ab \neq 0$.

Then, if $1 = 1_{M_k(F)}$, $1a = a, b1 = b$ and $a1 = 1b = 0$. It follows that $St_3(1, a, b) = 2ab + ba$. Since $ab \in J_{10} J_{01} \subseteq J_{11}$, $ba \in J_{01} J_{10} \subseteq J_{00}$, and $J_{11} \cap J_{00} = 0$, we obtain that $St_3 \notin \mathrm{Id}(R)$. The other cases $J_{01} J_{10} \neq 0, J_{10} J_{00} \neq 0, J_{00} J_{01} \neq 0$ are dealt with similarly.

Suppose now that $k \geq 2$ and let $J_{01} J_{10} \neq 0$. If $u \in J_{01}, v \in J_{10}$ are such that $uv \neq 0$, then there exists $e_{ii} \in M_k(F)$ such that $u e_{ii} v \neq 0$, for some $i \in \{1, \ldots, k\}$. Let $j \neq i$ and, as in the proof of Lemma 9.7.2, pick matrix units $a_1, \ldots, a_{2k-1} \in$

$M_k(F)$ such that $St_{2k-1}(a_1,\ldots,a_{2k-1}) = e_{ij}$. Then, since $M_k(F)u = vM_k(F) = 0$, we obtain
$$St_{2k+1}(ue_{ii}, a_1, \ldots, a_{2k-1}, e_{ji}v)$$
$$= ue_{ii}St_{2k-1}(a_1,\ldots,a_{2k-1})e_{ji}v + f(a_1,\ldots,a_{2k-1}, e_{ji}vue_{ii}),$$
where f is a suitable multilinear polynomial in $a_1,\ldots,a_{2k-1}, e_{ji}vue_{ii}$. Since
$$ue_{ii}St_{2k-1}(a_1,\ldots,a_{2k-1})e_{ji}v = ue_{ii}e_{ij}e_{ji}v = ue_{ii}v$$
is a non-zero element of $J_{01}J_{10} \subseteq J_{00}$, $vu \in J_{10}J_{01} \subseteq J_{11}$ and $J_{00} \cap J_{11} = 0$, it follows that $St_{2k+1} \notin \mathrm{Id}(R)$. In case $J_{01}J_{10} = 0$ and $J_{10}J_{01} \neq 0$, let $u \in J_{10}, v \in J_{01}$ be such that $uv \neq 0$. There exist $e_{ii}, e_{jj} \in M_k(F)$ such that $e_{ii}uve_{jj} \neq 0$. Since $e_{ii}uve_{jj} = e_{ik}e_{ki}e_{ii}uve_{jj}e_{jk}e_{kj}$, by replacing u with $e_{ki}u$ and v with ve_{jk}, we may assume that $e_{kk}u = u$ and $ve_{kk} = v$.

We now compute
$$St_{2k+1}(e_{11}, e_{12}, e_{22}, \ldots, e_{k-1,k}, e_{kk}, u, v) = 2e_{1k}uv \neq 0,$$
since $vu \in J_{01}J_{10} = 0$. Hence $St_{2k+1} \notin \mathrm{Id}(R)$ in this case. Similarly one can show that the same conclusion holds in case $J_{10}J_{00} \neq 0$ or $J_{00}J_{01} \neq 0$. □

LEMMA 9.7.5. *Suppose that $R = A + J(R)$ satisfies S_{2k+1}. If $J_{01}J_{10} = J_{10}J_{01} = J_{10}J_{00} = J_{00}J_{01} = 0$, then $\mathrm{var}(R) = \mathrm{var}(A_1 \oplus A_2 \oplus J_{00})$ where $A_1 = A + J_{10}$ and $A_2 = A + J_{01}$.*

PROOF. Clearly $\mathrm{Id}(R) \subseteq \mathrm{Id}(A_1 \oplus A_2 \oplus J_{00})$. Now let $f = f(x_1,\ldots,x_n)$ be a multilinear polynomial such that $f \notin \mathrm{Id}(R)$.

Suppose first that
$$f \in \mathrm{Id}(A + J_{11} + J_{10}) \cap \mathrm{Id}(A + J_{11} + J_{01}) \cap \mathrm{Id}(J_{00})$$
and let $b_1,\ldots,b_n \in R$ be such that $f(b_1,\ldots,b_n) \neq 0$. We may assume by linearity that $b_1,\ldots,b_n$ belong to $A \cup J_{10} \cup J_{01} \cup J_{11} \cup J_{00}$. By the assumption, $b_1,\ldots,b_n$ do not belong, at the same time, to $A \cup J_{11} \cup J_{10}$ or to $A \cup J_{11} \cup J_{01}$ or to J_{00}. Thus there exist b_i, b_j, $i \neq j$, such that one of the following three possibilities occurs: 1) $b_i \in J_{10}$ and $b_j \in J_{01}$. 2) $b_i \in J_{10}$ and $b_j \in J_{00}$. 3) $b_i \in J_{01}$ and $b_j \in J_{00}$. Since the J_{kl}'s are A-bimodules, $J_{01}J_{10} = J_{10}J_{01} = J_{10}J_{00} = J_{00}J_{01} = 0$ and, by Lemma 9.7.1, $J_{01}J_{00} = J_{00}J_{10} = J_{00}J_{11} = J_{11}J_{00} = J_{01}J_{01} = J_{10}J_{10} = 0$, we have that each of the above three cases leads to $b_{\sigma(1)} \cdots b_{\sigma(n)} = 0$, for all $\sigma \in S_n$. Thus $f \in \mathrm{Id}(R)$, contrary to the assumption.

We have proved that
$$\mathrm{Id}(R) \supseteq \mathrm{Id}(A + J_{11} + J_{10}) \cap \mathrm{Id}(A + J_{11} + J_{01}) \cap \mathrm{Id}(J_{00}).$$
If we prove that $\mathrm{Id}(A + J_{11} + J_{10}) = \mathrm{Id}(A + J_{10})$ and $\mathrm{Id}(A + J_{11} + J_{01}) = \mathrm{Id}(A + J_{01})$, we would get that $\mathrm{Id}(R) \supseteq \mathrm{Id}(A_1) \cap \mathrm{Id}(A_2) \cap \mathrm{Id}(J_{00})$ and the proof would be complete.

In order to prove that $\mathrm{Id}(A+J_{11}+J_{10}) = \mathrm{Id}(A+J_{10})$, suppose that there exists $f(x_1,\ldots,x_n) \notin \mathrm{Id}(A + J_{11} + J_{10})$ and let f be multilinear. Since $J_{11} = AN$, A commutes with N and N is commutative by Lemma 9.7.3, we have that for all $b_1,\ldots,b_m \in A + J_{11} + J_{10}$, $a \in A, d \in N$,
$$b_1 \cdots b_k adb_{k+1} \cdots b_m = db_1 \cdots b_k ab_{k+1} \cdots b_m.$$
It follows that if $b_1,\ldots,b_n \in A \cup J_{11} \cup J_{10}$ are such that $f(b_1,\ldots,b_n) \neq 0$, then we can write
$$f(b_1,\ldots,b_n) = d'f(b'_1,\ldots,b'_n),$$

for some $d' \in N$, $b'_1, \ldots, b'_n \in A \cup J_{10}$. Thus $f \notin \mathrm{Id}(A + J_{10})$ and $\mathrm{Id}(A + J_{11} + J_{10}) = \mathrm{Id}(A + J_{10})$ follows. Similarly one can show that $\mathrm{Id}(A + J_{11} + J_{01}) = \mathrm{Id}(A + J_{01})$. This completes the proof of the lemma. □

9.8. Asymptotics for the standard and the Capelli identities

In this section we compute the asymptotics of the variety determined by the standard polynomial or the Capelli polynomial of the type Cap_{k^2+1}.

Recall that for a polynomial f, $\mathrm{var}(f)$ denotes the variety whose T-ideal is generated by f. Also, two functions $g(n)$ and $h(n)$ are asymptotically equal and we write $g(n) \simeq h(n)$ if $\lim_{n \to \infty} g(n)/h(n) = 1$. If f denotes the standard or the Capelli polynomial, our aim is to find a suitable algebra A such that for all n, $c_n(f) \simeq c_n(A)$ and the codimensions of A are computable.

Our main tool will be a simplified form of Theorem 9.4.3 giving the decomposition of a variety into varieties generated by the Grassmann envelope of reduced superalgebras. In fact, in case of finite dimensional algebras, Theorem 9.4.3 has the following form. Recall that any algebra can be regarded as a superalgebra with trivial grading.

COROLLARY 9.8.1. *Let A be a finite dimensional algebra over the algebraically closed field F. Then there exist a finite number of reduced algebras $B_1, \ldots, B_t$ and a finite dimensional algebra D such that $\mathrm{var}(A) = \mathrm{var}(B_1 \oplus \cdots \oplus B_t \oplus D)$ and $\exp(A) = \exp(B_1) = \ldots = \exp(B_t)$, $\exp(D) < \exp(A)$.*

We need to introduce the following algebras.

DEFINITION 9.8.2. For $k, l \geq 1$, $M_{k \times l}(F)$ denotes the algebra of $(k+l) \times (k+l)$ matrices having the last l rows and the last k columns equal to zero.

It is clear that the algebra $M_{k \times l}(F)$ is of the type $M_k(F) + J$ and we studied such algebras in the previous section. We have

THEOREM 9.8.3 ([**GZ8**]). *Let F be algebraically closed. Then*

1) $\mathrm{var}(St_{2k}) = \mathrm{var}(M_k(F) \oplus B)$ *for some finite dimensional algebra B such that $\exp(B) < k^2$. In particular,*

$$c_n(St_{2k}) \simeq c_n(M_k(F)).$$

2) $\mathrm{var}(St_{2k+1}) = \mathrm{var}(M_{k \times 2k}(F) \oplus M_{2k \times k}(F) \oplus B)$ *for some finite dimensional algebra B with $\exp(B) < k^2$. In particular,*

$$c_n(St_{2k+1}) \simeq c_n(M_{k \times 2k}(F) \oplus M_{2k \times k}(F)).$$

PROOF. By Theorem 9.1.3, $\exp(St_{2k}) = \exp(St_{2k+1}) = k^2$. Also, by Theorem 7.1.4, for any $q \geq 1$, $\mathrm{var}(St_q)$ is generated by a finite dimensional algebra. Thus, by Corollary 9.8.1, there exists a finite number of finite dimensional reduced algebras $B_1, \ldots, B_t$ and a finite dimensional algebra D such that

(9.24) $$\mathrm{var}(St_q) = \mathrm{var}(B_1 \oplus \cdots \oplus B_t \oplus D)$$

and $\exp(B_1) = \cdots = \exp(B_t) = k^2$, $\exp(D) < k^2$.

We shall first analyze the structure of a finite dimensional reduced algebra R such that $St_q \in \mathrm{Id}(R)$ for fixed $q \geq 1$. Next we shall split the proof into the two cases $q = 2k$ and $q = 2k + 1$. In each case we shall plug into (9.24) the structure of each reduced algebra and we will deduce the desired result.

Let R be a finite dimensional reduced algebra satisfying $St_q \equiv 0$ ($q = 2k$ or $q = 2k+1$). Write $R = A_1 \oplus \cdots \oplus A_m + J$ where $A_i \cong M_{d_i}(F)$, $1 \le i \le m$ and $J = J(R)$. By the definition of reduced algebra, $A_1 J A_2 J \cdots J A_m \ne 0$. Hence by Theorem 8.2.1, R contains a minimal superalgebra with trivial grading isomorphic to the upper block triangular matrix algebra $UT(d_1, \ldots, d_m)$. Moreover,

$$\exp(R) = \exp(UT(d_1, \ldots, d_m)) = d_1^2 + \cdots + d_m^2.$$

Clearly $UT(d_1, \ldots, d_m)$ does not satisfy St_{2d-1} where $d = d_1 + \cdots + d_m$. Hence $UT(d_1, \ldots, d_m)$ satisfies St_q for $q = 2k$ or $q = 2k+1$ only if $d_1 + \cdots + d_m \le k$. On the other hand, $d_1^2 + \cdots + d_m^2 < k^2$ for any $m > 1$. It follows that if R is a reduced algebra with $\exp(R) = k^2$ satisfying $St_q \equiv 0$ ($q = 2k$ or $q = 2k+1$), then $R \cong M_k(F) + J$.

We now consider the cases $q = 2k$ and $q = 2k+1$.

Case 1. Suppose that $q = 2k$ and let $R = A + J$ be a reduced algebra as above where $A = M_k(F)$. Write $J = J_{00} + J_{01} + J_{10} + J_{11}$ as in Lemma 9.7.1. Since $St_{2k} \in \mathrm{Id}(R)$, by Lemmas 9.7.2 and 9.7.3, $J_{10} + J_{01} = 0$ and $J_{11} = AN$ where N is commutative. Since $AJ_{00} = J_{00}A = J_{11}J_{00} = J_{00}J_{11} = 0$ and $J_{10} = J_{01} = 0$, we obtain that $R = (A + J_{11}) \oplus J_{00}$ with $J_{00} \subseteq J$ a nilpotent ideal of R. Moreover, by Lemma 9.7.1, $A + J_{11} = A + AN \cong A \otimes_F N^\sharp$ where $N^\sharp$ is the algebra obtained from N by adjoining a unit element. Since $N^\sharp$ is commutative, it follows that $A + J_{11}$ and A satisfy the same identities. Thus $\mathrm{var}(R) = \mathrm{var}(A \oplus J_{00})$.

We have proved that if R is any reduced algebra such that $St_{2k} \in \mathrm{Id}(R)$ and $\exp(R) = k^2$, then $\mathrm{var}(R) = \mathrm{var}(M_k(F) \oplus J_{00})$ with J_{00} a finite dimensional nilpotent algebra. But then, by recalling the decomposition given in (9.24), we get that $\mathrm{var}(St_{2k}) = \mathrm{var}(M_k(F) \oplus D')$ where D' is a finite dimensional algebra with $\exp(D') < \exp(M_k(F)) = k^2$.

Case 2. Suppose now that $q = 2k+1$ and let $R = A + J$ be a reduced algebra with $A = M_k(F)$ and R satisfies St_{2k+1}. In this case, by Lemma 9.7.4, we get that $J_{01}J_{10} = J_{10}J_{01} = J_{10}J_{00} = J_{00}J_{01} = 0$. Hence, by Lemma 9.7.5, $\mathrm{var}(R) = \mathrm{var}((A + J_{10}) \oplus (A + J_{01}) \oplus J_{00})$ and J_{00} is a nilpotent algebra. Now, the left A-module J_{10} is isomorphic to $t \ge 1$ copies of a left ideal of $M_k(F)$. Since $J_{10}A = J_{10}J_{10} = 0$ and $A = M_{k \times k}(F)$, then $A + J_{10}$ as an F-algebra is isomorphic to $M_{k \times (k+t)}(F)$ and it is clear that $A + J_{10}$ has the same identities as $M_{k \times 2k}(F)$. Similarly one shows that $A + J_{01}$ satisfies the same identities as $M_{2k \times k}(F)$.

We have proved that if R is any reduced algebra such that $St_{2k+1} \in \mathrm{Id}(R)$, then $\mathrm{var}(R) = \mathrm{var}(M_{k \times 2k}(F) \oplus M_{k \times 2k}(F) \oplus J_{00})$ with J_{00} a finite dimensional nilpotent algebra. By invoking the decomposition given in (9.24), we get that

$$\mathrm{var}(St_{2k+1}) = \mathrm{var}(M_{k \times 2k}(F) \oplus M_{2k \times k}(F) \oplus B)$$

where $\exp(B) \le k^2 - 1$. $\square$

COROLLARY 9.8.4. $c_n(St_{2k}) \simeq \alpha n^{-\frac{k^2-1}{2}} k^{2n}$ and

$$\frac{\alpha}{k^2} n^{\frac{3-k^2}{2}} k^{2n} \lesssim c_n(St_{2k+1}) \lesssim \frac{2\alpha}{k^2} n^{\frac{3-k^2}{2}} k^{2n}$$

where $\alpha = \left(\frac{1}{\sqrt{2\pi}}\right)^{k-1} \left(\frac{1}{2}\right)^{\frac{1}{2}(k^2-1)} \cdot 1!2! \cdots (k-1)! k^{\frac{1}{2}(k^2+4)}$.

PROOF. The precise asymptotics for $c_n(St_{2k})$ follow from Theorem 5.10.4, and the relation $c_n(St_{2k}) \simeq c_n(M_k(F))$.

Let $I_1 = \mathrm{Id}(M_{k \times 2k}(F))$ and $I_2 = \mathrm{Id}(M_{2k \times k}(F))$. By Theorem 7.5.3, it follows that $I_1 = \mathrm{Id}(M_k(F)) \cdot F\langle X \rangle$, $I_2 = F\langle X \rangle \cdot \mathrm{Id}(M_k(F))$ and $c_n(I_1) = c_n(I_2) = nc_{n-1}(M_k(F))$. Since $\langle St_{2k+1} \rangle_T = I_1 \cap I_2 \cap I_3$ where, by Theorem 9.8.3, I_3 is a T-ideal with $c_n(I_3) \lesssim (k^2-1)^n$, we get
$$c_n(I_1) \le c_n(St_{2k+1}) \le c_n(I_1) + c_n(I_2) + c_n(I_3) \simeq 2c_n(I_1).$$
The second part of the corollary now follows from the asymptotics of $c_n(M_k(F))$. □

In Theorem 9.1.5 we proved that $\exp(Cap_{m+1}) = m, m-1, m-2$ or $m-3$. Also, $\exp(Cap_{m+1}) = m$ if and only if $m = k^2$ is a square.

In the next theorem we compute the asymptotics of the codimensions of Cap_{m+1} and the asymptotics of the corresponding codimensions if m is a square ([**GZ8**]).

THEOREM 9.8.5. *Let F be algebraically closed. Then*
$$\mathrm{var}(Cap_{k^2+1}) = \mathrm{var}(M_k(F) \oplus B),$$
for some finite dimensional algebra B such that $\exp(B) < k^2$.

PROOF. By Theorem 7.1.4, $\mathrm{var}(Cap_{k^2+1})$ is generated by a finite dimensional algebra and, by Theorem 9.1.5, $\exp(Cap_{k^2+1}) = k^2$. Thus, by Corollary 9.8.1, there exist finite dimensional algebras $B_1, \ldots, B_t, D$ such that
$$(9.25) \qquad \mathrm{var}(Cap_{k^2+1}) = \mathrm{var}(B_1 \oplus \cdots \oplus B_t \oplus D)$$
where $B_1, \ldots, B_t$ are reduced algebras and $\exp(B_1) = \cdots = \exp(B_t) = k^2$, $\exp(D) < k^2$. Next we analyze the structure of any such reduced algebra.

Let R be a finite dimensional reduced algebra such that $Cap_{k^2+1} \in \mathrm{Id}(R)$ and $\exp(R) = k^2$. As in the proof of the previous theorem, R contains a subalgebra of the type $UT(d_1, \ldots, d_t)$ with $\exp(R) = d_1^2 + \cdots + d_t^2 = k^2$. By Lemma 9.1.4 the algebra $UT(d_1, \ldots, d_t)$ does not satisfy $Cap_{d_1^2+\cdots+d_t^2+t-1}$. Hence $t = 1, d_1 = k$ and we may write $R = A + J$ where $A = M_k(F), J = J(R)$. Since by Lemma 9.7.2, $J_{01} = J_{10} = 0$, then $J = J_{00} + J_{11}$ and, as noticed in the proof of Theorem 9.8.3, we have that $R = (A + J_{11}) \oplus J_{00}$ with J_{00} a nilpotent ideal of R. Also, by Lemma 9.7.3, $J_{11} = AN$ where N is a nilpotent commutative subalgebra centralizing A in R. Hence, as in the proof of Theorem 9.8.3, A and $A + J_{11}$ have the same identities and $\mathrm{var}(R) = \mathrm{var}(A \oplus J_{00})$ follows.

By the decomposition given in (9.25), it follows that $\mathrm{var}(Cap_{k^2+1})$ is generated by $M_k(F) \oplus B$ where B is a finite dimensional algebra and $\exp(B) < k^2$. □

COROLLARY 9.8.6.
$$c_n(Cap_{k^2+1}) \simeq c_n(St_{2k}).$$

CHAPTER 10

G-Identities and $G \wr S_n$-Action

In this chapter we extend the asymptotic methods developed in the previous chapters to algebras with an additional structure of superalgebra or more generally of group graded algebra or of algebra with involution. Our intent is twofold: to generalize methods and results to a weaker class of identities and to construct finer invariants that can be related to the old ones.

By using a unified approach in this chapter we deal with algebras A on which a finite group G of automorphisms and antiautomorphisms acts. We study the corresponding G-identitites and we relate them to the ordinary identities of A. We show that there is a natural action of the group wreath product $G \wr S_n$ on the space of multilinear G-polynomials in n variables extending the permutation action of S_n. We exploit this action in order to study the G-identities of the algebra A. To this end we define the notions of G-codimension, G-cocharacter and we compare them to the ordinary notions of codimension and cocharacter. As an application of the efficiency of the asymptotic method, we prove that any algebra satisfying an essential G-identity satisfies an ordinary identity. As a corollary we deduce a celebrated theorem of Amitsur stating that any algebra with involution $*$ satisfying a non-trivial $*$-identity is a PI-algebra.

Later in this chapter we specialize our discussion to the following significant cases: when G is a finite abelian group of automorphisms or when G is a group of order two. In the first case, as we have seen in Chapter 3, the algebra A is G-graded and we study the corresponding graded identities. We then give a brief account of the ordinary representation theory of the group $G \wr S_n$ over an algebraically closed field of characteristic zero in case G is an abelian group. Through this theory we are able to characterize G-graded algebras whose sequence of G-codimensions is polynomially bounded.

When $G = \mathbb{Z}_2$ has order two, then either A has a structure of superalgebra or of algebra with involution if φ is an automorphism or an antiautomorphism of order two, respectively. By extending some of the techniques of Chapter 6, we prove that for any finite dimensional algebra, the $\mathbb{Z}_2$-codimensions have an integer exponential growth and we give a recipe for computing the corresponding $\mathbb{Z}_2$-exponent. In the next chapter we shall split the two cases (of superalgebras and algebras with involution) and we shall deal with them separately.

Throughout this chapter the ground field F is assumed to be of characteristic zero.

10.1. G-identities, G-codimensions and $G \wr S_n$-action

We recall some notation and terminology from Chapter 3. For an F-algebra A, $\text{Aut}^*(A)$ denotes the group of automorphisms and antiautomorphisms of A and $\text{Aut}(A)$ is the subgroup of automorphisms of A.

Throughout this section G is a finite subgroup of $\mathrm{Aut}^*(A)$ and $F\langle X|G\rangle$ is the free algebra on X with G-action. Recall that the algebra $F\langle X|G\rangle$ is the algebra freely generated on the set $\{x^g = g(x) \mid x \in X, g \in G\}$ and its elements are called G-polynomials. A G-polynomial $f(x_1^{g_1},\ldots,x_n^{g_n})$ is a G-identity for A, and we write $f \equiv 0$ on A, if $f(a_1^{g_1},\ldots,a_n^{g_n}) = 0$ for all $a_1,\ldots,a_n \in A$. Also,

$$\mathrm{Id}^G(A) = \{f \in F\langle X|G\rangle \mid f \equiv 0 \text{ on } A\}$$

is the ideal of G-identities of A. Notice that $\mathrm{Id}^G(A)$ is invariant under all endomorphisms of the free algebra commuting with the G-action.

By naturally extending the ordinary case (when G is trivial), the degree of a monomial M in a variable $x \in X$, is defined as the number of times the variables x^g appear in M (regardless of the exponent $g \in G$).

Thus, for every $n \geq 1$, we define the space of multilinear G-polynomials

$$P_n^G = \mathrm{span}\{x_{\sigma(1)}^{g_1} \cdots x_{\sigma(n)}^{g_n} \mid \sigma \in S_n, g_1,\ldots,g_n \in G\}.$$

Clearly, $P_n^G \cap \mathrm{Id}^G(A)$ is the set of all multilinear G-identities of degree n in the first n variables.

Since $\mathrm{char}\, F = 0$, the corresponding generalization of the multilinearization process applies and as in Chapter 1, it follows that any G-identity is equivalent to a system of multilinear G-identities. Hence the subspaces $P_n^G \cap \mathrm{Id}^G(A)$, $n \geq 1$, carry enough information on the G-identities of A.

DEFINITION 10.1.1. The dimension

$$c_n^G(A) = \dim \frac{P_n^G}{P_n^G \cap \mathrm{Id}^G(A)}$$

is called the nth G-codimension of A.

We now want to compare the sequence of G-codimensions and the sequence of ordinary codimensions of an algebra. When $G = 1$, $F\langle X|G\rangle = F\langle X\rangle$ is the free algebra on X and $\mathrm{Id}^G(A) = \mathrm{Id}(A)$ is the T-ideal of $F\langle X\rangle$ of ordinary polynomial identities of A. In this case $P_n^G = P_n$ and $c_n^G(A) = c_n(A)$, the nth codimension of A.

If we identify any polynomial $f(x_1,\ldots,x_n) \in P_n$ with the corresponding polynomial $f(x_1^1,\ldots,x_n^1) \in P_n^G$, then we have $P_n \subseteq P_n^G$ and

$$P_n \cap \mathrm{Id}^G(A) = P_n \cap \mathrm{Id}(A).$$

Observe that $\dim P_n^G = |G|^n n!$ versus $\dim P_n = n!$. Next lemma holds for any algebra A.

LEMMA 10.1.2. For all $n \geq 1$, $c_n(A) \leq c_n^G(A)$.

PROOF. We have

$$\frac{P_n}{P_n \cap \mathrm{Id}(A)} = \frac{P_n}{P_n \cap (P_n^G \cap \mathrm{Id}^G(A))} \simeq \frac{P_n + (P_n^G \cap \mathrm{Id}^G(A))}{P_n^G \cap \mathrm{Id}^G(A)} \subseteq \frac{P_n^G}{P_n^G \cap \mathrm{Id}^G(A)}.$$

$\square$

In the case of PI-algebras the G-codimensions can also be bounded from above.

LEMMA 10.1.3. *Let A be a PI-algebra satisfying a non-trivial G-identity, with $G \subseteq \mathrm{Aut}^*(A)$ a finite group. Then*
$$c_n^G(A) \le |G|^n c_n(A) \le |G|^n (d-1)^{2n},$$
where d is the degree of an ordinary identity satisfied by A.

PROOF. Let $\{f_1(x_1,\ldots,x_n),\ldots,f_k(x_1,\ldots,x_n)\}$ be a basis of $P_n \pmod{\mathrm{Id}(A)}$. Then for any $\sigma \in S_n$,
$$x_{\sigma(1)} \cdots x_{\sigma(n)} \equiv \sum_{i=1}^{k} \alpha_{i,\sigma} f_i(x_1,\ldots,x_n) \pmod{\mathrm{Id}(A)}.$$
For any fixed n-ple $(g_1,\ldots,g_n) \in G^n$, we have
$$x_{\sigma(1)}^{g_{\sigma(1)}} \cdots x_{\sigma(n)}^{g_{\sigma(n)}} \equiv \sum_{i=1}^{k} \alpha_{i,\sigma} f_i(x_1^{g_1},\ldots,x_n^{g_n}) \pmod{\mathrm{Id}^G(A)}.$$
Since $\mathrm{Id}(A) \subseteq \mathrm{Id}^G(A)$, we obtain that
$$\{f_i(x_1^{g_1},\ldots,x_n^{g_n}) \mid (g_1,\ldots,g_n) \in G^n, 1 \le i \le k\}$$
spans $P_n^G \pmod{\mathrm{Id}^G(A)}$. Thus $c_n^G(A) \le |G|^n c_n(A)$.
The inequality $|G|^n c_n(A) \le |G|^n (d-1)^{2n}$ follows from Theorem 4.2.4. $\square$

Since by Theorem 4.2.4, an algebra A is a PI-algebra if and only if $c_n(A)$ is exponentially bounded, as an immediate consequence of the above lemma we obtain.

THEOREM 10.1.4. *Let A be an F-algebra satisfying a non-trivial G-identity where $G \subseteq \mathrm{Aut}^*(A)$ is a finite group. Then A is a PI-algebra if and only if its sequence of G-codimensions $c_n^G(A)$, $n = 1, 2, \ldots$, is exponentially bounded.*

We shall apply this result in Section 10.3 in order to prove a celebrated theorem of Amitsur stating that if an algebra with involution satisfies a $*$-identity, then A satisfies an ordinary identity.

Next we wish to extend the permutation action of the symmetric group S_n on the space of multilinear polynomials in $x_1,\ldots,x_n$ (see Chapter 2) to the case of G-polynomials. It turns out that the analogous role in this setting can be played by the group $G \wr S_n$, the wreath product of G and S_n ([**GR**]). In fact we shall see below how one can naturally describe the spaces of multilinear G-identities through the representations of this group.

Recall (see for instance [**JK**]) that the wreath product of G and S_n is the group defined by
$$G \wr S_n = \{(g_1,\ldots,g_n;\sigma) \mid g_1,\ldots,g_n \in G, \sigma \in S_n\}$$
with multiplication given by
$$(g_1,\ldots,g_n;\sigma)(h_1,\ldots,h_n;\tau) = (g_1 h_{\sigma^{-1}(1)},\ldots,g_n h_{\sigma^{-1}(n)};\sigma\tau).$$
There is a convenient way of describing the group algebra $F(G \wr S_n)$ which is suitable for our purpose as follows.

Let us denote $\underbrace{FG \otimes \cdots \otimes FG}_{n} = FG^{\otimes n}$. We let the group S_n act on $FG^{\otimes n}$ by permuting coordinates: if $\sigma \in S_n$, $\mathbf{b} = b_1 \otimes \cdots \otimes b_n \in FG^{\otimes n}$, then
$$\sigma(\mathbf{b}) = \sigma(b_1 \otimes \cdots \otimes b_n) = b_{\sigma^{-1}(1)} \otimes \cdots \otimes b_{\sigma^{-1}(n)}.$$

The group algebra $F(G \wr S_n)$, that we shall simply write as $FG \wr S_n$, can then be identified with the vector space $FG^{\otimes n} \otimes FS_n$ with multiplication induced by

$$(\mathbf{b} \otimes \sigma)(\mathbf{c} \otimes \tau) = \mathbf{b} \cdot \sigma(\mathbf{c}) \otimes \sigma\tau$$

where $\mathbf{b}, \mathbf{c} \in FG^{\otimes n}$, $\sigma, \tau \in S_n$. In this terminology, the element $(g_1, \ldots, g_n; \sigma)$ of the wreath product $G \wr S_n$ is written as $\mathbf{g} \otimes \sigma$ where $\mathbf{g} = g_1 \otimes \cdots \otimes g_n \in G^{\otimes n}$.

Our next goal is to define an action of $G \wr S_n$ on P_n^G preserving the ideals of G-identities. Following [**GR**], in order to accomplish this, we notice that the map

$$g_1 \otimes \cdots \otimes g_n \otimes \sigma \to x_{\sigma(1)}^{g_{\sigma(1)}^{-1}} \cdots x_{\sigma(n)}^{g_{\sigma(n)}^{-1}}$$

where $g_1, \ldots, g_n \in G$, $\sigma \in S_n$, when extended by linearity to $FG \wr S_n$, gives a linear isomorphism

$$\psi : FG \wr S_n \to P_n^G.$$

This in turn defines a left action of $FG \wr S_n$ on P_n^G given by $a\psi(b) = \psi(ab)$, for $a, b \in FG \wr S_n$. As a consequence we obtain the following lemma.

LEMMA 10.1.5. *If $f(x_1, \ldots, x_n) \in P_n^G$ and $g_1 \otimes \cdots \otimes g_n \otimes \sigma \in G \wr S_n$, then*

(10.1) $$(g_1 \otimes \cdots \otimes g_n \otimes \sigma)f(x_1, \ldots, x_n) = f(x_{\sigma(1)}^{g_{\sigma(1)}^{-1}}, \ldots, x_{\sigma(n)}^{g_{\sigma(n)}^{-1}}).$$

PROOF. Let $\mathbf{g}, \mathbf{h} \in G^{\otimes n}, \sigma, \tau \in S_n$, and denote

$$M_{\mathbf{h} \otimes \tau}(x_1, \ldots, x_n) = x_{\tau(1)}^{h_1} \cdots x_{\tau(n)}^{h_n} = \psi(\mathbf{h} \otimes \tau).$$

Clearly it is enough to prove that

$$(\mathbf{g} \otimes \sigma)M_{\mathbf{h} \otimes \tau}(x_1, \ldots, x_n) = M_{\mathbf{h} \otimes \tau}(x_{\sigma(1)}^{g_{\sigma(1)}^{-1}}, \ldots, x_{\sigma(n)}^{g_{\sigma(n)}^{-1}}).$$

We have

$$(\mathbf{g} \otimes \sigma)M_{\mathbf{h} \otimes \tau}(x_1, \ldots, x_n) = (\mathbf{g} \otimes \sigma)\psi((\mathbf{h} \otimes \tau)) = \psi((\mathbf{g} \otimes \sigma)(\mathbf{h} \otimes \tau))$$
$$= \psi(\mathbf{g} \cdot \sigma(\mathbf{h}) \otimes \sigma\tau) = \psi(\mathbf{k} \otimes \sigma\tau)$$

where $\mathbf{k} = k_1 \otimes \cdots \otimes k_n$ and $k_i = g_i h_{\sigma^{-1}(i)}$, for $1 \leq i \leq n$. Now

$$\psi(\mathbf{k} \otimes \sigma\tau) = x_{\sigma\tau(1)}^{k_{\sigma\tau(1)}^{-1}} \cdots x_{\sigma\tau(n)}^{k_{\sigma\tau(n)}^{-1}}$$

and

$$k_{\sigma\tau(i)}^{-1} = (g_{\sigma\tau(i)} h_{\sigma^{-1}\sigma\tau(i)})^{-1} = h_{\tau(i)}^{-1} g_{\sigma\tau(i)}^{-1}.$$

Hence, since

$$x_{\sigma\tau(i)}^{k_{\sigma\tau(i)}^{-1}} = x_{\sigma\tau(i)}^{h_{\tau(i)}^{-1} g_{\sigma\tau(i)}^{-1}} = (x_{\sigma\tau(i)}^{g_{\sigma\tau(i)}^{-1}})^{h_{\tau(i)}^{-1}},$$

we obtain

$$\psi(\mathbf{k} \otimes \sigma\tau) = (x_{\sigma\tau(1)}^{g_{\sigma\tau(1)}^{-1}})^{h_{\tau(1)}^{-1}} \cdots (x_{\sigma\tau(n)}^{g_{\sigma\tau(n)}^{-1}})^{h_{\tau(n)}^{-1}} = M_{\mathbf{h} \otimes \tau}(x_{\sigma(1)}^{g_{\sigma(1)}^{-1}}, \ldots, x_{\sigma(n)}^{g_{\sigma(n)}^{-1}}),$$

and the proof is complete. $\square$

An important consequence of the previous lemma is that, the ideal $\mathrm{Id}^G(A)$ of G-identities of the algebra A is left invariant under the $G \wr S_n$-action. This makes $P_n^G/(P_n^G \cap \mathrm{Id}^G(A))$ a left $G \wr S_n$-module.

At this stage we need to describe the representation theory of $G \wr S_n$. We do this in Section 10.4 in the case of abelian groups G.

10.2. Decomposable monomials

In this section we introduce the notion of decomposable monomial which generalizes the notion of bad permutation given in Chapter 4. This will be an essential tool for proving the existence of polynomial identities in the next sections.

We first introduce an ordering on the variables x_i by requiring that $x_i < x_j$ if $i < j$; we then extend this ordering lexicographically to all monomials (words) in P_n by comparing them from left to right.

Following [**R4**] we introduce the following.

DEFINITION 10.2.1. A monomial $w \in P_n$ is m-decomposable if it can be represented in the form $w = aw_m w_{m-1} \cdots w_1 b$ where w_i, $i = 1, \ldots, m$, are non-empty monomials such that

a) the left variable in the monomial w_i is greater than any other variable in this monomial, $i = 1, \ldots, m$;

b) the first variable in the monomial w_{i+1} is greater than the first variable in the monomial w_i, $i = 1, \ldots, m-1$.

If w has no m-decompositions, then we say that w is m-indecomposable.

Here we want to find an upper bound on the number of m-indecomposable monomials in $x_1, \ldots, x_n$. We start with the following.

LEMMA 10.2.2. *Let $w = a x_n b$ be a multilinear monomial in $x_1, \ldots, x_n$. Then w is m-indecomposable if and only if a is m-indecomposable and b is $(m-1)$-indecomposable.*

PROOF. Clearly an m-decomposition of a gives an m-decomposition of w. If b has an $(m-1)$-decomposition, then we easily construct an m-decomposition for w since $x_n > x_1, \ldots, x_{n-1}$.

Suppose now that a is m-indecomposable, b is $(m-1)$-indecomposable but $w = u w_1 \cdots w_m \theta$ is an m-decomposition for w. This means that $w_1 > \cdots > w_m$ in the left lexicographic order and any w_i has the form $w_i = x_{t_i} u_i$ where x_{t_i} is greater than any indeterminate in u_i.

We examine the position of x_n in w. Write $w = a x_n b = u w_1 \cdots w_m \theta$. If $x_n \in \theta$ then, $\theta = \theta' x_n \theta''$ and $w = u w_1 \cdots w_m \theta' x_n \theta'' = a x_n b$, that is, $b = \theta''$, $a = u w_1 \cdots w_m \theta'$ and $w_1, \ldots, w_m$ gives an m-decomposition of a. Similarly, if $x_n \in u$, then $u = u' x_n u''$, $w = u' x_n u'' w_1 \cdots w_m \theta = a x_n b$, hence $b = u'' w_1 \cdots w_m \theta$ is m-decomposable.

Finally, suppose that $x_n \in w_i$. Since x_n is the greatest indeterminate, it follows that $i = 1$ and $w_1 = x_n w_1'$. Then $w = a x_n b = u x_n w_1' w_2 \cdots w_m \theta$ and $b = w_1' w_2 \cdots w_m \theta$. In particular, the subwords $w_2, \ldots, w_m$ give an $(m-1)$-decomposition of b. This contradiction completes the proof of the lemma. □

Denote by $a_m(n)$ the number of m-indecomposable words in $x_1, \ldots, x_n$, $n \geq 1$, and set $a_m(0) = 1$. We then have the following recursive formula.

LEMMA 10.2.3. $a_m(n) = \sum_{i=0}^{n-1} \binom{n-1}{i} a_m(i) a_{m-1}(n-1-i)$.

PROOF. Denote by Q the set of all m-indecomposable monomials in the n indeterminates $x_1, \ldots, x_n$. Then

$$Q = Q_0 \cup Q_1 \cup \ldots \cup Q_{m-1}$$

where
$$Q_i = \{w \in Q \mid w = w' x_n w'', \deg w' = i\}.$$
According to Lemma 10.2.2, $w' x_n w''$ is m-indecomposable if and only if w' is m-indecomposable and w'' is $(m-1)$-indecomposable in the corresponding indeterminates. Hence
$$|Q_i| = \binom{n-1}{i} a_m(i) a_{m-1}(n-1-i)$$
and
$$a_m(n) = |Q_0| + \cdots + |Q_{n-1}| = \sum_{i=0}^{n-1} \binom{n-1}{i} a_m(i) a_{m-1}(n-1-i).$$
$\square$

We now set
$$b_m(n) = \frac{a_m(n)}{n!}.$$
It can be shown that asymptotically $b_m(n)$ is smaller than $\left(\frac{1}{c}\right)^n$ for any fixed c. In the sequel we shall need an explicit estimate of $b_m(n)$, for n depending only on c and m.

Let $t = m + [\log_2 c]$ and $N = 2^{t 2^{t+1}}$. We denote by $p_j, j \geq 3$, an integer for which
$$\underbrace{\log_N \ldots \log_N}_{j-2} p_j = p_2$$
where $p_2 = 2^{t 2^t}$. We remark that with such a choice of p_2 the inequality
$$(10.2) \qquad n! p_2 > 2^{tn}$$
is satisfied for every natural n. We set $f(m, c) = \log_2 p_m$.

LEMMA 10.2.4. *If $n \geq f(m, c)$, then $b_m(n) < \left(\frac{1}{c}\right)^n$.*

PROOF. From Lemma 10.2.3 it follows that $b_k(n)$ satisfies the following recursive relation
$$(10.3) \qquad b_k(n) = \frac{1}{n} \sum_{i=0}^{n-1} b_k(i) b_{k-1}(n-1-i),$$
where $b_k(0) = 1$. It is easy to see that $b_2(n) = \frac{1}{n!}$. In what follows we assume that $k \geq 2$.

First, we show that if the conditions
$$(10.4) \qquad b_k(j) < p \varepsilon^j, \ j = 0, 1, \ldots$$
and
$$(10.5) \qquad b_{k+1}(j) < q (2\varepsilon)^j, \ j = 0, 1, \ldots, n-1,$$
with $\varepsilon < \frac{1}{2}$, are satisfied, then
$$(10.6) \qquad b_{k+1}(n) < \frac{2pq}{n} (2\varepsilon)^{n-1}.$$

In fact, it follows from (10.3) that

$$b_{k+1}(n) < \frac{1}{n}\sum_{i=0}^{n-1} q(2\varepsilon)^i p\varepsilon^{n-1-i} = \frac{pq}{n}\varepsilon^{n-1}\sum_{i=0}^{n-1} 2^i < \frac{2pq}{n}(2\varepsilon)^{n-1}.$$

We assume now that (10.4) holds, $n_0 > \frac{p}{\varepsilon}$ and $q \geq \left(\frac{1}{2\varepsilon}\right)^{n_0} > \left(\frac{1}{2\varepsilon}\right)^{\frac{p}{\varepsilon}}$. Then the inequality (10.5) is satisfied for all $j \leq n_0$, since $b_k(n) \leq 1$ for all k and n. But then, by (10.6), $b_{k+1}(j) < q(2\varepsilon)^j$ also for $j = n_0 + 1$, since $\frac{2p}{n_0} \leq 2\varepsilon$. It follows that with such a choice of q, the inequality (10.5) holds for all j. So, we have shown that if

$$b_k(j) < q_k \varepsilon_k^j$$

for all j, then

$$b_{k+1}(j) < q_{k+1}\varepsilon_{k+1}^j$$

also holds for all j with $\varepsilon_{k+1} = 2\varepsilon_k$ and

(10.7) $$q_{k+1} \geq \left(\frac{1}{\varepsilon_{k+1}}\right)^{\frac{q_k}{\varepsilon_k}}.$$

Set $\varepsilon_2 = 2^{-t}$. From (10.2) it follows that $b_2(j) < p_2\varepsilon_2^j$ for all j. Then $\varepsilon_{k+1} = 2^{-r}$ and $r \leq t$. Hence

$$\left(\frac{1}{\varepsilon_{k+1}}\right)^{\frac{q_k}{\varepsilon_k}} = (2^{r2^{r+1}})^{q_k} \leq N^{q_k}$$

and the inequality (10.7) is satisfied by all numbers of the form $q_2 = p_2, \ldots, q_k = p_k, q_{k+1} = N^{p_k} = p_{k+1}$. Hence, for $k = m$ we have

$$b_m(n) < p_m \varepsilon_m^n \quad \text{and} \quad \varepsilon_m = 2^{m-2}\varepsilon_2 = 2^{m-t-2}.$$

By the choice of the number t the following inequality holds:

$$t \geq m + \log_2 c - 1 = m + \log_2 2c - 2.$$

Hence $\varepsilon_m \leq \frac{1}{2c}$ and

(10.8) $$b_m(n) < p_m \left(\frac{1}{2c}\right)^n.$$

By the hypothesis, $n \geq \log_2 p_m$. Hence $p_m \left(\frac{1}{2}\right)^n \leq 1$ and from (10.8) we obtain the required estimate. $\square$

10.3. Essential G-identities. Amitsur's theorem on $*$-identities

Let A be an algebra and $G \leq \text{Aut}^*(A)$ a finite group. In this section we want to prove that under some circumstances a G-identity satisfied by A implies an ordinary identity of A. Notice that in general, an algebra A might satisfy a non-trivial G-identity, and still A is not a PI-algebra. For instance, let $A = B \oplus C$ be the algebra direct sum of the algebras B and C where $B \cong F\langle X \rangle$ and $C^2 = 0$. Then $\varphi \in \text{Aut}(A)$ such that $\varphi(b+c) = b - c$ for $b \in B, c \in C$ has order two and let $G = \langle \varphi \rangle$. Then A satisfies the G-identity $(x_1 - x_1^\varphi)(x_2 - x_2^\varphi) \equiv 0$ and still A is not a PI-algebra.

For positive results, when G is a group of automorphisms and A has no $|G|$-torsion, Kharchenko [**Ka**] proved that a polynomial identity in the ring of invariants $A^G = \{a \in A \mid a^g = a, \text{ for all } g \in G\}$ forces the existence of a polynomial identity in A. Notice that since for all $a \in A$, $\sum_{g \in G} a^g \in A^G$, then clearly an identity satisfied by A^G implies a G-identity on A.

On the other hand, Amitsur in [**A3**] and [**A4**] proved that with no further hypotheses on A, if $G = \{1, *\}$ where $*$ is an involution on A, then a G-identity (a $*$-identity) on A implies an ordinary identity on A.

We should remark that, if G is arbitrary, by combining the results of Amitsur and Kharchenko it is easy to prove that A^G PI forces A PI when there is no $|G|$-torsion.

In this section as a consequence of a more general theorem, we shall prove Amitsur's theorem and we shall also provide information on the degree of the polynomial identity satisfied by A.

Since by Lemma 10.1.2 $c_n(A) \leq c_n^G(A)$, in order to obtain a polynomial identity for the algebra A we shall bound the G-codimensions from above by making use of the m-indecomposable words introduced in the previous section.

Recall that P_n^G is the space of multilinear G-polynomials in $x_1, \ldots, x_n$. Let $G^n = G \times \cdots \times G$ and $\mathbf{g} = (g_1, \ldots, g_n) \in G^n$. Denote by

$$P_n^{\mathbf{g}} = \operatorname{span}_F \{x_{\sigma(1)}^{g_{\sigma(1)}} \cdots x_{\sigma(n)}^{g_{\sigma(n)}} \mid \sigma \in S_n\}$$

the space of multilinear polynomials in $F\langle X | G \rangle$ in the variables $x_1^{g_1}, \ldots, x_n^{g_n}$. In particular, for $\mathbf{1} = (1, \ldots, 1)$ we have $P_n^{\mathbf{1}} = P_n$. Notice that $P_n^G = \sum_{\mathbf{g} \in G^n} P_n^{\mathbf{g}}$ is the space of multilinear G-polynomials in $x_1, \ldots, x_n$.

We introduce a partial ordering on the variables x_i^g by requiring $x_i^g < x_j^h$ as soon as $i < j$. We then extend this ordering left lexicographically to all monomials in P_n^G as in the previous section.

DEFINITION 10.3.1. A G-identity $f \in P_n^G$ will be called an essential G-identity if it is of the form

$$f = x_1^1 \cdots x_n^1 + \sum_{\substack{1 \neq \sigma \in S_n \\ g = (g_1, \ldots, g_n) \in G^n}} \alpha_{\sigma, g} x_{\sigma(1)}^{g_1} \cdots x_{\sigma(n)}^{g_n}.$$

Let $f(m, c)$ be the function defined in the previous section. We can now prove the following result of Bahturin, Giambruno and Zaicev.

THEOREM 10.3.2 ([**BGZ1**]). *Let A be an algebra over a field F and G a finite subgroup of $\operatorname{Aut}^*(A)$. Suppose that A satisfies some mutilinear essential G-identity of degree d. Then for n sufficiently large we have $c_n^G(A) \leq |G|^n (f(d, |G|) - 1)^{2n}$ and A satisfies a non-trivial polynomial identity, whose degree is bounded by the function $f(d, |G|)$. If $G \leq \operatorname{Aut}(A)$, then $c_n^G(A) \leq |G|^n (d-1)^{2n}$ and A satisfies a non-trivial polynomial identity, whose degree is bounded by $3|G|(d-1)^2$.*

PROOF. We first claim that P_n^G is spanned (mod $\operatorname{Id}^G(A)$) by the d-indecomposable monomials.

In fact, suppose by contradiction that the claim is false. Then there exists a counterexample $w = x_{i_1}^{s_1} \cdots x_{i_n}^{s_n}$ which is minimal in the left lexicographic order defined above.

By hypothesis, A satisfies an essential identity of degree d, hence

$$(10.9) \qquad x_1^1 \cdots x_d^1 \equiv \sum_{\substack{1 \neq \sigma \in S_d \\ g = (g_1, \ldots, g_d) \in G^d}} \alpha_{\sigma, g} x_{\sigma(1)}^{g_1} \cdots x_{\sigma(d)}^{g_d} \pmod{\operatorname{Id}^G(A)}$$

for some $\alpha_{\sigma,g} \in F$.

Since $\mathrm{Id}^G(A)$ is invariant under all endomorphisms of $F\langle X|G\rangle$ commuting with the G-action, (10.9) implies that

$$(10.10) \qquad t_1 \cdots t_d \equiv \sum_{\substack{1 \neq \sigma \in S_d \\ g=(g_1,\ldots,g_d) \in G^d}} \alpha_{\sigma,g} t_{\sigma(1)}^{g_1} \cdots t_{\sigma(d)}^{g_d}, \quad (\mathrm{mod}\ \mathrm{Id}^G(A))$$

for any $t_1, \ldots, t_n \in F\langle X|G\rangle$.

Since w is d-decomposable, there exist indices $j_1, \ldots, j_d$ which determine a d-decomposition on w. We will write $y_1 = x_{i_1}^{s_1}, \ldots, y_n = x_{i_n}^{s_n}$ for convenience.

We denote by a_0 the product $y_1 \cdots y_{j_1-1}$, if $j_1 > 1$. If $j_1 = 1$, we simply assume that a_0 is the empty word. Similarly, we set $a_{d+1} = y_{j_d+1} \cdots y_n$ if $j_d < n$, and set a_{d+1} the empty word as soon as $j_d = n$. For all $k = 1, \ldots, d-1$ we set

$$a_k = y_{j_k} y_{j_k+1} \cdots x_{j_{k+1}-1},$$

and $a_d = y_{j_d}$. Then $w = a_0 a_1 \cdots a_d a_{d+1}$. But from (10.10) it follows that, modulo $\mathrm{Id}^G(A)$, we can express w as a linear combination of products of the form $w_{\sigma,g} = a_0 a_{\sigma(1)}^{g_1} \cdots a_{\sigma(d)}^{g_d} a_{d+1}$ where $\sigma \in S_d$ and $\sigma \neq 1$. Since w is a minimal counterexample and all $w_{\sigma,g}$ are smaller than w, we obtain $(\mathrm{mod}\ \mathrm{Id}^G(A))$ an expression of w as a linear combination of d-indecomposable monomials, a contradiction.

We have proved that P_n^G is spanned $(\mathrm{mod}\ \mathrm{Id}^G(A))$ by d-indecomposable monomials. Hence

$$c_n(A) \leq c_n^G(A) = \dim \frac{P_n^G}{P_n^G \cap \mathrm{Id}^G(A)} \leq |G|^n a_d(n).$$

Invoking Lemma 10.2.4, for $n \geq f(d, |G|)$ we obtain

$$c_n(A) \leq |G|^n b_d(n) n! < |G|^n \left(\frac{1}{|G|}\right)^n n! = n!$$

and A satisfies an identity of degree $f(d, |G|)$. By Lemma 10.1.3 it follows that $c_n^G(A) \leq |G|^n c_n(A) \leq |G|^n (f(d, |G|) - 1)^{2n}$.

If $G \leq \mathrm{Aut}(A)$, one can give an estimate of the degree of an identity on A which is better than the one given by the function $f(d, |G|)$ above. To accomplish this, observe that the space P_n^G can be spanned $(\mathrm{mod}\ \mathrm{Id}^G(A))$ by the d-good monomials in the sense of Definition 4.2.1 (see the proof of Theorem 4.2.4). In this case since by Lemma 4.2.3, the number of d-good monomials is $\leq (d-1)^{2n}$ it follows that $c_n(A) \leq c_n^G(A) \leq |G|^n (d-1)^{2n}$.

Consider now Euler's gamma function

$$\Gamma(x) = \int_0^\infty t^{x-1} e^{-t}\, dt.$$

The following inequality holds (see for instance [**FR**, page 105])

$$\left(\frac{x}{e}\right)^x < \frac{\Gamma(x+1)}{\sqrt{2\pi x}} < \Gamma(x+1),$$

for all $x \geq 1$. Substituting n for x where n is the least integer greater than $e|G|(d-1)^2$ yields

$$c_n(A) \leq (|G|(d-1)^2)^n < \left(\frac{n}{e}\right)^n < \Gamma(n+1) = n!$$

and A satisfies a PI of degree $\leq e|G|(d-1)^2 \leq 3|G|(d-1)^2$. $\square$

We can now deduce Amitsur's theorem on identities with involution ([**A3**], [**A4**]). If $G = \{1, *\}$ where $*$ is an involution, G-polynomials and G-identities are called $*$-polynomials and $*$-identities, respectively. Also, $c_n^G(A) = c_n^*(A)$ is called the nth $*$-codimension of A.

THEOREM 10.3.3. *Let A be an algebra with involution $*$ over a field F satisfying a non-trivial $*$-identity of degree d. Then for n sufficiently large we have $c_n^*(A) \le 2^n(f(2d,2) - 1)^{2n}$ and A satisfies a non-trivial polynomial identity whose degree is bounded by the function $f(2d,2)$.*

PROOF. By applying the usual linearization process to the given $*$-identity of A, we get that A satisfies a $*$-identity of the form

$$\sum_{s \in G^d} \alpha_s x_1^{s_1} \cdots x_d^{s_d} + \sum_{\substack{1 \ne \sigma \in S_d \\ q \in G^d}} \beta_{\sigma,q} x_{\sigma(1)}^{q_1} \cdots x_{\sigma(d)}^{q_d}$$

where $s = (s_1, \ldots, s_d), q = (q_1, \ldots, q_d) \in G^d$ and $G = \{1, *\}$. Also, without loss of generality, we may assume that $\alpha_1 \ne 0$ for $1 = (1, \ldots, 1)$. Replacing x_i by $x_{2i-1} x_{2i}$ for all $i = 1, \ldots d$, we obtain a $*$-identity of the form

$$x_1 \cdots x_{2d} + \sum_{\substack{1 \ne \sigma \in S_{2d} \\ q \in G^{2d}}} \gamma_{\sigma,q} x_{\sigma(1)}^{q_1} \cdots x_{\sigma(2d)}^{q_{2d}}.$$

Since this is an essential G-identity on A, Theorem 10.3.2 gives the desired conclusion. □

10.4. Representations of wreath products

In this section we describe all the irreducible representations of the group $G \wr S_n$ for any finite abelian group G over an algebraically closed field of characteristic zero. For such groups the representation theory is closely related to that of symmetric groups and we shall show how to reduce the problem of decomposing $P_n^G/(P_n^G \cap \mathrm{Id}^G(A))$ into the sum of $G \wr S_n$-irreducibles to a similar problem for the direct product of symmetric groups.

Throughout this section we assume that G is a finite abelian group of order k and F is an algebraically closed field of characteristic zero. We start by describing, following [**Re5**], how to construct irreducible representations for the wreath product $G \wr S_n$.

DEFINITION 10.4.1. A multipartition $\langle \lambda \rangle$ of n is a finite sequence of partitions $\langle \lambda \rangle = (\lambda(1), \ldots, \lambda(k))$, such that $\lambda(1) \vdash n_1 \ge 0, \ldots, \lambda(k) \vdash n_k \ge 0$ and $n = n_1 + \cdots + n_k$.

For $k \ge 1$, write $n = n_1 + \cdots + n_k$ where $n_i \ge 0$, $1 \le i \le k$, and let S_{n_i} be the symmetric group acting on the set $\{n_1 + \cdots + n_{i-1} + 1, \ldots, n_1 + \cdots + n_i\}$, where we set $n_0 = 0$. For every partition $\lambda(i) \vdash n_i$, $i = 1, \ldots, k$, choose a tableau of shape $\lambda(i)$ and let $e_{\lambda(i)}$ be the corresponding minimal essential idempotent of FS_{n_i}.

Let $f_1, \ldots, f_k$ be the minimal idempotents of FG and identify $F(S_{n_1} \times \cdots \times S_{n_k})$ with $FS_{n_1} \otimes \cdots \otimes FS_{n_k}$. For a fixed multipartion $\langle \lambda \rangle = (\lambda(1), \ldots, \lambda(k))$, $\lambda(i) \vdash n_i$, and corresponding minimal essential idempotents $e_{\lambda(i)}$ we define

$$e_{\langle \lambda \rangle} = f_1^{\otimes n_1} \otimes \cdots \otimes f_k^{\otimes n_k} \otimes e_{\lambda(1)} \otimes \cdots \otimes e_{\lambda(k)}$$

and
$$L_{\langle\lambda\rangle} = f_1^{\otimes n_1} \otimes \cdots \otimes f_k^{\otimes n_k} \otimes FS_{n_1}e_{\lambda(1)} \otimes \cdots \otimes FS_{n_k}e_{\lambda(k)}.$$

Notice that $e_{\langle\lambda\rangle}$ is an essential idempotent since $e_{\lambda(1)}, \ldots, e_{\lambda(k)}$ are essential idempotents of FS_n and $f_1, \ldots, f_k$ are orthogonal idempotents of FG.

For $n = n_1 + \cdots + n_k$, let
$$\binom{n}{n_1, \ldots, n_k} = \frac{n!}{n_1! \cdots n_k!}$$

denote the corresponding generalized binomial coefficient. Next theorem describes the irreducible representations of $G \wr S_n$.

THEOREM 10.4.2. *Let F be an algebraically closed field of characteristic zero and let G be a finite abelian group. Then there is a one-to-one correspondence between multipartitions $\langle\lambda\rangle$ of n and non-equivalent irreducible representations $N_{\langle\lambda\rangle}$ of $G \wr S_n$. Moreover,*

1) *for any multipartition $\langle\lambda\rangle$ of n, $e_{\langle\lambda\rangle}$ is a minimal essential idempotent of $FG \wr S_n$ and*
$$N_{\langle\lambda\rangle} \cong (FG \wr S_n)e_{\langle\lambda\rangle} = \bigoplus_{\tau \in \Gamma} \tau L_{\langle\lambda\rangle},$$
where Γ is a left transversal of $S_{n_1} \times \cdots \times S_{n_k}$ in S_n.

2) $\dim N_{\langle\lambda\rangle} = \binom{n}{n_1, \ldots, n_k} \cdot d_{\lambda(1)} \cdots d_{\lambda(k)}$ *where* $d_{\lambda(i)} = \dim(FS_{n_i}e_{\lambda(i)}) = \deg \chi_{\lambda(i)}.$

For any multipartition $\langle\lambda\rangle$ of n, let us denote by $\chi_{\langle\lambda\rangle}$ the irreducible $G \wr S_n$-character corresponding to $\langle\lambda\rangle$. Recall that according to (10.1) the group $G \wr S_n$ acts on the space P_n^G of multilinear G-polynomials and by Lemma 10.1.5, $P_n^G \cap \mathrm{Id}^G(A)$ is invariant under this action. Thus $P_n^G/(P_n^G \cap \mathrm{Id}^G(A))$ has a natural structure of $G \wr S_n$-module and we make the following.

DEFINITION 10.4.3. *The nth G-cocharacter of the algebra A, denoted $\chi_n^G(A)$, is the character of the $G \wr S_n$-module*
$$P_n^G(A) = \frac{P_n^G}{P_n^G \cap \mathrm{Id}^G(A)}.$$

Since $\mathrm{char}\, F = 0$, by complete reducibility, we can write $\chi_n^G(A)$ as a sum of irreducible characters
$$\chi_n^G(A) = \sum_{\langle\lambda\rangle \vdash n} m_{\langle\lambda\rangle} \chi_{\langle\lambda\rangle},$$
where $m_{\langle\lambda\rangle} \geq 0$ denotes the corresponding multiplicity.

Recall from Section 3.2, that for any finite abelian group $G \subseteq \mathrm{Aut}^*(A)$, the free algebra with G-action is G-graded as a vector space and if we denote by $E = \{f_1, \ldots, f_k\}$ the set of minimal idempotents of FG, then
$$F\langle X|G\rangle = F\langle x^{f_i} \mid x \in X, f_i \in E\rangle,$$
i.e., $F\langle X|G\rangle$ is freely generated by the elements x^{f_i}, where $x \in X$ and $f_i \in E$.

For any $\mathbf{e} = (e_1, \ldots, e_n) \in E^n$, we define
$$P_n^{\mathbf{e}} = \mathrm{span}\{x_{\sigma(1)}^{e_{\sigma(1)}} \cdots x_{\sigma(n)}^{e_{\sigma(n)}} | \sigma \in S_n\}.$$

Then it is clear that P_n^G has the decomposition
$$P_n^G = \bigoplus_{\mathbf{e} \in E^n} P_n^{\mathbf{e}}.$$

Since the variables $x_i^{f_j}$ are free generators then, by writing $h \in P_n^G \cap \mathrm{Id}^G(A)$ as $h = \sum_{\mathbf{e} \in E^n} h^{\mathbf{e}}$ with $h^{\mathbf{e}} \in P_n^{\mathbf{e}}$, we have that $h^{\mathbf{e}} \in \mathrm{Id}^G(A)$ for all $\mathbf{e} \in E^n$. This says that
$$P_n^G \cap \mathrm{Id}^G(A) = \bigoplus_{\mathbf{e} \in E^n} (P_n^{\mathbf{e}} \cap \mathrm{Id}^G(A)).$$

Now fix $\mathbf{e} \in E^n$ and suppose that in $P_n^{\mathbf{e}}$, n_1 variables x_i have exponent f_1, ..., n_k variables x_i have exponent f_k. Clearly $n_1 + \cdots + n_k = n$.

Let the group $S_{n_1} \times \cdots \times S_{n_k}$ act on the left on the space $P_n^{\mathbf{e}}$ by permuting each set of variables with the same exponent. Since $\mathrm{Id}^G(A)$ is invariant under this permutation action, $P_n^{\mathbf{e}}/(P_n^{\mathbf{e}} \cap \mathrm{Id}^G(A))$ becomes an $S_{n_1} \times \cdots \times S_{n_k}$-module.

For the representation theory of $S_{n_1} \times \cdots \times S_{n_k}$, there is a one-to-one correspondence between non-equivalent irreducible $S_{n_1} \times \cdots \times S_{n_k}$-representations and multipartitions of n, $\langle \lambda \rangle = (\lambda(1), \ldots, \lambda(k))$ such that $\lambda(1) \vdash n_1, \ldots, \lambda(k) \vdash n_k$. Moreover, for any $i = 1, \ldots, k$, consider a partition $\lambda(i) \vdash n_i$ and a tableau of shape $\lambda(i)$. If $e_{\lambda(i)}$ is the corresponding minimal essential idempotent of the group algebra FS_{n_i}, then
$$e_{\lambda(1)} \otimes \cdots \otimes e_{\lambda(k)}$$
is a minimal essential idempotent of the algebra
$$FS_{n_1} \otimes \cdots \otimes FS_{n_k} \equiv F(S_{n_1} \times \cdots \times S_{n_k}).$$

From this, it is clear that there is a close connection between irreducible $G \wr S_n$-representations and irreducible $S_{n_1} \times \cdots \times S_{n_k}$-representations. We shall state this precisely in terms of characters in the next theorem.

Let $\mathbf{e} = (e_1, \ldots, e_n)$ and $\mathbf{h} = (h_1, \ldots, h_n)$ be two n-tuples in E^n. If f_i appears n_i times in $\mathbf{e} = (e_1, \ldots, e_n)$ and n'_i times in $\mathbf{h} = (h_1, \ldots, h_n)$, then clearly,

(10.11)
$$\frac{P_n^{\mathbf{e}}}{P_n^{\mathbf{e}} \cap \mathrm{Id}^G(A)} \cong \frac{P_n^{\mathbf{h}}}{P_n^{\mathbf{h}} \cap \mathrm{Id}^G(A)}$$

if $n_i = n'_i$, $i = 1, \ldots k$.

If $n_1 + \cdots + n_k = n$ and $\mathbf{e} = \{\underbrace{f_1, \ldots, f_1}_{n_1}, \ldots, \underbrace{f_k, \ldots, f_k}_{n_k}\}$, we introduce the notation

(10.12)
$$P_{n_1,\ldots,n_k} = P_n^{\mathbf{e}}$$

and
$$P_{n_1,\ldots,n_k}(A) = \frac{P_{n_1,\ldots,n_k}}{P_{n_1,\ldots,n_k} \cap \mathrm{Id}^G(A)}.$$

DEFINITION 10.4.4. The character of the $S_{n_1} \times \cdots \times S_{n_k}$-module
$$P_{n_1,\ldots,n_k}(A) = \frac{P_n^{\mathbf{e}}}{P_n^{\mathbf{e}} \cap \mathrm{Id}^G(A)}$$
is denoted $\chi_{n_1,\ldots,n_k}(A)$.

For any partition $\lambda \vdash r$, let χ_λ be the irreducible S_r-character corresponding to λ. Then $\chi_{\lambda(1)} \otimes \cdots \otimes \chi_{\lambda(k)}$ is the irreducible $S_{n_1} \times \cdots \times S_{n_k}$-character associated to the multipartition $\langle \lambda \rangle = (\lambda(1), \ldots, \lambda(k))$. By complete reducibility we can write

$$\chi_{n_1,\ldots,n_k}(A) = \sum_{\langle \lambda \rangle \vdash n} m'_{\langle \lambda \rangle} \chi_{\lambda(1)} \otimes \cdots \otimes \chi_{\lambda(k)},$$

where $\langle \lambda \rangle = (\lambda(1), \ldots, \lambda(k))$ is a multipartition of n such that $\lambda(1) \vdash n_1, \ldots, \lambda(k) \vdash n_k$ and $m'_{\langle \lambda \rangle} \geq 0$. The connection between $G \wr S_n$-characters and $S_{n_1} \times \cdots \times S_{n_k}$-characters is given in the following.

THEOREM 10.4.5. *If $P_n^G(A)$ has $G \wr S_n$-character*

$$\chi_n^G(A) = \sum_{\langle \lambda \rangle \vdash n} m_{\langle \lambda \rangle} \chi_{\langle \lambda \rangle}$$

and $P_{n_1,\ldots,n_k}(A)$ has $S_{n_1} \times \cdots \times S_{n_k}$-character

$$\chi_{n_1,\ldots,n_k}(A) = \sum_{\langle \lambda \rangle \vdash n} m'_{\langle \lambda \rangle} \chi_{\lambda(1)} \otimes \cdots \otimes \chi_{\lambda(k)},$$

then $m_{\langle \lambda \rangle} = m'_{\langle \lambda \rangle}$, for all multipartitions $\langle \lambda \rangle = (\lambda(1), \ldots, \lambda(k))$ where $\lambda(1) \vdash n_1, \ldots, \lambda(k) \vdash n_k$, $n_1 + \cdots + n_k = n$.

PROOF. Fix a multipartition $\langle \lambda \rangle = (\lambda(1), \ldots, \lambda(k))$ of the desired form. Then by Theorem 10.4.2,

$$(FG \wr S_n)e_{\langle \lambda \rangle}(x_1 \cdots x_n)$$
$$= \bigoplus_{\tau \in \Gamma} \tau F S_{n_1} e_{\lambda(1)} \otimes \cdots \otimes F S_{n_k} e_{\lambda(k)}(x_1^{f_1} \cdots x_{n_1}^{f_1} \cdots x_{n_1+\cdots+n_{k-1}+1}^{f_k} \cdots x_n^{f_k})$$

is an irreducible $G \wr S_n$-module in P_n^G, where Γ is a left transversal of $S_{n_1} \times \cdots \times S_{n_k}$ in S_n.

For every $\tau \in \Gamma$, the corresponding summand of $(FG \wr S_n)e_{\langle \lambda \rangle}(x_1 \cdots x_n)$ consists of polynomials in n_1 variables x with exponent f_1, ..., n_k variables x with exponent f_k. Also these multisets of variables are distinct for different $\tau \in \Gamma$. Each summand is isomorphic to an irreducible $S_{n_1} \times \cdots \times S_{n_k}$-module with character $\chi_{\lambda(1)} \otimes \cdots \otimes \chi_{\lambda(k)}$ where $S_{n_1}, \cdots, S_{n_k}$ act on the variables with exponent $f_1, \ldots, f_k$, respectively. This implies that the multiplicities of the irreducible characters in $\chi_n(A)$ and in $\chi_{n_1,\ldots,n_k}(A)$ are the same. $\square$

Notice that as a consequence of the previous theorem we rediscover the formula given in Theorem 10.4.2 in the following form:

$$(10.13) \qquad c_n^G(A) = \sum_{n_1+\cdots+n_k=n} \binom{n}{n_1,\ldots,n_k} \dim P_{n_1,\ldots,n_k}(A).$$

10.5. Graded identities and polynomial growth

Throughout this section we assume that $A = \oplus_{g \in G} A^{(g)}$ is an algebra graded by a finite group G.

We remind the reader that if G is an abelian group, and F is algebraically closed, by Section 3.2, a G-grading on A uniquely determines G as a group of automorphisms of A and viceversa.

Recall from Section 3.3, the definition of the free G-graded algebra on a countable set X. We write $X = \bigcup_{g \in G} X_g$, where $X_g = \{x_1^{(g)}, x_2^{(g)}, \ldots\}$ are disjoint sets.

If $F\langle X \rangle^{(g)}$ is the subspace of the free algebra $F\langle X \rangle$ generated by all monomials having homogeneous degree g, then $F\langle X \rangle = \bigoplus_{g \in G} F\langle X \rangle^{(g)}$ is the free G-graded algebra on X, denoted $F\langle X \rangle^{gr}$.

Now for any G-graded algebra A, $\mathrm{Id}^{gr}(A)$ denotes the ideal of $F\langle X \rangle^{gr}$ of G-graded identities of A. As we have already remarked in the case of G-action, since $\mathrm{char}\, F = 0$, the ideal $\mathrm{Id}^{gr}(A)$ is determined by its multilinear polynomials. Hence, for every $n \geq 1$, we consider the space of multilinear G-graded polynomials

$$P_n^{gr} = \mathrm{span}\{x_{\sigma(1)}^{(g_1)} \cdots x_{\sigma(n)}^{(g_n)} \mid \sigma \in S_n, g_1, \ldots, g_n \in G\}$$

and $P_n^{gr} \cap \mathrm{Id}^{gr}(A)$ is the set of all multilinear G-graded identities of degree n in the first n variables.

DEFINITION 10.5.1. The nth G-graded codimension of A is defined as

$$c_n^{gr}(A) = \dim \frac{P_n^{gr}}{P_n^{gr} \cap \mathrm{Id}^{gr}(A)}.$$

Notice that the proof of Lemma 10.1.2 and Lemma 10.1.3 carry over, with the due changes, to the graded case and one easily gets the following:

1) for any G-graded algebra A, $c_n(A) \leq c_n^{gr}(A)$;
2) if A is a PI-algebra and A is G-graded, then $c_n^{gr}(A) \leq |G|^n c_n(A)$.

As in the case of G-polynomials there is a convenient decomposition of the space of multilinear graded polynomials. In fact, if for $\mathbf{g} = (g_1, \ldots, g_n) \in G^n$, we define

$$P_n^{\mathbf{g}} = \mathrm{span}\{x_{\sigma(1)}^{(g_{\sigma(1)})} \cdots x_{\sigma(n)}^{(g_{\sigma(n)})} \mid \sigma \in S_n\},$$

then clearly

(10.14) $$P_n^{gr} = \bigoplus_{\mathbf{g} \in G^n} P_n^{\mathbf{g}}.$$

Write $G = \{\gamma_1, \ldots, \gamma_k\}$ and let $\mathbf{g} = \{\underbrace{\gamma_1, \ldots, \gamma_1}_{n_1}, \ldots, \underbrace{\gamma_k, \ldots, \gamma_k}_{n_k}\}$ with $n_1 + \cdots + n_k = n$. We then define $P_{n_1, \ldots, n_k} = P_n^{\mathbf{g}}$ and

$$P_{n_1, \ldots, n_k}(A) = \frac{P_n^{\mathbf{g}}}{P_n^{\mathbf{g}} \cap \mathrm{Id}^{gr}(A)}.$$

Hence $P_{n_1, \ldots, n_k}(A)$ is the space of multilinear elements of the relatively free algebra $F\langle X \rangle^{gr}/\mathrm{Id}^{gr}(A)$ which are polynomials in the variables $\bar{y}_1, \ldots, \bar{y}_n$ where the first n_1 variables are homogeneous of degree γ_1, next n_2 variables are homogeneous of degree γ_2 and so on.

Our objective in this section is to give a characterization of G-graded algebras A whose sequence of codimensions is polynomially bounded when G is any finite group. This result states that a graded algebra has polynomially bounded graded codimensions if and only if it satisfies a certain set of graded identities. Here we shall give the proof of the necessity of the condition. We refer the reader to the original paper of Giambruno, Mishchenko and Zaicev ([**GMZ1**]) for a detailed proof of the sufficiency. The theorem is the following.

THEOREM 10.5.2 ([**GMZ1**]). *Let $A = \bigoplus_{g \in G} A^{(g)}$ be an F-algebra graded by a finite group G. Then the sequence of graded codimensions of A is polynomially* bounded if and only if for some positive integers m and r the algebra A satisfies the following graded identities:

1) $[x_1, x_2] \cdots [x_{2m-1}, x_{2m}] \equiv 0$, for all $x_1, \ldots, x_{2m} \in A^{(1)}$;
2) $x_1 x_3 \cdots x_{2r-1} y x_2 x_4 \cdots x_{2r}$
$$- \sum_{1 \neq \tau \in H} a_\tau x_{\tau(1)} x_{\tau(3)} \cdots x_{\tau(2r-1)} y x_{\tau(2)} x_{\tau(4)} \cdots x_{\tau(2r)} \equiv 0$$

for all $x_1, \ldots, x_{2r} \in A^{(1)}, y \in A^{(g)}$, where H is the subgroup of S_{2r} generated by the transpositions $(1\,2), (3\,4), \ldots, (2r-1\,2r)$, and $a_\tau \in F$ depends on $g \in G$;

3) the ideal of A generated by $\oplus_{g \neq 1} A^{(g)}$ is nilpotent.

PROOF. We shall only prove that 1), 2) and 3) hold provided A has polynomially bounded G-graded codimensions.

Let $|G| = k, G = \{\gamma_1, \ldots, \gamma_k\}$. From the definition of graded codimension it follows that for all $n_1, \ldots, n_k$, the space $P_{n_1, \ldots, n_k}(A)$ has polynomially bounded dimension. If we take $\gamma_1 = 1$, the unit of G, $n_1 = n$, $n_2 = \cdots = n_k = 0$, then $c_{n,0\ldots,0} = \dim P_{n,0\ldots,0}(A)$ is the ordinary nth codimension of the algebra $A^{(1)}$, the unit component of A. Hence the sequence $c_n(A^{(1)}), n = 1, 2, \ldots$, is bounded by a polynomial function of n. But then from Theorem 7.2.12, it follows that $A^{(1)}$ satisfies an identity of type 1).

Now take $n_1 = n-1, n_2 = 1$ and $n = 2r+1$ odd. If H denotes the subgroup of S_{2r} generated by $(1\,2), \ldots, (2r-1\,2r)$, we consider all monomials of the form

(10.15) $\qquad x_{\tau(1)} x_{\tau(3)} \cdots x_{\tau(2r-1)} y x_{\tau(2)} x_{\tau(4)} \cdots x_{\tau(2r)}$

where $x_1, \ldots, x_{2r} \in F\langle X \rangle$, $\tau \in H$, and y is a homogeneous variable from the component $F\langle X \rangle^{(\gamma_2)}$, $\gamma_2 = g \neq 1$. Notice that there are only $|H| = 2^r = 2^{(n-1)/2}$ monomials of the type (10.15). Since $c_{n-1,1,0\ldots,0}$ is polynomially bounded and $(\sqrt{2})^n$ grows faster than any polynomial, the monomials in (10.15) must be linearly dependent modulo $\text{Id}^{gr}(A)$. This says that A satisfies an identity of type 2). Hence A satisfies a graded identity of type 2).

We are only left with the proof that the ideal of A generated by $\oplus_{g \neq 1} A^{(g)}$ is nilpotent.

Now, for $n_1 + \cdots + n_k = n$, set $c_{n_1, \ldots, n_k} = \dim P_{n_1, \ldots, n_k}(A)$. Then, recalling the decomposition given in (10.14), we obtain

(10.16) $\qquad c_n(A)^{gr} = \sum_{n_1 + \cdots + n_k = n} \binom{n}{n_1, \ldots, n_k} c_{n_1, \ldots, n_k} \leq B n^r,$

for some constants B and r. Suppose that there exists $i \in \{1, \ldots, k\}$ such that $n_i > r$ and $\sum_{j \neq i} n_j > r$.

Then
$$\binom{n}{n_1, \ldots, n_k} = \frac{n!}{n_1! \ldots n_k!} > \frac{n!}{n_i! (\sum_{j \neq i} n_j)!} = \binom{n}{n_i}$$

and $\binom{n}{n_i}$ is asymptotically greater than $B n^r$. It follows that there exists $N \geq 0$ such that for all $n \geq N$, $c_{n_1, \ldots, n_k} = 0$, as soon as at least two integers among $n_1, \ldots, n_k$ are greater than r. In particular, any product of $N+1$ homogeneous elements of A containing $r+1$ elements of $A^{(g)}$ and $r+1$ elements of $A^{(h)}$ with $g \neq h$ must be zero.

We claim that for any $g \neq 1$, $(A^{(g)})^n = 0$ as soon as $n \geq N + (k+1)(r+1)$. In fact, any product $b_1 \cdots b_n$ with $b_1, \ldots, b_n \in A^{(g)}$ can be written as

$$w = a_1 \cdots a_{r+1} y_1 \cdots y_{N+r+1} b'$$

where $y_1, \ldots, y_{N+r+1} \in A^{(g)}, b' \in A$ and

$$a_1 = b_1 \cdots b_k, \quad \ldots, \quad a_{r+1} = b_{kr+1} \cdots b_{k(r+1)}.$$

Since $|G| = k$, then $a_1, \ldots, a_{r+1} \in A^{(1)}$. Hence, w contains $r+1$ factors from $A^{(1)}$, $r+1$ factors from $A^{(g)}, g \neq 1$, and the total degree of w is greater than N. By the above, $w = 0$ and the claim is established.

Next we prove that $w = b_1 \cdots b_n = 0$ for any $b_1, \ldots, b_n \in \cup_{g \neq 1} A^{(g)}$ whenever $n > M = (N + (k+1)(r+1))(r+1) + r$. Recall that if there exist $r+1$ factors b_i belonging to $A^{(g)}$ and $r+1$ factors b_j belonging to $A^{(h)}$ with $h \neq g$, then $w = 0$. Since $n > rk$, we may assume that at least $r+1$ factors come from the same component $A^{(g)}$. Also the number of all factors b_i belonging to all other components $A^{(h)}$, $h \neq g$, cannot be greater than r. Hence, we can write w as

$$w = y_0 a_1 y_1 a_2 \cdots a_m y_m$$

where $a_1, \ldots, a_m \in \cup_{h \neq g} A^{(h)}$ and $y_0, \ldots, y_m$ are products of some b_i's belonging to $A^{(g)}$. Besides, $m \leq r$. Hence the degree of some y_j is at least $q = \frac{M-r}{r+1} \geq N + (k+1)(r+1)$. By the previous claim we get $y_j \in (A^{(g)})^q = 0$ and $w = 0$ follows.

Now let I be the ideal generated $\oplus_{g \neq 1} A^{(g)}$. Then any product of M elements of I can be written as a linear combination of products of M factors from $\cup_{g \neq 1} A^{(g)}$ since any $A^{(g)}$, $g \neq 1$, is an $A^{(1)}$-bimodule. Therefore $I^M = 0$ and the proof is complete. □

We now consider the special case when G is abelian and F is algebraically closed. As mentioned above, by Section 3.2 there is a duality between G-gradings and G-actions. Hence G acts on A as a group of automorphisms, i.e., $G \subseteq \text{Aut}(A)$ (here we identify G and $\hat{G}$).

The free algebra with G-action $F\langle X|G\rangle$ is G-graded as an algebra and by Proposition 3.3.6, is the free G-graded algebra of countable rank. Moreover, $F\langle X|G\rangle$ is freely generated by the elements x^{f_i} which are homogeneous in the G-grading, where $x \in X$ and $f_1, \ldots, f_k$ are the minimal idempotents of FG. For convenience we rename the generators $\{x_i^{f_j} | i \geq 1, 1 \leq j \leq k\}$ of $F\langle X|G\rangle$ as $\{x_i^{(g)} | i \geq 1, g \in G\}$. Hence we view $F\langle X|G\rangle$ as the free G-graded algebra with homogeneous set of generators $x_i^{(g)}$, $\text{Id}^G(A) = \text{Id}^{gr}(A)$ and the notation is consistent with the above.

In this case, Theorem 10.5.2 is a result about G-codimensions.

COROLLARY 10.5.3. *Let A be an algebra over the algebraically closed field F and let G be a finite abelian group acting by automorphisms on A. Then the sequence of G-codimensions of A is polynomially bounded if and only if A satisfies the conditions 1), 2), 3) of Theorem 10.5.2.*

If one applies the representation theory of the group $G \wr S_n$, as in the previous section, it is possible to deduce a theorem describing when $c_n^G(A)$ is polynomially bounded in terms of properties of the G-cocharacter sequence of A.

10.5. GRADED IDENTITIES AND POLYNOMIAL GROWTH

LEMMA 10.5.4. *Let G be a finite group and let A be a G-graded algebra. If the G-graded codimensions of A are polynomially bounded, then A satisfies a Capelli identity.*

PROOF. By the analogue of Lemma 10.1.2, for all $n \geq 1$, $c_n(A) \leq c_n^{gr}(A)$. Hence the ordinary codimensions of A are polynomially bounded. But then, by Theorem 7.2.14, the S_n-characters appearing with non-zero multiplicity in $\chi_n(A)$ have corresponding diagram of height bounded by a constant. By Theorem 4.6.1, this says that A satisfies a Capelli identity. $\square$

LEMMA 10.5.5. *Let $\lambda = (\lambda_1, \lambda_2)$ be a partition of n with $\lambda_2 \geq q+2$. Then $d_\lambda = \deg \chi_\lambda \geq n^q$, for large enough n.*

PROOF. From the hook formula (Proposition 2.2.8) we obtain

$$d_\lambda = \frac{n!}{(n-k)!k!\frac{n-k+1}{n-2k+1}} = \binom{n}{k}\frac{n-2k+1}{n-k+1}$$

where $k = \lambda_2 \geq q+2$. Since $q+2 \leq k \leq \frac{1}{2}n$, then $\binom{n}{k} > \binom{n}{q+2}$. Hence

$$d_\lambda \geq \binom{n}{q+2}\frac{n-2k+1}{n-k+1} \geq \binom{n}{q+2}\frac{1}{n} \geq \frac{n^{q+1}}{C},$$

where C is a constant not depending on n. $\square$

LEMMA 10.5.6. *Let $\lambda \vdash n$ and let t, q be fixed integers. If $h(\lambda) \leq t$, i.e., λ lies in a strip of height t, and $\lambda_2 \geq q+4$, then $d_\lambda \geq n^{q+1}$, for large enough n.*

PROOF. Let $\lambda = (\lambda_1, \ldots, \lambda_m)$ and consider the partition $\lambda' = (\lambda_1, \lambda_2)$ of $n' = \lambda_1 + \lambda_2$. Clearly $d_{\lambda'} \leq d_\lambda$. On the other hand, $n' > \frac{n}{t}$ since $\lambda_1 \geq \cdots \geq \lambda_m$ and $m \leq t$.

Hence by the previous lemma

$$d_\lambda \geq d_{\lambda'} \geq \left(\frac{n}{t}\right)^{q+2}$$

and the asymptotic inequality $d_\lambda \gtrsim n^{q+1}$ holds. $\square$

Recall that if $G \subseteq \text{Aut}(A)$ is a finite abelian group and F is algebraically closed, by Section 10.4, the nth G-cocharacter of A has the decomposition

$$\chi_n^G(A) = \sum_{\langle\lambda\rangle \vdash n} m_{\langle\lambda\rangle} \chi_{\langle\lambda\rangle},$$

where $\chi_{\langle\lambda\rangle}$ is the irreducible $G \wr S_n$-character associated to the multipartition $\langle\lambda\rangle = (\lambda(1), \ldots, \lambda(k))$. We write $\lambda(i) = (\lambda(i)_1, \lambda(i)_2, \ldots)$ and $|\lambda(i)| = m$, when $\lambda(i) \vdash m$. We can now prove the following characterization.

THEOREM 10.5.7 ([**GMZ3**]). *Let A be an algebra over the algebraically closed field F and let $G \subseteq \text{Aut}(A)$ be a finite abelian group of order k. Then $c_n^G(A) \leq Cn^t$, for some constants C and t, if and only if there exists a constant q such that the nth G-cocharacter of A has the following decomposition:*

$$\chi_n^G(A) = \sum_{\substack{\langle\lambda\rangle \vdash n \\ |\lambda(1)| - \lambda(1)_1 \leq q \\ |\lambda(2)| + \cdots + |\lambda(k)| \leq q}} m_{\langle\lambda\rangle} \chi_{\langle\lambda\rangle}.$$

PROOF. Suppose first that $c_n^G(A) \leq Cn^t$, for some constants C, t. By Theorem 10.5.2, the ideal I generated by $\oplus_{g \neq 1} A^{(g)}$ is nilpotent and let $I^N = 0$. Let $n = n_1 + \cdots + n_k$ and suppose that $n_2 + \cdots + n_k \geq N$. By Theorem 10.4.5, the $S_{n_1} \times \cdots \times S_{n_k}$-character of $P_{n_1,\ldots,n_k}(A)$ has the decomposition

$$\chi_{n_1,\ldots,n_k}(A) = \sum_{\langle\lambda\rangle \vdash n} m_{\langle\lambda\rangle} \chi_{\lambda(1)} \otimes \cdots \otimes \chi_{\lambda(k)}$$

where $\langle\lambda\rangle = (\lambda(1), \ldots, \lambda(k))$, $\lambda(1) \vdash n_1, \ldots, \lambda(k) \vdash n_k$, and the multiplicities $m_{\langle\lambda\rangle}$ are the same as in $\chi_n^G(A) = \sum_{\langle\lambda\rangle \vdash n} m_{\langle\lambda\rangle} \chi_{\langle\lambda\rangle}$.

Since $I^N = 0$ and $n_2 + \cdots + n_k \geq N$, it follows that $P_{n_1,\ldots,n_k}(A) = 0$. Hence $m_{\langle\lambda\rangle} = 0$ for any $\langle\lambda\rangle$ with $n_2 + \cdots + n_k \geq N$.

It remains to show that if $\langle\lambda\rangle = (\lambda(1), \ldots, \lambda(k))$, $\lambda(1) \vdash n_1, \ldots, \lambda(k) \vdash n_k$, is such that $m_{\langle\lambda\rangle} \neq 0$, then $n_1 - \lambda(1)_1$ is bounded by a constant.

By Lemma 10.5.4, the algebra A satisfies the Capelli identity of some rank $q + 1$. Hence any multilinear polynomial alternating on $q + 1$ variables vanishes in A. Such a polynomial is clearly still zero if we evaluate its variables in $A^{(1)}$. By Theorem 4.6.1, $\lambda(1)$ must lie in a strip of height q.

We shall prove next that $\lambda(1)_2 < t + 4$. This together with the above will say that $n_1 - \lambda(1)_1 \leq q(t+4)$, the desired conclusion.

Suppose to the contrary that $\lambda(1)_2 \geq t + 4$. Then, by Lemma 10.5.6, $\chi_{\lambda(1)}(1)$ is asymptotically greater than n_1^{t+1}. Since, by what we proved above, $n_2 + \cdots + n_k < N$, we must have that $n_1 = n - n_2 - \cdots - n_k > n - N$. Hence, $c_n^G(A) = \dim P_{n_1,\ldots,n_k}(A) \geq \chi_{\lambda(1)}(1) \cdots \chi_{\lambda(k)}(1) \geq (n-N)^{t+1}$, a contradiction.

Conversely, suppose that there exists a constant C such that $|\lambda(2)| + \cdots + |\lambda(k)| + |\lambda(1)| - \lambda(1)_1 \leq C$, for all multipartitions $\langle\lambda\rangle = (\lambda(1), \ldots, \lambda(k))$ with $m_{\langle\lambda\rangle} \neq 0$. Then, for any n, we have that the number of non-zero spaces $P_{n_1,\ldots,n_k}(A)$ does not exceed the constant C^{k-1}. Hence, by recalling the formula (10.13) relating the G-codimensions to the dimensions of the spaces $P_{n_1,\ldots,n_k}(A)$, we get that, in the sum (10.13), the number of non-zero summands is bounded by a constant.

Now let $W = P_{n_1,\ldots,n_k}(A)$ be any non-zero subspace. Any irreducible $S_{n_1} \times \cdots \times S_{n_k}$-submodule of W is of the type $W_1 \otimes \cdots \otimes W_k$, where every W_i is an irreducible S_{n_i}-module corresponding to a partition $\lambda(i) \vdash n_i$, $i = 1, \ldots, k$. Moreover, $\dim W_i < C!$ for any $i = 2, \ldots, k$. For $i = 1$, from the hook formula it easily follows that $\dim W_1 \leq \frac{n!}{(n-C)!} \leq n^C$. Since the dimensions of all S_{n_1}-modules are polynomially bounded, their multiplicities are also polynomially bounded, both in FS_{n_1} and in W. Hence the dimension of any $P_{n_1,\ldots,n_k}(A)$ is polynomially bounded and, by (10.13), $c_n^G(A)$ is polynomially bounded. $\square$

Notice that the above theorem says that in order for the G-codimensions to be polynomially bounded, in the decomposition $n = |\lambda(1)| + \cdots + |\lambda(k)|$, the integers $|\lambda(1)| - \lambda(1)_1, |\lambda(2)|, \ldots, |\lambda(k)|$ should be bounded by some common constant, i.e., all the Young diagrams $\lambda(2), \ldots, \lambda(k)$ contain only a bounded number of boxes, and the diagram $\lambda(1)$ contains only a bounded number of boxes out of the first row.

10.6. The $\mathbb{Z}_2 \wr S_n$-action

In this and in the next section we only consider algebras with G-action in the special case when $G = \mathbb{Z}_2$ is a group of order two. In this case A has a structure of superalgebra or algebra with involution and in both cases we are able to make significant progress on the study of the corresponding codimensions.

Our first objective is to prove that if A is any such finite dimensional algebra, then the corresponding sequence of $\mathbb{Z}_2$-codimensions has an exponential rate of growth which is an integer. As in the ordinary case, we will be able to explicitly construct the corresponding exponent.

We recall some notation from 3.3. Suppose that $\mathbb{Z}_2 = \langle \varphi \rangle$ is the multiplicative group of order 2. Then $\frac{1+\varphi}{2}$ and $\frac{1-\varphi}{2}$ are the minimal idempotents of the group algebra $F\mathbb{Z}_2$ and the free algebra with $\mathbb{Z}_2$-action is freely generated by the elements $x_i + x_i^\varphi$ and $x_i - x_i^\varphi$, $i = 1, 2, \ldots$. For every i, we write $x_i + x_i^\varphi = y_i$ and $x_i - x_i^\varphi = z_i$. Then $F\langle X|G\rangle = F\langle Y, Z\rangle$ is the free associative algebra on the two sets $Y = \{y_1, y_2, \ldots\}$ and $Z = \{z_1, z_2 \ldots\}$. Notice that for every i, $y_i^\varphi = y_i$ and $z_i^\varphi = -z_i$.

Recall that if φ is an automorphism, $F\langle Y, Z\rangle$ is the free superalgebra on $Y \cup Z$ where the variables from Y have homogeneous degree 0 and the variables from Z have homogeneous degree 1. If φ is an involution, we write $\varphi = *$ and $F\langle X, *\rangle = F\langle Y, Z\rangle$ is the free algebra with involution $*$ on X. In this case the elements of Y are just symmetric variables, and the elements of Z are skew-symmetric variables.

For any algebra A with φ-action let us define

$$A^+ = \{a \in A \mid a^\varphi = a\} \quad \text{and} \quad A^- = \{a \in A \mid a^\varphi = -a\}.$$

Since for every i, $x_i + x_i^\varphi = y_i$ and $x_i - x_i^\varphi = z_i$, then according to (10.12) for $0 \leq r \leq n$, $P_{r,n-r}$ is the space of multilinear polynomials in the variables $y_1, \ldots, y_r, z_{r+1}, \ldots, z_n$. In order to simplify the notation we make the following.

DEFINITION 10.6.1. For $0 \leq r \leq n$, $P_{r,n-r}$ denotes the space of multilinear polynomials in the variables $y_1, \ldots, y_r, z_1, \ldots, z_{n-r}$.

Moreover, if A is an F-algebra with φ-action, then

$$P_{r,n-r}(A) = \frac{P_{r,n-r}}{P_{r,n-r} \cap \mathrm{Id}^{\mathbb{Z}_2}(A)}$$

and $\chi_{r,n-r}(A)$ is its $S_r \times S_{n-r}$-character.

If $\lambda \vdash r$ and $\mu \vdash n - r$, we write

$$\chi_{\langle\lambda,\mu\rangle} = \chi_{\lambda,\mu}$$

for the irreducible $\mathbb{Z}_2 \wr S_n$ character corresponding to the pair (λ, μ).

Recall that the connection between $\mathbb{Z}_2 \wr S_n$-characters and $S_r \times S_{n-r}$-characters is given in Theorem 10.4.5 which we now restate in this setting.

COROLLARY 10.6.2. *If $P_n^{\mathbb{Z}_2}(A)$ has $\mathbb{Z}_2 \wr S_n$-character*

$$\chi_n^{\mathbb{Z}_2}(A) = \sum_{|\lambda|+|\mu|=n} m_{\lambda,\mu} \chi_{\lambda,\mu}$$

and $P_{r,n-r}(A)$ has $S_r \times S_{n-r}$-character

$$\chi_{r,n-r}(A) = \sum_{\lambda \vdash r, \mu \vdash n-r} m'_{\lambda,\mu} \chi_\lambda \otimes \chi_\mu,$$

then $m_{\lambda,\mu} = m'_{\lambda,\mu}$, for all $\lambda \vdash r, \mu \vdash n-r$, $r = 0, \ldots, n$. In particular,

(10.17) $$c_n^{\mathbb{Z}_2}(A) = \sum_{r=0}^{n} \binom{n}{r} \dim P_{r,n-r}(A).$$

In the next section we shall also use the following result which we state here without proof. The reader can look at the original paper of Berele ([**Be3**]) for a proof that holds for a wider class of algebras.

THEOREM 10.6.3 ([**Be3**]). *Let A be an algebra with φ-action. Then in the nth $\mathbb{Z}_2$-cocharacter*

$$\chi_n^{\mathbb{Z}_2}(A) = \sum_{|\lambda|+|\mu|=n} m_{\lambda,\mu} \chi_{\lambda,\mu},$$

the multiplicities $m_{\lambda,\mu}$ are polynomially bounded.

10.7. Finite dimensional algebras with φ-action

Throughout this and the next section, A will be a finite dimensional algebra over F with an automorphism or antiautomorphism φ of order two. Following [**GZ3**] and [**BGP**] we shall prove that the $\mathbb{Z}_2$-codimensions of A have an integral exponential growth. According to Theorem 3.4.4, we can write $A = B + J$, where B is a maximal semisimple subalgebra of A such that $B^\varphi = B$ and $J = J(A)$ is the Jacobson radical of A. Also $J^\varphi = J$ and we can write $B = B_1 \oplus \cdots \oplus B_m$ where $B_1, \ldots, B_m$ are φ-simple algebras.

Inspired by the ordinary case (see Chapter 6), we define an integer $d = d(A)$ in the following way: we consider all possible non-zero products of the type

(10.18) $$C_1 J C_2 J \cdots C_{k-1} J C_k \neq 0,$$

where $C_1, \ldots, C_k$ are distinct subalgebras taken from the set $\{B_1, \ldots, B_m\}$ and $k \geq 1$. The case $k = 1$ means that $C_1 = B_i$ for some $1 \leq i \leq n$. We then define

(10.19) $$d = d(A) = \max_{C_1, \ldots, C_k} \dim(C_1 \oplus \cdots \oplus C_k)$$

where $C_1, \ldots, C_k \in \{B_1, \ldots, B_m\}$ are distinct and satisfy (10.18).

In the next section we shall prove that for the sequence of $\mathbb{Z}_2$-codimensions of A, $c_n^{\mathbb{Z}_2}(A)$, $n = 1, 2, \ldots$, the limit $\lim_{n \to \infty} \sqrt[n]{c_n^{\mathbb{Z}_2}(A)}$ exists and is an integer. Also, if F is algebraically closed, this integer coincides with d.

As in Lemma 6.2.1, the following remark holds.

REMARK 10.7.1. *If $B_{i_1}, \ldots, B_{i_t}$ are not necessarily distinct among the B_j's and $B_{i_1} J B_{i_2} J \cdots J B_{i_t} \neq 0$, then $\dim(B_{i_1} + \cdots + B_{i_t}) \leq d$.*

We start with the following.

LEMMA 10.7.2. *Let $t \geq 0$, $a + b > d$ and let $f(y_1, \ldots, y_a, z_1, \ldots, z_b, x_1, \ldots, x_t)$ be a multilinear polynomial alternating on $\{y_1, \ldots, y_a\}$ and on $\{z_1, \ldots, z_b\}$. If $\bar{y}_1, \ldots, \bar{y}_a \in B^+$, $\bar{z}_1, \ldots, \bar{z}_b \in B^-$ and $\bar{x}_1, \ldots, \bar{x}_t \in A$, then*

$$f(\bar{y}_1, \ldots, \bar{y}_a, \bar{z}_1, \ldots, \bar{z}_b, \bar{x}_1, \ldots, \bar{x}_t) = 0.$$

PROOF. Since the polynomial f is multilinear, it is enough to evaluate f on a basis of A. For each $i = 1, \ldots, m$, let $E_i = E_i^+ \cup E_i^-$ be a basis of B_i with $E_i^+ \subseteq A^+$ and $E_i^- \subseteq A^-$. Then $E = \cup E_i = E^+ \cup E^-$ is a basis of B where $E^+ = \cup E_i^+$ and $E^- = \cup E_i^-$.

Now, if all the variables in f are evaluated into elements of the given basis of B then, since $B_i B_j = 0$ for $i \neq j$, we will get a zero value unless all elements come from one φ-simple component, say B_t. In this case, since $\dim B_t < d$ and $a + b > d$, then either $a > |E_t^+|$ or $b > |E_t^-|$. Since f is alternating in the y_i's and the z_i's then the value of f will still be zero. Therefore, in order to get a non-zero value of f we should evaluate at least one variable in J.

Evaluate the variables $y_1, \ldots, y_a, z_1, \ldots, z_b$ into distinct basis elements $\bar{y}_1 \in D_1, \ldots, \bar{y}_a \in D_a, \bar{z}_1 \in D_{a+1}, \ldots, \bar{z}_b \in D_{a+b}$ respectively, where $D_1, \ldots, D_{a+b} \in \{B_1, \ldots, B_n\}$ are not necessarily distinct φ-simple components. Since for $i \neq j$ we have $B_i B_j = 0$, it follows that any monomial in f takes value in a subspace of one of the following types:

$$D_{i_1} J D_{i_2} J \cdots J D_{i_{m-1}} J D_{i_m}, \quad J D_{i_1} J D_{i_2} J \cdots J D_{i_{m-1}} J D_{i_m},$$

$$D_{i_1} J D_{i_2} J \cdots J D_{i_{m-1}} J D_{i_m} J, \quad J D_{i_1} J D_{i_2} J \cdots J D_{i_{m-1}} J D_{i_m} J.$$

Since $a_1, \ldots, a_{d+1}$ are linearly independent,

$$\dim(D_{i_1} + \cdots + D_{i_m}) = \dim(D_1 + \cdots + D_{d+1}) \geq d+1$$

and, by Remark 10.7.1, all the above products are zero. Hence f vanishes on these elements. □

Recall that if $\lambda \vdash r, \mu \vdash n-r$ and $W_{\lambda,\mu}$ is a left irreducible $S_r \times S_{n-r}$-module, then $W_{\lambda,\mu} \cong F(S_r \times S_{n-r}) e_{T_\lambda} e_{T_\mu}$ where S_r and S_{n-r} act on disjoint sets of integers and T_λ, T_μ are tableaux of shape λ and μ, respectively. Recall also that for a partition λ, $\lambda' = (\lambda_1', \lambda_2', \ldots)$ denotes the conjugate partition of λ.

REMARK 10.7.3. Let $\lambda \vdash r, \mu \vdash n-r$ and let $W_{\lambda,\mu} \subseteq P_{r,n-r}$ be a left irreducible $S_r \times S_{n-r}$-module. Then there exists $f \in W_{\lambda,\mu}$ such that

$$f = f(y_1^1, \ldots, y_{\lambda_1'}^1, y_1^2, \ldots, y_{\lambda_2'}^2, \ldots, z_1^1, \ldots, z_{\mu_1'}^1, z_1^2, \ldots, z_{\mu_2'}^2, \ldots) \neq 0$$

and f is alternating on each set of variables $\{y_1^i, \ldots, y_{\lambda_i'}^i\}, \{z_1^j, \ldots, z_{\mu_j'}^j\}$, $1 \leq i \leq \lambda_1, 1 \leq j \leq \mu_1$.

PROOF. If $f_0 \in W_{\lambda,\mu}$ is a non-zero polynomial, there exist tableaux T_λ, T_μ such that $g = e_{T_\lambda} e_{T_\mu} f_0 \neq 0$. But then

$$f = \sum_{\tau \in C_{T_\lambda}} \sum_{\rho \in C_{T_\mu}} (\operatorname{sgn} \tau)(\operatorname{sgn} \rho) \, g$$

is a polynomial with the prescribed property by Lemma 2.5.1. □

LEMMA 10.7.4. Let $\lambda \vdash r, \mu \vdash n-r$ and let $W_{\lambda,\mu} \subseteq P_{r,n-r}$ be a left irreducible $S_r \times S_{n-r}$-module. If $W_{\lambda,\mu} \not\subseteq P_{r,n-r} \cap Id^{\mathbb{Z}_2}(A)$, then

$$h(\lambda) \leq \dim A^+, \; h(\mu) \leq \dim A^- \text{ and } \lambda_{l+1}' + \mu_{l+1}' \leq d,$$

where $J^{l+1} = 0$. Moreover, $\dim W_{\lambda,\mu} \leq n^a (\lambda_{l+1}')^r (\mu_{l+1}')^{n-r}$, for some $a \geq 1$.

PROOF. Let f be the polynomial described in the previous remark. Clearly $W_{\lambda,\mu} = F(S_r \times S_{n-r})f$. Since f is alternating on $\{y_1^1, \ldots, y_{\lambda_1'}^1\}$, it follows that $\lambda_1' = h(\lambda) \leq \dim A^+$. Similarly, $h(\mu) \leq \dim A^-$.

Suppose by contradiction that $\lambda_{l+1}' + \mu_{l+1}' > d$. This implies that $\lambda_i' + \mu_i' > d$ for $i = 1, \ldots, l$. Since by hypothesis f does not vanish on A, then, by the previous lemma, we must have that each set $\{y_1^i, \ldots, y_{\lambda_i'}^i, z_1^i, \ldots, z_{\mu_i'}^i\}$, $1 \leq i \leq l+1$, must contain at least one variable that evaluates to an element of J. But $J^{l+1} = 0$ says that f vanishes on A, a contradiction.

Let $\dim A^+ = p$ and $\dim A^- = q$. From the hook formula it follows that $\chi_\lambda(1) \leq n^{pl}(\lambda'_{l+1})^r$ and $\chi_\mu(1) \leq n^{ql}(\mu'_{l+1})^{n-r}$. Therefore

$$\dim W_{\lambda,\mu} = \chi_\lambda(1)\chi_\mu(1) \leq n^{pl+ql}(\lambda'_{l+1})^r(\mu'_{l+1})^{n-r}.$$

□

In the next lemma we find an upper bound for the $\mathbb{Z}_2$-codimensions of any finite dimensional algebra.

LEMMA 10.7.5. *Let A be a finite dimensional algebra over F with φ-action and let d be the integer defined in (10.19). Then $c_n^{\mathbb{Z}_2}(A) \leq an^t d^n$, for some constants a, t.*

PROOF. By Lemma 10.7.4 we can write

$$\chi_{r,n-r}(A) = \sum_{0 \leq \lambda'_{l+1}+\mu'_{l+1} \leq d} \sum_{\substack{\lambda \vdash r \\ \mu \vdash n-r}} m_{\lambda,\mu}(\chi_\lambda \otimes \chi_\mu).$$

Hence

$$c_{r,n-r}(A) \leq \sum_{0 \leq \lambda'_{l+1}+\mu'_{l+1} \leq d} \sum_{\substack{\lambda \vdash r \\ \mu \vdash n-r}} m_{\lambda,\mu} n^\alpha (\lambda'_{l+1})^r (\mu'_{l+1})^{n-r},$$

for some constant α. Recall that by Theorem 10.6.3 the multiplicities $m_{\lambda,\mu}$ are polynomially bounded. Hence by Corollary 10.6.2 we obtain

$$c_n^{\mathbb{Z}_2}(A) = \sum_{r=0}^n \binom{n}{r} c_{r,n-r}(A) \leq an^b \sum_{\substack{t_1+t_2 \leq d \\ t_1, t_2 \geq 0}} \sum_{r=0}^n \binom{n}{r} t_1^r t_2^{n-r}$$

$$= an^b \sum_{\substack{t_1+t_2 \leq d \\ t_1, t_2 \geq 0}} (t_1+t_2)^n \leq an^b d^{n+2},$$

where t_1, t_2 are integers. □

10.8. The $\mathbb{Z}_2$-exponent of a finite dimensional algebra

In this section we shall compute the exponential rate of growth of the $\mathbb{Z}_2$-codimensions of any finite dimensional algebra with φ-action. We make the following:

DEFINITION 10.8.1. Let A be an F-algebra with φ-action and let $c_n^{\mathbb{Z}_2}(A)$, $n = 1, 2, \ldots$, be its sequence of $\mathbb{Z}_2$-codimensions. Then define

$$\mathbb{Z}_2\text{-}\underline{\exp}(A) = \liminf_{n \to \infty} \sqrt[n]{c_n^{\mathbb{Z}_2}(A)}, \quad \mathbb{Z}_2\text{-}\overline{\exp}(A) = \limsup_{n \to \infty} \sqrt[n]{c_n^{\mathbb{Z}_2}(A)}.$$

In case of equality,

$$\mathbb{Z}_2\text{-}\exp(A) = \lim_{n \to \infty} \sqrt[n]{c_n^{\mathbb{Z}_2}(A)},$$

is the $\mathbb{Z}_2$-exponent of A.

The existence of multialternating central polynomials for $k \times k$ matrices proved in Theorem 5.7.4, easily leads to the following.

LEMMA 10.8.2. *Let C be a finite dimensional central φ-simple algebra over F, $\dim C^+ = p$ and $\dim C^- = q$. For all $m \geq 1$, there exists a multilinear polynomial*
$$f = f(y_1^1, \ldots, y_p^1, \ldots, y_1^{2m}, \ldots, y_p^{2m}, z_1^1, \ldots, z_q^1, \ldots, z_1^{2m}, \ldots, z_q^{2m})$$
such that:
1) *f is alternating on each set of variables $\{y_1^i, \ldots, y_p^i\}$, $1 \leq i \leq 2m$, and $\{z_1^j, \ldots, z_q^j\}$, $1 \leq j \leq 2m$.*
2) *There exist $\bar{y}_i^j \in C^+$, $\bar{z}_i^j \in C^-$ such that $f(\bar{y}_1^1, \ldots, \bar{y}_p^{2m}, \bar{z}_1^1, \ldots, \bar{z}_q^{2m}) = 1_C$.*

PROOF. Suppose first that C is simple. The polynomial constructed in Theorem 5.7.4, is alternating in two distinct sets of variables of order $(\dim C)^2 = (p+q)^2$. By taking the product of $m \geq 1$ of such polynomials in distinct sets of variables, we obtain the existence of a multilinear polynomial
$$f(x_1^1, \ldots, x_{p+q}^1, \ldots, x_1^{2m}, \ldots, x_{p+q}^{2m})$$
alternating on each set of variables $\{x_1^i, \ldots, x_{p+q}^i\}$, $1 \leq i \leq 2m$, and f is a central polynomial for C. Since $p+q = \dim C$, it is clear that f can be viewed as alternating on $2m$ disjoint sets of symmetric variables y_i or skewsymmetric variables z_i.

If C in not simple, according to Theorem 3.4.4, $C = C_1 \oplus C_1^\varphi$ with C_1 simple. In this case $p = q = \dim C_1$ and the polynomial
$$f(y_1^1, \ldots, y_p^1, \ldots, y_1^{2m}, \ldots, y_p^{2m}) f(z_1^1, \ldots, z_p^1, \ldots, z_1^{2m}, \ldots, z_p^{2m})$$
is the required one. $\square$

The next step is to find multialternating polynomials of suitable degree not vanishing in a finite dimensional algebra. This is achieved by "gluing together" the polynomials found in the previous lemma.

LEMMA 10.8.3. *Let F be algebraically closed and let A be a finite dimensional algebra over F. Let $C_1 J C_2 J \cdots C_{k-1} J C_k \neq 0$ where $C_1, \ldots, C_k$ are distinct φ-simple subalgebras of A and $C_1 + \cdots + C_k = C_1 \oplus \cdots \oplus C_k$. If $p = \dim(C_1 + \cdots + C_k)^+$, $q = \dim(C_1 + \cdots + C_k)^-$, then for all $m \geq 1$, there exists a multilinear polynomial*
$$f = f(y_1^1, \ldots, y_p^1, \ldots, y_1^{2m}, \ldots, y_p^{2m}, z_1^1, \ldots, z_q^1, \ldots, z_1^{2m},$$
$$\ldots, z_q^{2m}, y_1, \ldots, y_{k_1}, z_1, \ldots, z_{k_2})$$
where $k_1 + k_2 = 2k - 1$, such that:
1) *f is alternating on each set of variables $\{y_1^i, \ldots, y_p^i\}$, $1 \leq i \leq 2m$, and $\{z_1^j, \ldots, z_q^j\}$, $1 \leq j \leq 2m$.*
2) *There exist $\bar{y}_i^j \in (C_1 + \cdots + C_k)^+, \bar{z}_i^j \in (C_1 + \cdots + C_k)^-, \bar{y}_i \in A^+, \bar{z}_i \in A^-$, such that $f(\bar{y}_1^1, \ldots, \bar{y}_p^{2m}, \bar{z}_1^1, \ldots, \bar{z}_q^{2m}, \bar{y}_1, \ldots, \bar{y}_{k_1}, \bar{z}_1, \ldots, \bar{z}_{k_2}) \neq 0$.*

PROOF. For every $i = 1, \ldots, k$, let $p_i = \dim C_i^+, q_i = \dim C_i^-$ and let
$$f_i = f_i(y_{i,1}^1, \ldots, y_{i,p_i}^1, \ldots, y_{i,1}^{2m}, \ldots, y_{i,p_i}^{2m}, z_{i,1}^1, \ldots, z_{i,q_i}^1, \ldots, z_{i,1}^{2m}, \ldots, z_{i,q_i}^{2m})$$
be the polynomial constructed in Lemma 10.8.2. We let
$$\tilde{f} = \mathcal{A}_1 \cdots \mathcal{A}_{2m} \mathcal{A}'_1 \cdots \mathcal{A}'_{2m} x_1 f_1 x'_1 x_2 f_2 \cdots x_{k-1} f_{k-1} x'_{k-1} x_k f_k$$
where $\mathcal{A}_i$ means alternation on the p variables $y_{1,1}^i, \ldots, y_{1,p_1}^i, \ldots, y_{k,1}^i, \ldots, y_{k,p_k}^i$ and $\mathcal{A}'_i$ means alternation on the q variables $z_{1,1}^i, \ldots, z_{1,q_1}^i, \ldots, z_{k,1}^i, \ldots, z_{k,q_k}^i$.

We remark that each polynomial f_i corresponds to a pair of tableaux (P_i, Q_i) where P_i is a $2m \times p_i$ rectangle, Q_i a $2m \times q_i$ rectangle and the variables in each column of P_i (resp. Q_i) are alternating. Now, $\tilde{f}$ corresponds to the pair (P, Q) obtained by gluing the rectangles P_i (resp. Q_i) one on top of the other and by alternating the variables in the columns. Hence $\mathcal{A}_j$ is alternation on the variables in the jth column of P and $\mathcal{A}'_j$ is alternation on the variables in the jth column of Q.

Since $C_1 J C_2 J \cdots C_{k-1} J C_k \neq 0$, there exist $c_i \in C_i$, $(1 \leq i \leq k)$, $b_1, \ldots, b_{k-1} \in J$ such that $c_1 b_1 c_2 b_2 \cdots b_{k-1} c_k \neq 0$.

For every $i = 1, \ldots, k$, let $\bar{y}_{i,j}^t \in C_i^+, \bar{z}_{i,j}^t \in C_i^-$ be such that

$$f_i(\bar{y}_{i,1}^1, \ldots, \bar{y}_{i,p_i}^{2m}, \bar{z}_{i,1}^1, \ldots, \bar{z}_{i,q_i}^{2m}) = 1_{C_i}.$$

Notice that, since $C_i C_j = 0$ for $i \neq j$, when evaluating the variables y's and z's on the C_i's, alternation on the columns of the rectangle P (resp. Q) can be replaced with alternation on the columns of each subrectangle P_i (resp. Q_i). Hence

$$\tilde{f}(\bar{y}_{1,1}^1, \ldots, \bar{y}_{k,p_k}^{2m}, \bar{z}_{1,1}^1, \ldots, \bar{z}_{k,q_k}^{2m}, c_1, \ldots, c_k, b_1, \ldots, b_{k-1})$$
$$= (p_1! \cdots p_k! q_1! \cdots q_k!)^{2m} c_1 f_1 b_1 c_2 f_2 b_2 \cdots c_{k-1} f_{k-1} b_{k-1} c_k f_k$$
$$= (p_1! \cdots p_k! q_1! \cdots q_k!)^{2m} c_1 b_1 c_2 \cdots b_{k-1} c_k \neq 0.$$

We may clearly assume that $c_1, b_1, \ldots, b_{k-1}, c_k \in A^+ \cup A^-$; suppose that k_1 of them belong to A^+ and k_2 of them belong to A^-, $k_1 + k_2 = 2k - 1$. Then

$$f = \tilde{f}(y_{1,1}^1, \ldots, y_{k,p_k}^{2m}, z_{1,1}^1, \ldots, z_{k,q_k}^{2m}, y_1, \ldots, y_{k_1}, z_1, \ldots, z_{k_2})$$

does not vanish in A and is the desired polynomial. $\square$

We are now in a position to find the lower bound for the exponential growth of the $\mathbb{Z}_2$-codimensions.

THEOREM 10.8.4 ([**GZ3**], [**BGP**]). *Let A be a finite dimensional algebra with φ-action over the algebraically closed field F and let d be the integer defined in (10.19). Then*

$$a_1 n^{b_1} d^n \leq c_n^{\mathbb{Z}_2}(A) \leq a_2 n^{b_2} d^n,$$

for some non-zero constants a_1, a_2, b_1, b_2. Hence $\mathbb{Z}_2$-$\exp(A) = d$.

PROOF. The upper bound for $c_n^{\mathbb{Z}_2}(A)$ was found in Lemma 10.7.5. In order to find the lower bound, we proceed as follows.

Let $C_1, \ldots, C_k$ be distinct φ-simple subalgebras of B such that

$$C_1 J C_2 J \cdots C_{k-1} J C_k \neq 0.$$

Let $p = \dim(C_1^{(0)} + \cdots + C_k^{(0)}), q = \dim(C_1^{(1)} + \cdots + C_k^{(1)})$ and $d = p + q$.

Take any $n \geq 2d + (k_1 + k_2)$, where k_1 and k_2 are as in Lemma 10.8.3, and divide $n - (k_1 + k_2)$ by $2d$. Then we can write

$$n = 2m(p + q) + (k_1 + k_2) + t$$

for some m, t, where $0 \leq t < 2d$.

Set $s = 2mp + k_1 + t, n - s = 2mq + k_2$. Let f be the polynomial constructed in the previous lemma of degree $2mp + 2mq + k_1 + k_2$, and set $g = f y_{k_1+1} \cdots y_{k_1+t} \in P_{s,n-s}$. It is easy to check that g does not vanish on A.

Let the group $G = S_{2mp} \times S_{2mq}$ act on g (S_{2mp} acts on the variables y_i^l and S_{2mq} acts on the variables z_i^l) and let M be the G-submodule of $P_{s,n-s}$ generated by g. By complete reducibility M contains an irreducible G-submodule of the form $W_{\lambda,\mu} = FGe_{T_\lambda}e_{T_\mu}(g)$ for some tableaux T_λ and T_μ of shape $\lambda \vdash 2mp$ and $\mu \vdash 2mq$, respectively.

Let $l(\lambda)$ be the length of the first row of λ. Now, for every $\tau \in S_{2mp}$, $\tau(g)$ is still alternating on $2m$ disjoint sets of even variables and $\sum_{\sigma \in R_{T_\lambda}} \sigma$ acts by symmetrizing $l(\lambda)$ variables. Thus, if $l(\lambda) > 2m$, we would get $e_{T_\lambda}(g) = 0$, a contradiction. Similarly for $l(\mu)$. The outcome of this is that $l(\lambda) \leq 2m$ and $l(\mu) \leq 2m$.

Recall that for a partition $\nu \vdash n$, $h(\nu)$ denoted the height of the corresponding Young tableau. Notice that from the above, since $l(\lambda) \leq 2m$ and $l(\mu) \leq 2m$, then $h(\lambda) \geq p$ and $h(\mu) \geq q$.

Suppose that either $h(\lambda) > p$ or $h(\mu) > q$ and that the total number of boxes out of the first p rows of the diagram D_λ and out of the first q rows of the diagram D_μ is at least $l+1$ where $J^{l+1} = 0$.

Since $FGe_{T_\lambda}e_{T_\mu}$ is a minimal left ideal of FG, then $FG\bar{C}_{T_\lambda}\bar{C}_{T_\mu}e_{T_\lambda}e_{T_\mu} = FGe_{T_\lambda}e_{T_\mu}$ where $\bar{C}_{T_\lambda} = \sum_{\sigma \in C_{T_\lambda}} (\operatorname{sgn} \sigma)\sigma$ and $\bar{C}_{T_\mu} = \sum_{\sigma \in C_{T_\mu}} (\operatorname{sgn} \sigma)\sigma$. We need to evaluate the polynomial $\bar{g} = \bar{C}_{T_\lambda}\bar{C}_{T_\mu}e_{T_\lambda}e_{T_\mu}(g)$ on A.

Let
$$\lambda' = (p + r_1, \ldots, p + r_u, \lambda'_{u+1}, \ldots, \lambda'_m)$$
and
$$\mu' = (q + s_1, \ldots, q + s_v, \mu'_{v+1}, \ldots, \mu'_n)$$
be the conjugate partitions of λ and μ, respectively. Here
$$r_1 + \cdots + r_u + s_1 + \cdots + s_v \geq l+1, \quad \lambda'_{u+1}, \ldots, \lambda'_m \leq p \quad \text{and} \quad \mu'_{v+1}, \ldots, \mu'_n \leq q.$$

Now, the polynomial $\bar{g}$ is alternating on each of the u sets of symmetric variables y_i of order $p + r_1, \ldots, p + r_u$ and on each of the v sets of skew-symmetric variables z_i of order $q + s_1, \ldots, q + s_v$. If we substitute on any of these sets of variables only elements from the semisimple subalgebra $C = C_1 \oplus \cdots \oplus C_k$, we would get zero since $\dim C^+ = p$, $\dim C^- = q$. It follows that we have to substitute into these sets of variables at least
$$r_1 + \cdots + r_u + s_1 + \cdots + s_v \geq l+1$$
elements from the Jacobson radical J. Since $J^{l+1} = 0$, we get that $\bar{g}$ vanishes in A, a contradiction.

We have proved that D_λ and D_μ must contain at most a total number of l boxes out of the first p and q rows, respectively. Since $l(\lambda) \leq 2m$ and $l(\mu) \leq 2m$, the outcome is that λ must contain the rectangle $\nu = ((2m-l)^p)$ and μ must contain the rectangle $\nu' = ((2m-l)^q)$. Recall that as $m \to \infty$,
$$(\deg \chi_\nu)(\deg \chi_{\nu'}) \simeq a((2m-l)p)^b((2m-l)q)^{b'} p^{(2m-l)p} q^{(2m-l)q} \geq n^u p^{2mp} q^{2mq},$$
for some constants a, b, b', u. Hence, since
$$\dim W_{\lambda,\mu} = (\deg \chi_\lambda)(\deg \chi_\mu) \geq (\deg \chi_\nu)(\deg \chi_{\nu'}),$$
we obtain
$$c_{s,n-s}(A) \geq \dim W_{\lambda,\mu} \geq n^u p^{2mp} q^{2mq}.$$
Therefore by Corollary 10.6.2,
$$c_n^{\mathbb{Z}_2}(A) = \sum_{r=0}^n \binom{n}{r} c_{r,n-r}(A) \geq \binom{n}{s} c_{s,n-s}(A) \geq n^u \frac{n!}{s!(n-s)!} p^{2mp} q^{2mq}.$$

Recalling that $s = 2mp + k_1 + t$ and $n - s = 2mq + k_2$, we have that
$$\frac{n!}{s!(n-s)!} \geq \frac{(2mp+2mq)!}{(2mp)!(2mq)!}$$
and by Stirling formula we get
$$c_n^{\mathbb{Z}_2}(A) \geq n^\alpha \frac{(2mp+2mq)^{2mp+2mq}}{(2mp)^{2mp}(2mq)^{2mq}} p^{2mp} q^{2mq} = n^\alpha (p+q)^{2mp+2mq}$$
$$= an^\alpha (p+q)^n = an^\alpha d^n$$
where $a = (p+q)^{(-k_1 - k_2 - t)}$ is a constant. □

In light of the previous theorem we summarize the results obtained so far on the $\mathbb{Z}_2$-exponent.

COROLLARY 10.8.5. *Let A be a finite dimensional algebra with φ-action over the algebraically closed field F and let $B = B^\varphi$ be a maximal semisimple subalgebra of A. Then*
$$\mathbb{Z}_2\text{-}\exp(A) = \max_i \dim_F(C_1^{(i)} + \cdots + C_{t_i}^{(i)})$$
where $C_1^{(i)}, \ldots, C_{t_i}^{(i)}$ are distinct φ-simple subalgebras of B and
$$C_1^{(i)} J C_2^{(i)} J \cdots J C_{t_i}^{(i)} \neq 0.$$

Since by an obvious extension of Theorem 4.1.9, any extension of the base field does not effect the $\mathbb{Z}_2$-codimensions of an algebra, from the previous theorem we immediately get.

COROLLARY 10.8.6. *If A is a finite dimensional algebra over the field F with φ-action, then $\mathbb{Z}_2\text{-}\exp(A)$ exists and is an integer $\leq \dim_F A$.*

10.9. Simple and semisimple φ-algebras

Let $\bar{F}$ be the algebraic closure of the field F. For an algebra A over F with φ-action, we let $Z = Z(A)$ be the center of A and $Z^+ = Z(A)^+ = Z \cap A^+$ the symmetric center of A. The next two results give an exact estimate of $\mathbb{Z}_2\text{-}\exp(A)$ in the case of simple or semisimple algebras ([**GZ3**], [**BGP**]).

COROLLARY 10.9.1. *Let A be a finite dimensional algebra with φ-action over F.*

1) *If A is φ-simple, $\mathbb{Z}_2\text{-}\exp(A) = \dim_{Z^+} A$.*
2) *If A is semisimple and $A = \oplus_i A_i$ is the decomposition of A into φ-simple subalgebras, then*
$$\mathbb{Z}_2\text{-}\exp(A) = \max_i \dim_{Z_i^+} A_i$$
   where $Z_i^+ = Z(A_i) \cap A_i^+$ is the symmetric center of A_i.
3) *$\mathbb{Z}_2\text{-}\exp(A) = \dim_F A$ if and only if A is φ-simple and $F = Z^+$.*

PROOF. Suppose first that A is simple.
If $Z^+ = Z$, then
$$A \otimes_F \bar{F} \cong \bigoplus_{i=1}^{[Z:F]} (A \otimes_Z \bar{F}_i)$$

where $\bar{F}_i \cong \bar{F}$ and $A \otimes_Z \bar{F}_i$ is a central simple algebra over $\bar{F}$ with induced φ-action. Moreover,

$$[Z:F]\dim_Z A = \dim_F A = \dim_{\bar{F}}(A \otimes_F \bar{F}) = [Z:F]\dim_{\bar{F}}(A \otimes_Z \bar{F}_i).$$

From Corollary 10.8.5 it follows that

$$\dim_Z A = \dim_{\bar{F}}(A \otimes_Z \bar{F}_i) = \mathbb{Z}_2\text{-}\exp(A \otimes_F \bar{F}) = \mathbb{Z}_2\text{-}\exp(A).$$

If $Z^+ \neq Z$, then $Z = Z^+(\alpha)$ with $\alpha^\varphi = -\alpha \in Z$ is a quadratic extension of Z^+ and $Z \otimes_{Z^+} \bar{F} \cong \bar{F} \oplus \bar{F}$. Hence

$$A \otimes_{Z^+} \bar{F} \cong A \otimes_Z Z \otimes_{Z^+} \bar{F} \cong A \otimes_Z (\bar{F} \oplus \bar{F}) \cong (A \otimes_Z \bar{F}) \oplus (A \otimes_Z \bar{F})$$

and φ acts as $(a \otimes f_1 + b \otimes f_2)^\varphi = b^\varphi \otimes f_1 + a^\varphi \otimes f_2$, when φ is an involution and $(a \otimes f_1 + b \otimes f_2)^\varphi = a^\varphi \otimes f_1 + b^\varphi \otimes f_2$, when φ is an automorphism of order two. As above, it follows that

$$A \otimes_F \bar{F} \cong A \otimes_{Z^+} (\bigoplus_{i=1}^{[Z^+:F]} \bar{F}_i) \cong \bigoplus_{i=1}^{[Z^+:F]} (A \otimes_{Z^+} \bar{F}_i)$$

$$\cong \bigoplus_{i=1}^{[Z^+:F]} (A \otimes_Z \bar{F}_i \oplus A \otimes_Z \bar{F}_i)$$

where, for $i = 1, \ldots, [Z^+ : F]$, $\bar{F}_i \cong \bar{F}$ and $A \otimes_Z \bar{F}_i \oplus A \otimes_Z \bar{F}_i$ is a φ-simple algebra. As in the previous case it follows that

$$\dim_{Z^+} A = \dim_{\bar{F}}(A \otimes_Z \bar{F}_i \oplus A \otimes_Z \bar{F}_i) = \mathbb{Z}_2\text{-}\exp(A \otimes_F \bar{F}) = \mathbb{Z}_2\text{-}\exp((A).$$

This proves 1, when A is simple.

Suppose now that A is φ-simple but not simple. Then $A \cong C \oplus C^\varphi$ where C is simple. In this case since $Z^+ = Z(A)^+ \cong Z(C)$, we get

$$A \otimes_F \bar{F} \cong \bigoplus_{i=1}^{[Z(C):F]} (C \otimes_{Z(C)} \bar{F}_i) \oplus (C^\varphi \otimes_{Z(C)} \bar{F}_i),$$

where $\bar{F}_i \cong \bar{F}$ and $(C \otimes_{Z(C)} \bar{F}_i) \oplus (C^\varphi \otimes_{Z(C)} \bar{F}_i)$ is φ-simple over $\bar{F}$. Thus as before we get

$$\dim_{Z^+} A = \dim_{\bar{F}}(A \otimes_Z \bar{F}_i \oplus A \otimes_Z \bar{F}_i) = \mathbb{Z}_2\text{-}\exp(A \otimes_F \bar{F}) = \mathbb{Z}_2\text{-}\exp(A).$$

This proves the first part of the corollary.

Suppose now that A is semisimple. Then $A \otimes_F \bar{F} \cong \oplus(A_i \otimes_F \bar{F})$ and by part 1) of the corollary, $A_i \otimes_F \bar{F} \cong B_{i1} \oplus \cdots \oplus B_{it_i}$ where $B_{i1} \cong \cdots \cong B_{it_i}$ are φ-simple algebras central over $\bar{F}$ and $t_i = [Z_i^+ : F]$. It follows that

$$\mathbb{Z}_2\text{-}\exp(A) = \mathbb{Z}_2\text{-}\exp(A \otimes_F \bar{F}) = \max_i \dim_{\bar{F}} B_{i1}.$$

Since $\dim_{Z_i^+} A_i = \dim_{\bar{F}} B_{i1}$, the conclusion of the second part of the corollary now follows.

In light of part 1), in order to prove 3), we only need to show that $\mathbb{Z}_2\text{-}\exp(A) = \dim_F A$ implies A φ-simple and $F = Z^+$.

Let $\bar{A} = A \otimes_F \bar{F}$ with induced φ-action. Then

$$\dim_{\bar{F}} \bar{A} = \dim_F A = \mathbb{Z}_2\text{-}\exp(A) = \mathbb{Z}_2\text{-}\exp(\bar{A}).$$

If $\bar{A}$ is nilpotent, then $\mathbb{Z}_2\text{-}\exp(\bar{A}) = 0$, a contradiction. Hence $\bar{A}$ contains a maximal semisimple subalgebra $B = B^\varphi$ and by part 2), $\dim_{\bar{F}} \bar{A} = \mathbb{Z}_2\text{-}\exp(\bar{A}) = \dim_{\bar{F}} B_i$

for a suitable φ-simple subalgebra B_i of B. Hence $\bar{A} = B_i$ is φ-simple. This implies that A is φ-simple and by the proof of 1), $\dim_F A = \mathbb{Z}_2\text{-}\exp(A) = \dim_{Z^+} A$ implies $F = Z^+$. □

CHAPTER 11

Superalgebras, ∗-Algebras and Codimension Growth

This chapter is devoted to the study of superalgebras, algebras with involution and their identities. As we have shown in the previous chapter for any such finite dimensional algebra the corresponding sequence of $\mathbb{Z}_2$-codimensions (now called supercodimensions or ∗-codimensions if A has a structure of superalgebra or of algebra with involution ∗) has integral exponential growth. Here we consider arbitrary algebras not necessarily finite dimensional and we prove a characterization when the $\mathbb{Z}_2$-codimensions are polynomially bounded. As a consequence of this result we prove that ∗-varieties and supervarieties of intermediate growth do not exist. In particular, we exhibit a complete list of superalgebras and of algebras with involution having almost polynomial growth of the corresponding codimensions. It is worth noting that while such a list of superalgebras includes G and UT_2 with various $\mathbb{Z}_2$-gradings, only two algebras with involution of small dimension have this property.

We complete our discussion on the codimension growth of superalgebras and algebras with involution by comparing such codimensions with the ordinary ones and by giving several examples of different behavior.

Throughout this chapter the base field is assumed to be of characteristic zero.

11.1. Notation and more

As we have already observed, an automorphism or antiautomorphism φ of order two on an algebra A determines uniquely a structure of superalgebra or of an algebra with involution on A. Since φ can be regarded as a unary operation on the algebra A, one can naturally speak of varieties of superalgebras, denoted supervarieties, and of varieties of algebras with involution, denoted ∗-varieties.

Recall that if A is a superalgebra, we denote by supvar(A) the supervariety generated by A, by $\mathrm{Id}^{gr}(A)$ the ideal of $F\langle Y, Z \rangle$ of graded identities of A, and we use the following notation:

- $P_n^{gr} = P_n^{\mathbb{Z}_2}$, is the space of multilinear polynomials in the even variables $y_1, \ldots, y_n$ and the odd variables $z_1, \ldots, z_n$,
- $\chi_n^{gr}(A) = \chi_n^{\mathbb{Z}_2}(A)$ is the nth graded cocharacter of A,
- $c_n^{sup}(A) = c_n^{gr}(A) = \deg \chi_n^{gr}(A)$, is the corresponding nth graded codimension or supercodimension of A.

Recall that the nth graded cocharacter of A is the $\mathbb{Z}_2 \wr S_n$-character of the module $P_n^{gr}/(P_n^{gr} \cap \mathrm{Id}^{gr}(A))$ and has the decomposition

$$\chi_n^{gr}(A) = \sum_{|\lambda|+|\mu|=n} m_{\lambda,\mu} \chi_{\lambda,\mu}.$$

Moreover, $P_{r,n-r}$ denotes the space of multilinear polynomials in the variables $y_1, \ldots, y_r, z_1, \ldots, z_{n-r}$ and $\chi_{r,n-r}(A)$ is the $S_r \times S_{n-r}$-character of the module $P_{r,n-r}/(P_{r,n-r} \cap \mathrm{Id}^{gr}(A))$. If $\chi_{r,n-r}(A)$ has the decomposition

$$\chi_{r,n-r}(A) = \sum_{\lambda \vdash r, \mu \vdash n-r} m'_{\lambda,\mu} \chi_\lambda \otimes \chi_\mu, \tag{11.1}$$

then, according to Corollary 10.6.2, we have that $m_{\lambda,\mu} = m'_{\lambda,\mu}$, for all $\lambda \vdash r, \mu \vdash n-r$, $r = 0, \ldots, n$.

Also, if $\mathcal{V}$ is a variety of superalgebras and A is a generating superalgebra, we write $c_n^{sup}(A) = c_n^{sup}(\mathcal{V})$.

In order to have a consistent notation, we also write

$$\mathbb{Z}_2\text{-}\exp(A) = \mathrm{supexp}(A)$$

(in case it is well defined) and we call it the superexponent of the superalgebra A. According to Corollary 10.8.6, $\mathrm{supexp}(A)$ exists and is an integer for any finite dimensional superalgebra A.

We recall that according to Theorem 4.8.3 for any finitely generated superalgebra A, there exists a finite dimensional superalgebra B such that $\mathrm{supvar}(A) = \mathrm{supvar}(B)$. Hence in light of this result, Corollary 10.8.6, implies the following.

COROLLARY 11.1.1. *If A is a finitely generated superalgebra over the field F, then $\mathrm{supexp}(A)$ exists and is an integer.*

If A is an algebra with involution $*$, we denote by $*$-$\mathrm{var}(A)$ the $*$-variety generated by A and by $\mathrm{Id}^*(A)$ the ideal of $F\langle Y, Z \rangle$ of $\mathbb{Z}_2$-identities or $*$-identities of A. Also, we use the following notation:

- $P_n^* = P_n^{\mathbb{Z}_2}$, is the space of multilinear polynomials in the symmetric variables $y_1, \ldots, y_n$ and the skew variables $z_1, \ldots, z_n$,
- $\chi_n^*(A) = \chi_n^{\mathbb{Z}_2}(A)$ is the nth $*$-cocharacter of A,
- $c_n^*(A) = c_n^{\mathbb{Z}_2}(A) = \deg \chi_n^*(A)$, is the corresponding nth $\mathbb{Z}_2$-codimension or $*$-codimension of A.

Recall that the nth $*$-cocharacter of A is the $\mathbb{Z}_2 \wr S_n$-character of the module $P_n^*/(P_n^* \cap \mathrm{Id}^*(A))$ and has the decomposition

$$\chi_n^*(A) = \sum_{|\lambda|+|\mu|=n} m_{\lambda,\mu} \chi_{\lambda,\mu}.$$

As above, $P_{r,n-r}$ denotes the space of multilinear polynomials in the variables $y_1, \ldots, y_r, z_1, \ldots, z_{n-r}$ and $\chi_{r,n-r}(A)$ is the $S_r \times S_{n-r}$-character of the module $P_{r,n-r}/(P_{r,n-r} \cap \mathrm{Id}^*(A))$. Also in this case, if $\chi_{r,n-r}(A)$ has the decomposition given in (11.1), $m_{\lambda,\mu} = m'_{\lambda,\mu}$, for all $\lambda \vdash r, \mu \vdash n-r$, $r = 0, \ldots, n$.

If $\mathcal{V}$ is a variety of $*$-algebras and A is a generating algebra, we shall write $c_n^*(A) = c_n^*(\mathcal{V})$. Also we denote

$$\mathbb{Z}_2\text{-}\exp(A) = *\text{-}\exp(A)$$

(in case it is well defined) and we call it the $*$-exponent of the algebra A.

In the next sections we shall characterize the supervarieties and the $*$-varieties having polynomially bounded $\mathbb{Z}_2$-codimensions. By analogy with the ordinary case we make the following.

11.2. *-varieties of almost polynomial growth

DEFINITION 11.1.2. If a variety $\mathcal{V}$ of superalgebras (of algebras with involution) has sequence of the supercodimensions (*-codimensions resp.) polynomially bounded, we say that $\mathcal{V}$ has polynomial growth. We say that $\mathcal{V}$ has almost polynomial growth if $\mathcal{V}$ does not have polynomial growth, but every proper subvariety $\mathcal{U}$ of $\mathcal{V}$ has polynomial growth.

Next we shall be looking for a minimal list of superalgebras $A_1, \ldots, A_n$, having the following property. Given a supervariety $\mathcal{V}$, then $\mathcal{V}$ has polynomial growth if and only if $A_i \notin \mathcal{V}$, for all $i = 1, \ldots, n$. It is clear that each supervariety $\mathrm{supvar}(A_i)$ will have almost polynomial growth. The same remark applies with the due changes to *-varieties.

In what follows, if A is a superalgebra, by writing $\mathrm{supexp}(A) = d$ we mean tacitly that $\mathrm{supexp}(A)$ exists and equals d.

11.2. *-varieties of almost polynomial growth

In this section we shall characterize T-ideals with polynomial growth of the *-codimensions. We shall also study two algebras with involution generating *-varieties of almost polynomial growth.

Our first example is the algebra $D = F \oplus F$ with exchange involution given by $(a, b)^* = (b, a)$. Notice that D can be regarded as the superalgebra $D = F \oplus cF$ where $c^2 = 1$ with grading $D^{(0)} = F$, $D^{(1)} = cF$. In fact, we can write

$$D = (\frac{1+c}{2})F \oplus (\frac{1-c}{2})F$$

and $*$ is given by $(\frac{1+c}{2}a + \frac{1-c}{2}b)^* = \frac{1+c}{2}b + \frac{1-c}{2}a$.

Next we determine the *-codimensions and the *-cocharacter of the algebra D. Recall that $\chi_n^*(D)$ is the $\mathbb{Z}_2 \wr S_n$-cocharacter of D and $\chi_{r,n-r}(D)$ is the $S_r \times S_{n-r}$-cocharacter of D.

LEMMA 11.2.1. For every $n \geq 1$,

$$\chi_n^*(D) = \sum_{r=0}^{n} \chi_{(r),(n-r)}.$$

Hence $c_n^*(D) = 2^n$.

PROOF. For every $r \geq 0$ write

$$\chi_{r,n-r}(D) = \sum_{\substack{\lambda \vdash r \\ \mu \vdash n-r}} m_{\lambda,\mu}(\chi_\lambda \otimes \chi_\mu).$$

Since $\dim D^+ = \dim D^- = 1$, by Lemma 10.7.2, it follows that $m_{\lambda,\mu} = 0$ whenever $h(\lambda) > 1$ or $h(\mu) > 1$. Also the commutativity of D implies that $m_{(r),(n-r)} \leq 1$.

Since the polynomial $y^r z^{n-r}$ does not vanish in D, we get that $m_{(r),(n-r)} = 1$. Thus $\chi_{r,n-r}(D) = \chi_{(r)} \otimes \chi_{(n-r)}$ and by Corollary 10.6.2, we get the desired conclusion. Also $c_n^*(D) = \sum_{r=0}^{n} \binom{n}{r} = 2^n$. $\square$

LEMMA 11.2.2. Let $\mathcal{V}$ be a *-variety. Then $D \notin \mathcal{V}$ if and only if $z^d \in \mathrm{Id}^*(\mathcal{V})$, for some $d \geq 1$, where z is a skew variable.

PROOF. Since $z^d \notin \mathrm{Id}^*(D)$, one implication is obvious. Suppose now that $D \notin \mathcal{V}$. Then $\mathrm{Id}^*(\mathcal{V}) \not\subseteq \mathrm{Id}^*(D)$ and let $f = f(y_1, \ldots, y_r, z_1, \ldots, z_{n-r}) \in \mathrm{Id}^*(\mathcal{V})$, $f \notin \mathrm{Id}^*(D)$. By the standard multilinearization process we may clearly assume that f is multilinear; hence f does not vanish on a basis of D. If we write $D = F \oplus cF$, $c^2 = 1$, clearly $\{1\}$ and $\{c\}$ are bases of D^+ and D^-, respectively. Since $f \notin \mathrm{Id}^*(D)$, then

$$0 \neq f(1, \ldots, 1, c, \ldots, c) = f(c^2, \ldots, c^2, c, \ldots, c) = \alpha c^{n+r}$$

where $\alpha \neq 0$ is the sum of all the coefficients of f. Since z^2 is a symmetric variable, it follows that $f(z^2, \ldots, z^2, z, \ldots, z) = \alpha z^{n+r} \in \mathrm{Id}^*(\mathcal{V})$ and, since $\alpha \neq 0$, we get $z^{n+r} \in \mathrm{Id}^*(\mathcal{V})$. □

The proof of the following remark is left as an exercise.

REMARK 11.2.3. $\mathrm{Id}^*(D)$ is generated by the polynomials $[y_1, y_2], [y_1, z_1]$ and $[z_1, z_2]$.

LEMMA 11.2.4. *Let A be an algebra with involution such that $c_n^*(A) \leq \alpha n^t$ for some α. Then there exists a constant β such that*

(11.2) $$\chi_n^*(A) = \sum_{|\lambda|+|\mu|=n} m_{\lambda,\mu} \chi_{\lambda,\mu},$$

where $m_{\lambda,\mu} = 0$ if either $|\lambda| - \lambda_1 > \beta$ or $|\mu| - \mu_1 > \beta$.

PROOF. We decompose the ordinary nth cocharacter of A into irreducible S_n-characters

(11.3) $$\chi_n(A) = \sum_{\nu \vdash n} m_\nu \chi_\nu$$

where for $\nu \vdash n$, m_ν is the multiplicity of χ_ν in $\chi_n(A)$. By Lemma 10.1.2, $c_n(A) \leq c_n^*(A) \leq \alpha n^t$; hence by Theorem 7.2.2, there exists a constant β such that, in (11.3), $m_\nu = 0$ whenever $|\nu| - \nu_1 > \beta$.

Suppose by contradiction that there exist $\lambda \vdash r$, $\mu \vdash n - r$ such that $|\lambda| - \lambda_1 > \beta$ (or $|\mu| - \mu_1 > \beta$) and $m_{\lambda,\mu} \neq 0$. Then there exist tableaux T_λ, T_μ such that $e_{T_\lambda} e_{T_\mu}$ has a non-trivial action on $P_{r,n-r}(A)$. This says that there exists a non-trivial polynomial $f \in e_{T_\lambda} e_{T_\mu} P_{r,n-r}$ such that $f = f(y_1, \ldots, y_r, z_1, \ldots, z_{n-r})$ is not a $*$-identity of A. In particular, this implies that the polynomial

$$f(x_1, \ldots, x_r, x_{r+1}, \ldots, x_n)$$

is not an ordinary (without $*$) identity of A. Moreover, $F(S_r \otimes S_{n-r}) f(x_1, \ldots, x_n)$ is an irreducible $S_r \otimes S_{n-r}$-module where S_r acts on $x_1, \ldots, x_r$ and S_{n-r} acts on $x_{r+1}, \ldots, x_n$.

Let $M = F(S_r \otimes S_{n-r}) f \uparrow S_n$ be the S_n-module obtained by inducing $F(S_r \otimes S_{n-r}) f$ up to S_n. If $\chi^{S_n}(M)$ denotes the S_n-character of M, then we can decompose

(11.4) $$\chi^{S_n}(M) = \sum_{\nu \vdash n} m'_\nu \chi_\nu.$$

Since $|\lambda| - \lambda_1 \geq \beta$ (or $|\mu| - \mu_1 \geq \beta$), by the Littlewood-Richardson rule (Theorem 2.3.9) we get that $m'_\nu = 0$ for all $\nu \vdash n$ such that $|\nu| - \nu_1 \leq \beta$.

Since $M \subseteq P_n(A)$, then, by comparing (11.3) and (11.4) we get that $m'_\nu \leq m_\nu$, for all $\nu \vdash n$. Hence $m'_\nu = 0$ for all $\nu \vdash n$ and this contradicts $M \neq 0$. This completes the proof of the lemma. □

Recall that by Theorem 10.4.2, if $d_{\lambda,\mu} = \chi_{\lambda,\mu}(1)$ is the degree of the irreducible $\mathbb{Z}_2 \wr S_n$-character corresponding to the pair (λ, μ), then $d_{\lambda,\mu} = \binom{n}{r} d_\lambda d_\mu$ where $d_\lambda = \chi_\lambda(1)$ and $d_\mu = \chi_\mu(1)$.

LEMMA 11.2.5. *Let A be such that $c_n^*(A) \leq \alpha n^t$ for some α. Then there exists a constant γ such that, in (11.2), $m_{\lambda,\mu} = 0$ if $|\lambda| > \gamma$ and $|\mu| > \gamma$.*

PROOF. Let $n > 2t$. If r is such that $r > t$ and $n - r > t$, then $\binom{n}{r} \geq \binom{n}{t+1}$. Since $\binom{n}{t+1}$ is a polynomial in n of degree $t+1$ and $c_n^*(A) \leq \alpha n^t$, there exists N such that for $n > N$, $\binom{n}{r} \geq \binom{n}{t+1} > c_n(A, *)$. Let $\gamma = \max\{N, t\}$ and suppose that $r = |\lambda| > \gamma$ and $n - r = |\mu| > \gamma$. In case $m_{\lambda,\mu} \neq 0$, we get

$$c_n^*(A) \geq d_{\lambda,\mu} = \binom{n}{r} d_\lambda d_\mu \geq \binom{n}{r} > c_n^*(A),$$

a contradiction. Hence $m_{\lambda,\mu} = 0$ and the lemma is proved. □

We are now able to characterize the *-cocharacter sequence of algebras with polynomially bounded *-codimensions. The analogous result for superalgebras is Theorem 10.5.7.

THEOREM 11.2.6. *Let A be an algebra with involution. Then $c_n^*(A) \leq \alpha n^t$ for some α, t if and only if there exists a constant δ such that*

$$\chi_n^*(A) = \sum_{|\lambda|+|\mu|=n} m_{\lambda,\mu} \chi_{\lambda,\mu}$$

and $m_{\lambda,\mu} = 0$ whenever either $|\lambda| - \lambda_1 > \delta$ or $|\mu| > \delta$.

PROOF. Since the *-codimensions of D have exponential growth, then $D \notin {}*\text{-var}(A)$. Hence by Lemma 11.2.2 there exists an integer $d \geq 1$ such that $z^d \in \text{Id}^*(A)$. Let $\delta = d(\beta + \gamma + 1)$ where β and γ are the integers determined in Lemma 11.2.4 and Lemma 11.2.5, respectively.

Let μ be such that $|\mu| > \delta$ and suppose that for some λ, $\chi_{\lambda,\mu}$ appears with non-zero multiplicity in $\chi_n^*(A)$, i.e., $m_{\lambda,\mu} \neq 0$. But then, by Lemma 11.2.5 we get $|\lambda| \leq \gamma$ and by Lemma 11.2.4 we get $|\mu| - \mu_1 \leq \beta$.

Let T_λ and T_μ be any two tableaux of shape λ and μ, respectively. For any polynomial $f \in e_{T_\lambda} e_{T_\mu} P_{r,n-r}$ we denote by g the polynomial obtained from f by identifying all the skew variables corresponding to the first row of μ (denote this variable by z). Recall that since char $F = 0$, $f \equiv 0$ and $g \equiv 0$ are equivalent *-identities; it follows that in order to prove that f vanishes in A it is enough to show that g does also.

Since $n = |\lambda| + |\mu| > \delta = d(\beta + \gamma + 1)$ and $n - \mu_1 = |\lambda| + |\mu| - \mu_1 \leq \beta + \gamma$, then $\mu_1 \geq (d-1)(\beta + \gamma + 1) + 1$. From this it follows that every monomial of g contains z^d as a submonomial. Since z^d vanishes in A, we get that $g \in \text{Id}^*(A)$ and, so, $m_{\lambda,\mu} = 0$, a contradiction. This completes the proof of the theorem. □

As an immediate consequence we get the following.

PROPOSITION 11.2.7. *The variety $*\text{-var}(D)$ has almost polynomial growth and $*\text{-exp}(D) = 2$.*

PROOF. Let $\mathcal{U}$ be a proper subvariety of $*$-var(D). By hypothesis, for any λ and μ, if $m_{\lambda,\mu}$ is the multiplicity of $\chi_{\lambda,\mu}$ in $\chi_n^*(\mathcal{U})$ and $m'_{\lambda,\mu}$ is the multiplicity of $\chi_{\lambda,\mu}$ in $\chi_n^*(D)$, then $m_{\lambda,\mu} \leq m'_{\lambda,\mu}$. Hence by Lemma 11.2.1,

$$\chi_n^*(\mathcal{U}) = \sum_{r=0}^{n} m_r \chi_{(r),(n-r)},$$

where $m_r \in \{0,1\}$.

Since $\mathcal{U}$ is a proper subvariety of $*$-var(D), by Lemma 11.2.2, $z^d \in \mathrm{Id}^*(\mathcal{U})$ for some $d \geq 1$. Also, since all symmetric and skew variables commute modulo $\mathrm{Id}^*(D)$, it follows that $m_r = 0$ for any r such that $n - r \geq d$. This means that the nth $*$-cocharacter of the $*$-variety $\mathcal{U}$ satisfies one of the equivalent conditions of Theorem 11.2.6. Hence $c_n^*(\mathcal{U})$ is polynomially bounded and the proposition is proved. $\square$

We remark that the above results hold with the due changes for the algebra D viewed as a superalgebra. In particular, in the proof of Proposition 11.2.7 one should invoke Theorem 10.5.7, instead of Theorem 11.2.6. We summarize the corresponding result in the following.

PROPOSITION 11.2.8 ([**GM1**], [**GM2**]). *For the superalgebra $D = F \oplus cF$ the following holds.*

1) *For every $n \geq 1$,*

$$\chi_n^{gr}(D) = \sum_{r=0}^{n} \chi_{(r),(n-r)}.$$

2) *Let $\mathcal{V}$ be a supervariety. Then $D \notin \mathcal{V}$ if and only if $z^d \in \mathrm{Id}^{gr}(\mathcal{V})$, for some $d \geq 1$, where z is an odd variable.*

3) *The variety* supvar(D) *has almost polynomial growth and* supexp$(D) = 2$.

We conclude this section with another example of algebra with involution. Define the following subalgebra of $M_4(F)$,

$$M = F(e_{11} + e_{44}) \oplus Fe_{12} \oplus F(e_{22} + e_{33}) \oplus Fe_{34}$$

endowed with the involution $*$ obtained by reflecting a matrix along its secondary diagonal, i.e.,

$$\begin{pmatrix} u & r & 0 & 0 \\ 0 & s & 0 & 0 \\ 0 & 0 & s & v \\ 0 & 0 & 0 & u \end{pmatrix}^* = \begin{pmatrix} u & v & 0 & 0 \\ 0 & s & 0 & 0 \\ 0 & 0 & s & r \\ 0 & 0 & 0 & u \end{pmatrix},$$

for some $u, r, s, v \in F$.

Clearly M is a 4-dimensional algebra and if we set $e_{11} + e_{44} = a$, $e_{22} + e_{33} = b$, $e_{12} = c$, $e_{34} = c^*$, then $M = \mathrm{span}_F\{a, b, c, c^*\}$ with the following multiplication table

	a	b	c	c^*
a	a	0	c	0
b	0	b	0	c^*
c	0	c	0	0
c^*	c^*	0	0	0

Hence $M^+ = \text{span}\{a, b, c + c^*\}$ and $M^- = \text{span}\{c - c^*\}$. Also, it is easily checked that $z_1 z_2 \equiv 0$ is a $*$-identity of M.

Next we state the two main results about the algebra M. We refer the reader to the original paper of Mishchenko and Valenti for a detailed proof. We remark that since $\dim M < \infty$, by Corollary 10.8.5, $*$-$\exp(M) = 2$.

THEOREM 11.2.9 ([**MV**]). *Let*
$$\chi_n^*(M) = \sum_{|\lambda|+|\mu| \vdash n} m_{\lambda,\mu} \chi_{\lambda,\mu}$$
be the nth $*$-cocharacter of the algebra with involution M. Then $m_{\lambda,\mu} = q + 1$ if either

1) $\lambda = (p+q, p), \mu = (1),$ for all $p \geq 0, q \geq 0,$ or
2) $\lambda = (p+q, p), \mu = \emptyset$ for all $p \geq 1, q \geq 0,$ or
3) $\lambda = (p+q, p, 1), \mu = \emptyset$ for all $p \geq 1, q \geq 0.$

In all other cases $m_{\lambda,\mu} = 0$, except the case $m_{(n),\emptyset} = 1$.

THEOREM 11.2.10 ([**MV**]). *The variety $*$-$\text{var}(M)$ has almost polynomial growth and $*$-$\exp(M) = 2$. Moreover, the identity $z_1 z_2 \equiv 0$ is a basis of $\text{Id}^*(M)$.*

11.3. Supervarieties of almost polynomial growth

In this section we present a list of superalgebras generating corresponding varieties of almost polynomial growth. Later in this chapter we shall prove that they generate, together with the superalgebra D, the only supervarieties of almost polynomial growth.

Recall that if A is any PI-algebra, then A can be viewed as a superalgebra with trivial grading, i.e., $A = A^{(0)} \oplus A^{(1)}$ where $A^{(0)} = A$ and $A^{(1)} = 0$. Thus, $\text{var}(A)$ can be viewed as the supervariety defined by all the identities of the algebra A in even variables and by the identity $z \equiv 0$ with z odd.

Let us consider the infinite dimensional Grassmann algebra G with its natural $\mathbb{Z}_2$-grading $G = G^{(0)} \oplus G^{(1)}$, and with the trivial grading $G = G^{(0)}$. In order to distinguish between the two, we make the following.

DEFINITION 11.3.1. G^{gr} denotes the Grassmann algebra with its natural $\mathbb{Z}_2$-grading and G denotes the Grassmann algebra with trivial grading.

Another useful example is the algebra UT_2, of 2×2 upper triangular matrices over the field F. In this case it can be easily proved (see [**V**]) that UT_2 can have only two possible $\mathbb{Z}_2$-gradings: the natural grading $UT_2 = (UT_2)^{(0)} \oplus (UT_2)^{(1)}$ where $(UT_2)^{(0)} = Fe_{11} + Fe_{22}$ is the subspace of diagonal matrices, and $(UT_2)^{(1)} = Fe_{12}$, and the trivial grading $UT_2 = (UT_2)^{(0)}$. We then set the following.

DEFINITION 11.3.2. UT_2^{gr} denotes the algebra UT_2 with its natural $\mathbb{Z}_2$-grading and UT_2 denotes the algebra UT_2 with trivial grading.

We start by examining the structure of the superidentities of the above four algebras. The first result concerns G and UT_2 with trivial grading.

PROPOSITION 11.3.3. *If G and UT_2 are endowed with the trivial $\mathbb{Z}_2$-grading, then $\text{supvar}(G)$ and $\text{supvar}(UT_2)$ are varieties with almost polynomial growth and $\text{supexp}(G) = \text{supexp}(UT_2) = 2$.*

PROOF. Let A denote the algebra G or the algebra UT_2. Since $z_1 \in \mathrm{Id}^{gr}(A)$, then $P_{r,n-r}(A) = 0$ for all $n \geq 1$ and $r \neq n$. This says that we are dealing with ordinary (not graded) identities. Then the conclusion now follows from 4.1.8 and 4.1.5. Clearly in this case the superexponent and the ordinary exponent coincide. □

Now we treat the case of G^{gr}. Recall that if $r \geq 0$, $\chi_{r,n-r}(G^{gr})$ is the $S_r \times S_{n-r}$-cocharacter of the algebra G^{gr}. Also, if $f_1, \ldots, f_n \in F\langle Y, Z\rangle$, we write $\langle f_1, \ldots, f_n\rangle_{T_2}$ for the T_2-ideal of $F\langle Y, Z\rangle$ generated by the polynomials $f_1, \ldots, f_n$.

In the next proposition we describe $\chi_{r,n-r}(G^{gr})$ and we exhibit a set of generators for $\mathrm{Id}^{gr}(G^{gr})$.

PROPOSITION 11.3.4. *Let $G^{gr} = G^{(0)} \oplus G^{(1)}$. Then, for every $r \geq 0$,*

$$\chi_{r,n-r}(G^{gr}) = \chi_{(r)} \otimes \chi_{(1^{n-r})}$$

and

$$\mathrm{Id}^{gr}(G^{gr}) = \langle [y_1, y_2], z_1 z_2 + z_2 z_1, [y_1, z_1]\rangle_{T_2}.$$

PROOF. Let $I = \langle [y_1, y_2], z_1 z_2 + z_2 z_1, [y_1, z_1]\rangle_{T_2}$ and let $\mathcal{U}$ be the variety of superalgebras determined by the ideal I. Clearly $G^{gr} \in \mathcal{U}$. Also, it is easy to see that for any $n \geq 1$ and $r \geq 0$ the space $P_{r,n-r}$ is generated, modulo $I \cap P_{r,n-r}$, by the monomial $y_1 \cdots y_r z_1 \cdots z_{n-r}$. Since this monomial does not vanish in G^{gr} for any $r \geq 0$, we get that $I = \mathrm{Id}^{gr}(G^{gr})$ and $\dim P_{r,n-r}(\mathcal{U}) = 1$.

It follows that $P_{r,n-r}(\mathcal{U})$, as an $S_r \times S_{n-r}$-module, has character $\chi_{(r)} \otimes \chi_{(1^{n-r})}$. □

In the next lemma we deal with proper subvarieties of $\mathrm{supvar}(G^{gr})$.

LEMMA 11.3.5. *Let $\mathcal{U}$ be a subvariety of $\mathrm{supvar}(G^{gr})$. Then $\mathcal{U}$ is a proper subvariety if and only if there exists $N \geq 1$ such that $z_1 \cdots z_N \in \mathrm{Id}^{gr}(\mathcal{U})$.*

PROOF. Since $z_1 \cdots z_N \notin \mathrm{Id}^{gr}(G^{gr})$ for all $N \geq 1$, one direction is obvious. Now let $\mathcal{U}$ be a proper subvariety of $\mathrm{supvar}(G^{gr})$. Then $\mathcal{U}$ satisfies some additional multilinear identity $f = f(y_1, \ldots, y_r, z_1, \ldots, z_{n-r})$. By the previous proposition, f is a monomial $y_1 \cdots y_r z_1 \cdots z_{n-r}$. If we make the substitution $y_1 = z_{n-r+1} z_{n-r+2}, \ldots, y_r = z_{n+r-1} z_{n+r}$, we get the desired conclusion. □

Now we prove the following.

REMARK 11.3.6. If $g(z_1, \ldots, z_N) \in P_{0,n}(G^{gr})$, then

$$\sum_{\sigma \in S_n} (\mathrm{sgn}\,\sigma) g(z_{\sigma(1)}, \ldots, z_{\sigma(n)}) = 0$$

in the free algebra $F\langle Y, Z\rangle$.

PROOF. Since $\chi_{0,n}(G^{gr}) = \emptyset \otimes \chi_{(1^n)}$, by complete reducibility we have that the S_n-module $P_{0,n} \cap \mathrm{Id}^{gr}(G^{gr})$ has character $\chi = \sum_{\substack{\lambda \vdash n \\ \lambda \neq (1^n)}} d_\lambda \chi_\lambda$, where $d_\lambda = \deg \chi_\lambda$. This says that if $g \in P_{0,n} \cap \mathrm{Id}^{gr}(G^{gr})$, then $e_{T_{(1^n)}} \cdot g(z_1, \ldots, z_n) = 0$. □

As a consequence of the previous lemmas we can now characterize the varieties not containing G^{gr} in terms of graded identities and we can prove that $\mathrm{supvar}(G^{gr})$ has almost polynomial growth ([**GMZ2**]).

COROLLARY 11.3.7. *Let $\mathcal{V}$ be a variety of superalgebras. Then $G^{gr} \notin \mathcal{V}$ if and only if $St_N(z_1, \ldots, z_N) \in \mathrm{Id}^{gr}(\mathcal{V})$ for some $N \geq 1$.*

PROOF. Let $St_N(z_1, \ldots, z_N) \in \mathrm{Id}^{gr}(\mathcal{V})$. Since $St_N(z_1, \ldots, z_N) \notin \mathrm{Id}^{gr}(G^{gr})$, then $G^{gr} \notin \mathcal{V}$.

Suppose now $G^{gr} \notin \mathcal{V}$. Then $\mathcal{V} \cap \mathrm{supvar}(G^{gr}) \subsetneq \mathrm{supvar}(G^{gr})$ and, by the previous lemma there exists $N \geq 1$ such that
$$z_1 \cdots z_N \in \mathrm{Id}^{gr}(\mathcal{V} \cap \mathrm{supvar}(G^{gr})) = \mathrm{Id}^{gr}(\mathcal{V}) + \mathrm{Id}^{gr}(G^{gr}).$$
It follows that there exists $g \in \mathrm{Id}^{gr}(G^{gr})$ such that $z_1 \cdots z_N + g \in \mathrm{Id}^{gr}(\mathcal{V})$. Also, by the multihomogeneity of T_2-ideals we may assume that $g = g(z_1, \ldots, z_N)$.

Now if we alternate $z_1 \cdots z_N + g$ with respect to the variables $z_1, \ldots, z_N$ and we apply the previous remark, we get
$$\sum_{\sigma \in S_N} (\mathrm{sgn}\,\sigma) z_{\sigma(1)} \cdots z_{\sigma(N)} \in \mathrm{Id}^{gr}(\mathcal{V}).$$
□

PROPOSITION 11.3.8. *The variety $\mathrm{supvar}(G^{gr})$ has almost polynomial growth and $\mathrm{supexp}(G^{gr}) = 2$.*

PROOF. By Proposition 11.3.4, for all $n \geq 1$ and $0 \leq r \leq n$ we have that $\dim P_{r,n-r}(G^{gr}) = \dim P_{r,n-r}/(P_{r,n-r} \cap Id^{gr}(G^{gr})) = 1$. Hence, since
$$c_n^{sup}(G^{gr}) = \sum_{r=0}^n \binom{n}{r} \dim P_{r,n-r}(G^{gr}),$$
we obtain $c_n^{sup}(G^{gr}) = \sum_{r=0}^n \binom{n}{r} = 2^n$ and $\mathrm{supvar}(G^{gr})$ has exponential growth.

Let $\mathcal{U}$ be a proper subvariety of $\mathrm{supvar}(G^{gr})$. By Lemma 11.3.5, $z_1 \cdots z_N \in \mathrm{Id}^{gr}(\mathcal{U})$, for some $N \geq 1$. Since $[y_1, z_1] \in \mathrm{Id}^{gr}(G^{gr})$, it follows that $P_{r,n-r} \subseteq \mathrm{Id}^{gr}(\mathcal{U})$ as soon as $n - r \geq N$. Hence, for all n, we have
$$c_n^{sup}(\mathcal{U}) = \sum_{r=0}^n \binom{n}{r} (\dim_F P_{r,n-r}) \leq \sum_{n-r<N} \binom{n}{r} \leq \alpha n^N,$$
for some constant $\alpha \geq 1$. □

Now we consider the superalgebra $(UT_2)^{gr}$. We start with the following.

REMARK 11.3.9. $\mathrm{Id}^{gr}(UT_2^{gr}) = \langle [y_1, y_2], z_1 z_2 \rangle_{T_2}$.

PROOF. Write $I = \langle z_1 z_2, [y_1, y_2] \rangle_{T_2}$. We first claim that for any variable $x = y + z$, $z_1 x z_2 \in \langle z_1 z_2, [y_1, y_2] \rangle_{T_2}$.

In fact, since $z_1 y$ has homogeneous degree 1, it follows that $z_1 y z_2 \in \langle z_1 z_2 \rangle_{T_2}$. Hence $z_1(y+z)z_2, \in \langle z_1 z_2 \rangle_{T_2}$ and, so, $z_1 x z_2 \in \langle z_1 z_2 \rangle_{T_2}$.

Now let $f(y_1, \ldots, y_t, z_1, \ldots, z_t)$ be a multilinear polynomial in $\mathrm{Id}^{gr}(A)$. We wish to show that, modulo I, f is the zero polynomial. From the above it is clear that we can write
$$f(y_1, \ldots, y_t, z_1, \ldots, z_t) = f_1(y_1, \ldots, y_s) + f_2(z, y_1, \ldots, y_s) \pmod{I}$$

and, by the multihomogeneity of T_2-ideals, it follows that f_1 and f_2 are both identities of A. Since $[y_1, y_2] \in I$, we obtain that $f_1 = \alpha y_1 \cdots y_s$. But then, by substituting $y_1 = \ldots = y_s = e_{11}$ we obtain $\alpha = 0$ and, so, $f_1 = 0$ (mod I). Write

$$f_2 = \sum \alpha y_{i_1} \cdots y_{i_t} z y_{j_1} \cdots y_{j_{n-t}},$$

where $i_1 < \ldots < i_t$ and $j_1 < \ldots < j_{n-t}$. Fix one non-zero monomial of f_2, let it be $\alpha y_1 \cdots y_s z y_{s+1} \cdots y_n$. By substituting $y_1 = \ldots = y_s = e_{11}, y_{s+1} = \ldots = y_n = e_{22}, z = e_{12}$ we get $f = \alpha e_{12}$. Hence $\alpha = 0$, a contradiction. It follows that $f_2 = 0$ (mod I) and we are done. $\square$

Next we record a result of Valenti giving all multiplicities of the graded cocharacter of the superalgebra $(UT_2)^{gr}$. She also proved that $(UT_2)^{gr}$ generates a variety of almost polynomial growth. We remark that in light of Corollary 10.8.5, we already know that supexp(UT_2^{gr}) = 2.

PROPOSITION 11.3.10 ([**V**]). *The variety* supvar(UT_2^{gr}) *has almost polynomial growth and* supexp(UT_2^{gr}) = 2. *Moreover, if*

$$\chi_n^{gr}(UT_2^{gr}) = \sum_{|\lambda|+|\mu|=n} m_{\lambda,\mu} \chi_{\lambda,\mu}$$

is its nth graded cocharacter, we have
1) $m_{\lambda,(1)} = q + 1$, *if* $\lambda = (p+q, p)$,
2) $m_{(n),\emptyset} = 1$ *and*
3) $m_{\lambda,\mu} = 0$, *in all other cases.*

11.4. Capelli identities on superalgebras

In this section we shall characterize the supervarieties satisfying a Capelli identity. The next lemma allows us to glue Capelli and standard identities on a superalgebra.

LEMMA 11.4.1. *Let* $A = A^{(0)} \oplus A^{(1)}$ *be a superalgebra. If* $A^{(0)}$ *satisfies a Capelli identity and* $A^{(1)}$ *satisfies a standard identity, then* A *satisfies a Capelli identity.*

PROOF. Suppose that $A^{(0)}$ satisfies the Capelli identity of rank m and let $St_m(z_1, \ldots, z_m) \in \mathrm{Id}^{gr}(A)$. Set $M = 4m^2$. We claim that A satisfies the standard identity of degree M.

In fact, it is clear from the linearity of the standard polynomial that it is enough to show that $St_M(y_1, \ldots, y_r, z_1, \ldots, z_{M-r}) \in \mathrm{Id}^{gr}(A)$ for all $0 \le r \le M$. Suppose first that $r < 4m$. Then every monomial in $St_M(y_1, \ldots, y_r, z_1, \ldots, z_{M-r})$ is of the form

$$w_1 y_{i_1} w_2 y_{i_2} \cdots w_r y_{i_r} w_{r+1},$$

where $w_1, \ldots, w_{r+1}$ are monomials in odd variables and, since $r < 4m$, there exists $j \in \{1, \ldots, r+1\}$ such that $|w_j| \ge m$. But then, since $St_m(z_1, \ldots, z_m) \in \mathrm{Id}^{gr}(A)$, we get that $St_M(y_1, \ldots, y_r, z_1, \ldots, z_{M-r}) \in \mathrm{Id}^{gr}(A)$, the desired conclusion.

Suppose now that $r \ge 4m$ and write $St_M(y_1, \ldots, y_r, z_1, \ldots, z_{M-r})$ as a linear combination (with coefficients ± 1) of polynomials of the type

$$f_j = \sum_{\sigma \in S_r} (\mathrm{sgn}\, \sigma) y_{\sigma(1)} \cdots y_{\sigma(j_1)} z_{i_1} y_{\sigma(j_1+1)} \cdots y_{\sigma(j_1+j_2)} z_{i_2} \cdots$$
$$\cdots z_{i_{M-r}} y_{\sigma(j_1+\cdots+j_{M-r-1}+1)} \cdots y_{\sigma(j_1+\cdots+j_{M-r})},$$

where $j_1, \ldots, j_{M-r} \in \{1, \ldots, r\}$ and $j_1 + \cdots + j_{M-r} = r$.

Since
$$\sum_{k=1}^{[\frac{M-r-2}{2}]} j_{2k+1} + \sum_{k=1}^{[\frac{M-r-1}{2}]} j_{2k} + j_1 + j_{M-r} = r \geq 4m,$$

we must have that at least one of the summands on the left-hand side is greater or equal to m.

If either $j_1 \geq m$ or $j_{M-r} \geq m$, then $f_j \in \mathrm{Id}^{gr}(A)$, since $A^{(0)}$ satisfies the Capelli identity of rank m. Suppose that $\sum_{k=1}^{[\frac{M-r-1}{2}]} j_{2k} = N > m$. In this case we write f_j as a linear combination of polynomials of the form

$$f'_j = a_0 \sum_{\tau \in S_N} y_{\tau(j_1+1)} \cdots y_{\tau(j_1+j_2)} a_1 y_{\tau(j_1+j_2+j_3+1)} \cdots y_{\tau(j_1+\cdots+j_4)} a_2 \cdots,$$

where τ acts on the set $\{j_1+1, \ldots j_1+j_2, j_1+j_2+j_3+1, \ldots j_1+\cdots+j_4, \ldots\}$
$a_0 = y_{p_1} \cdots y_{p_{j_1}} z_{i_1}$ and $a_s = z_{i_{2s}} y_{q_1} \cdots y_{q_t} z_{i_{2s+1}}$ where $t = j_{2s+1}$ for $s \geq 1$.

Notice that $a_1, a_2, \ldots$ are of homogeneous degree zero. Hence, since $A^{(0)}$ satisfies the Capelli identity of rank m and τ acts on $N > m$ indices, we get that $f'_j \in \mathrm{Id}^{gr}(A)$. Thus $f_j \in \mathrm{Id}^{gr}(A)$ and we are done in this case.

If $\sum_{k=1}^{[\frac{M-r-2}{2}]} j_{2k+1} > m$ the proof is similar. In this case we gather together the monomials of the type $z_{i_{2s+1}} y_{q_1} \cdots y_{q_t} z_{i_{2s+2}}$, where $t = j_{2s}$ for $s \geq 0$, and we get the claim.

We have proved that A satisfies the standard identity of degree M. By Theorem 7.1.4, A satisfies the Capelli identity of some rank. □

THEOREM 11.4.2. *Let A be a superalgebra. Then $G, G^{gr} \notin \mathrm{supvar}(A)$ if and only if A satisfies a Capelli identity.*

PROOF. Let $A = A^{(0)} \oplus A^{(1)}$ and suppose that $G, G^{gr} \notin \mathrm{supvar}(A)$. Since $G \notin \mathrm{supvar}(A)$, then $G \notin \mathrm{var}(A^{(0)})$. By Theorem 7.1.4, $A^{(0)}$ satisfies a Capelli identity.

Also, since $G^{gr} \notin \mathrm{supvar}(A)$, by Corollary 11.3.7, we get that $St_m(z_1, \ldots, z_m) \in \mathrm{Id}^{gr}(A)$, for some $m \geq 1$. The conclusion now follows from the previous lemma. The converse holds since G and G^{gr} do not satisfy any Capelli identity of any rank. □

The following result generalizes 3) → 2) of Theorem 7.1.4.

THEOREM 11.4.3. *Let $\mathcal{V}$ be a variety of superalgebras. If $\mathcal{V}$ satisfies the Capelli identity of some rank, then $\mathcal{V} = \mathrm{supvar}(A)$, for some finitely generated superalgebra A.*

PROOF. Suppose that $\mathcal{V}$ satisfies $Cap_{d+1} \equiv 0$ and denote by A the relatively free algebra of the variety $\mathcal{V}$ with d even and d odd generators. We shall prove that $\mathcal{V} = \mathrm{supvar}(A)$, i.e., any graded identity $f \equiv 0$ of A is a graded identity of $\mathcal{V}$.

Since $\mathrm{char}\, F = 0$, we may assume that f is multilinear and let
$$f = f(y_1, \ldots, y_r, z_1, \ldots, z_{n-r})$$

with $y_1, \ldots, y_r$ even variables and $z_1, \ldots, z_{n-r}$ odd variables. Let M be the $S_r \times S_{n-r}$-module generated by f and let $M = M_1 \oplus \cdots \oplus M_m$ be its decomposition into irreducible components. Let M_i be generated by f_i as an $S_r \times S_{n-r}$-module, $1 \leq i \leq m$. If $f_i \equiv 0$ is an identity of $\mathcal{V}$ for all $i = 1, \ldots, m$, then also $f \equiv 0$

is an identity of $\mathcal{V}$. Hence, without loss of generality, we may assume that M is irreducible.

Let $\chi_\lambda \otimes \chi_\mu$ be the character of M, where $\lambda \vdash r, \mu \vdash n-r$. Let $e_{T_\lambda} = \left(\sum_{\tau \in R_{T_\lambda}} \tau\right)\left(\sum_{\sigma \in C_{T_\lambda}} (\operatorname{sgn} \sigma)\sigma\right)$ and $e_{T_\mu} = \left(\sum_{\tau \in R_{T_\mu}} \tau\right)\left(\sum_{\sigma \in C_{T_\mu}} (\operatorname{sgn} \sigma)\sigma\right)$ be corresponding essential idempotents. Then

$$h = h(y_1, \ldots, y_r, z_1, \ldots, z_{n-r}) = \left(\sum_{\tau \in C_{T_\lambda}} (\operatorname{sgn} \tau)\tau\right)\left(\sum_{\tau' \in C_{T_\mu}} (\operatorname{sgn} \tau')\tau'\right) f$$

and

$$g = g(y_1, \ldots, y_r, z_1, \ldots, z_{n-r}) = \left(\sum_{\tau \in R_{T_\lambda}} \tau\right)\left(\sum_{\tau' \in R_{T_\mu}} \tau'\right) f$$

are graded identities of $\mathcal{V}$ if and only if f is a graded identity of $\mathcal{V}$ since M is irreducible and h, g are non-zero by Lemma 2.5.1 and Lemma 2.5.2.

Suppose first that $\lambda_{d+1} \neq 0$, i.e., λ does not lie in the strip $H(d, 0)$ of width d. Then h is alternating on at most $d+1$ even variables, so $h \equiv 0$ is an identity of $\mathcal{V}$ since $\mathcal{V}$ satisfies $Cap_{d+1} \equiv 0$. Similarly, we obtain that $h \equiv 0$ is an identity of $\mathcal{V}$ if $\mu_{d+1} \neq 0$.

Therefore, in order to finish the proof, we may assume that λ and μ lie in the strip $H(d, 0)$. In this case g is symmetric on at most d disjoint subsets $Y_1, \ldots, Y_d$ of $\{y_1, y_2, \ldots\}$, and is symmetric on at most d disjoint subsets $Z_1, \ldots, Z_d$ of $\{z_1, z_2, \ldots\}$.

If we identify all variables of Y_1 with y_1, all variables of Z_1 with z_1, all variables of Y_2 with y_2, and so on, we obtain a homogeneous polynomial $t = t(y_1, \ldots, y_d, z_1, \ldots, z_d)$ which is still a graded identity of A. From the definition of relatively free algebra, it follows that for any superalgebra $R = R^{(0)} \oplus R^{(1)} \in \mathcal{V}$, $t(a_1, \ldots, a_d, b_1, \ldots, b_d) = 0$, for any $a_1, \ldots, a_d \in R^{(0)}$, $b_1, \ldots, b_d \in R^{(0)}$. That is, $t \equiv 0$ is a graded identity of $\mathcal{V}$. But the complete linearization of t on all even and odd variables is equal to $\gamma g(y_1, \ldots, y_r, z_1, \ldots, z_{n-r})$ where

$$\gamma = \lambda_1! \cdots \lambda_d! \mu_1! \cdots \mu_d! \neq 0.$$

Hence $g \equiv 0$ and $f \equiv 0$ are identities of $\mathcal{V}$ and the proof is complete. $\square$

11.5. Superalgebras and polynomial growth

In this section we shall prove that the five superalgebras described in Section 11.3 characterize supervarieties of polynomial growth. Recall that for an algebra A, the commutator ideal of A is the ideal generated by $[A, A]$.

PROPOSITION 11.5.1. *Let A be a superalgebra. If G, G^{gr}, $F \oplus cF$, $UT_2 \notin \operatorname{supvar}(A)$, then the commutator ideal of A and the ideal generated by $A^{(1)}$ are both nilpotent.*

PROOF. Suppose that $G, G^{gr}, F \oplus cF \notin \operatorname{supvar}(A)$. By Theorem 11.4.2, A satisfies the Capelli identity of some rank. Hence by Theorem 11.4.3, $\operatorname{supvar}(A) = \operatorname{supvar}(B)$, for some finitely generated superalgebra B. Also, by Theorem 4.8.3 $\operatorname{supvar}(B) = \operatorname{supvar}(C)$, for some finite dimensional superalgebra C. Hence we may assume that A itself is finite dimensional over F.

Write $A = A_1 \oplus \cdots \oplus A_t + J$ where $A_1, \ldots, A_t$ are $\mathbb{Z}_2$-graded simple algebras over F and $J = J(A)$ is the Jacobson radical. By extending the scalars we may assume that F is algebraically closed. Hence, since z^k is a graded identity for A_i, for all i, from the classification of the $\mathbb{Z}_2$-graded simple algebras (see Chapter 3) it follows that A_i must have trivial grading, i.e., $A_i = A_i^{(0)}$ for all $i = 1, \ldots, t$. Hence $A_1 \oplus \cdots \oplus A_t \subseteq A^{(0)}$, so $A^{(1)} \subseteq J$. It follows that the two-sided ideal generated by $A^{(1)}$ is nilpotent. Moreover, $A_1, \ldots, A_t$ are commutative algebras since $UT_2 \notin \mathcal{V}$. Hence $[A, A] \subseteq J$ and also the commutator ideal of A is nilpotent. $\square$

The following theorem characterizes supervarieties of polynomial growth.

THEOREM 11.5.2 ([**GMZ2**]). *Let $\mathcal{V}$ be a variety of superalgebras. Then $\mathcal{V}$ has polynomial growth if and only if $G, UT_2, G^{gr}, UT_2^{gr}, F \oplus cF \notin \mathcal{V}$.*

PROOF. Since each of the above algebras generates a variety of exponential growth one direction is obvious.

Suppose that $G, UT_2, G^{gr}, UT_2^{gr}, F \oplus tF \notin \mathcal{V}$. If $z \equiv 0$ is an identity for $\mathcal{V}$, then $\mathcal{V}$ is an ordinary variety of associative algebras and $c_n^{sup}(\mathcal{V}) = c_n(\mathcal{V})$ is the nth codimension of $\mathcal{V}$. Since $G, UT_2 \notin \mathcal{V}$, then by Theorem 7.2.5, $\mathcal{V}$ has polynomial growth.

Suppose now that $z \notin \mathrm{Id}^{gr}(\mathcal{V})$ and let $\mathcal{V} = \mathrm{supvar}(A)$, where $A = A^{(0)} \oplus A^{(1)}$. Since the codimensions are not affected by any extension of the scalars, we may assume that F is algebraically closed. Hence, in order to prove that $\mathcal{V}$ has polynomial growth we need only to prove that A verifies conditions 1), 2), and 3) of Theorem 10.5.2. Since $G, G^{gr}, F \oplus cF, UT_2 \notin \mathcal{V}$, by Proposition 11.5.1 conditions 1) and 3) are clearly satisfied. Next we prove that 2) holds.

Since $UT_2^{gr} \notin \mathcal{V}$, then $\mathcal{W} = \mathcal{V} \cap \mathrm{supvar}(UT_2^{gr})$ is a proper subvariety of $\mathrm{supvar}(UT_2^{gr})$. By Proposition 11.5.1, $\mathcal{W}$ has polynomial growth hence, by Theorem 10.5.2, $f_r(y_1, \ldots, y_{2r}, x)$ is an identity for $\mathcal{W}$ for some $r \geq 1$ where x is either an even or a odd variable. It follows that A satisfies the identity $f_r(y_1, \ldots, y_{2r}, x)$ modulo $\mathrm{Id}^{gr}(UT_2^{gr})$. By Remark 11.3.9, the polynomials $[y_1, y_2]$ and $z_1 z_2$ are a basis of the graded identities of UT_2^{gr}. Hence, we can write

$$(11.5) \qquad f_r(y_1, \ldots, y_{2r}, x) \equiv \sum_{i,j} a_{ij}[y_i, y_j]b_{ij} \pmod{\mathrm{Id}^{gr}(A)}$$

where a_{ij}, b_{ij} are polynomials in the set of variables $\{x, y_1, \ldots, y_{2r}\} \setminus \{y_i, y_j\}$. Now define the polynomial

$$g_t(y_1, \ldots, y_{2t}, x) = \sum_{\tau \in H_t} (\mathrm{sgn}\,\tau) y_{\tau(1)} y_{\tau(3)} \cdots y_{\tau(2t-1)} x y_{\tau(2t)} \cdots y_{\tau(4)} y_{\tau(2)}$$

where H_t is the subgroup of S_{2t} generated by the transpositions $(12), (34), \ldots, (2t-1\ 2t)$.

From (11.5) and the obvious properties of commutators it follows that

$$(11.6) \qquad g_r(y_1, \ldots, y_{2r}, x) \equiv \sum_{i,j} a'_{ij}[y_i, y_j]b'_{ij} \pmod{\mathrm{Id}^{gr}(A)}$$

where $a'_{ij} b'_{ij}$ are still polynomials in $\{x, y_1, \ldots, y_{2r}\} \setminus \{y_i, y_j\}$.

Notice that the polynomials g_t satisfy the recurrence relation

$$(11.7) \qquad g_{t+s}(y_1, \ldots, y_{2(t+s)}, x)$$
$$= \sum_{\tau \in H_t} (\mathrm{sgn}\,\tau) y_{\tau(1)} y_{\tau(3)} \cdots y_{\tau(2t-1)} g_s(y_{2t+1}, \ldots, y_{2t+2s}, x) y_{\tau(2t)} \cdots y_{\tau(4)} y_{\tau(2)}.$$

Now let $m \geq 1$ be the index of nilpotence of the commutator ideal U of A. By an easy induction it follows from (11.6) and (11.7) that g_{rm} can be written (mod $\mathrm{Id}^{gr}(A)$) as a linear combination of terms each containing at least m commutators in the y_i's. Since such terms evaluate to zero in A we get that $g_{rm} \in \mathrm{Id}^{gr}(A)$.

Now take $t = rm^2 + 2(m-1)$. By recalling that $U^m = 0$, we obtain that $y_1 \cdots y_t$ can be written (mod $\mathrm{Id}^{gr}(A)$) as a linear combination of products of the type

$$y_{i_1+j_1} \cdots y_{i_1+1} y_{i_1} w_1 y_{i_2+j_2} \cdots y_{i_2+1} y_{i_2} w_2 \cdots w_{k-1} y_{i_k+j_k} \cdots y_{i_k+1} y_{i_k}$$

where all the w_i's are of the form $[y_u, y_v]$, $k-1 \leq m-1$ and $j_a \geq rm$ for some $a \geq 1$. Since $g_{rm} \in \mathrm{Id}^{gr}(A)$, this implies that $f_t \in \mathrm{Id}^{gr}(A)$ and the proof is complete. □

As an immediate consequence we obtain the following two corollaries.

COROLLARY 11.5.3. *There is no variety of superalgebras of intermediate growth.*

COROLLARY 11.5.4. $\mathrm{supvar}(G)$, $\mathrm{supvar}(UT_2)$, $\mathrm{supvar}(G^{gr})$, $\mathrm{supvar}(UT_2^{gr})$ *and* $\mathrm{supvar}(F \oplus cF)$ *are the only varieties of superalgebras of almost polynomial growth.*

11.6. $*$-algebras and the Nagata-Higman theorem

Throughout this section A will be an F-algebra with involution $*$. Notice that $A^- = L$ has a structure of Lie algebra under the bracket operation; we then denote with $L^{(i)}$ the $(i+1)$th term of the solvable series of L, i.e., $L^{(0)} = L$ and $L^{(i)} = [L^{(i-1)}, L^{(i-1)}]$ for $i \geq 1$.

LEMMA 11.6.1. *Let $A^- = L$ and suppose that $L^{(i_1)} \cdots L^{(i_M)} = 0$, for some $M \geq 1$ and $i_j \geq 0$. Then, for any (eventually empty) associative words $w_1, \ldots, w_{M-1}$ in skew elements, we have*

$$L^{(i_1)} w_1 L^{(i_2)} w_2 \cdots w_{M-1} L^{(i_M)} = 0.$$

PROOF. Since $L^{(i)}$ is a Lie ideal of L, $[L^{(i)}, L] \subseteq L^{(i)}$; hence for any $k \in L$, $kL^{(i)} \subseteq L^{(i)}k + L^{(i)}$. This says that for any linear combination f of monomials in elements of L, $fL^{(i)} \subseteq L^{(i)}g$ where g is still a linear combination of monomials in elements of L. It follows that $L^{(i_1)} w_1 L^{(i_2)} w_2 \cdots w_{M-1} L^{(i_M)} \subseteq L^{(i_1)} \cdots L^{(i_M)} h$ where h is still an associative polynomial evaluated in L. □

LEMMA 11.6.2. *If A satisfies the $*$-identity $z^m \equiv 0$ for some $m \geq 1$, then A^- is Lie nilpotent.*

PROOF. The Lie algebra $L = A^-$ satisfies the Engel condition $[z_1, \underbrace{z_2, \ldots, z_2}_{2m}] \equiv 0$. By a celebrated result of Zelmanov [**Z**] L is Lie nilpotent. □

LEMMA 11.6.3. *Suppose that A satisfies the $*$-identity $z^m \equiv 0$ and let $i \geq 1$. If there exists M_i such that $\underbrace{L^{(i)} \cdots L^{(i)}}_{M_i} = 0$, then $\underbrace{L^{(i-1)} \cdots L^{(i-1)}}_{M_{i-1}} \equiv 0$ where $M_{i-1} = M_i m$.*

PROOF. By Lemma 11.6.1, for every $k_{j_1}, \ldots, k_{j_m} \in L^{(i-1)}$ we have

$$\underbrace{L^{(i)} \cdots L^{(i)}}_{M_i - 1} k_{j_1} \cdots k_{j_{t-1}} [k_{j_t}, k_{j_{t+1}}] k_{j_{t+2}} \cdots k_{j_m} = 0.$$

This says that in a monomial in skew elements of length m lying on the right-hand side of $(L^{(i)})^{M_i-1}$ we can permute any two consecutive elements.

Since $z^m \equiv 0$ is a $*$-identity of A, by complete linearization we get that $\sum_{\sigma \in S_m} z_{\sigma(1)} \cdots z_{\sigma(m)} \equiv 0$ is also a $*$-identity of A. But then, by the above we can write

$$\underbrace{L^{(i)} \cdots L^{(i)}}_{M_i-1}(m! k_1 \cdots k_m) \subseteq \underbrace{L^{(i)} \cdots L^{(i)}}_{M_i-1} \sum_{\sigma \in S_m} k_{\sigma(1)} \cdots k_{\sigma(m)} = 0.$$

Since char $F = 0$, we obtain that $\underbrace{L^{(i)} \cdots L^{(i)}}_{M_i-1} \underbrace{L^{(i-1)} \cdots L^{(i-1)}}_{m} = 0$.

If we now apply Lemma 11.6.1 to the last equality, we can write

$$\underbrace{L^{(i)} \cdots L^{(i)}}_{M_i-2} k_{j_1} \cdots k_{j_{t-1}}[k_{j_t}, k_{j_{t+1}}] k_{j_{t+2}} \cdots k_{j_m} \underbrace{L^{(i-1)} \cdots L^{(i-1)}}_{m} = 0$$

for all $k_{j_s} \in L^{(i-1)}$. It is clear that a repeated application of the argument above leads to the desired conclusion. □

LEMMA 11.6.4. *If A satisfies the $*$-identity $z^m \equiv 0$ for some $m \geq 1$, then there exists $M \geq 1$ such that $z_1 z_2 \cdots z_M \equiv 0$ is a $*$-identity of A.*

PROOF. By Lemma 11.6.2, $A^- = L$ is Lie nilpotent, hence Lie solvable. Let $L^{(c)} = 0, L^{(c-1)} \neq 0$. We repeatedly apply Lemma 11.6.3 by starting with $i = c$ and $M_i = 1$. Then $M_{c-1} = m$, $M_{c-2} = m^2$ and $M_0 = m^c$. This says that $k_1 k_2 \cdots k_{m^c} = 0$ for all $k_i \in L$ and the lemma is proved. □

A well-known theorem of Nagata and Higman (see [**D10**]) states that if A is an algebra over a field of characteristic zero and A is nil of bounded exponent, then A is nilpotent. In other words, the identity $x^n \equiv 0$ for some $n \geq 1$, implies the identity $x_1 \cdots x_N \equiv 0$ for some $N \geq 1$. In the next theorem ([**GM2**]) we prove the skew analogue of the Nagata-Higman theorem for algebras with involution.

THEOREM 11.6.5. *Let A be an algebra with involution and $m \geq 1$. If for all $k \in A^-$, $k^m = 0$, then there exists $M \geq 1$ such that for all $k_1, \ldots, k_M \in A^-$,*

$$k_1 w_1 k_2 w_2 \cdots w_{M-1} k_M = 0$$

where $w_1, \ldots, w_{M-1}$ are (eventually empty) words in elements of A.

PROOF. By Lemma 11.6.4 there exists M such that $k_1 \cdots k_M = 0$ for all $k_i \in A^-$. Since for any $a \in A$, $a = s + k$, $s \in A^+, k \in A^-$, it is enough to prove the theorem for words w_i in elements of A^+ and A^-. Now, for $s \in A^+, k \in A^-$ we have that $sk + ks \in A^-$; hence $ks = -sk + k'$ for some $k' \in A^-$. This says that we can write $k_1 w_1 k_2 w_2 \cdots w_{M-1} k_M$ as a linear combination of words each containing at least M consecutive skew elements to the right-hand side. The proof is completed by recalling that the product of M skew elements of A is zero. □

We can easily get the symmetric analogue of the previous theorem.

THEOREM 11.6.6. *Let A be an algebra with involution. if $y^m \equiv 0$ is a $*$-identity for A, then there exists $N \geq 1$ such that $x_1 x_2 \cdots x_N \equiv 0$ is an ordinary identity for A.*

PROOF. Since for $k \in A^-$, $k^2 \in A^+$, we get that $k^{2m} = 0$. Also, the commutator of two symmetric elements is a skew element. Hence by the previous theorem we get that for some $M \geq 1$,

(11.8) $\qquad [s_1, s_2]w_1[s_3, s_4]w_2 \cdots w_M[s_{2M-1}, s_{2M}] = 0,$

for all $s_1, \ldots, s_{2m} \in A^+$ and $w_1, \ldots, w_M$ any words in elements of A.

Let $a, b \in A$. Next we prove that if $aw[s, s']w'b = 0$, for all $s, s' \in A^+$ and for any words w, w' in A, then $as_1 \cdots s_m b = 0$ for all $s_1, \ldots, s_m \in A^+$.

Now, by the hypothesis of the claim, for any $s_1, \ldots, s_m \in A^+$ we get that $as_{i_1} \cdots s_{i_{t-1}}[s_{i_t}, s_{i_{t+1}}]s_{i_{t+2}} \cdots s_m b = 0$ for $t = 1, \ldots, m-1$. Since $y^m \equiv 0$ is an identity of A, by complete linearization we get that $\sum_{\sigma \in S_m} s_{\sigma(1)} \cdots s_{\sigma(m)} = 0$ for all $s_1, \ldots, s_m \in A^+$. Hence, by the above,

$$0 = a \sum_{\sigma \in S_m} s_{\sigma(1)} \cdots s_{\sigma(m)} b = m! a s_1 \cdots s_m b,$$

so $as_1 \cdots s_m b = 0$ as claimed. A repeated application of this result into equality (11.8) implies that $s_1 s_2 \cdots s_{mM} = 0$ for all $s_i \in A^+$. Take $n = M^2 m + 1$.

We claim that $a_1 a_2 \cdots a_n = 0$ for all $a_i \in A$. In fact, write each a_i as a sum of a symmetric and a skew element and expand the product $a_1 \cdots a_n$. It is clearly enough to prove that $a_1 \cdots a_n = 0$ where, for each i, $a_i \in A^+$ or $a_i \in A^-$. If a product contains at least M skew elements, then by Theorem 11.6.5 it will be zero. If, on the other hand, a product contains less than M skew elements, then it must contain at least Mm consecutive symmetric elements and we will get zero also in this case since $s_1 \cdots s_{mM} = 0$ for all $s_i \in A^+$. $\square$

11.7. Polynomial growth of the $*$-codimensions

Our aim in this section is to prove that the two algebras with involution $D = F \oplus cF$ and M studied in Section 11.2 generate the only two $*$-varieties of almost polynomial growth and to characterize $*$-varieties with polynomial growth by proving that $\mathcal{V}$ is such a variety if and only if $D, M \notin \mathcal{V}$. To this end we first study $*$-varieties that do not contain D. We start with the following consequence of Theorem 11.6.5.

LEMMA 11.7.1. *Let* $\chi_n^*(A) = \sum_{|\lambda|+|\mu|=n} m_{\lambda,\mu} \chi_{\lambda,\mu}$ *be the nth $*$-cocharacter of the algebra A. If $z^m \in \mathrm{Id}^*(A)$, for some $m \geq 1$, then there exists $t \geq 1$ such that $m_{\lambda,\mu} = 0$ whenever $|\mu| \geq t$.*

PROOF. By Theorem 11.6.5 there exists $t \geq 1$ such that $z_1 w_1 z_2 w_2 \cdots w_{t-1} z_t \in \mathrm{Id}^*(A)$ where $w_1, \ldots, w_{t-1}$ are (eventually empty) words in symmetric or skew variables. This says that if $n - r \geq t$, then $P_{r, n-r} \subseteq \mathrm{Id}^*(A)$; hence for all $\lambda \vdash r$, $m_{\lambda, \mu} = 0$ as soon as $|\mu| \geq t$. $\square$

THEOREM 11.7.2. *Let $\mathcal{V}$ be a $*$-variety. Then $D \notin \mathcal{V}$ if and only if*

$$\chi_n^*(\mathcal{V}) = \sum_{\substack{|\lambda|+|\mu|=n \\ |\mu|<t}} m_{\lambda,\mu} \chi_{\lambda,\mu},$$

for some constant $t \geq 0$.

PROOF. Suppose first that $D \notin \mathcal{V}$. Then by Lemma 11.2.2 $z^m \in \mathrm{Id}^*(\mathcal{V})$, for some $m \geq 1$, and the conclusion of the first part follows from the previous lemma.

Suppose now that $D \in \mathcal{V}$ and let
$$\chi_n^*(\mathcal{V}) = \sum_{|\lambda|+|\mu|=n} m_{\lambda,\mu} \chi_{\lambda,\mu}.$$

Since by Lemma 11.2.1,
$$\chi_n^*(D) = \sum_{r=0}^n \chi_{(r),(n-r)},$$

then $1 \leq m_{(r),(n-r)}$ for all $r = 0, \ldots, n$. On the other hand, by hypothesis if $|\mu| \geq t$, then $m_{\lambda,\mu} = 0$. This leads to the contradiction $1 \leq m_{(r),(n-r)} = 0$, as soon as $n - r \geq t$. □

LEMMA 11.7.3. *Let $\mathcal{V}$ be a $*$-variety. If $D \notin \mathcal{V}$, then there exists $t \geq 1$ such that $[x_1, x_2] \cdots [x_{2t-1}, x_{2t}] \equiv 0$ is an ordinary identity of $\mathcal{V}$.*

PROOF. Since $D \notin \mathcal{V}$, by Lemma 11.2.1 $z^m \in \mathrm{Id}^*(\mathcal{V})$, for some m. If we now apply Theorem 11.7.2 above, we get that any monomial in symmetric and skew variables containing at least t skew variables must lie in $\mathrm{Id}^*(\mathcal{V})$.

Notice that in order to prove that $[x_1, x_2] \cdots [x_{2t-1}, x_{2t}] \equiv 0$ is an ordinary identity of $\mathcal{V}$, it is enough to prove that $[w_1, w_2] \cdots [w_{2t-1}, w_{2t}] \equiv 0$ is a $*$-identity of $\mathcal{V}$ where the w_i's are either symmetric or skew variables. But each commutator $[w_{2i-1}, w_{2i}]$ either evaluates to a skew element (if both w_{2i-1} and w_{2i} are symmetric variables) or contains at least one skew variable. In any case $g = [w_1, w_2] \cdots [w_{2t-1}, w_{2t}]$ evaluates to a linear combination of monomials each containing at least t skew elements. By Theorem 11.7.2, $g \in \mathrm{Id}^*(\mathcal{V})$. □

Next, we compare the $*$-identities and the ordinary identities of an algebra with involution. We start by describing the T-ideal of ordinary identities of the algebra M. Recall that by Theorem 4.1.5, for the algebra of 2×2 upper triangular matrices UT_2, we have $\mathrm{Id}(UT_2) = \langle [x_1, x_2][x_3, x_4] \rangle_T$.

REMARK 11.7.4. $\mathrm{Id}(M) = \mathrm{Id}(UT_2) = \langle [x_1, x_2][x_3, x_4] \rangle_T$.

PROOF. It is easily seen that the subalgebra of M generated by a, b and c is isomorphic to UT_2. Hence $\mathrm{Id}(M) \subseteq \mathrm{Id}(UT_2)$. Since $f - [x_1, x_2][x_3, x_4] \in \mathrm{Id}(M)$ and $\langle f \rangle_T = \mathrm{Id}(UT_2)$, we get the equality $\mathrm{Id}(M) = \mathrm{Id}(UT_2)$. □

Notice that if $\mathcal{V}$ is a $*$-variety $\mathrm{Id}(\mathcal{V}) \subseteq \mathrm{Id}^*(\mathcal{V})$, where $\mathrm{Id}(\mathcal{V})$ is the T-ideal of the free associative algebra (without involution) of ordinary identities of $\mathcal{V}$. We then denote by $\hat{\mathcal{V}}$ the variety of algebras without involution corresponding to $\mathrm{Id}(\mathcal{V})$.

LEMMA 11.7.5. *Let $\mathcal{V}$ be a $*$-variety and suppose that $M \notin \mathcal{V}$ and $D \notin \mathcal{V}$. Then there exists an ordinary polynomial f such that $f \in \mathrm{Id}(\mathcal{V})$ and $f \notin \mathrm{Id}(UT_2)$.*

PROOF. Let $\mathcal{W} = \mathcal{V} \cap *\text{-var}(M)$. Since $M \notin \mathcal{V}$, then $\mathcal{W}$ is a proper subvariety of $*\text{-var}(M)$. Hence by Theorem 11.2.10, $\mathcal{W}$ has polynomial growth. But then by Theorem 11.2.6, there exists $k \geq 0$ such that if
$$\chi_n^*(\mathcal{W}) = \sum_{|\lambda|+|\mu|=n} m_{\lambda,\mu} \chi_{\lambda,\mu}$$

is the nth $*$-cocharacter of $\mathcal{W}$, then $m_{\lambda,\mu} = 0$ whenever $|\lambda| - \lambda_1 > k$ or $|\mu| > k$. It follows that if $r > k$, then $e_{T_\lambda} e_{T_\mu} y_1 \cdots y_{2r+1} \in \mathrm{Id}^*(\mathcal{W})$, for the pair of tableaux

$$(T_\lambda, T_\mu) = \left(\begin{array}{|c|c|c|c|c|} \hline 1 & 2 & \cdots & r & r+1 \\ \hline r+2 & r+3 & \cdots & 2r+1 \\ \cline{1-4} \end{array} , \emptyset \right).$$

Also, $e_{T_\lambda} e_{T_\mu} y_1 \cdots y_r z y_{r+2} y_{2r+1} \in \mathrm{Id}^*(\mathcal{W})$, for the pair of tableaux

$$(T_\lambda, T_\mu) = \left(\begin{array}{|c|c|c|c|} \hline 1 & 2 & \cdots & r \\ \hline r+2 & r+3 & \cdots & 2r+1 \\ \hline \end{array} , \begin{array}{|c|} \hline r+1 \\ \hline \end{array} \right).$$

We deduce that

$$f(y_1, \ldots, y_{2r}, x) = \sum_{\sigma, \tau \in H} (\mathrm{sgn}\, \sigma\tau) y_{\sigma(1)} \cdots y_{\tau(r)} x y_{\tau(r+1)} \cdots y_{\sigma(2r)} \in \mathrm{Id}^*(\mathcal{W})$$

for any symmetric or skew variable x, where H is the subgroup of S_{2r} generated by the transpositions $(1, 2r), (2, 2r-1), \ldots, (r, r+1)$. We can write

$$f(y_1, \ldots, y_{2r}, x) \equiv g(y_1, \ldots, y_{2r}, x) \pmod{\mathrm{Id}^*(\mathcal{V})}$$

where $g \in \mathrm{Id}^*(M) = \langle z_1 z_2 \rangle$ is a linear combination of terms each containing at least one commutator of the form $[y_i, y_j]$. By Lemma 11.2.2 and Theorem 11.6.5, there exists t such that any monomial of P_n^* containing at least t skew variables lies in $\mathrm{Id}^*(\mathcal{V})$. Hence an easy induction shows that $f(y_1, \ldots, y_{2T}, x) \in \mathrm{Id}^*(\mathcal{V})$, for any $T \geq kt$. By writing each variable $x_i = y_i + z_i$, and by using again the same property, it follows that $f(x_1, \ldots, x_{2N}, x) \in \mathrm{Id}(\mathcal{V})$ for $N = t - 1 + t(kt - 1)$. Clearly $f \notin \mathrm{Id}(UT_2)$. $\square$

REMARK 11.7.6. Let A be an algebra with involution and $t \geq 1$. If every monomial of P_n^* containing at least t skew variables lies in $\mathrm{Id}^*(A)$, then $c_n^*(A) \leq n^t c_n(A)$.

PROOF. Let $B = \{w_i(x_1, \ldots, x_n) \mid 1 \leq i \leq p\}$ be a basis of $P_n \pmod{\mathrm{Id}(A)}$. By specializing the variables, we get that $\{w_i(y_1, \ldots, y_r, z_{r+1}, \ldots, z_n) \mid 1 \leq i \leq p\}$ generates $P_{r,n-r} \pmod{\mathrm{Id}^*(A)}$. By counting we get

$$c_n^*(A) \leq \left(\binom{n}{0} + \binom{n}{1} + \cdots + \binom{n}{t-1} \right) c_n(A) \leq n^t c_n(A).$$

$\square$

The next theorem characterizes $*$-varieties of polynomial growth.

THEOREM 11.7.7 ([**GM2**]). *Let $\mathcal{V}$ be a $*$-variety. Then $\mathcal{V}$ has polynomial growth if and only if $F \oplus cF, M \notin \mathcal{V}$.*

PROOF. Since by Proposition 11.2.7 and Theorem 11.2.9 D and M have exponential growth of the $*$-codimensions, it is clear that if $\mathcal{V}$ has polynomial growth, then $D, M \notin \mathcal{V}$.

Conversely, since $D \notin \mathcal{V}$, by Lemma 11.7.3, we get that

$$f = [x_1, x_2] \cdots [x_{2t-1}, x_{2t}] \in \mathrm{Id}(\mathcal{V}),$$

for some $t \geq 1$. But then, since f is not an identity for the Grassmann algebra G, we obtain that $G \notin \hat{\mathcal{V}}$. Also, since $M \notin \mathcal{V}$, by Lemma 11.7.5, $UT_2 \notin \hat{\mathcal{V}}$. Hence $G, UT_2 \notin \hat{\mathcal{V}}$ and by Theorem 7.2.5, $\hat{\mathcal{V}}$ has polynomial growth. Let $c_n(\mathcal{V}) \leq \alpha n^q$.

Since $D \notin \mathcal{V}$, as above, by applying Theorem 11.6.5, we get that there exists $t \geq 1$ such that every monomial of P_n^* containing at least t skew variables lies in

Id$^*(\mathcal{V})$. But then by Remark 11.7.6 we get that $c_n^*(A) \leq n^t c_n(A) \leq \alpha n^t n^q$ and $\mathcal{V}$ has polynomial growth. □

As an immediate consequence we get.

COROLLARY 11.7.8. *There is no $*$-variety of intermediate growth.*

COROLLARY 11.7.9. *$*$-var$(F \oplus cF)$ and $*$-var(M) are the only two $*$-varieties of almost polynomial growth.*

11.8. Supervarieties of exponent 2

In this section we shall characterize the supervarieties generated by a finitely generated superalgebra having superexponent equal to two. We remark that an analogous characterization of $*$-varieties generated by finite dimensional algebra with involution can be found in [**Pi**]. Recall that a characterization of supervarieties of polynomial growth was given in Section 11.5. We first need to introduce some notation.

Recall that UT_n is the algebra of $n \times n$ upper triangular matrices over the field F. We define the following four $\mathbb{Z}_2$-gradings on UT_3:

1) UT_3^{gr0} is the algebra UT_3 with trivial grading $(UT_3, \{0\})$;

2) UT_3^{gr1} is the algebra UT_3 with grading $\left(\begin{pmatrix} F & F & 0 \\ 0 & F & 0 \\ 0 & 0 & F \end{pmatrix}, \begin{pmatrix} 0 & 0 & F \\ 0 & 0 & F \\ 0 & 0 & 0 \end{pmatrix} \right)$;

3) UT_3^{gr2} is the algebra UT_3 with grading $\left(\begin{pmatrix} F & 0 & 0 \\ 0 & F & F \\ 0 & 0 & F \end{pmatrix}, \begin{pmatrix} 0 & F & F \\ 0 & 0 & 0 \\ 0 & 0 & 0 \end{pmatrix} \right)$;

4) UT_3^{gr3} is the algebra UT_3 with grading $\left(\begin{pmatrix} F & 0 & F \\ 0 & F & 0 \\ 0 & 0 & F \end{pmatrix}, \begin{pmatrix} 0 & F & 0 \\ 0 & 0 & F \\ 0 & 0 & 0 \end{pmatrix} \right)$.

When the field F is algebraically closed and char $F = 0$, in [**VZ**] the authors classified all possible G-gradings on UT_n, for G an arbitrary finite abelian group. It turns out that the above are the only possible $\mathbb{Z}_2$-gradings on UT_3.

We shall also need the following algebras:

1) $A_1 = \begin{pmatrix} F \oplus cF & F \oplus cF \\ 0 & F \end{pmatrix}$, with grading $\left(\begin{pmatrix} F & F \\ 0 & F \end{pmatrix}, \begin{pmatrix} cF & cF \\ 0 & 0 \end{pmatrix} \right)$;

2) $A_2 = \begin{pmatrix} F & F \oplus cF \\ 0 & F \oplus cF \end{pmatrix}$, with grading $\left(\begin{pmatrix} F & F \\ 0 & F \end{pmatrix}, \begin{pmatrix} 0 & cF \\ 0 & cF \end{pmatrix} \right)$;

We start by examining some special cases in the next two lemmas. If A is a finite dimensional superalgebra, we write $A = B + J$ where B is a maximal semisimple subalgebra of A with induced $\mathbb{Z}_2$-grading and $B = B_1 \oplus \cdots \oplus B_m$ where $B_1, \ldots, B_m$ are $\mathbb{Z}_2$-graded simple subalgebras of A.

LEMMA 11.8.1. *Let A be a finite dimensional superalgebra over an algebraically closed field F and suppose that supexp$(A) > 2$. If there exist three graded simple components $B_i \cong B_k \cong B_l \cong F$ with grading $(F, \{0\})$ such that $B_i J B_k J B_l \neq 0$, then $UT_3^{gri} \in$ supvar(A) for some $i \in \{0, 1, 2, 3\}$.*

PROOF. Let e_1, e_2, e_3 be the unit elements of B_i, B_k, B_l, respectively. It is clear that $e_p^2 = e_p$, $e_p \in B_p^{(0)}$ and $e_r e_s = \delta_{rs} e_r$ for $r, s = 1, 2, 3$ and $p \in \{i, k, l\}$. Since

$B_iJB_kJB_l \neq 0$, there exist $j_1 = j_1^{(0)} + j_1^{(1)}$, $j_2 = j_2^{(0)} + j_2^{(1)} \in J$ with $j_1^{(0)}, j_2^{(0)} \in J^{(0)}$, $j_1^{(1)}, j_2^{(1)} \in J^{(1)}$ such that

$$e_1 j_1 e_2 j_2 e_3 = e_1(j_1^{(0)} + j_1^{(1)})e_2(j_2^{(0)} + j_2^{(1)})e_3 \neq 0.$$

It follows that one of the following four inequalities must hold:

$$e_1 j_1^{(0)} e_2 j_2^{(0)} e_3 \neq 0, \ e_1 j_1^{(0)} e_2 j_2^{(1)} e_3 \neq 0, \ e_1 j_1^{(1)} e_2 j_2^{(0)} e_3 \neq 0, \ e_1 j_1^{(1)} e_2 j_2^{(1)} e_3 \neq 0.$$

Suppose first that $e_1 j_1^{(0)} e_2 j_2^{(0)} e_3 \neq 0$. Then clearly $e_1 j_1^{(0)} e_2 \neq 0$, $e_2 j_2^{(0)} e_3 \neq 0$ and let D_1 be the subalgebra of A generated by the elements

$$e_1, e_2, e_3, e_1 j_1^{(0)} e_2, e_2 j_2^{(0)} e_3, e_1 j_1^{(0)} e_2 j_2^{(0)} e_3.$$

D_1 is a superalgebra with induced $\mathbb{Z}_2$-grading $(D_1, \{0\})$. Moreover, it is easy to check that D_1 is isomorphic to the superalgebra UT_3^{gr0}. In fact, one can build up the isomorphism of superalgebras

$$\varphi: D_1 \to UT_3^{gr0}$$

induced by setting $\varphi(e_i) = e_{ii}$, $i = 1, 2, 3$, $\varphi(e_1 j_1^{(0)} e_2) = e_{12}$, $\varphi(e_2 j_2^{(0)} e_3) = e_{23}$ and $\varphi(e_1 j_1^{(0)} e_2 j_2^{(0)} e_3) = e_{13}$. Therefore $UT_3^{gr0} \in \mathrm{supvar}(A)$ in this case.

Suppose now that $e_1 j_1^{(0)} e_2 j_2^{(1)} e_3 \neq 0$. Then $e_1 j_1^{(0)} e_2 \neq 0$, $e_2 j_2^{(1)} e_3 \neq 0$ and we construct the subalgebra D_2 of A generated by the elements

$$e_1, e_2, e_3, e_1 j_1^{(0)} e_2, e_2 j_2^{(1)} e_3, e_1 j_1^{(0)} e_2 j_2^{(1)} e_3.$$

D_2 is a superalgebra with induced $\mathbb{Z}_2$-grading $(D_2^{(0)}, D_2^{(1)})$, where

$$D_2^{(0)} = \mathrm{span}_F\{e_1, e_2, e_3, e_1 j_1^{(0)} e_2\} \quad \text{and} \quad D_2^{(1)} = \mathrm{span}_F\{e_2 j_2^{(1)} e_3, e_1 j_1^{(0)} e_2 j_2^{(1)} e_3\}.$$

It is not difficult to show that D_2 is isomorphic as a superalgebra to UT_3^{gr1}. Hence $UT_3^{gr1} \in \mathrm{supvar}(A)$.

In the remaining cases, i.e., if either $e_1 j_1^{(1)} e_2 j_2^{(0)} e_3 \neq 0$ or $e_1 j_1^{(1)} e_2 j_2^{(1)} e_3 \neq 0$, one can construct two subalgebras of A isomorphic to either UT_3^{gr2} or UT_3^{gr3}. $\square$

LEMMA 11.8.2. *Let A be a finite dimensional superalgebra over an algebraically closed field F and suppose that $\mathrm{supexp}(A) > 2$. If there exist two graded simple components $B_i \cong F \oplus cF$, where $c^2 = 1$, with grading (F, cF) and $B_k \cong F$ with grading $(F, \{0\})$ such that either $B_i J B_k \neq 0$ or $B_k J B_i \neq 0$, then A_1 or $A_2 \in \mathrm{supvar}(A)$.*

PROOF. Suppose first that $B_i J B_k \neq 0$. Let $e_1 \in B_i$ and $e_2 \in B_k$ be the unit elements of B_i and B_k, respectively. From $B_i J B_k \neq 0$ it follows that there exists some element $j \in J$ such that $e_1 j e_2 \neq 0$. By eventually multiplying by c on the left, we may assume that $e_1 j^{(0)} e_2 \neq 0$ for some $j^{(0)} \in J^{(0)}$. Let D be the subalgebra of A generated by the elements

$$e_1, ce_1, e_2, e_1 j^{(0)} e_2, ce_1 j^{(0)} e_2.$$

Since all gradings are direct, it is easy to check that these five elements are linearly independent over F. Thus $D = D^{(0)} \oplus D^{(1)}$ is a superalgebra with induced $\mathbb{Z}_2$-grading, where

$$D^{(0)} = \mathrm{span}_F\{e_1, e_2, e_1 j^{(0)} e_2\} \quad \text{and} \quad D^{(1)} = \mathrm{span}_F\{ce_1, ce_1 j^{(0)} e_2\}.$$

We claim that D is isomorphic to the superalgebra A_1. In fact, it is enough to consider the homomorphism
$$\varphi : D \to A_1$$
induced by setting $\varphi(e_1) = e_{11}$, $\varphi(ce_1) = ce_{11}$, $\varphi(e_2) = e_{22}$, $\varphi(e_1 j^{(0)} e_2) = e_{12}$ and $\varphi(ce_1 j^{(0)} e_2) = ce_{12}$. Thus it follows that if $B_i J B_k \neq 0$, then $A_1 \in \mathrm{supvar}(A)$.

A similar argument shows that if $B_k J B_i \neq 0$, then $A_2 \in \mathrm{supvar}(A)$. □

We can now prove the main result of this section. Let us denote by M_2 the algebra of 2×2 matrices over F with trivial grading. Recall that $M_{1,1} = M_{1,1}(F)$ denotes the algebra $M_2(F)$ with the only non-trivial $\mathbb{Z}_2$-grading when F is algebraically closed.

THEOREM 11.8.3 ([**BGP**]). *Let A be a finitely generated superalgebra over F satisfying an ordinary polynomial identity. Then $\mathrm{supexp}(A) > 2$ if and only if one of the superalgebras $UT_3^{gr0}, UT_3^{gr1}, UT_3^{gr2}, UT_3^{gr3}, M_2, M_{1,1}, A_1, A_2$, belongs to $\mathrm{supvar}(A)$.*

PROOF. As in the proof of Theorem 7.6.1, we may assume that F is algebraically closed. Also, by Theorem 4.8.3 we may assume that A is a finite dimensional superalgebra over F.

Let $A = B_1 \oplus \cdots \oplus B_m + J$ be the Wedderburn-Malcev decomposition of A as above. Recall that by Theorem 3.5.3, a simple finite dimensional superalgebra B_i over F is isomorphic to either $M_{k,l}(F)$, $k \geq l \geq 0, k \geq 1$, or to $M_n(F \oplus cF)$ where $c^2 = 1$.

Now, if for some i, $B_i \cong M_{k,l}(F)$ with $k + l \geq 2$, then $M_2(F)$ with a suitable grading lies in $\mathrm{supvar}(A)$ and we are done in this case.

On the other hand, if $B_i \cong M_n(F \oplus cF)$ and $n \geq 2$, then $M_2(F)$ with trivial grading lies in $\mathrm{supvar}(A)$ and we are done.

Recall that by the basic property of the superexponent, since $\mathrm{supexp}(A) > 2$, it follows that there exist distinct simple components $B_{i_1}, \ldots, B_{i_n}$ such that $B_{i_1} J \cdots J B_{i_n} \neq 0$ and $\dim_F(B_{i_1} + \cdots + B_{i_n}) > 2$. Therefore we may assume that one of the following possibilities occurs:

(1) for some $i \neq k$, $B_i J B_k \neq 0$ where $B_i \cong F \oplus cF$ with $c^2 = 1$ and $B_k \cong F$;
(2) for some $i \neq k$, $B_i J B_k \neq 0$ where $B_i \cong F$ and $B_k \cong F \oplus cF$ with $c^2 = 1$;
(3) there exist distinct B_i, B_k, B_l such that $B_i J B_k J B_l \neq 0$ and $B_i \cong B_k \cong B_l \cong F$.

If either (1) or (2) holds, then by Lemma 11.8.2, A contains a superalgebra isomorphic to A_1 or A_2, respectively. On the other hand, if case (3) occurs, then by Lemma 11.8.1 the superalgebra A contains a superalgebra isomorphic to UT_3 with some grading. In any case the proof is complete. □

It can be checked that the above list of algebras cannot be reduced. In fact, if B and C are any two such algebras, it is easy to see that $\mathrm{Id}^{gr}(B) \not\subseteq \mathrm{Id}^{gr}(C)$ (see the original paper [**BGP**]). More generally, generators for the T_2-ideal and the decomposition of the graded cocharacter has been determined for each of them. We refer the reader to [**Di**] for the algebra $M_{1,1}(F)$, to [**DiL**] for A_1 and A_2 and to [**DiD**] and [**DiN**] for the algebras UT_3^{gr1}, UT_3^{gr2} and UT_3^{gr3}.

As a consequence of Theorem 11.8.3 and Theorem 11.5.2, we can now characterize finitely generated superalgebras of superexponent 2.

COROLLARY 11.8.4. *Let A be a finitely generated superalgebra over F satisfying an ordinary polynomial identity. Then* $\operatorname{supexp}(A) = 2$ *if and only if*

$$UT_3^{gr0}, UT_3^{gr1}, UT_3^{gr2}, UT_3^{gr3}, M_2, M_{1,1}, A_1, A_2 \notin \operatorname{supvar}(A)$$

and either UT_2 or UT_2^{gr} or $F \oplus cF$ belongs to $\operatorname{supvar}(A)$.

PROOF. By Theorem 11.5.2 $\operatorname{supexp}(A) > 1$ if and only if one of the algebras $G, UT_2, G^{gr}, UT_2^{gr}, F \oplus cF$ lies in $\operatorname{supvar}(A)$.

Since A is finitely generated, by Theorem 7.1.4, $G, G^{gr} \notin \operatorname{supvar}(A)$.

Now let $St_n(x_1, \ldots, x_n) = \sum_{\sigma \in S_n} (\operatorname{sgn} \sigma) x_{\sigma(1)} \cdots x_{\sigma(n)}$ be the standard polynomial of degree n. Notice that for any $n \geq 1$, if $e_1, \ldots, e_n \in G^{(1)}$ are generating elements of G, then $St_n(e_1, \ldots, e_n) = n! e_1 \cdots e_n \neq 0$. Hence $St_n(z_1, \ldots, z_n) \notin \operatorname{Id}^{gr}(G_2)$, for any $n \geq 1$. On the other hand, since A is a finite dimensional superalgebra, $St_n(z_1, \ldots, z_n) \in \operatorname{Id}^{gr}(A)$ for any $n > \dim_F A^{(1)}$. It follows that $\operatorname{Id}^{gr}(G_1) \not\supseteq \operatorname{Id}^{gr}(A)$. This establishes the claim.

The proof of the corollary now follows from Theorem 11.8.3. □

11.9. Further properties

In this section we collect further properties of finite dimensional superalgebras and their superexponent.

We start by comparing $\exp(A)$ and $\operatorname{supexp}(A)$. By Lemma 10.1.3, for any superalgebra A satisfying an ordinary polynomial identity we have that

$$c_n(A) \leq c_n^{sup}(A) \leq 2^n c_n(A).$$

If A is finitely generated, these inequalities imply the following.

LEMMA 11.9.1. *If A is a finitely generated superalgebra satisfying an ordinary polynomial identity, then*

$$\exp(A) \leq \operatorname{supexp}(A) \leq 2 \exp(A).$$

It is natural to wonder how large can the gap be between $\exp(A)$ and $\operatorname{supexp}(A)$. In other words, given any positive integers d and $0 \leq k \leq d$ does there exist a finitely generated superalgebra A such that $\exp(A) = d$ and $\operatorname{supexp}(A) = d + k$? The answer is yes as the following proposition shows.

THEOREM 11.9.2. *Given integers $d \geq k \geq 0$ there exists a finitely generated superalgebra A such that $\exp(A) = d$ and $\operatorname{supexp}(A) = d + k$.*

PROOF. Let us denote by F_2 the algebra $F_2 = \left\{ \begin{pmatrix} a & b \\ b & a \end{pmatrix} \mid a, b \in F \right\}$. This algebra is $\mathbb{Z}_2$-graded, $F_2 = F_2^{(0)} \oplus F_2^{(1)}$ where $F_2^{(0)} = F(e_{11}+e_{22})$, $F_2^{(1)} = F(e_{12}+e_{21})$ and clearly $F_2 \cong F \oplus cF$.

We write $d + k = 2k + (d - k)$ and we let $U(k, d - k)$ be the algebra of upper block triangular matrices over F of the type $\begin{pmatrix} A & B \\ 0 & C \end{pmatrix}$, where

$$A = \begin{pmatrix} F_2 & & * \\ & \ddots & \\ 0 & & F_2 \end{pmatrix} \subseteq M_{2k}(F)$$

is an algebra of upper block triangular matrices over F, $C = UT_{d-k}(F)$ and B is the space of $2k \times (d-k)$ matrices over F. The Wedderburn-Malcev decomposition of this algebra is

$$U(k, d-k) = \underbrace{F_2 \oplus \cdots \oplus F_2}_{k} \oplus \underbrace{F \oplus \cdots \oplus F}_{d-k} + J$$

where J consists of all strictly upper block triangular matrices.

Recalling the defining properties of $\exp(A)$ and $\operatorname{supexp}(A)$ it is easy to check that $\exp(U(k,d-k)) = k+d-k = d$ and $\operatorname{supexp}(U(k,d-k)) = 2k+d-k = d+k$. $\square$

Next we characterize finite dimensional superalgebras whose sequence of graded codimensions is polynomially bounded (i.e., $\operatorname{supexp}(A) \leq 1$). We have

THEOREM 11.9.3. *Let A be a finite dimensional superalgebra over an algebraically closed field F. Then $\operatorname{supexp}(A) \leq 1$ if and only if:*

1) $\exp(A) \leq 1$;
2) $A = B + J$ where B is a maximal commutative semisimple subalgebra with trivial induced $\mathbb{Z}_2$-grading.

PROOF. Suppose that $\operatorname{supexp}(A) \leq 1$. Then for all n, $c_n(A) \leq c_n^{sup}(A) \leq \alpha n^t$ for some constants α, t, and the sequence of codimensions is polynomially bounded. By the definition of the superexponent it follows that for any maximal semisimple superalgebra $B \subseteq A$, then $B = B_1 \oplus \cdots \oplus B_m$ and, for all i, $B_i \cong F$ has trivial $\mathbb{Z}_2$-grading. This proves the first part of the theorem.

Conversely, suppose that $A = B + J$ where $B = B_1 \oplus \cdots \oplus B_m$ and, for all i $B_i \cong F$ with trivial $\mathbb{Z}_2$-grading. In this case if $a \in A$, we write $a = b + j$, $b \in B = B^{(0)}$, $j \in J$. Then $a - a^{(0)} = j - j^{(0)} \in J$ and $A^{(1)} \subseteq J$ follows.

Notice that if we assume that $c_n(A)$ is polynomially bounded, then $c_{r,n-r}(A) \leq c_n(A) \leq \alpha n^t$, for some α, t, for all $r \geq 0$. Let $J^q = 0$. Since $A^{(1)} \subseteq J$, then $V_{r,n-r} \cap \operatorname{Id}^{gr}(A) = V_{r,n-r}$ and $c_{r,n-r}(A) = 0$ follows, for all $r \leq n - q$. Hence, for all n, we obtain

$$c_n^{sup}(A) = \sum_{r=0}^{n} \binom{n}{r} c_{r,n-r}(A) \leq \alpha n^t \sum_{r=n-q+1}^{n} \binom{n}{r}$$

$$= \alpha n^t \sum_{r=0}^{q-1} \binom{n}{r} \leq \alpha n^{t+q}.$$

and $c_n^{sup}(A)$ is polynomially bounded. $\square$

CHAPTER 12

Lie Algebras and Non-associative Algebras

In this chapter we shall give an account of some of the phenomena occurring in the study of the codimension growth of non-associative algebras.

We shall see that the most important properties of the codimension growth of associative algebras fail in general. For instance, the codimension sequence of a non-associative PI-algebra is not in general exponentially bounded. Even in case of algebras whose codimensions have exponential growth, the rate of growth of this exponential function may be non-integer and the colength growth may be overpolynomial. We shall mostly be concerned with Lie algebras and Lie superalgebras. Even in this case we shall see that there exist PI-algebras with overexponential codimension growth or with overpolynomial colength growth or with exponential growth but with non-integer exponent. In this chapter we shall prove some of the results and we shall survey on the most important phenomena. As a consequence this will show the exceptional role played in PI-theory by the associative algebras.

12.1. Introduction to Lie algebras

Throughout this chapter, unless otherwise stated, F will always be a field of characteristic zero.

DEFINITION 12.1.1. Let L be a (non-associative) algebra over F. Then L is a Lie algebra if L satisfies the relations

(12.1) $$xy = -yx,$$

(12.2) $$(xy)z + (yz)x + (zx)y = 0,$$

for all $x, y, z \in L$.

The relations (12.1) and (12.2) are called anticommutativity and Jacobi identity, respectively. Notice that if char $F = 2$, (12.1) should be replaced by $x^2 = 0$. Here we shall use the standard notation of writing the product of two elements $x, y \in L$ by $[x, y]$. This product is usually called the commutator of x and y. For all $n \geq 3$ and for any $x_1, \ldots, x_n \in L$, we define inductively the right-normed commutator

$$[x_1, \ldots, x_n] = [x_1, [x_2, \ldots, x_n]].$$

EXAMPLE 12.1.2. Let A be an associative algebra. Denote by $A^{(-)}$ the vector space A with the product $[a, b] = ab - ba$. Straightforward computation shows that $A^{(-)}$ is a Lie algebra.

EXAMPLE 12.1.3. Let R be any (non-associative) algebra. Denote by Der$R$ the set of all derivations of R, i.e., all linear transformations $\delta : R \to R$ such that $\delta(ab) = \delta(a)b + a\delta(b)$. It is easy to check that the commutator $[\delta, \mu] = \delta\mu - \mu\delta$ is a derivation as soon as $\delta, \mu \in \operatorname{Der} R$. Since all linear transformations of R form an associative algebra, by the previous example $\operatorname{Der} R$ satisfies (12.1) and (12.2), hence $\operatorname{Der} R$ is a Lie algebra.

EXAMPLE 12.1.4. Let $R = F[t]$ be the polynomial ring in the indeterminate t. Let $d \in \operatorname{Der} R$. Since $d(t^n) = nt^{n-1}d(t)$ and $d(1) = 0$, then d is uniquely determined by its value $d(t)$. In other words, d acts on $F[t]$ as $f(t)\frac{d}{dt}$:
$$d(h(t)) = f(t)\frac{dh}{dt}.$$
It follows that $W_1 = \operatorname{Der} F[t]$ is an infinite dimensional Lie algebra. In fact, the derivations
$$e_{-1} = \frac{d}{dt}, e_0 = t\frac{d}{dt}, \ldots, e_n = t^{n+1}\frac{d}{dt}, \ldots$$
form a basis of W_1 with multiplication given by $[e_i, e_j] = (j-i)e_{i+j}$.

More generally, the Lie algebra $W_n = \operatorname{Der} F[t_1, \ldots, t_n]$ is spanned by all derivations of the form $f(t_1, \ldots, t_n)\frac{\delta}{\delta t_i}$, $1 \leq i \leq n$, with product
$$[f\frac{\delta}{\delta t_i}, g\frac{\delta}{\delta t_j}] = f\frac{\delta g}{\delta t_i} \cdot \frac{\delta}{\delta t_j} - g\frac{\delta f}{\delta t_j} \cdot \frac{\delta}{\delta t_i}.$$

For any subspaces A, B of a Lie algebra L define the commutator $[A, B]$ as the subspace of all linear combinations of products $[a, b]$, $a \in A$, $b \in B$.

A subspace $M \subseteq L$ is called an ideal of L if $[M, L] \subseteq M$. The chain of ideals
$$L = L^1 \supseteq L^2 \supseteq \ldots \supseteq L^n \supseteq \ldots$$
defined inductively by $L^2 = [L, L]$, $L^{n+1} = [L^n, L]$ for $n \geq 2$ is called the lower derived series of L. The chain of ideals
$$L = L^{(0)} \supseteq L' \supseteq L'' \supseteq \ldots \supseteq L^{(n)} \supseteq \ldots$$
defined by $L' = [L, L]$, $L^{(n)} = [L^{(n-1)}, L^{(n-1)}]$ for $n \geq 2$, is called the derived series of L.

DEFINITION 12.1.5. A Lie algebra L is nilpotent of step $k \geq 1$ if $L^k \neq 0$ and $L^{k+1} = 0$.

DEFINITION 12.1.6. A Lie algebra L is solvable of step $k \geq 1$ if $L^{(k-1)} \neq 0$ and $L^{(k)} = 0$.

A Lie algebra L is nilpotent (resp. solvable) if it is nilpotent of some step k (resp. solvable of some length). A Lie algebra L is called abelian if $L^2 = L' = 0$, i.e., L is a vector space with zero multiplication. An algebra L is called metabelian if $L'' = 0$ and $L' \neq 0$.

DEFINITION 12.1.7. A non-abelian Lie algebra L is simple if it has no ideals different from 0 and L.

Next we record some well-known results about Lie algebras which can be found in many standard references (see, for instance, [**GG**]).

A linear map $l : L \to A$ from a Lie algebras L to an associative algebra A is called a homomorphism if $f([x, y]) = f(x)f(y) - f(y)f(x)$. In other words, f is a homomorphism of the Lie algebras L and $A^{(-)}$.

Let V be a vector space and let $\operatorname{End} V$ be the associative algebra of all linear transformations of V.

DEFINITION 12.1.8. A homomorphism $f : L \to \text{End}V$ is called a representation of L.

The dimension of V over F is called the dimension of the representation of f.

Denote by ad$x$ the left multiplication by $x \in L$, i.e., ad$x(y) = [x, y]$, for all $y \in L$. Then clearly, ad$x$ is a linear transformation of L and ad $: L \to \text{End}L$ is a homomorphism.

DEFINITION 12.1.9. The representation ad $: L \to \text{End}L$ is called the adjoint representation of L. Its kernel
$$Z = Z(L) = \{x \in L | [x, L] = 0\}$$
is called the center of L.

DEFINITION 12.1.10. Let L be a Lie algebra. An associative algebra $\mathcal{U} = \mathcal{U}(L)$ with 1 is called a universal enveloping algebra of L if
1) there exists a canonical homomorphism $\epsilon : L \to \mathcal{U}$ such that $\mathcal{U}$ as an associative algebra is generated by $\epsilon(L)$;
2) for any associative algebra A and for any homomorphism $f : L \to A$, there exists a uniquely defined homomorphism of associative algebras $h : \mathcal{U} \to A$ such that $h(\epsilon(x)) = f(x)$ for all $x \in L$.

THEOREM 12.1.11. (Poincare-Birkhoff-Witt). *For any Lie algebra L over F there exists a universal enveloping algebra $\mathcal{U}(L)$ and the canonical homomorphism $\epsilon : L \to \mathcal{U}(L)$ is injective. If we choose a linearly order basis $B \subseteq L$ and we identify $\epsilon(x) \in \mathcal{U}(L)$ with x for all $x \in L$, then 1 and all the products*
$$b_1 b_2 \cdots b_n, \quad b_i \in B, \quad b_1 \leq b_2 \leq \ldots \leq b_n, \ n = 1, 2, \ldots$$
form a basis of L.

As a consequence of the Poincare-Birkhoff-Witt theorem, any Lie algebra L is embeddable into some associative algebra. Note that $\dim \mathcal{U}(L) = \infty$ for any $L \neq 0$.

For a finite dimensional Lie algebra one can find an embedding into a finite dimensional associative algebra.

THEOREM 12.1.12. (Ado-Iwasawa) *For any finite dimensional Lie algebra L there exists a finite dimensional representation $f : L \to \text{End}V$ with $\ker f = 0$.*

Recall that $\rho : L \to \text{End}V$ is called an irreducible representation if V has no non-trivial $\rho(L)$-invariant subspaces.

THEOREM 12.1.13. (Lie). *Let L be a finite dimensional solvable Lie algebra and let $f : L \to \text{End}V$ be an irreducible representation of L. Then $f(L^2) = 0$. In particular, L^2 is a nilpotent Lie algebra.*

We remark that Theorem 12.1.13 holds since char $F = 0$. In positive characteristic a finite dimensional Lie algebra L may have non-nilpotent commutator subalgebra L^2.

12.2. Identities of Lie algebras

In order to define a polynomial identity for a Lie algebra we need to define the free objects in the class of Lie algebras. Throughout we let X be a countable set.

DEFINITION 12.2.1. The Lie algebra $\mathcal{L}(X)$ generated by the set X over F is called a free Lie algebra on X if for any Lie algebra L and for any set-theoretical map $\varphi : X \to L$ there exists a homomorphism $\overline{\varphi} : \mathcal{L}(X) \to L$ such that $\overline{\varphi}(x) = \varphi(x)$ for all $x \in X$. The elements of X are called the free generators of $\mathcal{L}(X)$.

THEOREM 12.2.2. *Let $A = F\langle X \rangle$ be the free associative algebra over F. Then the Lie subalgebra of $A^{(-)}$ generated by X is the free Lie algebra on X.*

PROOF. Denote by $\mathcal{L}(X)$ the Lie subalgebra of $A^{(-)}$ generated by X and let L be an arbitrary Lie algebra. By the Poincare-Birkhoff-Witt theorem, L is embedded into $\mathcal{U} = \mathcal{U}(L)$. Let $\varphi : X \to L$ be any map. Then there exists a homomorphism of associative algebras $\rho : A \to \mathcal{U}$ such that $\rho(x) = \varphi(x)$ for any $x \in X$. Clearly, ρ preserves Lie commutators, i.e., ρ is a homomorphism of $A^{(-)}$ into $\mathcal{U}^{(-)}$ and the image of the Lie subalgebra of $A^{(-)}$ generated by X lies in the Lie subalgebra generated by $\rho(X) = \varphi(X)$, i.e., $\rho(\mathcal{L}(X)) \subseteq L$. Hence the restriction of ρ to $\mathcal{L}(X)$ is the required homomorphism of Lie algebras $\mathcal{L}(X) \to L$. □

The elements of $\mathcal{L}(X)$ are called Lie polynomials in X, any commutator of elements of X is called Lie monomial. Unlike the associative case, Lie monomials may be linearly dependent. For example,

$$[[x_1, x_2], [x_3, x_4]] = [x_1, x_2, x_3, x_4] - [x_2, x_1, x_3, x_4].$$

Nevertheless, any Lie monomial f in $x_1, \ldots, x_n$ is a multihomogeneous associative polynomial as an element of $F\langle X \rangle$. Hence $\deg f$ and $\deg_{x_i} f$, $1 \leq i \leq n$, are all well-defined.

DEFINITION 12.2.3. A non-zero polynomial $f = f(x_1, \ldots, x_n) \in \mathcal{L}(X)$ is called an identity of a Lie algebra L, if $f(a_1, \ldots, a_n) = 0$ for all $a_1, \ldots, a_n \in L$. The Lie identity f is homogeneous and $\deg f = k$, if all Lie monomial of f are of degree k. An identity f is multihomogeneous if all its Lie monomials are multihomogeneous of the same multidegree. In particular, $f = f(x_1, \ldots, x_n)$ is multilinear if $\deg_{x_i} f = 1$ for all $i = 1, \ldots, n$.

As in the associative case all identities $\mathrm{Id}(L)$ of L form an ideal of $\mathcal{L}(X)$ stable under all endomorphisms of $\mathcal{L}(X)$. Although it is not the usual terminology we shall call T-ideals these ideals of $\mathcal{L}(X)$. The proofs of the following two lemmas follow the same patterns as in the associative case and we omit them here.

LEMMA 12.2.4. *Over a field of characteristic zero any identity is equivalent to a family of multilinear identities. In particular, any T-ideal of $\mathcal{L}(X)$ is completely determined by the multilinear polynomials it contains.*

LEMMA 12.2.5. *Let L be a Lie algebra and let $f = f(x_1, \ldots, x_n)$ be a Lie polynomial in $x_1, \ldots, x_n$. Then the linear span of all values $f(a_1, \ldots, a_n)$, $a_1, \ldots, a_n \in L$, is an ideal of L.*

In particular, the above lemma says that any subspace $I \subseteq \mathcal{L}(X)$ stable under all endomorphisms of $\mathcal{L}(X)$ is a T-ideal.

In order to study identities of Lie algebras we need to study the structure of the free Lie algebra $\mathcal{L}(X)$. The next proposition is well-known (see, for example, [**B1**]).

PROPOSITION 12.2.6. *Let $f = f(x_1, \ldots, x_n)$ be a homogeneous Lie polynomial, $\deg f = k$. Then*

1) f is a linear combination of right-normed monomials $[x_i, \ldots, x_{i_k}]$;

2) if f is multilinear, $k = n$, then f is a linear combination of the monomials

(12.3) $$[x_{\sigma(1)}, \ldots, x_{\sigma(n-1)}, x_n], \quad \sigma \in S_{n-1};$$

3) the elements (12.3) are linearly independent over F.

We begin our study of Lie identities by analyzing the most important examples. As in the associative case one can define the standard identity and the Capelli identity.

DEFINITION 12.2.7. The multilinear Lie polynomial
$$st_{n+1}(x_0, x_1, \ldots, x_n) = \sum_{\sigma \in S_n} [x_{\sigma(1)}, \ldots, x_{\sigma(n)}, x_0]$$
is called the standard Lie polynomial of degree $n+1$. A Lie algebra L satisfies the standard identity of degree n if $st_{n-1} \equiv 0$ in L.

Recall that in the associative case the Capelli identity is defined as a finite set of multilinear alternating polynomials. In the case of Lie algebras, the Capelli identities are defined in a similar way but with an infinite set of polynomials. It can be proved (but this requires a quite difficult proof) that this set is PI-equivalent to some finite subset.

DEFINITION 12.2.8. A Lie polynomial $f = f(x_1, x_2, \ldots)$ is skew-symmetric in x_1 and x_2 if f is linear on x_1 and on x_2 and $f(x_1, x_2, \ldots) = -f(x_1, x_2, \ldots)$.

DEFINITION 12.2.9. A Lie polynomial $f = f(x_1, \ldots, x_n, y_1, \ldots, y_m)$ is alternating on $x_1, \ldots, x_n$ if f is skew-symmetric on any pair x_i, x_j, $1 \leq i < j \leq n$.

For short we shall say that f is alternating on n indeterminates if f is alternating in some n of its variables.

DEFINITION 12.2.10. A Lie algebra L satisfies the Capelli identity of rank k if $f \equiv 0$ on L as soon as f is a polynomial alternating on k indeterminates.

Clearly, if L satisfies the Capelli identity of rank k, then st_{k+1} is an identity of L. In the associative case the standard identity implies the Capelli identity. In the Lie case the relation between standard Lie identities and Capelli identities is much more involved.

THEOREM 12.2.11. The Lie algebra $W_1 = \mathrm{Der} F[t]$ satisfies the standard identity $st_4(x_0, \ldots, x_5) \equiv 0$ and does not satisfy the Capelli identity of any rank k.

PROOF. First note that W_1 is a simple Lie algebra. Indeed, if $a = \alpha_{-1} e_{-1} + \alpha_0 e_0 + \cdots + \alpha_k e_k$ and $\alpha_k \neq 0$, then the ideal generated by a in W_1 contains the product
$$[\underbrace{e_{-1}, \ldots, e_{-1}}_{k+1}, a] = (-1)^{k+1} \alpha_k k! e_{-1} \neq 0.$$

Hence for all $j = 0, 1, \ldots$, $e_j = \frac{1}{j+1}[e_j, e_{-1}]$. Now denote by I the subspace of W_1 spanned by all values

(12.4) $$st_5(a_0, a_1, \ldots, a_4), \quad a_0, a_1, \ldots, a_4 \in W_1.$$

Since W_1 is simple, by Lemma 12.2.5 either $I = 0$ or $I = W_1$. On the other hand, by the linearity of st_5 we can take only $a_0, \ldots, a_4 \in \{e_{-1}, e_0, e_1, \ldots\}$ in (12.4). Fix five indices $i_0, i_1, \ldots, i_4 \geq 1$ and consider the value $st_4(e_{i_0}, \ldots, e_{i_4})$. If at least two among $i_1, \ldots, i_4$ coincide, then, by skew symmetry, this value is zero. But if all $i_1, \ldots, i_4$ are distinct, then $i_1 + \cdots + i_4 \geq -1 + 0 + 1 + 2 = 2$ and $i_0 + i_1 + \cdots + i_4 \geq 1$. By the multiplication rule of W_1, this implies that $st_4(e_{i_0}, \ldots, e_{i_4})$ is a scalar multiple of $e_{i_0 + \cdots + i_4}$ and all values $st_4(a_0, \ldots, a_4)$ lie in $\text{span}\{e_1, e_2, \ldots\}$. In particular, $I \neq W_1$, hence $I = 0$ and st_4 is an identity of W_1.

Now we prove that W_1 does not satisfy any Capelli identity of rank n for $n \geq 1$. Consider the polynomial

$$\sum_{\sigma \in S_n} (\text{sgn}\,\sigma)[[\underbrace{y, \ldots, y}_{k_1}, x_{\sigma(1)}], \ldots, [\underbrace{y, \ldots, y}_{k_n}, \sigma(x_n)]] = f(x_1, \ldots, x_n, y) = f,$$

where $k_1, \ldots, k_n$ are positive integers. Clearly, f is alternating on $x_1, \ldots, x_n$. We evaluate f on $y = e_0, x_1 = e_{z_1}, \ldots, x_n = e_{z_n}$. First compute the summand of f with $\sigma = 1$,

$$[[\underbrace{e_0, \ldots, e_0}_{k_1}, e_{z_1}], \ldots, [\underbrace{e_0, \ldots, e_0}_{k_n}, e_{z_n}]] = \varphi_1(z_1, \ldots, z_n) e_{z_1 + \cdots + z_n},$$

where

$$\varphi_1(z_1, \ldots, z_n) e_{z_1 + \cdots + z_n}$$
$$= \pm z_1^{k_1} \cdots z_n^{k_n}(z_n - z_{n-1})(z_n + z_{n-1} - z_{n-2}) \cdots (z_n + \cdots + z_2 - z_1).$$

Similarly, all other summands with $\sigma \neq 1$ give $e_{z_1 + \cdots + z_n}$ with coefficients

$$\varphi_\sigma(z_1, \ldots, z_n) = \pm z_{\sigma(1)}^{k_1} \cdots z_{\sigma(n)}^{k_n}(z_{\sigma(n)} - z_{\sigma(n-1)}) \cdots (z_{\sigma(n)} + \cdots + z_{\sigma(2)} - z_{\sigma(1)}).$$

Hence

$$f(e_{z_1}, \ldots, e_{z_n}, e_0) = h(z_1, \ldots, z_n) e_{z_1 + \cdots + z_n},$$

where

$$h(z_1, \ldots, z_n) = \sum_{\sigma \in S_n} \varphi_\sigma(z_1, \ldots, z_n).$$

Now let the integers $k_1, \ldots, k_n$ satisfy the inequalities

$$k_{j+1} - k_j > n, \quad j = 1, \ldots, n-1.$$

Then any polynomial $\varphi_\sigma(z_1, \ldots, z_n)$ on $z_1, \ldots, z_n$ with $\sigma \neq 1$ does not contain the monomial

$$z_1^{k_1} \cdots z_{n-1}^{k_{n-1}} z_n^{k_n + n - 1}$$

hence $h(z_1, \ldots, z_n)$ is a non-zero polynomial on $z_1, \ldots, z_n$. On the other hand, f is an identity of W_1 only if $h(z_1, \ldots, z_n)$ vanishes on all integers $z_1, \ldots, z_n \geq -1$. Since for a non-trivial polynomial it is impossible, we obtain $f \neq 0$ and the proof is complete. $\square$

Now we denote by V_n the space of multilinear Lie polynomials in $x_1, \ldots, x_n$, $V_n \subseteq \mathcal{L}(X)$. Then V_n is an S_n-module and $I \cap V_n$ is a submodule, for any T-ideal I. From Proposition 12.2.6 we immediately get the following.

PROPOSITION 12.2.12. 1) $\dim V_n = (n-1)!$,
2) $V_n \cong FS_{n-1}$ as left S_n-modules where S_{n-1} acts on $x_1, \ldots x_{n-1}$.

As in the associative case, given a Lie algebra L, one can consider the S_n-module $V_n/(V_n \cap \mathrm{Id}(L))$ and its character

$$\chi_n(L) = \chi\left(\frac{V_n}{V_n \cap \mathrm{Id}(L)}\right) = \sum_{\lambda \vdash n} m_\lambda \chi_\lambda$$

called the nth cocharacter of L. Accordingly, the nth codimension and the nth colength of L are defined as

$$c_n(L) = \dim \frac{V_n}{V_n \cap \mathrm{Id}(L)}, \quad l_n(L) = \sum_{\lambda \vdash n} m_\lambda.$$

Applying the same arguments as in the associative case we obtain the analogue of the strip theorem.

THEOREM 12.2.13. *Let L be a Lie algebra and let $\chi_n(L) = \sum_{\lambda \vdash n} m_\lambda \chi_\lambda$ be its nth cocharacter. Then L satisfies the Capelli identity of rank k if and only if $m_\lambda = 0$ wherever $h(\lambda) \geq k$.*

THEOREM 12.2.14. *Let L be a finite dimensional Lie algebra, $\dim L = k$. Then for any $n \geq 1$,*

$$\chi_n(L) = \sum_{\substack{\lambda \vdash n \\ h(\lambda) \leq k}} m_\lambda \chi_\lambda,$$

i.e, the nth cocharacter of L lies in a strip of height k.

We shall see the next section that for a Lie algebra L satisfying an identity the codimensions are not in general exponentially bounded. Nevertheless, we next show that in case $\chi_\lambda(L)$ lies in the strip $H(k,0)$ for all $n \geq 1$, the codimensions of L are exponentially bounded.

COROLLARY 12.2.15. *Let L be a Lie algebra such that*

$$\chi_n(L) = \sum_{\substack{\lambda \vdash n \\ h(\lambda) \leq k}} m_\lambda \chi_\lambda.$$

Then the sequence $c_n(L)$, $n = 1, 2, \ldots$, is exponentially bounded.

PROOF. Since V_n is an S_n-submodule of $P_n \subseteq F\langle X \rangle$ and P_n is isomorphic to FS_n as a left module, all multiplicities m_λ of $\chi_n(V_n)$ are bounded by $d_\lambda = \deg \chi_\lambda$. Hence by Theorems 12.2.13, 12.2.14 and Corollary 5.1.2, for a Lie algebra L we obtain

$$c_n(L) = \dim \frac{V_n}{V_n \cap \mathrm{Id}(L)} = \sum_{\substack{\lambda \vdash n \\ h(\lambda) \leq k}} m_\lambda d_\lambda \leq \sum_{\substack{\lambda \vdash n \\ h(\lambda) \leq k}} d_\lambda^2 \leq k^{2n}.$$

$\square$

COROLLARY 12.2.16. *For any finite dimensional Lie algebra L the sequence $c_n(L)$, $n = 1, 2, \ldots$, is exponentially bounded.*

12.3. Codimension growth of Lie algebras

In this section we study some classes of Lie algebras with exponentially bounded codimension growth. As we have seen above (see Corollary 12.2.16) $c_n(L)$ is exponentially bounded as soon as $\dim L < \infty$. Now we extend the class of finite dimensional algebras in the following way.

By the Ado-Iwasawa Theorem (Theorem 12.1.12) any finite dimensional Lie algebra L can be embedded into a finite dimensional associative algebra A which is clearly a PI-algebra.

By generalizing this situation we introduce a new class of Lie algebras.

DEFINITION 12.3.1. A Lie algebras L is called special (or SPI-algebra) if there exists an injective homomorphism $L \to A$ from L into an associative PI-algebra A.

In particular, if A is any associative PI-algebra, then the associative Lie algebra $A^{(-)}$ is SPI. Obviously, this new class is wider than that of finite dimensional algebras. We present here one example which will be used later.

EXAMPLE 12.3.2. Let G be the Grassmann algebra and let

$$A = \begin{pmatrix} G & G \\ 0 & 0 \end{pmatrix},$$

be the algebra of 2×2 matrices with entries from G in the first row and zero entries in the second row. Then A is a Lie algebra under the multiplication $[a,b] = ab - ba$. Since $[x,y,z] \equiv 0$ is an identity of G, we have that

$$[A, A, A] \subseteq \begin{pmatrix} 0 & G \\ 0 & 0 \end{pmatrix}.$$

Hence A satisfies the identity

(12.5) $$[[x_1, x_2, x_3], [y_1, y_2, y_3]] \equiv 0.$$

Notice that A is SPI, since it can be embedded into the associative PI-algebra $M_2(G)$.

PROPOSITION 12.3.3. *The Lie algebra* $A = \begin{pmatrix} G & G \\ 0 & 0 \end{pmatrix}$ *is solvable of length* 3 *and does not satisfy any standard identity.*

PROOF. Since for any Lie algebra L we have $L'' \subseteq L^3$, by (12.5) we have that

$$A''' \subseteq [A^3, A^3] = 0.$$

On the other hand, if $e_1, e_2, \ldots$ are the standard generators of G, A contains the elements

$$a = 2 \begin{pmatrix} e_1 e_2 & 0 \\ 0 & 0 \end{pmatrix} = \left[\begin{pmatrix} e_1 & 0 \\ 0 & 0 \end{pmatrix}, \begin{pmatrix} e_2 & 0 \\ 0 & 0 \end{pmatrix} \right],$$

$$b = \left[\begin{pmatrix} e_3 & 0 \\ 0 & 0 \end{pmatrix}, \begin{pmatrix} 0 & e_4 \\ 0 & 0 \end{pmatrix} \right] = \begin{pmatrix} 0 & e_3 e_4 \\ 0 & 0 \end{pmatrix}$$

and

$$[a, b] = \begin{pmatrix} 0 & e_1 e_2 e_3 e_4 \\ 0 & 0 \end{pmatrix} \neq 0.$$

Hence $A'' \neq 0$. Now, let $st_{n+1} = st_{n+1}(x_0, x_1, \ldots, x_n)$ be the Lie standard polynomial and let $St_n = St_n(x_1, \ldots, x_n)$ be the associative standard polynomial. If

$$a_0 = \begin{pmatrix} 0 & e_{n+1} \\ 0 & 0 \end{pmatrix}, a_1 = \begin{pmatrix} e_1 & 0 \\ 0 & 0 \end{pmatrix}, \ldots, a_n = \begin{pmatrix} e_n & 0 \\ 0 & 0 \end{pmatrix},$$

then

$$b = St_n(e_1, \ldots, e_n)e_{n+1} = n!e_1 \cdots e_{n+1} \neq 0$$

and st_{n+1} is not an identity of A. $\square$

Next we record some important properties of SPI-algebras. For a Lie algebra L, denote by AdL the associative subalgebra of the algebra of all linear transformation of L generated by all left multiplications ad$x : L \to L$, ad$x(y) = [x, y]$.

The next three theorems can be found in [**B1**]

THEOREM 12.3.4. *Let L be an SPI Lie algebra. Then* AdL *is a PI-algebra.*

THEOREM 12.3.5. *Let $Z = Z(L)$ be the center of the Lie algebra L. Then* AdL *is a PI-algebra if and only if L/Z is SPI.*

THEOREM 12.3.6. *Let L be an SPI-Lie algebra and let $\mathcal{V} = $ var(L). Then*

1) *any Lie algebra $H \in \mathcal{V}$ with zero center is SPI;*
2) *all relatively free algebras of $\mathcal{V}$ are SPI.*

Actually, if $L/Z(L)$ is SPI, then L is very close to an SPI-algebra but still it may not be SPI. In order to see this, we present one more example of Lie algebras.

EXAMPLE 12.3.7. Let H be a finite dimensional semisimple Lie algebra and let $F[t, t^{-1}]$ be the Laurent polynomial ring in the indeterminate t. Let

$$\widetilde{H} = H \otimes F[t, t^{-1}] + \langle z \rangle$$

and define the multiplication on $\widetilde{H}$ as $[\widetilde{H}, z] = 0$ and

$$[a \otimes t^m, b \otimes t^n] = [a, b] \otimes t^{m+n} + (a, b)m\delta_{m+n,0}z$$

where $a, b \in H$, (a, b) is the value of the Killing form on a, b and δ is the Kronecker delta.

Also let

$$\widehat{H} = \langle d \rangle + \widetilde{H}$$

and

$$[d, z] = 0, \quad [d, a \otimes f] = a \otimes t\frac{df}{dt}.$$

The algebras $\widetilde{H}$ and $\widehat{H}$ are, so-called, affine Kac-Moody algebras. We shall say that $\widetilde{H}$ and $\widehat{H}$ correspond to the semisimple Lie algebra H.

Clearly, $\widetilde{H}/\langle z \rangle$ is an SPI-algebra since by the Ado-Iwasawa theorem it can be embedded into $M_k(F) \otimes F[t, t^{-1}]$, for a suitable k. Nevertheless,

PROPOSITION 12.3.8. *For any semisimple Lie algebra H over an algebraically closed field F of characteristic zero, $\widetilde{H}$ is not SPI.*

This result was proved in [**Bl**] only for $H = sl_2(F)$ but it obviously extends to any H since any semisimple algebra contains a subalgebra isomorphic to $sl_2(F)$. Hence $\widetilde{H}$ contains $\widetilde{sl_2}(F)$.

DEFINITION 12.3.9. A Lie algebra L is called generalized SPI (or GSPI) if $L/Z(L)$ is SPI.

The structure of SPI- and GSPI-algebras is close to the structure of finite dimensional algebras. We record here only one result which is a combination of some theorems from [**Z4, R3**].

THEOREM 12.3.10. *Let L be a finitely generated generalized SPI Lie algebra over an infinite field F. Then*

1) *L has a maximal solvable ideal $R = \operatorname{Rad} L$ called the radical of L;*
2) *L/R is a subdirect product of finite dimensional simple algebras of bounded dimension;*
3) *If $\operatorname{char} F = 0$, then $[L, R]$ is a nilpotent ideal of L;*
4) *L satisfies the identities of some finite dimensional Lie algebra.*

The usual numerical invariants attached to the identities of an *SPI* or GSPI algebra have properties as the invariants of associative PI-algebras.

THEOREM 12.3.11. *Let L be a GSPI Lie algebra and let $\chi_n(L) = \sum_{\lambda \vdash n} m_\lambda \chi_\lambda$ be its nth cocharacter. Then*

1) *there exists an infinite hook $H(k, l)$ such that $\chi_n(L) \subseteq H(k, l)$ for all $n \geq 1$;*
2) *$c_n(L)$ is exponentially bounded;*
3) *$l_n(L) = \sum_{\lambda \vdash n} m_\lambda$ is polynomially bounded.*

PROOF. Suppose first that L is an SPI-algebra and let $L \subseteq A$ be an embedding of L into an associative PI-algebra A. Clearly, it is enough to prove the theorem only for the Lie algebra $A^{(-)}$.

Since $\mathcal{L}(X) \subseteq F\langle X \rangle$, we can consider the set V_n of multilinear Lie polynomials as a subspace of $P_n \subseteq F\langle X \rangle$. Thus V_n is viewed as an S_n-submodule. Let $I = \operatorname{Id}(A)$ be the T-ideal of polynomial identities of L. Then $V_n \cap I$ is the set of all multilinear Lie identities of the Lie algebra $A^{(-)}$.

Since

$$\frac{V_n}{V_n \cap I} \cong \frac{V_n + I}{I} \subseteq \frac{P_n + I}{I} \cong \frac{P_n}{P_n \cap I},$$

The statements 1), 2), and 3) follow from Theorems 4.5.1, 4.2.4 and 4.9.3, Chapter 4.

Now, let L be a Lie algebra with center $Z = Z(L)$. Since L/Z is SPI, by the first part of the proof, all cocharacters $\chi_n(L/Z)$ lie in a hook $H(k, l)$, $\{c_n(L/Z)\}_{n \geq 1}$ is exponentially bounded and $\{l_n(L/Z)\}_{n \geq 1}$ is polynomially bounded.

We first show that $\chi_n(L) \subseteq H(k+1, l+1)$. Suppose $\mu \vdash n$, $\mu \not\subset H(k+1, l+1)$. As in the associative case we need to prove that $e_{T_\mu} f \equiv 0$ in L for any Young tableau T_μ and for any multilinear polynomial $f = f(x_1, \ldots, x_n)$. Since $f = [x_1, g_1] + \cdots + [x_n, g_n]$ (see Proposition 12.2.6), it is sufficient to verify the relation $e_{T_\mu}[x_i, g_i] \equiv 0$ for any multilinear Lie polynomial g_i in the indeterminates $x_1, \ldots, x_{i-1}, x_{i+1}, \ldots, x_n$. Take, for instance, $i = n$ and consider the canonical embedding $S_{n-1} \subseteq S_n$ where S_{n-1} is the set of permutations fixing n. Denote by μ^- the set of all partitions of $n-1$ whose diagram is obtained from D_μ by deleting one box. Then by the Branching theorem (Theorem 2.3.1)

$$e_{T_\mu} e_{T_\lambda} = 0$$

for any $\lambda \vdash n-1$ as soon as $\lambda \notin \mu^-$. If $\lambda \in \mu^-$, then $\lambda \not\subset H(k,l)$ and then $e_{T_\lambda} g_n \equiv 0$ is an identity of L/Z. Hence
$$e_{T_\lambda}[x_n, g_n] = [x_n, e_{T_\lambda} g_n] \equiv 0$$
in L. Since the unit of FS_{n-1} is a linear combination of quasi-idempotents, we obtain
$$e_{T_\mu}[x_n, g_n] \equiv 0$$
in L. This implies the inclusion $\chi_n(L) \subseteq H(k+1, l+1)$ for all $n \geq 1$.

In order to prove 2), fix n and take a basis $\{f_1, \ldots, f_k\}$ of V_n modulo $\operatorname{Id}(L/Z)$. Then any multilinear polynomial of type $[x_{n+1}, g]$, $g \in V_n$, is a linear combination of $[x_{n+1}, f_i]$, $i = 1, \ldots, k$, modulo $\operatorname{Id}(L)$, and
$$c_{n+1}(L) \leq (n+1) c_n(L/Z).$$
Hence, since $c_n(L/Z)$ is exponentially bounded, we obtain that also $c_n(L)$ is exponentially bounded.

Finally, suppose that $l_n(L/Z)$ is polynomially bounded. Let $I = \operatorname{Id}(L/Z)$, $J = \operatorname{Id}(L)$, $I_n = I \cap V_n$, $J_n = J \cap V_n$.

Then V_n as an S_n-module decomposes into
$$V_n = I_n \oplus Q_n$$
where $l_n(Q_n) = l_n(L/Z)$. Since $J_{n+1} \supseteq [I_n, x_{n+1}]$, modulo J_{n+1}, we have
$$(12.6) \qquad [V_n, x_{n+1}] \equiv [Q_n, x_{n+1}] \pmod{J_{n+1}}.$$

Consider an irreducible S_n-submodule $M \subseteq Q_n$. Recall that all cocharacters $\chi_m(L)$ lie in a hook $H(k,l)$ and the total number of partitions $\nu \vdash n+1$ with $\nu \subset H(k,l)$ is less than $(n+1)^{k+l}$. Then the Branching theorem the length of the S_{n+1}-module generated by $[M, x_{n+1}]$ is bounded by $(n+1)^{k+l}$. Hence the length of the S_{n+1}-module generated by $[Q_n, x_{n+1}]$ is bounded by $(n+1)^{k+l} l_n(L/Z)$. Using (12.6) and applying the obvious relation $V_{n+1} = FS_{n+1}[V_n, x_{n+1}]$, we obtain the required upper bound
$$l_{n+1}(L) \leq (n+1)^{k+l} l_n(L/Z).$$
$\square$

Theorem 12.3.11 gives us a wide class of infinite dimensional Lie algebras with exponentially bounded codimension growth. In particular, any affine Kac-Moody algebra $\tilde{H}$ (see example 12.3.7) has exponentially bounded codimensions and polynomially bounded colength.

Another wide class of algebras of exponential codimension growth is given by Corollary 12.2.15. By generalizing this example we give the following definition ([**M3**]).

Let L be a Lie algebra with cocharacter
$$\chi_n(L) = \sum_{\lambda \vdash n} m_\lambda \chi_\lambda, \ n = 1, 2, \ldots.$$

DEFINITION 12.3.12. The algebra L is of associative type (or AT-algebra) if there exist $k, l \geq 0$ such that $\chi_n(L) \subseteq H(k,l)$, for all $n \geq 1$.

By Theorem 12.3.11 and Corollary 12.2.15 any generalized SPI-algebra and any Lie algebra with Capelli identity is of associative type. Next we show that GSPI is a proper subclass of AT.

EXAMPLE 12.3.13. Let $L = \langle h, e \rangle$ be the two-dimensional Lie algebra with multiplication $[h, e] = e$. Consider the left L-action on the polynomial ring $F[t]$ define by
$$h(f) = t\frac{df}{dt}, \ e(f) = tf$$
where $\frac{d}{dt}$ is the usual derivation by t. Then
$$B = L + F[t] = \langle h, e, 1, t, t^2, \ldots \rangle$$
is an infinite dimensional Lie algebra if we define the multiplication as follows:
$$[\lambda h + \mu e + f, \alpha h + \beta e + g] = (\lambda\beta - \mu\alpha)e + \lambda h(g) + \mu e(g) - \alpha h(f) - \beta e(f).$$
In particular, $F[t]$ is abelian ideal of B of codimension 2 and $L''' = 0$.

Note also that $B^2 = \langle e \rangle + F[t]$ is a non-nilpotent Lie algebra.

EXAMPLE 12.3.14. Let $H = \langle x, y, z \rangle$ be the so-called Heisenberg algebra, i.e., H is the three-dimensional vector space with basis $\{x, y, z\}$ and multiplication defined by $[x, y] = z$ all other commutators of the basis elements are zero. Consider the left H-action on $F[t]$:
$$x(f) = \frac{df}{dt} = f', \ y(f) = tf, \ z(f) = f$$
and define the multiplication on
$$C = H + F[t] = \langle x, y, z, 1, t, t^2, \ldots \rangle$$
as follows
$$[\alpha x + \beta y + \gamma z + f, \lambda x + \mu y + \nu z + g] =$$
$$(\alpha\mu - \beta\lambda)z + \alpha x(g) + \beta y(g) + z(g) - \lambda x(f) - \mu y(f) - z(f) =$$
$$(\alpha\mu - \beta\lambda)z + \alpha g' - \lambda f' + \beta tg - \mu tf + f - g.$$
Then C becomes a Lie algebra with $C''' = 0$. As in previous example, $C' = \langle z, 1, t, t^2, \ldots \rangle$ is not a nilpotent algebra.

These two Lie algebras have many properties in common.

PROPOSITION 12.3.15. *The algebras B and C are solvable of length 3 and do not satisfy the identities of any finite dimensional Lie algebra. Also for all $n = 1, 2, \ldots$, the following inclusions hold:*
$$\chi_n(B) \subseteq H(4; 0); \ \chi_n(BC) \subseteq H(5; 0).$$

PROOF. We have mentioned already that $B''' = 0$ and $C''' = 0$. It is not difficult to check that both B and C are finitely generated. Since B^2 and C^2 are not nilpotent, from Theorem 12.3.10 it follows that B and C are not generalized SPI. □

Although AT is a wider class than SPI, it has similar properties in the sense of codimensions and colength.

THEOREM 12.3.16 ([**ZM4**, **M2**]). *Let L be a Lie algebra of associative type. Then $c_n(L)$ is exponentially bounded and $l_n(L)$ is polynomially bounded.*

The class of Lie algebras with exponentially bounded codimensions is not exhausted by the algebras of exponential type. To show this we present below some counterexamples.

THEOREM 12.3.17 ([**M1**]). *Let $W_n = \mathrm{Der} F[t_1,\ldots,t_m]$ be the Lie algebra of derivations of the polynomial ring in n variables. Then $c_n(W_n)$ is exponentially bounded.*

THEOREM 12.3.18 ([**Z4**]). *Let H be any finite dimensional semisimple algebra and let $\widetilde{H}$ be the affine Kac-Moody algebra of Example 12.3.7. Then $c_n(\widetilde{H})$ is exponentially bounded.*

THEOREM 12.3.19 ([**Z4**]). *Let L be a finitely generated solvable Lie algebra of length of solvability equal to 3. Then $c_n(L)$ is exponentially bounded.*

Note that the algebras W_n and $\widetilde{H}$ of Theorems 12.3.17 and 12.3.18 are not of associative type. In general, a finitely generated Lie algebra L with $L''' = 0$ is also not AT.

Now we show that the codimension growth of a Lie algebra can be overexponential.

THEOREM 12.3.20. *Let $\mathcal{V}$ be the variety of Lie algebras defined by the identity*

$$(12.7) \qquad [[x_1, x_2, x_3], [x_4, x_5, x_6]] \equiv 0.$$

Then $c_n(\mathcal{V}) \geq (\frac{n-1}{2})!$, for all $n = 2m+1$.

PROOF. Let $H = H_\infty$ be the infinite dimensional Heisenberg Lie algebra with basis $a_i, b_i, c, i = 1, 2, \ldots$ and with only non-zero products

$$[a_i, b_i] = c, \ i = 1, 2, \ldots.$$

Denote by $\mathcal{U}$ the polynomial ring $F[t_1, t_2, \ldots]$ considered as a Lie algebra with zero multiplication. Define the multiplication between elements of H and $\mathcal{U}$ as follows,

$$[a_i, f] = \frac{\delta f}{\delta t_i}, \quad [b_i, f] = t_i f, \quad [c, f] = f,$$

for any $f = f(t_1,\ldots,t_n) \in \mathcal{U}$. Then $L = H + \mathcal{U}$ becomes a Lie algebra with $L^3 \subseteq \mathcal{U}$. Thus L satisfies (12.7).

We claim that L does not satisfy any identity of the type

$$f = f(x_1,\ldots,x_m, y_1,\ldots,y_m, z) = \sum_{\sigma \in S_m} \alpha_\sigma [[x_{\sigma(1)}, y_1],\ldots, [x_{\sigma(m)}, y_m], z].$$

Without loss of generality, we may assume that $\alpha_1 \neq 0$ where 1 is the unit of S_n. Then

$$[[a_1, b_1],\ldots, [a_m, b_m], h] = h,$$

for any $h = h(t_1,\ldots,t_k)$ and

$$[[a_{\sigma(1)}, b_1],\ldots, [a_{\sigma(m)}, b_m], h] = 0,$$

unless $\sigma \neq 1$. Hence f is not an identity of L and $\mathcal{V}$. Since all Lie monomials $[[x_{\sigma(1)}, y_1],\ldots, [x_{\sigma(m)}, y_m], z]$ are linearly independent modulo $\mathrm{Id}(L)$, we get

$$c_n(\mathcal{V}) \geq (\frac{n-1}{2})!$$

and the proof is complete. □

Since by Stirling formula

$$\left(\frac{n}{2}\right)! \simeq \frac{1}{\sqrt{\pi n}} \left(\frac{n}{2}\right)^{\frac{n}{2}} \left(\frac{1}{e}\right)^{\frac{n}{2}} > \sqrt{\frac{n}{2e}}^n \simeq \frac{\sqrt{n!}}{2^n},$$

the variety $\mathcal{V}$ of Theorem 12.3.20 has an overexponential codimension growth.

In one of the next sections we shall prove that there exist varieties of Lie algebras whose codimensions are asymptotically equal to $(n!)^{\frac{k-1}{k}}$ for any positive integer k. We complete this section by determining an upper bound of the type $n! \, a^{-n}, a > 1$.

Consider an n-tuple $(a_1, \ldots, a_n)$ of pairwise distinct integers. As in Section 10.2 we define the notion of m-decomposable and m-indecomposable sequences.

DEFINITION 12.3.21. The sequence $(a_1, \ldots, a_n)$ is m-decomposable if there exist indices $1 \leq i_1 < i_2 < \ldots < i_m \leq n$ such that $a_{i_{k+1}} > a_j$ for all $j = i_k, i_k + 1, \ldots, i_{k+1} - 1$, $k = 1, \ldots, m-1$.

From the definition it follows that $a_{i_1} < a_{i_2} < \ldots < a_{i_m}$. If the sequence $(a_1, \ldots, a_n)$ has no m-decompositions, then we call it m-indecomposable.

LEMMA 12.3.22. *Let L be a Lie algebra satisfying a multilinear identity of degree $m+1$. Then $V_{n+1} = V_{n+1}(x_0, x_1, \ldots, x_n)$ modulo $V_{n+1} \cap \mathrm{Id}(L)$ is a linear combination of right-normed monomials of the form $[x_{j_1}, \ldots, x_{j_n}, x_0]$ with $\{j_1, \ldots, j_n\} = \{1, \ldots, n\}$ such that the sequence $(j_1, \ldots, j_n)$ is m-indecomposable.*

PROOF. We shall prove the lemma by induction on n. Clearly, any n-tuple $(j_1, \ldots, j_n)$ is m-indecomposable as soon as $n < m$.

Now let $n \geq m$ and consider the right lexicographic order on n-tuples. Suppose the conclusion of the lemma is false. Then there exists a minimal counterexample, i.e., a right-normed monomial $u = [x_{j_1}, \ldots, x_{j_n}, x_0]$ where $(j_1, \ldots, j_n)$ is an m-decomposable sequence and any monomial $[x_{i_1}, \ldots, x_{i_n}, x_0]$ with $(i_1, \ldots, i_n) < (j_1, \ldots, j_n)$ is a linear combination of m-indecomposable monomials modulo $\mathrm{Id}(L)$. For short we shall call a monomial $[x_{\alpha_1}, \ldots, x_{\alpha_n}, x_0]$ m-decomposable or indecomposable according as $(\alpha_1, \ldots, \alpha_n)$ is m-decomposable or indecomposable, respectively. By the m-decomposability of $(j_1, \ldots, j_n)$ there exist $t_1 < \ldots < t_m$ such that

$$t_1, t_1 + 1, \ldots, t_2 - 1 < t_2,$$
$$t_2, t_2 + 1, \ldots, t_3 - 1 < t_3,$$
$$\ldots$$
$$t_{m-1}, t_{m-1} + 1, \ldots, t_m - 1 < t_m.$$

We first remark that for any right-normed monomials A and B the inequality $A < B$ implies

(12.8) $\qquad [x_l, \ldots, x_{l_s}, A] < [x_l, \ldots, x_{l_s}, B].$

Using the relation $[[a, b], c] = [a, b, c] - [b, a, c]$ we can also show that

(12.9) $\qquad [x_{i_1}, \ldots, x_{i_k}, x_j, x_\alpha, \ldots, x_\beta, x_0] - [[x_{i_1}, \ldots, x_{i_k}, x_j], x_\alpha, \ldots, x_\beta, x_0]$

is a linear combination of monomials $[x_{l_1}, \ldots, x_{l_{k+1}}, x_\alpha, \ldots, x_\beta, x_0]$ strictly smaller than $[x_{i_1}, \ldots, x_{i_k}, x_j, x_\alpha, \ldots, x_\beta, x_0]$ as soon as $i_1, \ldots, i_k < j$.

Now let
$$v_1 = t_1,$$
$$v_2 = [x_{t_1+1}, \ldots, x_{t_2}],$$
$$\vdots$$
$$v_m = [x_{t_{m-1}+1}, \ldots, x_{t_m}],$$
$$v_{m+1} = [x_{t_m+1}, \ldots, x_{j_n}, x_0],$$

and set $u' = [x_{j_1}, \ldots, x_{t_1-1}, v_1, \ldots, v_m, v_{m+1}]$. Then by the previous arguments, $u - u'$ is a linear combination of monomials strictly smaller than u. By the minimality of u, any monomial $w < u$ is a combination of m-indecomposable monomials.

Recall that L satisfies a multilinear identity of degree m. We can write it as

$$[y_1, \ldots, y_{m+1}, y_0] \equiv \sum_{\sigma \neq 1} \alpha_\sigma [y_{\sigma(1)}, \ldots, y_{\sigma(m)}, y_0] \quad (\text{mod } \text{Id}(L)).$$

Then

(12.10)
$$[x_{j_1}, \ldots, x_{t_1-1}, v_1, \ldots, v_m, v_{m+1}]$$
$$- \sum_{\sigma \neq 1} [x_{j_1}, \ldots, x_{t_1-1}, v_{\sigma(1)}, \ldots, v_{\sigma(m)}, v_{m+1}] \in \text{Id}(L).$$

Applying (12.8), (12.9) we see that the greatest monomial in any

$$[x_{j_1}, \ldots, x_{t_1-1}, v_{\sigma(1)}, \ldots, v_{\sigma(m)}, v_{m+1}]$$

in the decomposition into the sum of right-normed monomials is

(12.11) $$[x_{j_1}, \ldots, x_{t_{\sigma(1)}}, \ldots, x_{t_{\sigma(m)}}, x_{t_m+1}, \ldots, x_{j_n}, x_0]$$

where $x_{t_{\sigma(1)}}, \ldots, x_{t_{\sigma(m)}}$ are placed in the positions of $x_{t_1}, \ldots, x_{t_m}$, respectively. Hence, in the right lexicographic order (12.11) is strictly smaller than u and using (12.10) we obtain a decomposition of u into a linear combination of strictly smaller monomials and then into m-indecomposable monomials. This contradiction completes the proof of the lemma. □

Clearly the total number of m-indecomposable n-tuples coincides with the number of indecomposable monomials $a_m(n)$ given in Section 10.2 since the sequence $(i_1, \ldots, i_n)$ is m-decomposable if and only if the monomial $x_{i_n} \cdots x_{i_1}$ is m-decomposable. We now call the right-normed Lie monomial $[x_{i_1}, \ldots, x_{i_n}, x_0]$ m-decomposable or m-indecomposable if the corresponding sequence $(i_1, \ldots, i_n)$ is such.

Recall that if $a_m(n)$ denotes the number of m-indecomposable monomials in $x_1, \ldots, x_n$, $n \geq 1$, then by Lemma 10.2.3,

$$a_m(n) = \sum_{i=0}^{n-1} \binom{n-1}{i} a_m(i) a_{m-1}(n-1-i).$$

Now let $\{\alpha_i\}_{i \geq 0}$ and $\{\beta_i\}_{i \geq 0}$ be two infinite sequences of positive real numbers.

LEMMA 12.3.23. *Suppose that for any $b > 1$ there exists a constant $\mathcal{D} = \mathcal{D}(b)$ such that*

$$\beta_j < \mathcal{D} \frac{j!}{b^j}, \quad \text{for all } j = 0, 1, \ldots.$$

Also, let

$$\alpha_n = \sum_{i=0}^{n-1} \binom{n-1}{i} \alpha_i \beta_{n-1-i}.$$

Then for any $a > 1$ one can find $C = C(a)$ such that

$$\alpha_i < C \frac{i!}{a^i}, \quad \text{for all } i = 0, 1, \ldots.$$

PROOF. Given $a > 1$, first denote $C_i = i! a^{-i} \alpha_i^{-1}$. Then

$$\alpha_i = C_i \frac{i!}{a^i}, \quad i = 0, 1, \ldots.$$

Denote also $C_k^* = \max\{C_0, C_1, \ldots, C_{k-1}\}$ and let $\mathcal{D} = \mathcal{D}(2a)$. Then

$$\beta_j < \mathcal{D} \frac{j!}{(2a)^j}, \quad \text{for all } j \geq 0$$

and therefore

$$\alpha_n = \sum_{i=0}^{n-1} \frac{(n-1)!}{i!(n-1-i)!} \alpha_i \beta_{n-1-i} < \sum_{i=0}^{n-1} \frac{(n-1)!}{i!(n-1-i)!} C_i \frac{i!}{a^i} \mathcal{D} \frac{(n-1-i)!}{a^{n-1-i}} \left(\frac{1}{2}\right)^{n-1-i}$$

$$\leq C_{n-1}^* \mathcal{D} \frac{(n-1)!}{a^{n-1}} \sum_{i=0}^{n-1} \left(\frac{1}{2}\right)^{n-1-i} = C_{n-1}^* \mathcal{D} \frac{(n-1)!}{a^{n-1}} 2 \left(1 - \frac{1}{2^n}\right).$$

Hence

$$\alpha_n = C_n \frac{n!}{a^n} < \frac{2a\mathcal{D}}{n} \frac{n!}{a^n} C_{n-1}^*$$

and we get the inequality

$$C_n < \frac{2a\mathcal{D}}{n} \max\{C_0, C_1, \ldots, C_{n-1}\}.$$

This latter inequality shows that all $C_0, C_1, \ldots$ are bounded and the proof is complete. □

COROLLARY 12.3.24. *Let $a_m(n)$ be the number of m-indecomposable right-normed Lie monomials in $x_1, \ldots, x_n$. Then for any $a > 1$ there exists $C > 0$ such that*

$$a_m(n) < C \frac{n!}{a^n}.$$

PROOF. Induction on m. □

Combining Lemma 12.3.22 and Corollary 12.3.24 we get ([**Gi**]).

THEOREM 12.3.25. *Suppose that a Lie algebra L satisfies a non-trivial identity. Then for any $a > 1$ there exists $C > 0$ such that*

$$c_n(L) < C \frac{n!}{a^n}.$$

Note that if L does not satisfy any identity, then $c_n(L) = (n-1)!$ by Proposition 12.2.12.

12.4. Exponents of Lie algebras

In this section we compute the exponential growth of the codimensions of some Lie algebras and varieties of Lie algebras and we give a survey of known results of this area. We also present a Lie algebra with fractional exponent. We first improve Corollary 12.2.15 and we give a more precise upper bound for Lie algebras satisfying Capelli identities.

THEOREM 12.4.1. *Let L be a Lie algebra and let*

(12.12) $$\chi_n(L) = \sum_{\lambda \vdash n} m_\lambda \chi_\lambda$$

be its nth cocharacter. Suppose that there exist a constant C and an integer $k \geq 1$ such that $m_\lambda \neq 0$ in (12.12) only if $n - \lambda_1 - \cdots - \lambda_k \leq C$. Then there exists a polynomial $f = f(x)$ such that

$$c_n(L) \leq f(n) k^n$$

for all $n = 1, 2, \ldots$.

PROOF. Clearly, L satisfies a Capelli identity of rank $k + C + 1$. Hence, by Theorem 12.3.16, $l_n(L) = \sum_{\lambda \vdash n} m_\lambda$ is polynomially bounded and we need only to find an upper bound of the type $f(n)k^n$ for any degree $\deg \chi_\lambda$ with $m_\lambda \neq 0$ in (12.12). But this follows from Corollary 4.4.7, Chapter 4, and Lemma 6.2.4, Chapter 6. □

Using Theorem 12.4.1 we compute the exponent of some important examples. We start with a minimal non-solvable finite dimensional Lie algebra. Recall that $sl_n(F)$ denotes the algebra of $n \times n$ traceless matrices over F.

THEOREM 12.4.2. *Let $L = sl_2(F)$. Then the PI-exponent of L exists and is equal to $\dim L = 3$.*

PROOF. By Theorem 12.2.14 the cocharacter of $sl_2(F)$ lies in the strip $H(3,0)$ of width 3. Hence

(12.13) $$c_n(sl_2(F)) \leq f(n) 3^n$$

for some polynomial f. Now we fix the basis

$$h = \begin{pmatrix} 1 & 0 \\ 0 & -1 \end{pmatrix}, \; e = \begin{pmatrix} 0 & 1 \\ 0 & 0 \end{pmatrix}, \; f = \begin{pmatrix} 0 & 0 \\ 1 & 0 \end{pmatrix}$$

of L. Then $[h, e] = 2e$, $[h, f] = -2f$ and $[e, f] = h$. If $\text{ad} h, \text{ad} e, \text{ad} f$ are left multiplications by $e, f,$ and h, respectively, and $St_3(x_1, x_2, x_3)$ is the associative standard polynomial, then

$$st_4(e, e, f, h) = St_3(\text{ad} e, \text{ad} f, \text{ad} h)(e)$$
$$= 2[[e, f], h, e] - 2[[h, f], e, e] - 2[[e, h], f, e] = 16e.$$

Hence, the Lie polynomial

$$g_n = St_3(\text{ad} x_1^{(1)}, \text{ad} x_2^{(1)}, \text{ad} x_3^{(1)}) \cdots St_3(\text{ad} x_1^{(n)}, \text{ad} x_2^{(n)}, \text{ad} x_3^{(n)})(x_0)$$

of degree $3n + 1$ is not an identity of $sl_2(F)$ for any $n \geq 1$. If S_{3n} acts on $V_{3n+1} = V_{3n+1}(x_0, x_1, \ldots, x_{3n})$ preserving x_0, then, after renaming the variables,

g_n generates an irreducible S_{3n}-module with character χ_λ, $\lambda = (n,n,n)$. Applying Lemma 6.2.5, Chapter 6, we obtain

(12.14) $$c_{3n+1}(L) \geq \deg \chi_\lambda \geq 3^n.$$

Combining (12.13) and (12.14) we complete the proof of the theorem. $\square$

Next we compute the PI-exponent of the algebra A constructed in Example 12.3.2. Recall that $A = \begin{pmatrix} G & G \\ 0 & 0 \end{pmatrix}$ where G is the Grassmann algebra.

THEOREM 12.4.3. $\exp(A) = 2$.

PROOF. First consider the associative identities of the algebra A. Denote by I and J the T-ideals of $F\langle X \rangle$ of identities of G and A, respectively. If the polynomials $f_1, \ldots, f_N \in P_n$ form a basis of P_n modulo $I \cap P_n$ then any multilinear polynomial of the type gx_{n+1} is a linear combination of $f_i x_{n+1}$, $i = 1, \ldots, N$, modulo $J \cap P_{n+1}$. On the other hand, all polynomials

$$f_j(x_1, \ldots, x_{i-1}, x_{i+1}, \ldots, x_{n+1}) x_i, \quad 1 \leq i \leq n+1, \ 1 \leq j \leq N$$

are linearly independent modulo J. Hence, applying Theorem 4.1.8 we obtain

$$\dim \frac{P_n}{P_n \cap J} = n \dim \frac{P_{n-1}}{P_{n-1} \cap I} = n 2^{n-2}.$$

Now take multilinear associative polynomials $g_1, \ldots, g_k$ in $x_1, \ldots, x_n$ linearly independent modulo J. Then all

$$g_i(\operatorname{ad} x_1, \ldots, \operatorname{ad} x_n)(x_0), \ i = 1, \ldots, k$$

are multilinear Lie polynomials linearly independent modulo the Lie identities of A and we obtain the lower bound

$$c_{n+1}(A) \geq n 2^{n-2}$$

for the Lie codimension $c_{n+1}(A)$. On the other hand, the Lie codimension is less or equal to the associative codimension as it was shown in the proof of Theorem 12.3.11. Hence

$$c_{n+1}(A) \leq (n+1) 2^{n-1}$$

and $\exp(A) = 2$. $\square$

Other important Lie algebras were constructed in the Examples 12.3.13 and 12.3.14. Recall that

$$B = \langle h, e, 1, t, t^2, \ldots \rangle,$$

with

$$[h, e] = e, \ [h, t^n] = nt^n, \ [e, t^n] = t^{n+1}, \ [t_i, t_j] = 0$$

and

$$C = \langle x, y, z, 1, t, t^2, \ldots \rangle,$$

with

$$[x, y] = z, \ [x, z] = [y, z] = [t^i, t^j] = 0,$$
$$[x, t^n] = n t^{n-1}, \ [y, t^n] = t^{n+1}, \ [z, t^n] = t^n.$$

THEOREM 12.4.4. *The PI-exponent of the Lie algebras B and C exists and we have $\exp(B) = 2$, $\exp(C) = 3$.*

12.4. EXPONENTS OF LIE ALGEBRAS

PROOF. First consider the Lie algebra B. It contains the ideal $I = F[t]$ with zero multiplication of codimension 2. We claim that

$$(12.15) \qquad \chi_n(B) = \sum_{\substack{\lambda \vdash n \\ |\lambda| - \lambda_1 - \lambda_2 \leq 1}} m_\lambda \chi_\lambda.$$

We need to show that $e_{T_\lambda} f$ is an identity of B for any multilinear polynomial f as soon as $|\lambda| - \lambda_1 - \lambda_2 \geq 2$. Denote

$$e_{\tilde{T}_\lambda} = \Big(\sum_{\tau \in C_{T_\lambda}} (\operatorname{sgn} \tau) \tau \Big) e_{T_\lambda}.$$

Suppose $\lambda_4 \neq 0$. Then $g = e_{\tilde{T}_\lambda} f$ is alternating in 4 variables, say $x_1, \ldots, x_4$. Since g is multilinear, we can evaluate it only on basic elements of B. If we evaluate three x_i's on h, e, then the value will be zero by skew-symmetry. If we take at least two x_i's from I, then the value will be zero since $I^2 = 0$. Hence $g \equiv 0$ in B. Finally,

$$\Big(\sum_{\sigma \in C_{T_\lambda}} \sigma \Big) e_{\tilde{T}_\lambda} f = \alpha e_{T_\lambda} f$$

with $\alpha \neq 0, \alpha \in \mathbb{Q}$, and $e_{T_\lambda} f$ is an identity of B.

Similarly, if $\lambda_4 = 0$ but $\lambda_3 \geq 2$, then the polynomial $g = e_{\tilde{T}_\lambda} f$ contains two alternating sets of indeterminates of order 3. The same arguments show that g and $e_{T_\lambda} f$ are identities of B.

We have proved that $\lambda_4 = 0$ and $\lambda_3 \leq 1$ as soon as $m_\lambda \neq 0$. This implies (12.15). By Theorem 12.4.1 we obtain

$$c_n(B) \leq f(n) 2^n$$

for some polynomial f. Given $n \geq 1$, consider the multilinear Lie polynomial

$$f(x_0, x_1, \ldots, x_{2n}) = \sum_{\sigma \in R} \sum_{\tau \in C} (\operatorname{sgn} \tau)[[x_{\sigma\tau(1)}, x_{\sigma\tau(2)}], \ldots, [x_{\sigma\tau(2n-1)}, x_{\sigma\tau(2n)}], x_0]$$

where R and C are the subgroups of S_{2n},

$$R = \operatorname{Sym}\{1, 3, \ldots, 2n-1\} \times \operatorname{Sym}\{2, 4, \ldots, 2n\}$$

and C is generated by all transpositions $(12), (34), \ldots, (2n-1, 2n)$. Then

$$f(t^k, h, e, \ldots, h, e) = 2^n (n!)^2 t^{k+n} \neq 0,$$

for any $k \geq 0$, i.e., f is not an identity of B. Clearly, the S_{2n}-submodule of V_{2n+1} generated by f has irreducible character χ_λ with $\lambda = (n, n)$. Hence by Lemma 6.2.5, Chapter 6,

$$c_{2n+1}(B) \geq \deg \chi_{(n,n)} \geq a n^b 2^{2n},$$

for some constants a, b and $\exp(B) = 2$.

For the algebra C the same arguments give the decomposition

$$\chi_n(C) = \sum_{\substack{\lambda \vdash n \\ n - \lambda_1 - \lambda_2 - \lambda_3 \leq 1}} m_\lambda \chi_\lambda.$$

Hence by Theorem 12.4.1

$$c_n(C) \leq f(n) 3^n.$$

For finding the lower bound we take a polynomial

$$f = f(w, x_1, \ldots, x_n, y_1, \ldots, y_n, z_1, \ldots, z_n)$$

$$= \sum_{\sigma \in R} \sum_{\tau \in C'} (\operatorname{sgn} \tau) \sigma \tau [[x_1, y_1], z_1, [x_2, y_2], z_2, \ldots, [x_n, y_n], z_n, x_0]$$

where C' is the subgroup of S_{3n}, $C' \cong S_3 \times \cdots \times S_3 = S_3^n$ acting separately on $\{x_1, y_1, z_1\}, \ldots, \{x_n, y_n, z_n\}$ and $R \cong S_n \times S_n \times S_n$ acts separately on $\{x_1, \ldots, x_n\}$, $\{y_1, \ldots, y_n\}$, $\{z_1, \ldots, z_n\}$. Then

$$f(t^k, x, \ldots, x, y, \ldots, y, z, \ldots, z) = 2^n (n!)^3 t^k \neq 0$$

and f is not an identity of C. Under the S_{3n}-action f generates an irreducible module with characters χ_λ, $\lambda = (n, n, n)$. Hence

$$c_{3n+1}(C) \geq \deg \chi_{(n,n,n)} \geq a n^b 3^{3n}$$

for some constant a, b and the proof is complete. $\square$

Denote by $\mathcal{N}_t \mathcal{A}$ the variety of Lie algebras defined by the identity

(12.16) $\qquad [[x_1, x_2], \ldots, [x_{2t+1}, x_{2t+2}]] \equiv 0$

and let $\mathcal{V}_0 = \operatorname{var}(\operatorname{sl}_2(F))$, $\mathcal{V}_1 = \mathcal{N}_2 \mathcal{A}$, $\mathcal{V}_2 = \operatorname{var}(A)$, $\mathcal{V}_3 = \operatorname{var}(B)$, $\mathcal{V}_4 = \operatorname{var}(C)$, where A, B and C are the algebras of Examples 12.3.2, 12.3.13 and 12.3.14, respectively.

The next results characterize varieties of low growth.

THEOREM 12.4.5 ([**M2**]). *The varieties $\mathcal{V}_0, \mathcal{V}_1, \ldots, \mathcal{V}_4$ have almost polynomial growth. If $\mathcal{V}$ is a solvable variety of almost polynomial growth then $\mathcal{V} = \mathcal{V}_i$ for some $1 \leq i \leq 4$.*

THEOREM 12.4.6 ([**M2**]). *A variety $\mathcal{V}$ has polynomially bounded codimension growth if and only if $\mathcal{V} \subseteq \mathcal{N}_t \mathcal{A}$ and $\mathcal{V} \not\supseteq \mathcal{N}_2 \mathcal{A}$, for some $t \geq 1$.*

THEOREM 12.4.7 ([**M4**]). *Let $\mathcal{V}$ be a variety of Lie algebras. If $c_n(\mathcal{V})$ is asymptotically less than $(2 - \varepsilon)^n$ for some $\varepsilon > 0$, then $c_n(\mathcal{V})$ is polynomially bounded.*

There is a wide class of Lie algebras with integer PI-exponents.

THEOREM 12.4.8 ([**Z2**]). *Let L be a finite dimensional Lie algebra. Then $\exp(L)$ exists and is an integer.*

COROLLARY 12.4.9. *Let H be a finite dimensional semisimple Lie algebra and let $\tilde{H}$ be the affine Kac-Moody algebra constructed in Example 12.3.7. Then $\exp(\tilde{H})$ exists and is an integer.*

PROOF. It was proved in [**Z1**] that $\operatorname{var}(\tilde{H}) = \operatorname{var}(L)$ for a suitable finite dimensional algebra L and we apply Theorem 12.4.8. $\square$

It is not difficult to show that $\exp(L) \leq \dim L$ in the finite dimensional case. The equality is a characteristic property of simple algebras.

THEOREM 12.4.10 ([**GRZ**]). *Let L be a finite dimensional Lie algebra over F. If L is not simple, then $\exp(L) \leq \dim L - 1$. If L is simple, then*

$$\dim L = \exp(L)[\zeta(L) : F],$$

where $\zeta(L)$ is the centroid of L. In particular, if F is algebraically closed, then $\exp(L) = \dim L$.

By the Lie theorem any finite dimensional solvable Lie algebra satisfies the identity (12.16). Theorem 12.4.8 can be generalized to this class.

THEOREM 12.4.11 ([**MP1**]). *Let L be a Lie algebra with nilpotent commutator subalgebra. Then $\exp(L)$ exists and is an integer. For $\mathcal{N}_t\mathcal{A}$ we have $\exp(\mathcal{N}_t\mathcal{A}) = t$.*

No solvable Lie algebra has a nilpotent commutator subalgebra. Even if it is an algebra from the variety $\mathcal{AN}_2$ defined by the identity

(12.17) $$[[x_1, x_2, x_3], [x_4, x_5, x_6]] \equiv 0.$$

As it was shown in Theorem 12.3.20, the codimension growth of $\mathcal{AN}_2$ is overexponential. Nevertheless, any proper subvariety has exponential growth with integer exponent.

THEOREM 12.4.12 ([**MRZ2**]). *Let $\mathcal{V}$ be a proper subvariety of $\mathcal{AN}_2$. Then $\exp(\mathcal{V})$ exists and is an integer. If $L_m = L_m(\mathcal{AN}_2)$ is the relatively free algebra of $\mathcal{AN}_2$, then $\exp(L_m) = m + 1$ for m even and $\exp(L_m) = m$ for m odd.*

Nevertheless, the situation with asymptotic behavior is quite different in Lie and associative algebras even in case of exponential growth. Let $L = L(x_1, x_2, x_3)$ be the free metabelian Lie algebra in x_1, x_2, x_3. That is L is isomorphic to $\mathcal{L}(X)/\mathcal{L}''(X)$ where $\mathcal{L}(X)$ is the free Lie algebra with $X = \{x_1, x_2, x_3\}$. Any linear map $X \to L$ can be extended to a derivation of L. So, let $d : L \to L$ be a derivation such that $d(x_1) = x_2, d(x_2) = x_3, d(x_3) = x_1$. Denote by H the direct sum of vector spaces $\langle \alpha \rangle + L$ with multiplication $[d, a] = d(a)$, for all $a \in L$ and the initial products of all elements of L.

THEOREM 12.4.13 ([**ZM2**]). *The Lie algebra H is solvable of length 3, it satisfies the Capelli identity of rank 6 and*

$$3.1 < \underline{\exp}(L), \ \overline{\exp}(L) < 3.9.$$

12.5. Overexponential codimension growth

In this section we give a short survey of the phenomenon of overexponential growth. All results mentioned below may be found in [**P7**], [**P9**], [**GMZ1**], [**MZ**].

We have proved in Theorem 12.3.25 that $c_n(L)$ is asymptotically less than $n!/a^n$ for any $a > 1$. Improving this upper bound we record here the following result.

THEOREM 12.5.1. *Let L be a Lie algebra satisfying a non-trivial identity of degree $m > 3$. Then asymptotically*

$$c_n(L) < \frac{n!}{(\underbrace{\ln \ln \cdots \ln}_{m-3} n)^n}.$$

Given two varieties $\mathcal{U}$ and $\mathcal{V}$, the product variety $\mathcal{UV}$ is the class of all Lie algebras L such that L has an ideal $I \in \mathcal{U}$ satisfying $L/I \in \mathcal{V}$. Denote by $\mathcal{N}_t$ the variety of all nilpotent Lie algebras of class $\leq t$, i.e., the variety defined by the identity $[x_1, \ldots, x_{t+1}] \equiv 0$ and set $\mathcal{A} = \mathcal{N}_1$. Then, for example, $\mathcal{N}_t\mathcal{A}$ is the variety defined by (12.16) and $\mathcal{AN}_2$ is defined by (12.17).

We already know that $\exp(\mathcal{N}_t\mathcal{A}) = t$ and $\mathcal{AN}_2$ has overexponential growth.

THEOREM 12.5.2. *Let $t \geq 1$ and $s \geq 2$. Then asymptotically*

$$c_n(\mathcal{N}_t\mathcal{N}_s) \simeq (n!)^{\frac{s-1}{s}}.$$

THEOREM 12.5.3. *Let $q \geq 3$. Then asymptotically*
$$c_n(\mathcal{N}_{s_1}\cdots\mathcal{N}_{s_q}) \simeq \frac{n!}{(\underbrace{\ln\cdots\ln}_{q-2} n)^{\frac{n}{s_q}}}.$$

In particular, Theorem 12.5.3 shows that the upper bound given in Theorem 12.5.1 cannot be improved.

One of the minimal examples of overexponential codimension growth gives the first example of overpolynomially colength growth.

THEOREM 12.5.4. *The colength of the variety $\mathcal{AN}_2$ has an intermediate growth. Namely,*
$$\lim_{n\to\infty} \frac{\ln(l_n(\mathcal{AN}_2))}{\pi\sqrt{2n/3}} = 1.$$

This theorem means that asymptotically the colength $l_n(\mathcal{AN}_2)$ behaves like $e^{\sqrt{n}}$. Taking other products of nilpotent varieties we realize a faster colength behavior.

THEOREM 12.5.5. *Let $\mathcal{V} = \mathcal{N}_{s_1}\cdots\mathcal{N}_{s_q}$ be the product of $q \geq 2$ nilpotent varieties. Then*

1) *if $q = 2$, $s_2 = 2$, and $\mathcal{V} = \mathcal{N}_s\mathcal{N}_2$, then*
$$l_n(\mathcal{V}) \simeq (\sqrt{s})^n;$$

2) *if $q = 2$ and $\mathcal{V} = \mathcal{N}_s\mathcal{N}_t$ with $t \geq 3$, then*
$$l_n(\mathcal{V}) \geq (\sqrt{s})^n (n!)^{\frac{t-2}{2t}};$$

3) *if $q \geq 3$, then*
$$l_n(\mathcal{V}) \geq \frac{\sqrt{n!}}{(\underbrace{\ln\cdots\ln}_{q-2} n)^{\frac{n}{s_q}}}.$$

12.6. Lie superalgebras, alternative and Jordan algebras

A non-associative algebra $L = L^{(0)} \oplus L^{(1)}$ with $\mathbb{Z}_2$-grading is called a Lie superalgebra if for any homogeneous x, y, z,
$$[x,y] = -(-1)^{d(x)d(y)}[y,x],$$
$$[x,[y,z]] = [[x,y],z] + (-1)^{d(x)d(y)}[y,[x,z]].$$
Here $d(x) = 0$ for $x \in L^{(0)}$ and $d(x) = 1$ for $x \in L^{(1)}$. We also keep the notation $[x, y]$ for the product of two elements $x, y \in L$ and the notation $[x_1, \ldots, x_n]$ for the right-normed product of $x_1, \ldots, x_n$, as in the Lie case. Obviously, any Lie algebra L is a Lie superalgebra with $L^{(0)} = L$ and $L^{(1)} = 0$.

We consider varieties of Lie superalgebras, i.e., classes defined by graded identities. So, all indeterminates are homogeneous of degree zero (even) and one (odd). Clearly any non-graded multilinear identity of degree n can be considered as a family of 2^n graded identities. But in general a set of graded identities is not equivalent to any family non-graded identities. Nevertheless, we can study the non-graded codimensions of a variety $\mathcal{V}$ of Lie superalgebras assuming that $c_n(\mathcal{V}) = c_n(L_\infty(\mathcal{V}))$ where $L_\infty(\mathcal{V})$ is the relatively free superalgebra of $\mathcal{V}$ with countable sets of even and odd generators. Results of this section can be found in [**ZM1**].

THEOREM 12.6.1. *Let $\mathcal{V}$ a variety of Lie superalgebras. Then $c_n(\mathcal{V})$ is polynomially bounded if and only if*

1) $\mathcal{V}$ *satisfies the identity*

(12.18) $$[[z_1, z_2], \ldots, [z_{2q+1}, z_{2q+2}]] \equiv 0$$

for some $q \geq 0$,

2) *there exists a positive integer k such that any multilinear polynomial of degree $\geq k$ both on even and odd variables is an identity of $\mathcal{V}$,*

3) *for any $r = 0, 1, \ldots, q-1$, the variety $\mathcal{V}$ satisfies an identity of the type*

$$[[x_1, \ldots, x_m, z_1, z_2], t_1, \ldots, t_r, [x_{m+1}, \ldots, x_{2m}, z_3, z_4]]$$
$$\equiv \sum_{\sigma \in R} \alpha_{\sigma, r} [[x_{\sigma(1)}, \ldots, x_{\sigma(m)}, z_1, z_2], t_1, \ldots, t_r, [x_{\sigma(m+1)}, \ldots, x_{\sigma(2m)}, z_3, z_4]]$$

where R is the set of permutations $\sigma \in S_{2m}$ such that $\sigma(\{1, \ldots, m\}) \not\subseteq \{1, \ldots, m\}$, $\alpha_{\sigma, r} \in F$.

As in the Lie case, Lie superalgebras also do not admit an intermediate codimension growth but the answer is slightly different.

PROPOSITION 12.6.2. *Let $\mathcal{V}$ be a variety of Lie superalgebras. Then*

1) *if $c_n(\mathcal{V}) < 2^{\frac{n-2}{2}}$ for all n, then $\mathcal{V}$ satisfies an identity of type (12.18), for some $q \geq 1$;*
2) *if $\mathcal{V}$ satisfies (12.18) and $c_n(\mathcal{V}) < 2^{\frac{n}{4q}}$, then $c_n(\mathcal{V})$ is polynomially bounded.*

COROLLARY 12.6.3. *If $c_n(\mathcal{V})$ is less than any exponential function $a^n, a > 1$, then $c_n(\mathcal{V})$ is polynomially bounded.*

Proposition 12.6.2 and Corollary 12.6.3 do not give some fixed lower bound such that either $c_n(\mathcal{V}) \geq a^n$ or $c_n(\mathcal{V})$ is polynomially bounded.

There are very few results concerning codimension growth of other classes of non-associative algebras.

THEOREM 12.6.4 ([**BZ1**]). *Let S be a special Jordan PI-algebra. Then $c_n(\mathcal{V})$ is asymptotically less than $n!/a^n$, for any constant $a > 1$.*

Recall that if A is an associative algebra with involution $*$, then the subspace
$$A^{(+)} = \{a \in A |\ a^* = a\}$$
of all symmetric elements forms a Jordan algebra under multiplication $a \circ b = ab + ba$.

THEOREM 12.6.5 ([**GRZ**]). *Let $M_k(t), M_k(s)$ denote the $k \times k$ matrix algebra with transpose and symplectic involution, respectively. Then*

1) *for $k = 2, 3$, the PI-exponent of $M_k(t)^+$ exists and $\exp(M_k(t)^+) = \dim M_k(t)^+ = \frac{k(k+1)}{2}$;*
2) *for $k = 2, 4$, the PI-exponent of $M_k(s)^+$ exists and $\exp(M_k(s)^+) = \frac{k(k-1)}{2}$.*

THEOREM 12.6.6 ([**BZ2**]). *For any constant $a > 0$ and for any alternative PI-algebra A the codimensions $c_n(A)$ are asymptotically less than $n!/a^n$.*

Note that the proof of Theorems 12.6.4 and 12.6.6 is also based on counting the total number of m-indecomposable words as in Theorem 12.3.25.

12.7. The general non-associative case

Results of this section show that there are common phenomena in the finite dimensional case but in general associative or Lie algebras are very exceptional classes. First note that the total number of distinct arrangements of brackets on the non-associative monomial of degree n is the so-called $(n-1)$th Catalan number $\frac{1}{n}\binom{2n-2}{n-1}$.

PROPOSITION 12.7.1. *Denote by $\tilde{P}_n$ the subspace of all multilinear polynomials in $x_1,\ldots,x_n$ in the free non-associative algebra. Then*

$$\dim \tilde{P}_n = \frac{1}{n}\binom{2n-2}{n-1} n!.$$

Asymptotically Catalan numbers behaves like 4^n, hence for any partition $\lambda \vdash n$ the irreducible S_n-character χ_λ has multiplicity m_λ in the decomposition

$$\chi_n(\tilde{P}_n) = \sum_{\lambda \vdash n} m_\lambda \chi_\lambda$$

bounded by $4^n \deg \chi_\lambda$.

Suppose now that for some algebra A its cocharacter $\chi_n(A)$ lies in the infinite strip $H(d,0)$ of width d. For any such algebra, by using the previous remark, we obtain.

PROPOSITION 12.7.2. *Let A be an algebra such that*

$$\chi_n(A) = \sum_{\substack{\lambda \vdash n \\ h(\lambda) \leq d}} m_\lambda \chi_\lambda$$

for all $n = 1, 2, \ldots$. Then $l_n(A) \leq (4d)^n$ and $c_n(A) \leq (4d^2)^n$.

As in the associative case (see Theorem 4.6.2) it is not difficult to prove that if $\dim A = d < \infty$, then $m_\lambda = 0$ in $\chi_n(A)$ for all $\lambda \vdash n$ with $h(\lambda) \geq d+1$. In particular, $c_n(A)$ is exponentially bounded as soon as $\dim A < \infty$. A non-obvious remark is that the colength of any finite dimensional algebra is polynomially bounded.

THEOREM 12.7.3 ([**GMZ5**]). *Let A be a finite dimensional algebra $\dim A = d$. Then*

$$l_n(A) \leq d(n+1)^{d^2+d},$$

for all $n = 1, 2, \ldots$.

Another general property of all finite dimensional algebras is a gap between polynomial and exponential codimension growth.

THEOREM 12.7.4 ([**GMZ5**]). *Let A be a finite dimensional algebra, $\dim A = d$. Then either $c_n(A) > \frac{1}{n^2} 2^{\frac{n}{3d^2}}$ for all n sufficiently large, or $c_n(A)$ is polynomially bounded.*

We conclude the chapter by giving (without details) some interesting properties of the codimension behaviour of non-associative PI-algebras. By making use of the basic properties of periodic and Sturmian words, one can construct PI-algebras whose PI-exponent is any assigned real number greater than 1.

THEOREM 12.7.5 ([**GMZ4**]). *For any real number $\alpha > 1$ there exists a non-associative algebra A such that its PI-exponent exists and $\exp(A) = \alpha$.*

It is also possible to realize intermediate growth functions of the type $n^{n^\alpha}, 0 < \alpha < 1$.

THEOREM 12.7.6 ([**GMZ5**]). *For any $0 < \alpha < 1$ there exists an algebra A such that*
$$\lim_{n \to \infty} \log_n \log_n(c_n(A)) = \alpha.$$
Hence $c_n(A)$ grows as n^{n^α}.

APPENDIX A

The Generalized-Six-Square Theorem

In this appendix we present the proof of a theorem generalizing the classical Lagrange Theorem stating that any positive integer is the sum of at most four squares. The proof given here is due ro Regev and is based on the paper [**CR**].

A.1. The Theorem

We deal with the so-called *hyperbolic* (or *super*) integers $\mathcal{B}$, that is the elements of $\mathbb{Z} \times \mathbb{Z}$ with coordinatewise addition and the following multiplication:
$$(a,b)(c,d) = (ac+bd, ad+bc).$$

We start with the following definition.

DEFINITION A.1.1. The set of the generalized squares is
$$\mathcal{P} = \{(r^2, r^2), (r^2+s^2, 2rs) \mid r,s \in \mathbb{N}\}.$$
Here $\mathbb{N} = \{0, 1, 2, \ldots\}$. Note that in particular, $(r^2, 0)$, $(2r^2, 2r^2)$ as well as $(y^2+(y+k)^2, 2y(y+k))$ are in $\mathcal{P}$.

Let $r, s \geq 0$. If (r,s) is a sum of generalized squares (namely, of elements of $\mathcal{P}$), then $r \geq s$.

It can be shown that for instance $(10, 3)$ is not a sum of 5 generalized squares. The following is the generalized-six-square theorem.

THEOREM A.1.2. *Given $r \geq s \geq 0$ in $\mathbb{N}$, the pair (r,s) is always a sum of at most six elements in $\mathcal{P}$.*

The basic tool for proving Theorem A.1.2 is the following classical theorem due to Legendre and Lagrange (see [**D**, vol. 2, chapters VII, VIII]).

THEOREM A.1.3.
(1) *Every positive integer $m \in \mathbb{N}$ is a sum of at most four squares.*
(2) *Every positive integer $m \in \mathbb{N}$, which is not of the form $4^u(8k+7)$ with $u, k \in \mathbb{N}$, is a sum of three squares. Moreover, if $m \in \mathbb{N}$ is not divisible by 4, then m is a sum of three squares with no common factor: $m = x^2 + y^2 + z^2$ and $\gcd(x,y,z) = 1$.*

We shall also make use of the following theorem.

THEOREM A.1.4. *Every positive odd integer $m \in \mathbb{N}$ can be written in the form*
$$m = a^2 + b^2 + 2c^2.$$

Here we may assume that a is odd (hence $a \geq 1$) and that b is even. In addition, if $m > 1$, then it has such a presentation with either $b > 0$ or $c > 0$.

PROOF. Since 4 does not divide $2m$, by Theorem A.1.3 $2m = x^2 + y^2 + z^2$ and $gcd(x, y, z) = 1$. Since $2m \equiv 2 \pmod 4$, we may assume that x, y are odd, $x \geq y \geq 1$, and z even. Define
$$a = \frac{1}{2}(x+y), \quad b = \frac{1}{2}(x-y), \quad c = \frac{1}{2}z,$$
then it easily follows that
$$m = a^2 + b^2 + 2c^2.$$
Assume $m > 1$. If $b = c = 0$, it implies that $z = 0$ and that $x = y > 1$, a contradiction since $gcd(x, y, z) = 1$. □

THEOREM A.1.5 ([**D**]). *Every odd integer $s \in \mathbb{N}$ can be written as a sum of four squares of integers, of which two are consecutive:*
$$s = p^2 + q^2 + z^2 + (z+1)^2.$$

A.2. Basics

DEFINITION A.2.1. We call the following presentations of $s \in \mathbb{N}$ *quadratic-ternary* presentations.
  (1) $s = \varepsilon x^2 + \eta y^2 + \gamma z^2$, $\varepsilon, \eta, \gamma \in \{0, 1, 2\}$,
  (2) $s = \varepsilon x^2 + \eta y^2 + 2z(z+\ell)$, $\varepsilon, \eta, \in \{0, 1, 2\}$,
  (3) $s = \varepsilon x^2 + 2y(y+k) + 2z(z+\ell)$, $\varepsilon, \in \{0, 1, 2\}$.

The *length* of such a presentation is the number of its non-zero summands; its *shift* is 0 in case (1), ℓ^2 in case (2) and $k^2 + \ell^2$ in case (3). For example, the length of the presentation $23 = 3^2 + 2 \cdot 1 \cdot (1+2) + 2 \cdot 1 \cdot (1+3)$ is 3, and its shift is $2^2 + 3^2 = 13$.

LEMMA A.2.2. *Let $r \geq s \geq 0$ in $\mathbb{N}$.*
  (1) *Assume s has a quadratic-ternary presentation of length ≤ 2 with shift $\leq r - s$. Then (r, s) is a sum of six generalized squares, namely, (r, s) is the sum of six elements of $\mathcal{P}$.*
  (2) *Assume s has a quadratic-ternary presentation with shift d such that $d \leq r - s$ and $r - s - d$ is a sum of three squares in $\mathbb{N}$. Then (r, s) is a sum of six generalized squares.*

PROOF. 1. The proof here follows by expressing $r - s$–shift as a sum of four squares in $\mathbb{N}$. Here are the cases:

1.1. $s = \varepsilon x^2 + \eta y^2$, $\varepsilon, \eta \in \{0, 1, 2\}$ (hence shift $= 0$). In $\mathbb{N}$ let $r - s = q_1^2 + \cdots + q_4^2$; then
$$(r, s) = (s, s) + (r - s, 0) = (\varepsilon x^2, \varepsilon x^2) + (\eta y^2, \eta y^2) + (q_1^2, 0) + \cdots + (q_4^2, 0).$$

1.2. $s = \varepsilon x^2 + 2y(y+k)$, $\varepsilon, \in \{0, 1, 2\}$ and let $r - s$–shift$= r - s - k^2 = q_1^2 + \cdots + q_4^2$. Since $r = (s + k^2) + (r - s - k^2)$ and since $y^2 + (y+k)^2 = 2y(y+k) + k^2$, we have
$$(r, s) = (\varepsilon x^2, \varepsilon x^2) + (y^2 + (y+k)^2, 2y(y+k)) + (q_1^2, 0) + \cdots + (q_4^2, 0).$$

1.3. $s = 2y(y+k) + 2z(z+\ell)$. Now $r - s$–shift$= r - s - (k^2 + \ell^2) = (q_1^2, 0) + \cdots + (q_4^2, 0)$, and from $r = (s + k^2 + \ell^2) + (r - s - (k^2 + \ell^2))$ we deduce:
$$(r, s) = (y^2 + (y+k)^2, 2y(y+k)) + (z^2 + (z+\ell)^2, 2z(z+\ell)) + (q_1^2, 0) + \cdots + (q_4^2, 0).$$

2. The proof of part 2 is similar. Here are the details.

A.2. BASICS

2.1. $s = \varepsilon x^2 + \eta y^2 + \gamma z^2$, $\varepsilon, \eta, \gamma \in \{0, 1, 2\}$. Here shift$=d = 0$ while $r - s = (q_1^2, 0) + \cdots + (q_3^2, 0)$ and we get

$$(r, s) = (s, s) + (r - s, 0) = (\varepsilon x^2, \varepsilon x^2) + (\eta y^2, \eta y^2) + (\gamma z^2, \gamma z^2) + (q_1^2, 0) + \cdots + (q_3^2, 0).$$

2.2. $s = \varepsilon x^2 + \eta y^2 + 2z(z + \ell)$, $\varepsilon, \eta, \in \{0, 1, 2\}$, so $d = \ell^2$. By assumption $r - s - \ell^2 = (q_1^2, 0) + \cdots + (q_3^2, 0)$, and from $(r, s) = (s + \ell^2, s) + (r - s - \ell^2)$ we get

$$(r, s) = (\varepsilon x^2, \varepsilon x^2) + (\eta y^2, \eta y^2) + (z^2 + (z + \ell)^2, 2z(z + \ell)) + (q_1^2, 0) + \cdots + (q_3^2, 0).$$

2.3. $s = \varepsilon x^2 + 2y(y + k) + 2z(z + \ell)$, $\varepsilon, \in \{0, 1, 2\}$. By similar calculations

$$(r, s) = (s + k^2 + \ell^2, s) + (r - s - k^2 - \ell^2)$$
$$= (\varepsilon x^2, \varepsilon x^2) + (y^2 + (y+k)^2, 2y(y+k)) + (z^2 + (z+\ell)^2, 2z(z+\ell)) + (q_1^2, 0) + \cdots + (q_3^2, 0).$$

□

Note that if $M \leq 4$, then M is a sum of three squares in $\mathbb{N}$.

LEMMA A.2.3. *Let $5 \leq M \in \mathbb{N}$ and assume that none of $M, M - 1$ and $M - 4$ is a sum of three squares in $\mathbb{N}$. Then $M = 4^u(8\ell + 7)$ with $0 \leq \ell$ and $3 \leq u$, hence $M \geq 4^3 7 = 448$. (Note: at a crucial point in the proof of Theorem A.1.2 we shall use the fact that that number is > 149; see the case $a \geq 7$ in the last section here).*

PROOF. By Theorem A.1.3 $M = 4^u(8\ell + 7)$ with $0 \leq \ell, u$. Similarly, $M - 1 = 4^v(8k + 7)$, hence $4^u(8\ell + 7) = 4^v(8k + 7) + 1$. If $1 \leq v$, we must have $u = 0$ (otherwise 4 divides 1), hence l.h.s. is congruent to 3 modulo 4 while the r.h.s. is congruent to 1, a contradiction. Thus $v = 0$, so $4^u(8\ell + 7) = 8(k + 1)$, which implies that $2 \leq u$. Similarly, we have $M = M - 4 + 4 = 4^w(8m + 7) + 4$, therefore $4^u(8\ell + 7) = 4^w(8m + 7) + 4$; since $0 < u$, deduce that $1 \leq w$. By cancellation, $4^{u-1}(8\ell + 7) = 4^{w-1}(8m + 7) + 1$ and since $2 \leq u$, this implies that $w - 1 = 0$. Thus $4^{u-1}(8\ell + 7) = 8(m + 1)$ which implies that $2 \leq u - 1$. □

LEMMA A.2.4. *Let $5 \leq M \in \mathbb{N}$ and assume none of $M, M - 1$ and $M - 4$ is a sum of three squares. Let $t \geq 0$ be an integer that modulo 8 is congruent to either 2 or 5. Then $M - t$ is a sum of three squares.*

PROOF. If $M - t$ is not a sum of three squares, then $M = 4^u(8\ell + 7) = M - t + t = 4^q(8r + 7) + t$ for some $0 \leq \ell, q, r$ and with $3 \leq u$. Deduce a contradiction as follows. If $q = 0$, it implies that $t \equiv 1 \pmod 8$, a contradiction. If $q = 1$, then modulo 8, t is congruent to 4, a contradiction. Finally, if $2 \leq q$, then $t \equiv 0 \pmod 8$, again a contradiction. □

Note that the proof implies more: if modulo 8 t is not congruent to 0, 1 or 4 (hence, if modulo 8 t is congruent to 2, 3, 5, 6 or 7), then $M - t$ is a sum of three squares.

LEMMA A.2.5. *Let $9 \leq M \in \mathbb{N}$ and assume none of $M, M - 1$ and $M - 4$ is a sum of three squares. Then $M - 8$ is a sum of three squares.*

PROOF. Again, assume this is not the case and deduce a contradiction. Thus

$$M = 4^u(8\ell + 7) = M - 8 + 8 = 4^y(8s + 7) + 8$$

where $3 \leq u$ and $0 \leq \ell, k, s$. Since both sides are even, $1 \leq y$. Dividing by 4 we have

$$4^{u-1}(8\ell + 7) = 4^{y-1}(8s + 7) + 2.$$

Since $3 \leq u$, by parity we must have $2 \leq y$. Now reducing this equation modulo 4 yields $0 \equiv 2 \pmod{4}$, a contradiction. $\square$

A.3. Representations of integers

In this section we prove some lemmas on the representations of integers. These lemmas are then applied in the proof of Theorem A.1.2.

LEMMA A.3.1. *Every integer $s \in \mathbb{N}$ can be represented as $s = x^2 + y^2 + \varepsilon z^2$, $x, y, z \in \mathbb{N}$ and $\varepsilon \in \{0, 1, 2\}$.*

PROOF. The case s is odd is given by Theorem A.1.4 (with $\varepsilon = 2$), so let s be even. If $s \equiv 2 \pmod{4}$ then s cannot be of the form $s = 4^u(8\ell + 7)$, hence is a sum of three squares and we are done. So assume $s \equiv 0 \pmod{4}$. If s is a sum of three squares, we are done. Otherwise, $s = 4^u(8\ell + 7)$ with $1 \leq u$, and we can write
$$s = 2 \cdot 4^{u-1}(8\ell + 7) + 2 \cdot 4^{u-1}(8\ell + 7).$$
Since $2 \cdot 4^{u-1}(8\ell + 7)$ is not of the form $4^v(8k + 7)$, by Theorem A.1.2 it is a sum of three squares: $2 \cdot 4^{u-1}(8\ell + 7) = x^2 + y^2 + z^2$, which implies that
$$s = 2x^2 + 2y^2 + 2z^2 = (x+y)^2 + (x-y)^2 + 2z^2$$
as desired. $\square$

LEMMA A.3.2. *Every $s \in \mathbb{N}$ can be represented as $s = \varepsilon x^2 + y^2 + 2z(z+1)$, where $x, y, z \in \mathbb{N}$ and $\varepsilon \in \{0, 1, 2\}$.*

PROOF. Case 1: s is even. By Theorem A.1.5 $s + 1 = p^2 + q^2 + z^2 + (z+1)^2$, therefore $s = p^2 + q^2 + z^2 + 2z(z+1)$ and we are done.

Case 2: s is odd. Hence either $s = 4k + 1$ or $s = 4k + 3$. Assume first that $s = 4k+1$. By Theorem A.1.2 $8k+3$ is a sum of three squares: $8k+3 = a^2+b^2+c^2$. Reducing modulo 4 implies that a, b, c are all odd. Therefore we can write $8k + 4 = 1 + a^2 + b^2 + c^2$ as
$$8k + 4 = \frac{1}{2}\left((a-1)^2 + (a+1)^2 + (b-c)^2 + (b+c)^2\right).$$
Thus
$$4k + 2 = \left(\frac{a-1}{2}\right)^2 + \left(\frac{a+1}{2}\right)^2 + \left(\frac{b-c}{2}\right)^2 + \left(\frac{b+c}{2}\right)^2,$$
which implies that
$$s = 4k + 1 = 2\left(\frac{a-1}{2}\right)\left(\frac{a+1}{2}\right) + \left(\frac{b-c}{2}\right)^2 + \left(\frac{b+c}{2}\right)^2 = 2z(z+1) + \varepsilon x^2 + y^2,$$
where $\varepsilon = 1$.

Finally, consider the case $s = 4k+3$. The argument here is similar: By Theorem A.1.4, $8k + 7 = a^2 + b^2 + 2c^2$, where $a \geq 1$ is odd and $b \geq 0$ even. Thus
$$8k + 8 = 1 + a^2 + b^2 + 2c^2 = \frac{1}{2}\left((a-1)^2 + (a+1)^2\right) + b^2 + 2c^2.$$
This implies that
$$4k + 4 = \left(\frac{a-1}{2}\right)^2 + \left(\frac{a+1}{2}\right)^2 + \frac{1}{2}b^2 + c^2,$$

hence
$$s = 4k+3 = 2\left(\frac{a-1}{2}\right)\left(\frac{a+1}{2}\right) + 2\left(\frac{b}{2}\right)^2 + c^2$$
as desired. □

LEMMA A.3.3. *Every $s \in \mathbb{N}$ can be represented as $s = \varepsilon x^2 + y^2 + 2z(z+2)$, where $x, y, z \in \mathbb{N}$ and $\varepsilon \in \{0, 1, 2\}$.* .

PROOF. The proof is divided into several cases and subcases, and in each we show that s has the desired form.

Case 1. s is odd. By Theorem A.1.4 $s+2 = a^2 + b^2 + 2c^2$ with a odd, b even, and either $b > 0$ or $c > 0$.
Subcase 1.1: $c \neq 0$. It follows that $s = a^2 + b^2 + 2(c-1)(c+1)$ and we are done.
Subcase 1.2: $c = 0$, so $s+2 = a^2 + b^2$ and $b > 0$. Thus $b = 2k$, $k \geq 1$, so
$$s+2 = a^2 + (2k)^2 = a^2 + 2k^2 + 2k^2,$$
which implies that
$$s = a^2 + 2k^2 + 2(k-1)(k+1),$$
and we are done.

Case 2: s is even, hence $s+2 = 2^t v$ where $t \geq 1$ and v odd. By Theorem A.1.4, $v = p^2 + q^2 + 2r^2$ where, say, p is odd, so $p \neq 0$. Write
$$2^{2e}v = (2^e p)^2 + (2^e q)^2 + 2(2^e r)^2 = A^2 + B^2 + 2C^2$$
where $A = (2^e p)^2$ etc., and $A \neq 0$ since $p \neq 0$.
Subcase 2.1: $t = 2e+1$, i.e., odd. Then
$$s+2 = 2 \cdot 2^{2e}v = 2A^2 + 2B^2 + (2C)^2$$
with $A > 0$, so
$$s = 2(A-1)(A+1) + 2B^2 + (2C)^2$$
as desired.
Subcase 2.2: $t = 2e$, is even, with $e \geq 1$. Now
$$s+2 = 2^{2e}v = (2^e p)^2 + (2^e q)^2 + 2(2^e r)^2.$$
Subcase 2.2.1: $r \neq 0$. In that case
$$s = (2^e p)^2 + (2^e q)^2 + 2(2^e r - 1)(2^e r + 1)$$
is the desired presentation.
Subcase 2.2.2: $r = 0$. In that case $v = p^2 + q^2$ with $p > 0$. Here $s+2 = (2^e p)^2 + (2^e q)^2$ and $e \geq 1$. With $u = 2^{e-1}p \neq 0$ and $w = 2^e q$ we have
$$s+2 = 2u^2 + 2u^2 + w^2,$$
hence
$$s = 2(u-1)(u+1) + 2u^2 + w^2,$$
as required. □

COROLLARY A.3.4. *Let $r \geq s \geq 0$ be integers and let $M = r - s$. If at least one of M, $M-1$, $M-4$ is a sum of three squares in $\mathbb{N}$, then (r, s) is a sum of six generalized squares. In particular, if $r - s < 448$, then (r, s) is a sum of six generalized squares.*

PROOF. If M is a sum of three squares, apply Lemma A.3.1. Similarly, if $M-1$ is a sum of three squares, apply Lemma A.3.2, and if $M-4$ is a sum of three squares, apply Lemma A.3.3. For example, assume $M - 1 = q_1^2 + q_2^2 + q_3^2$ in $\mathbb{N}$. By Lemma A.3.2 $s = \varepsilon x^2 + y^2 + 2z(z+1)$, $\varepsilon \in \{0,1,2\}$, hence

$$(r,s) = (\varepsilon x^2, \varepsilon x^2) + (y^2, y^2) + (z^2 + (z+1)^2, 2z(z+1)) + (q_1^2, 0) + (q_2^2, 0) + (q_1^3, 0).$$

Similarly, assume $M - 4 = q_1^2 + q_2^2 + q_3^2$. By Lemma A.3.3, $s = \varepsilon x^2 + y^2 + 2z(z+2)$, $\varepsilon \in \{0,1,2\}$, hence

$$(r,s) = (\varepsilon x^2, \varepsilon x^2) + (y^2, y^2) + (z^2 + (z+2)^2, 2z(z+2)) + (q_1^2, 0) + (q_2^2, 0) + (q_1^3, 0).$$

The case M is a sum of three squares is left to the reader. $\square$

COROLLARY A.3.5. *Let $r \geq s \geq 0$ be integers with*

$$s = \varepsilon x^2 + 2y(y+1) + 2z(z+2), \quad \varepsilon \in \{0,1,2\},$$

then (r,s) is a sum of six generalized squares.

PROOF. Let $M = r - s$. If at least one of M, $M-1$, $M-4$ is a sum of three squares in $\mathbb{N}$, then, by Corollary A.3.4, (r,s) is a sum of six generalized squares. Thus, assume none of M, $M-1$, $M-4$ is a sum of three squares in $\mathbb{N}$. In particular, it implies that $M \geq 5$. By Lemma A.2.4 with $t = 5$, $M - 5 = r - s - 5 = q_1^2 + q_2^2 + q_3^2$. Now proceed as in Corollary A.3.4: $r = (s + 1 + 4) + (r - s - 5)$, and with s as in the lemma we get

$$(r,s) = (\varepsilon x^2, \varepsilon x^2) + (y^2 + (y+1)^2, 2y(y+1))$$
$$+ (z^2 + (z+2)^2, 2z(z+2)) + (q_1^2, 0) + (q_2^2, 0) + (q_1^3, 0).$$

$\square$

A.4. A crucial lemma

In Lemma A.4.1 below we show that most $s \in \mathbb{N}$ admit the presentation $s = \varepsilon x^2 + 2y(y+1) + 2z(z+2)$, $\varepsilon \in \{0,1,2\}$. Note that neither 3 nor 23 admit such a presentation.

LEMMA A.4.1. *Let $s \in \mathbb{N}$ and assume $2s+5$ cannot be written as $2s+5 = a^2 + 2c^2$ with both a and c odd. Then s can be written in the form*

$$s = \varepsilon x^2 + 2y(y+1) + 2z(z+2), \quad \varepsilon \in \{0,1,2\}.$$

PROOF. By Theorem A.1.4, $2s + 5 = a^2 + b^2 + 2c^2$, and we may assume that $a \geq 1$ is odd, $b \geq 0$ even, and either $b > 0$ or $c > 0$.

Case 1: $b \neq 0$. Since b is even, $b \geq 2$. It follows that

$$2s + 6 = 1 + a^2 + b^2 + 2c^2 = \frac{1}{2}\left((a-1)^2 + (a+1)^2 + 2b^2 + (2c)^2\right),$$

therefore

$$s + 3 = \left(\frac{a-1}{2}\right)^2 + \left(\frac{a+1}{2}\right)^2 + 2\left(\frac{b}{2}\right)^2 + c^2,$$

which implies that

$$s = 2\left(\frac{a-1}{2}\right)\left(\frac{a+1}{2}\right) + 2\left(\frac{b}{2} - 1\right)\left(\frac{b}{2} + 1\right) + c^2,$$

as required.

Case 2: Assume $b = 0$, then $c > 0$, and is even by the assumptions of the lemma. Thus
$$2s + 5 = a^2 + 2c^2,$$
$a \geq 1$ odd and $c \geq 2$ even. From
$$2s + 6 = \frac{1}{2}\left((a-1)^2 + (a+1)^2 + 4c^2\right)$$
it follows that
$$s + 3 = \left(\frac{a-1}{2}\right)^2 + \left(\frac{a+1}{2}\right)^2 + 4\left(\frac{c}{2}\right)^2,$$
and we conclude that
$$s = 2\left(\frac{a-1}{2}\right)\left(\frac{a+1}{2}\right) + 2\left(\frac{c}{2} - 1\right)\left(\frac{c}{2} + 1\right) + 2\left(\frac{c}{2}\right)^2,$$
which completes the proof. $\square$

COROLLARY A.4.2. *Let $r \geq s \geq 0$ and assume s satisfies the assumptions of Lemma A.4.1 (namely, $2s + 5$ cannot be written as $2s + 5 = a^2 + 2c^2$ with both a and c odd). Then (r, s) is a sum of six generalized squares.*

PROOF. Apply Lemma A.4.1 and Corollary A.3.5. $\square$

A.5. The proof of Theorem A.1.2

Recall that we want to prove that given $r \geq s \geq 0$ in $\mathbb{N}$, the pair (r, s) is always a sum of at most six elements in $\mathcal{P}$.

PROOF OF THEOREM A.1.2. Denote $M = r - s$. By Corollary A.3.4, if one of $M, M-1, M-4$ is a sum of three squares in $\mathbb{N}$, we are done. The same if $M < 448$. By Corollary A.4.2 the theorem holds if $2s+5$ cannot be written as $2s+5 = a^2 + 2c^2$ with both a and c odd. It therefore remains to prove the theorem in the following case:

None of $M, M - 1, M - 4$ is a sum of three squares in $\mathbb{N}$, $M \geq 448$, and $2s + 5 = a^2 + 2c^2$ with both $a, c \geq 1$ odd.

Case 1: $a \geq 7$. Then
$$2s - 44 = a^2 - 49 + 2c^2 = (a-7)(a+7) + 2c^2,$$
so
$$s - 22 = 2\left(\frac{a-7}{2}\right)\left(\frac{a+7}{2}\right) + c^2$$
which implies the presentation
$$s = 2 \cdot 1 \cdot 11 + 2\left(\frac{a-7}{2}\right)\left(\frac{a+7}{2}\right) + c^2.$$

The shift of this presentation is $10^2 + 7^2 = 149$; it is congruent to 5 modulo 8, and $149 < 448 \leq M$. By Lemma A.2.4 it implies that $M - 149$ is a sum of three squares in $\mathbb{N}$, hence, by part 2 of Lemma A.2.2, (r, s) is a sum of six generalized squares.

Case 2: $a = 5$. Then $2s + 5 = 25 + 2c^2$ so $s = 10 + c^2$ which implies the presentation
$$s = 2 \cdot 1 \cdot 2 + 2 \cdot 1 \cdot 3 + c^2.$$

Here the shift is $1^2 + 2^2 = 5$. By Lemma A.2.4 $r - s$–shift$= M - 5$ is a sum of three squares in $\mathbb{N}$, hence we are done by part (2) of Lemma A.2.2.

Case 3: $a = 3$. From $2s + 5 = 9 + 2c^2$ deduce that
$$s = 2 \cdot 1 + c^2.$$
The length of this presentation is 2, and by part (1) of Lemma A.2.2 we are done.

Case 3: $a = 1$. Then $2s + 5 = 1 + 2c^2$, so $2s + 4 = 2c^2$ and
$$s = c^2 - 2.$$
Since $s \geq 1$ and c is odd, it implies that $c \geq 3$, so we can write
$$s = c^2 - 2 = 1 + 2\left(\frac{c-1}{2}\right)\left(\frac{c+3}{2}\right) + 2\left(\frac{c+1}{2}\right)\left(\frac{c-3}{2}\right).$$
This is a presentation of the form
$$s = 1 + 2y(y+2) + 2z(z+2),$$
whose shift is 8. By Lemma A.2.5 $r - s$–shift$= M - 8$ is a sum of three squares in $\mathbb{N}$, and by part (2) of Lemma A.2.2 (r, s) is a sum of six generalized squares. This completes the proof. □

Bibliography

[A1] S. A. Amitsur, *The identities of PI-rings*, Proc. Amer. Math. Soc. **4** (1953), 27-34.

[A2] S. A. Amitsur, *The T-ideals of the free ring*, J. London Math. Soc. **30** (1955), 470-475.

[A3] S. A. Amitsur, *Rings with involution*, Israel J. Math., **6** (1968), 99-106.

[A4] S. A. Amitsur, *Identities in rings with involution*, Israel J. Math., **7** (1968), 63-68.

[AL] S. A. Amitsur and J. Levitzki, *Minimal identities for algebras*, Proc. Amer. Math. Soc. **1** (1950), 449-463.

[AR] S. A. Amitsur and A. Regev, *P.I. algebras and their cocharacters*, J. Algebra **78** (1982), 248-254.

[AS] S. A. Amitsur and L. W. Small, *Affine algebras with polynomial identities*, Recent developments in the theory of algebras with polynomial identities (Palermo, 1992). Rend. Circ. Mat. Palermo (2) Suppl. **31** (1993), 9-43.

[An] A. Z. Anan'in and A. R. Kemer, *Varieties of associative algebras whose lattices of subvarieties are distributive*, (Russian) Sibirsk. Mat. Z. **17** (1976), no. 4, 723-730.

[B1] Yu. A. Bahturin, Identical Relations in Lie algebras, Utrecht, VNU Science Press, 1987.

[B2] Yu. A. Bahturin, *Identities of algebras with actions of Hopf algebras*, Lect. Notes Pure Appl. Math., **198** (1998), 1-36.

[BD] Yu. A. Bahturin and V. Drensky, *Graded polynomial identities of matrices*, Linear Algebra Appl. **357** (2002), 15-34.

[BMPZ] Yu. A. Bahturin, A. V. Mikhalev, V. M. Petrogradsky and M. Zaicev, Infinite dimensional Lie superalgebras, Walter de Gruyter, Berlin, New York, 1992.

[BGR] Yu. A. Bahturin, A. Giambruno, D. Riley, *Group-graded algebras with polynomial identity*, Israel J. Math. **104** (1998), 145-155.

[BGZ1] Yu. A. Bahturin, A. Giambruno, M. Zaicev, *G-identities on associative algebras*, Proc. Amer. Math. Soc. **127** (1998), 63-69.

[BGZ2] Yu. A. Bahturin, A. Giambruno, M. V. Zaicev, *Symmetric identities in graded algebras*, Arch. Math **69** (1997), 461-464.

[BMR] Yu. A. Bahturin, S. Mishchenko and A. Regev, *On the Lie and associative codimensions growth*, Comm. Algebra **27** (1999), 4901-4908.

[BM] Yu. A. Bahturin, and S. Montgomery, *PI-envelopes of Lie superalgebras*, Proc. Amer. Math. Soc. **127** (1999), 2829-2839.

[BMZ] Yu. A. Bahturin, S. Montgomery and M. Zaicev, *Generalized Lie solvability of associative algebras*, Groups, rings, Lie and Hopf algebras (St. John's, NF, 2001), 1-23, Math. Appl., **555**, Kluwer Acad. Publ., Dordrecht, 2003.

[BZ1] Yu. A. Bahturin and M. Zaicev, *Identities of special Jordan algebras with finite grading* (Russian) Vestnik Moskov. Univ. Ser. I Mat. Mekh. 1998, no. 2, 26-29, 73; translation in Moscow Univ. Math. Bull. **53** (1998), no. 2, 28-31.

[BZ2] Yu. A. Bahturin and M. Zaicev, *Identities of graded alternative algebras*, Nonassociative algebra and its applications (São Paulo, 1998), 9-20, Lecture Notes in Pure and Appl. Math., **211**, Dekker, New York, 2000.

[BMM] K. I. Beidar, W. S. Martindale III and A. V. Mikhalev, Rings with generalized identities, Monographs and Textbooks in Pure and Applied Mathematics, **196** Marcel Dekker, Inc., New York, 1996.

[BR] A. Belov and L. H. Rowen, Computational aspects of polynomial identities, Research Notes in Mathematics **9**, A. K. Peters, Ltd., Wellesley, MA, 2005.

[BGP] F. Benanti, A. Giambruno and M. Pipitone, *Polynomial identities on superalgebras and exponential growth*, J. Algebra **269** (2003), 422-438.

[Be1] A. Berele, *Homogeneous polynomial identities*, Israel J. Math. **42** (1982), 258-272.

[Be2]   A. Berele, *Cocharacters of $Z/2Z$-graded algebras*, Israel J. Math. **61** (1988), 225-234.

[Be3]   A. Berele, *Cocharacter sequences for algebras with Hopf algebra actions*, J. Algebra **185** (1996), 869-885.

[Be4]   A. Berele, *Colength sequences for matrices*, J. Algebra **283** (2005), 700-710.

[BeGR]   A. Berele, A. Giambruno and A. Regev, *Involution codimensions and trace codimensions of matrices are asymptotically equal*, Israel J. Math., **96** (1996), 49-62.

[BR1]   A. Berele and A. Regev, *Applications of Hook Young diagrams to P.I. algebras*, J. Algebra, **82** (1983), 559-567.

[BR2]   A. Berele and A. Regev, *Hook Young diagrams with applications to combinatorics and to representations of Lie superalgebras*, Adv. Math. **64** (1987), 118-175.

[BR3]   A. Berele and A. Regev, *On the codimensions of the verbally prime P.I. algebras*, Israel J. Math. **91** (1995), 239-247.

[BR4]   A. Berele and A. Regev, *Codimensions of products and intersections of verbally prime T-ideals*, Israel J. Math. **103** (1998), 17-28.

[BR5]   A. Berele and A. Regev, *Exponential growth for codimensions of some p.i. algebras*, J. Algebra **241** (2001), 118-145.

[BC]   J. Bergen, M. Cohen, *Action of commutative Hopf algebras*, Bull. London Math. Soc **18** (1986), 159-164.

[B]   G. M. Bergman, *A ring primitive on the right but not on the left*, Proc. Amer. Math. Soc. **15** (1964), 473-475.

[BL]   G. M. Bergman and J. Lewin, *The semigroup of ideals of a fir is (usually) free*, J. London Math. Soc. (2) **11** (1975), 21-31.

[Bi]   N. L. Biggs, Discrete Mathematics, Clarendon Press, Oxford, 1989.

[Bl]   Yu. V. Billig, *A homomorphic image of a special Lie algebra* (Russian) Mat. Sb. (N.S.) **136(178)** (1988), no. 3, 320-323, 430; translation in Math. USSR-Sb. **64** (1989), no. 2, 319-322.

[Br]   G. Birkhoff, *On the structure of abstract algebras*, Proc. Camb. Philos. Soc. **31**, 433-454 (1935).

[Bo]   H. Boerner, Representations of groups. With special consideration for the needs of modern physics, North-Holland Publishing Co., Amsterdam-London; American Elsevier Publishing Co., Inc., New York, 1970.

[CR]   P. B. Cohen and A. Regev, *A six generalized squares theorem, with applications to polynomial identity algebras* J. Algebra **239** (2001), 174-190.

[C]   P. M. Cohn, Algebra. Vol. 1,2,3, John Wiley & Sons, London-New York-Sydney, 1977.

[CuR]   C. Curtis and I. Reiner, Representation Theory of Finite Groups and Associative Algebras, J. Wiley & Sons, New York, 1962.

[DR]   A. D'Amour and M. Racine, *$*$-polynomial identities of matrices with the transpose involution: the low degrees*, Trans. Amer. Math. Soc. **351** (1999), 5089-5106.

[DR2]   A. D'Amour and M. Racine, *$*$-polynomial identities of matrices with the symplectic involution: the low degrees*, Comm. Algebra **32** (2004), 895-918.

[De]   M. Dehn, *Über die Grundlagen der projektiven Geometrie und allgemeine Zahlsysteme*, (German) Math. Ann. **85**, 184-194 (1922).

[D]   L. E. Dickson, History of the theory of numbers. Vol. I, II, III, Chelsea Publishing Co., New York, 1966.

[Di]   O. M. Di Vincenzo, *On the graded identities of $M_{1,1}(E)$*, Israel J. Math. **80** (1992), 323-335.

[DiD]   O. M. Di Vincenzo and V. Drensky, *The basis of the graded polynomial identities for superalgebras of triangular matrices*, Comm. Algebra **24** (1996), 727-735.

[DKV]   O. M. Di Vincenzo, P. Koshlukov and A. Valenti, *Gradings on the algebra of upper triangular matrices and their graded identities*, J. Algebra **275** (2004), 550-566.

[DiL]   O. M. Di Vincenzo and R. La Scala, *Block-triangular matrix algebras and factorable ideals of graded polynomial identities*, J. Algebra **279** (2004), 260-279.

[DiN]   O. M. Di Vincenzo and V. Nardozza, *$\mathbb{Z}_2$-graded cocharacters for superalgebras of triangular matrices*, J. Pure Appl. Algebra **194** (2004), 193-211.

[D1]   V. Drensky, *Representations of the symmetric group and varieties of linear algebras*, (Russian), Mat. Sb. (N.S.) **115** (1981), no. 1, 98-115.

[D2]   V. Drensky, *A minimal basis for identities of a second-order matrix algebra over a field of characteristic 0*, (Russian), Algebra i Logika **20** (1981), no. 3, 282–290.

[D3] V. Drensky, *Codimensions of T-ideals and Hilbert series of relatively free algebras*, J. Algebra **91** (1984), 1-17.

[D4] V. Drensky, *Extremal varieties of algebras I*, (Russian), Serdica **13** (1987), 320-332.

[D5] V. Drensky, *Extremal varieties of algebras II*, (Russian), Serdica **14** (1988), 20-27.

[D6] V. Drensky, *Relations for the cocharacter sequences of T-ideals*, Proceedings of the International Conference on Algebra, Part 2 (Novosibirsk, 1989), 285–300, Contemp. Math., **131**, Part 2, Amer. Math. Soc., Providence, RI, 1992.

[D7] V. Drensky, *New central polynomials for the matrix algebra*, Israel J. Math. **92** (1995), 35-248.

[D8] V. Drensky, *Gelfand-Kirillov dimension of PI-algebras*, In: Methods in Ring Theory, Lect. Notes in Pure and Appl. Math., Vol. **198** (1998), 97-113.

[D9] V. Drensky, *New central polynomials for the matrix algebra*, Israel J. Math. **92** (1995), 235-248.

[D10] V. Drensky, Free Algebras and PI-Algebras, Graduate course in algebra, Springer-Verlag Singapore, Singapore, 2000.

[DF] V. Drensky and E. Formanek, Polynomial identity rings, Advanced Courses in Mathematics, CRM Barcelona, Birkhäuser Verlag, Basel, 2004.

[DG] V. Drensky and A. Giambruno, *Cocharacters, codimensions and Hilbert series of the polynomial identities for 2×2 matrices with involution*, Canadian J. Math., **46** (1994), 718-733.

[DK] V. Drensky and A. Kasparian, *A new central polynomial for 3×3 matrices*, Comm. Algebra **13** (1985), 745-752.

[DR] V. Drensky and A. Regev, *Exact asymptotic behaviour of the codimensions of some P.I. algebras*, Israel J. Math. **96** (1996), 231-242.

[FR] O.J. Farrell and B. Ross, Solved Problems: Gamma and Beta Functions, Legendre Polynomials, Bessel Functions, Dover Publications, New York, 1971.

[F1] E. Formanek, *Central polynomials for matrix rings*, J. Algebra **32** (1972), 129-132.

[F2] E. Formanek, *Invariants and the ring of generic matrices*, J. Algebra **89** (1984), 178-223.

[F3] E. Formanek, *A conjecture of Regev about the Capelli polynomial*, J. Algebra **109** (1987), 93-114.

[F4] E. Formanek, The Polynomial Identities and Invariants of $n \times n$ Matrices, CBMS Regional Conference Series in Mathematics, **78**, American Mathematical Society, Providence, RI, 1991.

[FHL] E. Formanek, P. Halpin and W. C. W. Li, *The Poincaré series of the ring of 2×2 generic matrices*, J. Algebra **69** (1981), 105-112.

[G]   A. Giambruno Ed., Recent developments in the theory of algebras with polynomial identities, Papers from the conference held in Palermo, June 15-18, 1992, Rend. Circ. Mat. Palermo (2) Suppl. **31**, Circolo Matematico di Palermo, Palermo, 1993.

[GL] A. Giambruno and D. La Mattina, *PI-algebras with slow codimension growth*, J. Algebra **284** (2005), 371-391.

[GM1] A. Giambruno and S. Mishchenko, *Polynomial growth of the $*$-codimensions and Young diagrams*, Comm. Algebra **29** (2001), 277-284.

[GM2] A. Giambruno and S. Mishchenko, *On star-varieties with almost polynomial growth*, Algebra Coll. **8** (2001), 33-42.

[GMZ1] A. Giambruno, S. Mishchenko and M. Zaicev, *On the colength of a variety of Lie algebras*, Internat. J. Algebra Comput. **9** (1999), 483-491.

[GMZ2] A. Giambruno, S. Mishchenko and M. Zaicev, *Polynomial identities on superalgebras and almost polynomial growth*, Special issue dedicated to Alexei Ivanovich Kostrikin, Comm. Algebra **29** (2001), 3787-3800.

[GMZ3] A. Giambruno, S. Mishchenko and M. Zaicev, *Group actions and asymptotic behavior of graded polynomial identities*, J. London Math. Soc. **66** (2002), 295-312.

[GMZ4] A. Giambruno, S. Mishchenko and M. Zaicev, *Codimensions of algebras and growth functions*, (preprint).

[GMZ5] A. Giambruno, S. Mishchenko and M. Zaicev, *Algebras with intermediate growth of the codimensions*, Adv. in Appl. Math. (to appear).

[GR] A. Giambruno and A. Regev, *Wreath products and P.I. algebras*, J. Pure Applied Algebra, **35** (1985), 133-149.

[GRZ]   A. Giambruno, A. Regev and M. V. Zaicev, *Simple and semisimple Lie algebras and codimension growth*, Trans. Amer. Math. Soc. **352** (2000), 1935-1946.

[GRZ2] A. Giambruno, A. Regev and M. V. Zaicev, Eds., Polynomial identities and combinatorial methods, Proceedings of the conference held on Pantelleria, September 2001, Lecture Notes in Pure and Applied Mathematics, **235**, Marcel Dekker, Inc., New York, 2003.

[GV]   A. Giambruno and A. Valenti, *Central polynomials and matrix invariants*, Israel J. Math. **96** (1996), 281-297.

[GZ1]   A. Giambruno and M. Zaicev, *On codimension growth of finitely generated associative algebras*, Adv. Math. **140** (1998), 145-155.

[GZ2]   A. Giambruno and M. Zaicev, *Exponential codimension growth of P.I. algebras: an exact estimate*, Adv. Math. **142** (1999), 221-243.

[GZ3]   A. Giambruno and M. Zaicev, *Involution codimensions of finite dimensional algebras and exponential growth*, J. Algebra **222** (1999), 471-484.

[GZ4]   A. Giambruno and M. Zaicev, *Minimal varieties of algebras of exponential growth*, Electron. Res. Announc. Amer. Math. Soc. **6** (2000), 40-44.

[GZ5]   A. Giambruno and M. Zaicev, *A characterization of varieties of associative algebras of exponent two*, Serdica Math. J. **26** (2000), 245-252.

[GZ6]   A. Giambruno and M. Zaicev, *A characterization of algebras with polynomial growth of the codimensions*, Proc. Amer. Math. Soc. **129** (2001), 59-67.

[GZ7]   A. Giambruno and M. Zaicev, *Minimal varieties of exponential growth*, Adv. Math. **174** (2003), 310–323.

[GZ8]   A. Giambruno and M. Zaicev, *Asymptotics for the standard and the Capelli identities*, Israel J. Math. **135** (2003), 125-145.

[GZ9]   A. Giambruno and M. Zaicev, *Codimension growth and minimal superalgebras*, Trans. Amer. Math. Soc. **355** (2003), 5091-5117.

[GG]   M. Goto and F. D. Grosshans, Semisimple Lie algebras, Lecture Notes in Pure and Applied Mathematics, Vol. **38**, Marcel Dekker, Inc., New York-Basel, 1978. vii+480 pp.

[Gi]   A. N. Grishkov, *Growth of varieties of Lie algebras*, (Russian), Mat. Zametki **44** (1988), no. 1, 51-54, 154; translation in Math. Notes **44** (1988), no. 1-2, 515-517 (1989).

[Gr]   M. L. Gromov, Geometric Group Theory, Vol 2 (Sussex, 1991), Cambridge Univ. Press Cambridge, 1993.

[Gu]   G. B. Gurevich, Foundation of the theory of algebraic invariants, Noordhoff, Groningen, 1964.

[GuR]   A. Guterman and A. Regev, *On the growth of identities*, In: Algebra (Moscow, 1998), 319-330, de Gruyter, Berlin, 2000.

[Ha]   M. Hall, Combinatorial Theory. Braisdess, London, 1967.

[Hl]   P. Halpin, *Central and weak identities for matrices*, Comm. Algebra **11** (1983), 2237-2248.

[H]   I. N. Herstein, Noncommutative Rings, Carus Monograph No. **15**, MAA Utreck, 1968.

[Ja]   N. Jacobson, PI-algebras. An introduction, Lecture Notes in Mathematics, Vol. **441**. Springer-Verlag, Berlin-New York, 1975.

[Ja2]   N. Jacobson, Basic algebra II, Second edition, W. H. Freeman and Company, New York, 1989.

[Jm]   G. D. James, The representation theory of the symmetric groups, Lecture Notes in Mathematics, **682** Springer, Berlin, 1978.

[JK]   G. James and A. Kerber, The Representation Theory of the Symmetric Group, Encyclopedia of Mathematics and its Applications, Vol. **16**, Addison-Wesley, London, 1981.

[K1]   I. Kaplansky, *Rings with a polynomial identity*, Bull. Amer. Math. Soc. **54** (1948), 496-500.

[K2]   I. Kaplansky, *Problems in the theory of rings, revisited*, Amer. Math. Monthly **77** (1970), 445-454.

[Ke1]   A. R. Kemer, *T-ideals with power growth of the codimensions are Specht*, (Russian), Sibirskii Matematicheskii Zhurnal **19** (1978), 54-69; English translation: Siberian Math. J. **19** (1978), 37-48.

[Ke2]   A. R. Kemer, *Varieties of finite rank*, Proc. 15-th All the Union Algebraic Conf., Krasnoyarsk, Vol **2**, p. 73, 1979, (Russian).

[Ke3]   A. R. Kemer, *Varieties of $\mathbb{Z}_2$-graded algebras*, (Russian), Izv. Akad. Nauk SSSR Ser. Mat. **48** (1984), no. 5, 1042-1059.

[Ke4] A. R. Kemer, *Finite basability of identities of associative algebras*, Algebra and Logic **26** (1987), no. 5, 362-397.

[Ke5] A. R. Kemer, *Solution of the problem as to whether associative algebras have a finite basis of identities*, (Russian), Dokl. Akad. Nauk SSSR **298** (1988), no. 2, 273-277; translation in Soviet Math. Dokl. **37** (1988), no. 1, 60-64.

[Ke6] A. R. Kemer, *Representability of reduced-free algebras*, (Russian), Algebra i Logika **27** (1988), no. 3, 274–294, 375; translation in Algebra and Logic **27** (1988), no. 3, 167–184 (1989).

[Ke7] A. R. Kemer, Ideals of Identities of Associative Algebras, AMS Translations of Mathematical Monograph, Vol. **87**, 1988.

[Ka] V. K. Kharchenko, *Galois extensions, and rings of quotients*, (Russian), Algebra i Logika **13** (1974), no. 4, 460-484, 488.

[Kn] D. E. Knuth, The Art of Computer Programming, Volume **3**, Sorting and searching, Addison-Wesley Series in Computer Science and Information Processing, Addison-Wesley Publishing Co., Reading, Mass.-London-Don Mills, Ont., 1973.

[Ko] P. Koshlukov, *Basis of the identities of the matrix algebra of order two over a field of characteristic $p \neq 2$*, J. Algebra **241** (2001), 410-434.

[Ks] B. Kostant, *A theorem of Frobenius, a theorem of Amitsur-Levitski and cohomology theory*, J. Math. Mech. **7** (1958), 237-264.

[Kt] A. I. Kostrikin, Around Burnside, Springer, 1990.

[KR] D. Krakowski and A. Regev, *The polynomial identities of the Grassmann algebra*, Trans. Amer. Math. Soc. **181** (1973), 429–438.

[KL] G. R. Krause and T. H. Lenagan, Growth of algebras and Gelfand-Kirillov dimension, Revised edition, Graduate Texts in Mathematics Vol. **22**, Amer. Math. Soc. Providence R.I., 2000.

[L] T. Y. Lam, A First Course in Noncommutative Rings, Second edition, Graduate Texts in Mathematics **131** Springer-Verlag, New York, 2001. xx+385 pp.

[L1] V. N. Latyshev, *On Regev's theorem on identities in tensor product of PI-algebras*, Ups. Mat. Nauk. **27** (1973), 213-214, (Russian).

[L2] V. N. Latyshev, *The complexity of nonmatrix varieties of associative algebras I*, (Russian), Algebra i Logika **16** (1977), no. 2, 149–183.

[L3] V. N. Latyshev, *The complexity of nonmatrix varieties of associative algebras II*, (Russian), Algebra i Logika **16** (1977), no. 2, 184–199.

[Le] J. Lewin, *A matrix representation for associative algebras I*, Trans. Amer. Math. Soc. **188** (1974), 293-308.

[LP] J. L. Loday and C. Procesi, *Homology of symplectic and orthogonal algebras*, Adv. Math. **69** (1988), 93-108.

[Lv] I. V. L'vov, *Maximality conditions in algebras with identity relations*, (Russian), Algebra i Logika **8** (1969), 449-459. English translation: Algebra and Logica **8** (1969), 258-263.

[M] I. G. Macdonald, Symmetric functions and Hall polynomials, Oxford Mathematical Monographs, Oxford Science Publications. The Clarendon Press, Oxford University Press, New York, 1995.

[Ma1] A. I. Mal'tsev, *Untersuchungen aus dem Gebiete der mathematischen Logik*, (German), Rec. Math. Moscou, n. Ser. 1, 323-335 (1936).

[Ma2] Yu. N. Malcev, *A basis for the identities of the algebra of upper triangular matrices*, (Russian), Algebra i Logika **10** (1971), 393-400; English translation: Algebra and Logic **10** (1971).

[Mt] H. Matsumura, Commutative Algebra, W. A. Benjamin, Inc., New York, 1970 xii+262 pp.

[M1] S. P. Mishchenko, *On the problem of the Engel property*, (Russian), Mat. Sb. (N.S.) **124(166)** (1984), no. 1, 56-67.

[M2] S. P. Mishchenko, *Growth of varieties of Lie algebras*, Uspekhi Mat. Nauk **45** (1990), 25-45, 189; English translation: Russian Math. Surveys **45** (1990), 27-52.

[M3] S. P. Mishchenko, *On some classes of Lie algebras*, (Russian), Vestnik Moskov. Univ. Ser. I Mat. Mekh. 1992, no. 3, 55-57; translation in Moscow Univ. Math. Bull. **47** (1992), no. 3, 27-28.

[M4]   S. P. Mishchenko, *Lower bounds on the dimensions of irreducible representations of symmetric groups and of the exponents of the exponential of varieties of Lie algebras*, (Russian), Sb. **187** (1996), 83-94; English translation: Sb. Math. **187** (1996), 81-92.

[MP1] S. P. Mishchenko and V. M. Petrogradsky, *Exponents of varieties of Lie algebras with a nilpotent commutator subalgebra*, Comm. Algebra **27** (1999), 2223-2230.

[MRZ1] S. P. Mishchenko, A. Regev and M. Zaicev, *A characterization of P.I. algebras with bounded multiplicities of the cocharacters*, J. Algebra **219** (1999), 356-368.

[MRZ2] S. P. Mishchenko, A. Regev and M. Zaicev, *Integrality of exponents of some abelian-by-nilpotent varieties of Lie algebras*, Comm. Algebra **28** (2000), 4105-4130.

[MRZ3] S. P. Mishchenko, A. Regev and M. Zaicev, *The exponential growth of codimensions for Capelli identities*, Israel J. Math. **115** (2000), 333-342.

[MV]   S. Mishchenko and A. Valenti, *A star-variety with almost polynomial growth*, J. Algebra, **223** (2000), 66-84.

[MZ]   S. P. Mishchenko and M. Zaicev, *Asymptotic behaviour of colength of varieties of Lie algebras*, Serdica Math. J. **26** (2000), 145-154.

[O]   A. Yu. Ol'shanskii, *On the distorsion of subgroups of finitely presented groups*, (Russian), Mat. Sb. **188** (1997), 51-98; translation in Sb. Math. **188** (1997), 1617-1664.

[OR]   J. B. Olson and A. Regev, *An application of representation theory to PI-algebras*, Proc. Amer. Math. Soc. **55** (1976), 253-257.

[P]   D. S. Passman, The Algebraic Structure of Group Rings, Pure and Applied Mathematics, Wiley-Interscience [John Wiley & Sons], New York-London-Sydney, 1977.

[P1]   V. M. Petrogradsky, *Exponents of subvarieties of upper triangular matrices over arbitrary fields are integral*, Serdica Math. J. **26** (2000), 167-176.

[Pe2] V. M. Petrogradsky, *Invariants of the action of a finite group on a free Lie algebra*, (Russian), Sibirsk. Mat. Zh. 41 (2000), 917-925; English translation: Siberian Math. J. **41** (2000), 763-770.

[P3]   V. M. Petrogradsky, *On growth of Lie algebras, generalized partitions, and analytic functions*, Formal power series and algebraic combinatorics (Vienna, 1997), Discrete Math. **217** (2000), 337-351.

[P4]   V. M. Petrogradsky, *On the numerical characteristics of subvarieties of three varieties of Lie algebras*, (Russian), Mat. Sb. 190 (1999),111-126; English translation: Sb. Math. **190** (1999), 887-902.

[P5]   V. M. Petrogradsky, *Growth of finitely generated polynilpotent Lie algebras and groups, generalized partitions, and functions analytic in the unit circle*, Internat. J. Algebra Comput. **9** (1999), 179-212.

[P6]   V. M. Petrogradsky, *Exponential Schreier's formula for free Lie algebras and its applications*, Algebra, 11. J. Math. Sci. (New York) **93** (1999), 939-950.

[P7]   V. M. Petrogradsky, *Scale for codimension growth of Lie algebras*, Methods in ring theory (Levico Terme, 1997), 213-222, Lecture Notes in Pure and Appl. Math., **198**, Marcel Dekker, New York, 1998.

[P8]   V. M. Petrogradsky, *Growth of polynilpotent varieties of Lie algebras, and rapidly increasing entire functions*, (Russian), Mat. Sb. **188** (1997), 119-138; English translation: Sb. Math. **188** (1997), 913-931.

[P9]   V. M. Petrogradsky, *On types of superexponential growth of identities in Lie PI-algebras*, (Russian), Fundam. Prikl. Mat. **1** (1995), 989-1007.

[Pi]   M. Pipitone, *Algebras with involution whose exponent of the $*$-codimensions is equal to two*, Comm. Algebra **30** (2002), no. 8, 3875–3883.

[Po]   E. C. Posner, *Prime rings satisfying a polynomial identity*, Proc. Amer. Math. Soc. **11** (1960), 180-183.

[Pr1] C. Procesi, *Non-commutative affine rings*, Atti Accad. Naz. Lincei, **8** (1967), 239-255.

[Pr2] C. Procesi, The invariant theory of $n \times n$ matrices, Adv. Math. **19** (1976), 306-381.

[Pr3] C. Procesi, Rings with polynomial identities, Pure and Applied Mathematics, **17**, Marcel Dekker, Inc., New York, 1973.

[R1]   Ju. P. Razmyslov, *A certain problem of Kaplansky*, (Russian) Izv. Akad. Nauk SSSR Ser. Mat. **37** (1973), 483-501.

[R2]   Ju. P. Razmyslov, *Trace identities of full matrix algebras over a field of characteristic zero*, (Russian), Izv. Akad. Nauk SSSR, Ser. Mat. **38** (1974), 723-756. Translation: Math USSR Izv. **8**, (1974), 727-760.

[R3] Yu. P. Razmyslov, *The Jacobson Radical in PI-algebras*, (Russian), Algebra i Logika **13** (1974), 337-360. English translation: Algebra and Logic **13** (1974), 192-204 (1975).

[R4] Yu. P. Razmyslov, Identities of algebras and their representations, Translations of Mathematical Monographs, **138**, American Mathematical Society, Providence, RI, 1994.

[Re1] A. Regev, *Existence of identities in $A \otimes B$*, Israel J. Math. **11** (1972), 131-152.

[Re2] A. Regev, *Algebras satisfying a Capelli identity*, Israel J. Math. **33** (1979), 149-154.

[Re3] A. Regev, *Asymptotic values for degrees associated with strips of Young diagrams*, Adv. Math., **41** (1981), 115-136.

[Re4] A. Regev, *Codimensions and trace codimensions of matrices are asymptotically equal*, Israel J. Math. **47** (1984), 246-250.

[Re5] A. Regev, *The representations of wreath products via double centralizing theorems*, J. Algebra **102** (1986), 423-443.

[Re6] A. Regev, *On the identities of subalgebras of matrices over the Grassmann algebra*, Israel J. Math. **58** (1987), 351-369.

[Re7] A. Regev, *Asymptotics of codimensions of some P.I. algebras*, In: Trends in Ring Theory, CMS Conference Proc. Vol. **22**, Amer. Math. Soc., Providence RI, 1998, 159-172.

[Rob] Robinson, G. de B. Representation theory of the symmetric group, Mathematical Expositions, No. 12. University of Toronto Press, Toronto 1961.

[R] S. Rosset, *A new proof of the Amitsur-Levitski identity*, Israel J. Math. **23** (1976), 187-188.

[Ro1] L. H. Rowen, *Some results on the center of a ring with polynomial identity*, Bull. Amer. Math. Soc. **79** (1973), 219-223.

[Ro2] L. H. Rowen, Polynomial Identities in Ring Theory, Academic Press, New York, 1980.

[Ro3] L. H. Rowen, Ring Theory, Vol. 1, 2, Academic Press, New York, 1988.

[Sa] B. E. Sagan, The symmetric group. Representations, combinatorial algorithms, and symmetric functions, Second edition, Graduate Texts in Mathematics, **203** Springer-Verlag, New York, 2001.

[S] I. Schur, *Uber die rationalen darstellungen der allgemeinen linearen gruppe (1927)*, in "Gesammelte Abhandlungen Band III", pp. 68-85, Springer-Verlag, Berlin, 1973.

[Sh] A. I. Shirshov, On rings with identity relations, (Russian) Mat. Sb. N. S. **43**(85) (1957), 277-283.

[Si] K. S. Siberskii, *Algebraic invariants for a set of matrices*, Sib. Mat. Zh. **9** (1968), 152-164.

[Sp] W. Specht, Gesetze in Ringen. I, Math. Z. **52** (1950), 557-589.

[Sr] T. A. Springer, Invariant Theory, Lecture Notes in Mathematics, Vol. **585**, Springer-Verlag, Berlin-New York, 1977.

[St] R. P. Stanley, Enumerative combinatorics. Volume 2, Cambridge Studies in Advanced Mathematics **62**, Cambridge: Cambridge University Press, 1999.

[SV] A. N. Stoyanova-Venkova, *Some lattices of varieties of associative algebras defined by identities of fifth degree* (Russian), C. R. Acad. Bulg. Sci. **35** (1982), 865-868.

[T] E. J. Taft, *Invariant Wedderburn factors*, Illinois J. Math. **1** (1957), 565-573.

[V] A. Valenti, *The graded identities of upper triangular matrices of size two*, J. Pure Appl. Algebra **172** (2002), 325-335.

[VZ] A. Valenti and M. Zaicev, *Abelian gradings on upper-triangular matrices*, Arch. Math. (Basel) **80** (2003), 12-17.

[Vo] I. B. Volichenko, *Varieties of Lie algebras with identity $[[X_1, X_2, X_3], [X_4, X_5, X_6]] = 0$ over a field of characteristic zero*, (Russian), Sibirsk. Mat. Zh. **25** (1984), 3, 40-54.

[Wa] W. Wagner, *Über die Grundlagen der projektiven Geometrie und allgemeine Zahlensysteme*, (German) Math. Ann. **113** (1936), 528-567.

[W] H. Weyl, The Classical Groups. Their Invariants and Representations, Princeton University Press, Princeton, N.J., 1939.

[Z1] M. Zaicev, *Identities of affine Kac-Moody algebras*, (Russian), Vestnik Moskov. Univ. Ser. I Mat. Mekh. 1996, no. 2, 33-36, 104; translation in Moscow Univ. Math. Bull. **51** (1996), no. 2, 29-31.

[Z2] M. Zaicev, *Integrality of exponents of growth of identities of finite-dimensional Lie algebras*, (Russian), Izv. Ross. Akad. Nauk Ser. Mat. **66** (2002), 23-48; English translation: Izv. Math. **66** (2002), 463-487.

[Z3] M. Zaicev, *A superrank of varieties of Lie algebras*, (Russian), Algebra i Logika, **37** (1998), 394-412; English translation: Algebra and Logic **37** (1998), 223-233.

[Z4]   M. Zaicev, *Varieties and identities of affine Kac-Moody algebras*, Methods in ring theory (Levico Terme, 1997), 303-314, Lecture Notes in Pure and Appl. Math., **198**, Marcel Dekker, New York, 1998.

[ZM1] M. Zaicev and S. P. Mishchenko, *A criterion for the polynomial growth of varieties of Lie superalgebras*, (Russian), Izv. Ross. Akad. Nauk Ser. Mat.**62** (1998), 103-116; English translation: Izv. Math. **62** (1998), 953-967.

[ZM2] M. Zaicev and S. P. Mishchenko, *An example of a variety of Lie algebras with a fractional exponent*, Algebra, 11. J. Math. Sci. (New York) **93** (1999), 977-982.

[ZM3] M. Zaicev and S. P. Mishchenko, *A new extremal property of the variety AN_2 of Lie algebras*, (Russian), Vestnik Moskov. Univ. Ser. I Mat. Mekh. 1999, no. 5, 15-18; English translation: Moscow Univ. Math. Bull. **54** (1999), 11-14.

[ZM4] M. Zaicev and S. P. Mishchenko, *On the polynomial growth of the colength of varieties of Lie algebras*, (Russian), Algebra i Logika, **38** (1999), 161-175; English translation: Algebra and Logic **38** (1999), 84-92.

[Z]   E. I. Zelmanov, *On Engel Lie algebras*, (Russian), Sibirsk. Mat. Zh. **29** (1988), no. 5, 112-117, 238; translation in Siberian Math. J. **29** (1988), no. 5, 777-781 (1989).

[ZSSS] K. A. Zhevlakov, A. M. Slinko, I. P. Shestakov and A.I. Shirshov, Rings that are nearly associative, Pure and Applied Mathematics, **104**, Academic Press, Inc. [Harcourt Brace Jovanovich, Publishers], New York-London, 1982.

Index

$*$-$\exp(A)$, 284
$*$-$\mathrm{var}(A)$, 284
$C(k,n)$, 125
$C(k,n)^{mult}$, 125
C_{T_λ}, 49
Cap_m, 13
D_λ, 47
FS_n, 46
$F\langle Y, Z\rangle$, 69
$F\langle X \mid G\rangle$, 65
$F\langle X, *\rangle$, 69
$F\langle X, \mathrm{Tr}\rangle$, 122
$F\langle X\rangle$, 1
$F\langle X\rangle^*$, 56
$F\langle X\rangle^{gr}$, 66
$F\{\xi\}$, 12
$G \wr S_n$, 257
G-identity, 66
G^{gr}, 289
$H(d, l)$, 58
I_λ, 47
$L_k(x; y)$, 129
MT_n, 122
$M \widehat{\otimes} N$, 50
$M_k(F)$, 11
$M_k(G)$, 83
$M_n(F \oplus cF)$, 75
$M_{k,l}(F)$, 75
$M_{k,l}(G)$, 83
$M_{k \times l}(F)$, 251
PT_n, 122
P_n, 53
$P_n(A)$, 54
$P_n(\mathcal{V})$, 54
P_n^G, 256
P_n^*, 284
P_n^{gr}, 268
$P_n^{\mathbb{Z}_2}$, 283
$P_n^{\mathbb{Z}_2}(A)$, 273
$P_{r,n-r}$, 273
R_{T_λ}, 49
$S(A)$, 84
$St_m(x_1, \ldots, x_m)$, 13

T_2-ideal, 80
T_λ, 47
$UT(d_1, \ldots, d_m)$, 24
$UT_2(F)$, 88
UT_2^{gr}, 289
$UT_n(F)$, 2
$[a, b]$, 2
$\Delta(\xi_1, \ldots, \xi_{k^2})$, 125
$\alpha \models n$, 50
$\chi \downarrow H$, 46
$\chi \uparrow G$, 46
$\chi_n^{mtr}(M_k(F))$, 135
χ_λ, 46
$\chi_n^G(A)$, 265
$\chi_{n_1, \ldots, n_k}(A)$, 266
$\deg f$, 1
$\deg_{x_i} f$, 1
$\exp(A)$, 144
$\exp(f)$, 215
$\hat{G}$, 63
$\lambda \vdash n$, 15
$\langle S\rangle_T$, 4
$\langle \lambda\rangle$, 264
$\mathcal{A}_{x_1, \ldots, x_r}$, 18
$\mathcal{V}$, 4
$\mathcal{V}^*$, 82
$\mathrm{Aut}(A)$, 63
$\mathrm{Aut}^*(A)$, 65
$\mathrm{End}_A(M)$, 29
$\mathrm{GKdim}(A)$, 39
$\mathrm{Id}(A)$, 3
$\mathrm{Id}^G(A)$, 66
$\mathrm{Id}^*(A)$, 284
$\mathrm{Id}^{gr}(A)$, 66
$\mathrm{supexp}(A)$, 284
$\mathrm{ad} x$, 309
$\det(a)$, 2
$\mathrm{tr\,deg}(K/F)$, 39
$\mathrm{tr}(a)$, 2
φ-ideals, 73
$c_n(A)$, 88
$c_n^G(A)$, 256
$c_n^{(h_1, \ldots, h_n)}(A)$, 98

$c_n^*(A)$, 284
$c_n^{gr}(A)$, 268
$c_n^{sup}(A)$, 283
$c_n^{\mathbb{Z}_2}(A)$, 273
d_λ, 47
e_λ, 47
e_{T_λ}, 49
$f \equiv 0$, 2
$h(d, l, t)$, 57
$l_n(A)$, 90
$\mathbb{Z}_2\text{-}\exp(A)$, 276

algebra
  algebraic of bounded degree, 14
  center, 31
  of generic elements, 11
  of generic matrices, 12
  of invariants, 124
  supercommutative, 83
  verbally prime, 83
Amitsur, 35, 36, 104, 106, 108, 215, 242–245
Amitsur identity, 109
Amitsur trick, 100
Amitsur-Levitzki Theorem, 18
antiautomorphism, 65
artinian ring, 29
AT-algebra, 317

Berele, 38, 117, 212, 274
Birkhoff Theorem, 4
block-triangular matrix algebra, 24
Branching Theorem, 50

Capelli identity, 13
Capelli polynomial, 13
central localization, 34
character, 45
  induced, 46
  inner product, 45
  irreducible, 45
cocharacter, 54
  ∗-cocharacter, 284
  G-cocharacter, 265
  graded cocharacter, 238
  mixed trace cocharacter, 135
  pure trace cocharacter, 123
codimension, 88
  ∗-codimension, 264, 284
  G-codimension, 256
  $\mathbb{Z}_2$-codimension, 273, 284
  graded codimension, 268, 283
  supercodimension, 283
  trace codimension, 123
colength, 90
column-stabilizer, 49
commuting ring, 29

decomposable monomial, 259

dense set, 29
Dilworth, 94
discriminant, 125
Drensky, 20, 28, 137, 185, 187, 207

element
  skew-symmetric, 69
  symmetric, 69
elementary symmetric function, 15
essential G-identity, 262
essential hook, 241
exponent, 144
  ∗-exponent, 284
  $\mathbb{Z}_2$-exponent, 276
  of a polynomial, 215
  superexponent, 284

First Fundamental Theorem, 125
Formanek, 27, 28, 40, 128, 134, 137
free G-graded algebra, 66
free algebra
  with G-action, 65
  with involution, 69
  with trace, 122
free associative algebra, 1
free Lie algebra, 310
free superalgebra, 69
free supercommutative algebra, 83
Frobenius reciprocity, 46

Gelfand-Kirillov dimension, 37
generalized square, 244
generic division ring, 40
generic element, 11
generic matrix, 12
graded algebra, 5, 61
  G-graded algebra, 61
  $\mathbb{Z}_2$-graded algebra, 65
graded identity, 66
graded subalgebra, 61
graded subspace, 61
Grassmann algebra, 2, 90
Grassmann envelope, 81
GSPI-algebra, 316
Gurevich, 125

Halpin, 28
Herstein, 29
homogeneous
  component, 61
  element, 61
hook, 58
Hook Formula, 48
hook number, 48
Hook Theorem, 105

idempotent, 45
  central, 45
  essential, 49

minimal, 45
    minimal graded, 194
indecomposable monomial, 259
induced module, 46
involution, 69
    exchange, 77
    symplectic, 77
    transpose, 77

Jacobson, 29

Kaplansky, 27
Kaplansky's Theorem, 31
Kasparian, 28
Kemer, 20, 83, 110, 112, 113, 169
Koshlukov, 20
Kostant, 120
Krull dimension, 39

lattice permutation, 51
Latyshev, 94
Lewin Theorem, 21
Lie algebra, 307
    abelian, 308
    adjoint representation, 309
    center, 309
    nilpotent, 308
    representation, 309
    simple, 308
    solvable, 308
    universal enveloping algebra, 309
Lie commutator, 2
Lie ideal, 9
Lie identity, 310
Littlewood-Richardson rule, 51
lower exponent, 144

Maschke's Theorem, 44
Mishchenko, 184, 268, 289
mixed trace polynomial, 122
monomial, 1
multilinearization process, 7
multipartition, 264

Newton's formulas, 15
Noether Normalization Theorem, 39

outer tensor product, 50

partition, 46
    conjugate, 47
permutation
    d-bad, 94
    d-good, 94
PI-algebra, 2
PI-exponent, 144
polynomial
    G-polynomial, 66
    alternating, 12
    central, 26
    consequence, 7
    equivalent, 7
    homogeneous, 5
    linear, 7
    multialternating, 138
    multihomogeneous, 5
    multihomogeneous component, 6
    multilinear, 7
polynomial growth, 171
polynomial identity, 2
Posner's Theorem, 34
power sums symmetric function, 15
prime ring, 34
primitive ring, 29
Procesi, 27, 40, 123, 125
product of varieties, 327
pure trace polynomial, 122

Razmyslov, 19, 20, 27, 28, 123
Razmyslow-Kemer-Braun Theorem, 35
Regev, 94–96, 108, 117, 128, 141, 184, 212
relatively free algebra, 4
representation, 43
    completely reducible, 44
    equivalent, 44
    irreducible, 44
    left regular, 44
Robinson-Schensted Correspondence, 102
Rosset, 18
row insertion algorithm, 101
row-stabilizer, 49
Rowen, 33

Schur, 120
Schur's Lemma, 29
Second Fundamental Theorem, 125
semiprime ring, 32
semistandard tableau, 51
Sibirskii, 125
skew-tableau, 51
Skolem-Noether Theorem, 78
SPI-algebra, 314
splitting field, 30, 45
stable identity, 10
standard Lie polynomial, 311
standard polynomial, 13
standard tableau, 47
Strip Theorem, 107
subdirect product, 33
superalgebra, 65
    minimal, 194
    reduced, 240
superenvelope, 84
supervariety, 80
symmetric algebra, 124
symmetric function, 15
symmetric polynomial, 15

T-ideal, 3

verbally prime, 82
trace identity, 122
trace polynomial, 122
transcendence degree, 39
trivial grading, 62

unirational, 40
unordered partition, 50
upper exponent, 144

Valenti, 128, 289, 292
Vandermonde matrix, 6
variety
   distributive, 177
   left noetherian, 176
   minimal, 205
   non-trivial, 4
   of almost polynomial growth, 171
   of polynomial growth, 171
   prime, 83
   proper, 4
variety of algebras, 4
Von Neumann Lemma, 50

Wedderburn, 29
Wedderburn-Artin Theorem, 29
Wedderburn-Malcev Theorem, 71
Weyl, 120
wreath product, 257

Young diagram, 47
Young tableau, 47
Young's Rule, 50
Young-Frobenius Formula, 48

Zorn, 30

Titles in This Series

122 **Antonio Giambruno and Mikhail Zaicev, Editors,** Polynomial identities and asymptotic methods, 2005

121 **Anton Zettl,** Sturm-Liouville theory, 2005

120 **Barry Simon,** Trace ideals and their applications, 2005

119 **Tian Ma and Shouhong Wang,** Geometric theory of incompressible flows with applications to fluid dynamics, 2005

118 **Alexandru Buium,** Arithmetic differential equations, 2005

117 **Volodymyr Nekrashevych,** Self-similar groups, 2005

116 **Alexander Koldobsky,** Fourier analysis in convex geometry, 2005

115 **Carlos Julio Moreno,** Advanced analytic number theory: L-functions, 2005

114 **Gregory F. Lawler,** Conformally invariant processes in the plane, 2005

113 **William G. Dwyer, Philip S. Hirschhorn, Daniel M. Kan, and Jeffrey H. Smith,** Homotopy limit functors on model categories and homotopical categories, 2004

112 **Michael Aschbacher and Stephen D. Smith,** The classification of quasithin groups II. Main theorems: The classification of simple QTKE-groups, 2004

111 **Michael Aschbacher and Stephen D. Smith,** The classification of quasithin groups I. Structure of strongly quasithin K-groups, 2004

110 **Bennett Chow and Dan Knopf,** The Ricci flow: An introduction, 2004

109 **Goro Shimura,** Arithmetic and analytic theories of quadratic forms and Clifford groups, 2004

108 **Michael Farber,** Topology of closed one-forms, 2004

107 **Jens Carsten Jantzen,** Representations of algebraic groups, 2003

106 **Hiroyuki Yoshida,** Absolute CM-periods, 2003

105 **Charalambos D. Aliprantis and Owen Burkinshaw,** Locally solid Riesz spaces with applications to economics, second edition, 2003

104 **Graham Everest, Alf van der Poorten, Igor Shparlinski, and Thomas Ward,** Recurrence sequences, 2003

103 **Octav Cornea, Gregory Lupton, John Oprea, and Daniel Tanré,** Lusternik-Schnirelmann category, 2003

102 **Linda Rass and John Radcliffe,** Spatial deterministic epidemics, 2003

101 **Eli Glasner,** Ergodic theory via joinings, 2003

100 **Peter Duren and Alexander Schuster,** Bergman spaces, 2004

99 **Philip S. Hirschhorn,** Model categories and their localizations, 2003

98 **Victor Guillemin, Viktor Ginzburg, and Yael Karshon,** Moment maps, cobordisms, and Hamiltonian group actions, 2002

97 **V. A. Vassiliev,** Applied Picard-Lefschetz theory, 2002

96 **Martin Markl, Steve Shnider, and Jim Stasheff,** Operads in algebra, topology and physics, 2002

95 **Seiichi Kamada,** Braid and knot theory in dimension four, 2002

94 **Mara D. Neusel and Larry Smith,** Invariant theory of finite groups, 2002

93 **Nikolai K. Nikolski,** Operators, functions, and systems: An easy reading. Volume 2: Model operators and systems, 2002

92 **Nikolai K. Nikolski,** Operators, functions, and systems: An easy reading. Volume 1: Hardy, Hankel, and Toeplitz, 2002

91 **Richard Montgomery,** A tour of subriemannian geometries, their geodesics and applications, 2002

90 **Christian Gérard and Izabella Łaba,** Multiparticle quantum scattering in constant magnetic fields, 2002

TITLES IN THIS SERIES

89 Michel Ledoux, The concentration of measure phenomenon, 2001
88 Edward Frenkel and David Ben-Zvi, Vertex algebras and algebraic curves, second edition, 2004
87 Bruno Poizat, Stable groups, 2001
86 Stanley N. Burris, Number theoretic density and logical limit laws, 2001
85 V. A. Kozlov, V. G. Maz'ya, and J. Rossmann, Spectral problems associated with corner singularities of solutions to elliptic equations, 2001
84 László Fuchs and Luigi Salce, Modules over non-Noetherian domains, 2001
83 Sigurdur Helgason, Groups and geometric analysis: Integral geometry, invariant differential operators, and spherical functions, 2000
82 Goro Shimura, Arithmeticity in the theory of automorphic forms, 2000
81 Michael E. Taylor, Tools for PDE: Pseudodifferential operators, paradifferential operators, and layer potentials, 2000
80 Lindsay N. Childs, Taming wild extensions: Hopf algebras and local Galois module theory, 2000
79 Joseph A. Cima and William T. Ross, The backward shift on the Hardy space, 2000
78 Boris A. Kupershmidt, KP or mKP: Noncommutative mathematics of Lagrangian, Hamiltonian, and integrable systems, 2000
77 Fumio Hiai and Dénes Petz, The semicircle law, free random variables and entropy, 2000
76 Frederick P. Gardiner and Nikola Lakic, Quasiconformal Teichmüller theory, 2000
75 Greg Hjorth, Classification and orbit equivalence relations, 2000
74 Daniel W. Stroock, An introduction to the analysis of paths on a Riemannian manifold, 2000
73 John Locker, Spectral theory of non-self-adjoint two-point differential operators, 2000
72 Gerald Teschl, Jacobi operators and completely integrable nonlinear lattices, 1999
71 Lajos Pukánszky, Characters of connected Lie groups, 1999
70 Carmen Chicone and Yuri Latushkin, Evolution semigroups in dynamical systems and differential equations, 1999
69 C. T. C. Wall (A. A. Ranicki, Editor), Surgery on compact manifolds, second edition, 1999
68 David A. Cox and Sheldon Katz, Mirror symmetry and algebraic geometry, 1999
67 A. Borel and N. Wallach, Continuous cohomology, discrete subgroups, and representations of reductive groups, second edition, 2000
66 Yu. Ilyashenko and Weigu Li, Nonlocal bifurcations, 1999
65 Carl Faith, Rings and things and a fine array of twentieth century associative algebra, 1999
64 Rene A. Carmona and Boris Rozovskii, Editors, Stochastic partial differential equations: Six perspectives, 1999
63 Mark Hovey, Model categories, 1999
62 Vladimir I. Bogachev, Gaussian measures, 1998
61 W. Norrie Everitt and Lawrence Markus, Boundary value problems and symplectic algebra for ordinary differential and quasi-differential operators, 1999
60 Iain Raeburn and Dana P. Williams, Morita equivalence and continuous-trace C^*-algebras, 1998

For a complete list of titles in this series, visit the
AMS Bookstore at **www.ams.org/bookstore/**.